PERTURBATIONS IN THE SPECTRA OF DIATOMIC MOLECULES

PERTURBATIONS IN THE SPECTRA OF DIATOMIC MOLECULES

Hélène Lefebvre-Brion

Laboratoire de Photophysique Moléculaire
Université de Paris-Sud
Orsay, France

Robert W. Field

Department of Chemistry
Massachusetts Institute of Technology
Cambridge, Massachusetts

1986

ACADEMIC PRESS, INC.

Harcourt Brace Jovanovich, Publishers

Orlando San Diego New York Austin
Boston London Sydney Tokyo Toronto

ACADEMIC PRESS, INC.
Orlando, Florida 32887

United Kingdom Edition published by
ACADEMIC PRESS INC. (LONDON) LTD.
24–28 Oval Road, London NW1 7DX

Library of Congress Cataloging in Publication Data

Lefebvre-Brion, Hélène.
Perturbations in the spectra of diatomic molecules.

Bibliography: p.
Includes index.
1. Molecular spectroscopy. 2. Perturbation (Quantum dynamics) I. Field, Robert W. II. Title.
QD96.M65L44 1986 539'.6 85-18589
ISBN 0–12–442690–5 (alk. paper)
ISBN 0–12–442691–3 (paperback)

PRINTED IN THE UNITED STATES OF AMERICA

86 87 88 89 9 8 7 6 5 4 3 2 1

Contents

Preface

Examined in sufficient detail, the spectrum of every diatomic molecule is full of surprises. These surprises or "perturbations" can be at least as interesting as the vast expanses of textbook spectra lying between the surprises. Perturbations are more than spectroscopic esoterica. We hope that our discussion of perturbations provides a useful and unified view of diatomic molecular structure. This is a book about the spectra of diatomic molecules, warts and all.

This book is for graduate students just beginning research, for theorists curious about what experimentalists actually measure, for experimentalists bewildered by theory, and for potential users of spectroscopic data in need of a user's guide. We have avoided abstract and elegant treatments (e.g., spherical tensors) wherever a simpler one (e.g., ladder operators) would suffice. We have worked through many examples rather than attempt to provide formulas and literature examples for all conceivable cases.

Chapters 1, 2, and 4 form the core of this book. Perturbations are defined and simple procedures for evaluating matrix elements of angular momentum operators are presented in Chapter 1. Chapter 2 deals with the troublesome terms in the molecular Hamiltonian that are responsible for perturbations. Particular attention is devoted to the reduction of matrix elements to separately evaluable rotational, vibrational, and electronic factors. Whenever possible the electronic factor is reduced to one- and two-electron orbital matrix elements. The magnitudes and physical interpretations of matrix elements are discussed in Chapter 4. In Chapter 3 the process of reducing spectra to molecular constants and the difficulty of relating empirical parameters to terms in the exact molecular Hamiltonian are described. Transition intensities, especially quantum mechanical interference effects, are discussed in Chapter 5. Also included in Chapter 5 are examples of experiments that illustrate, sample, or utilize perturbation effects. The phenomena of predissociation and autoionization are forms of perturbation and are discussed in Chapters 6 and 7.

We want to thank the following experts who have helped us by reading and criticizing portions of the manuscript: Prof. M. Alexander, Dr. R. F. Barrow, Dr. N. Bessis, Prof. M. Broyer, Dr. L. Brus, Dr. M. Chergui, Dr. M. Child, Dr. D. L. Cooper, Prof. K. Dressler, Dr. C. Effantin, Dr. D. Gauyacq, Dr. A. Giusti-Suzor, Dr. R. A. Gottscho, Dr. G. Gouedard, Dr. M. Graff, Prof. C. Green, Dr. J. T. Hougen, Dr. B. J. Howard, Prof. P. Houston, Dr. Ch. Jungen, Prof. A. Lagerqvist, Prof. C. Linton, Dr. S. McDonald, Prof. E. Miescher, Dr. J. Norman, Prof. S. Novick, Dr. I. Renhorn, Dr. J. Rostas, Prof. J. Schamps, Dr. H. Schweda, Prof. S. J. Silvers, Dr. J. Thoman, Dr. A. Tramer, Dr. P. Vaccaro, and Dr. J. F. Wyart. The errors and opacities that remain are ours, not theirs. We have not attempted to cite all relevant references; we apologize for our probably numerous failures to discuss the first and/or most important examples of phenomena, interpretations, and techniques. This book could never have been produced without the artistry of Mr. J. Lefèvre and the skillful and energetic word processing of Ms. V. Siggia, Ms. K. Garcia, and Ms. S. M. Moore.

RWF is grateful to the Laboratoire de Photophysique Moléculaire of the Université de Paris–Sud for a visiting professorship in 1981, to Professors S. Novick and B. Kohler of Wesleyan University and Professor R. J. Saykally of the University of California, Berkeley, for their hospitality, and to project MAC of MIT for a grant of computer time. A NATO international travel grant greatly accelerated and enriched the process of preparing this book. RWF owes his fascination with molecules to Prof. W. A. Klemperer and the late Prof. H. P. Broida.

HLB has been greatly stimulated by many discussions with the experimentalists in the groups led by Dr. S. Leach (Orsay) and Prof. J. d'Incan (Lyon). Their numerous questions led her and Dr. Leach to organize the French meetings of the GESEM,* which provided an opportunity to develop some of the material included in this book. HLB is also grateful to the late Prof. A. Kastler, who invited her to lecture at the International Winter College in 1973 in Trieste. The lecture notes written for this purpose were the first draft for some sections of this book. Finally, her interest in perturbations goes back to a problem presented to her in 1963 by Prof. E. Miescher. This was the starting point for a fruitful and pleasant scientific exchange that is still underway. This is why this book is dedicated to Prof. E. Miescher.

* Groupe d'Etudes de Spectroscopie Electronique Moléculaire (1969–1979).

Chapter 1

Introduction

1.1 What Is a Perturbation?

For many years the study of perturbations has been a main focus of the research of a small number of spectroscopists specializing in the electronic structure of diatomic molecules. The pioneering studies have been by Heurlinger (1917–1919), Kronig (1928), Van Vleck (1929, 1932, 1936), Mulliken (1930, 1931, 1932), Ittmann (1931), Dieke (1935), Schmid and Gerö (1935), Budó (1937), Budó and Kovács (1938), Kovács (1937, 1969), and Stepanov (1940, 1945). More recently, Lagerqvist, Barrow, Miescher, Herzberg, Ramsay, and Broida have been among the major workers in this field. The study of perturbations is an esoteric, usually avoided, but occasionally necessary pastime of a much larger group of spectroscopists, and an impenetrable mystery to most users of spectroscopic data. The label or footnote *perturbed*, when applied to a band in a data compilation (Herzberg, 1950; Huber and Herzberg, 1979; Rosen, 1970; Barrow, 1973, 1975, 1979, 1982; Pearse and Gaydon, 1976), has been like a quarantine poster warning against the use of that band because of a dangerous and perhaps contagious form of irregular

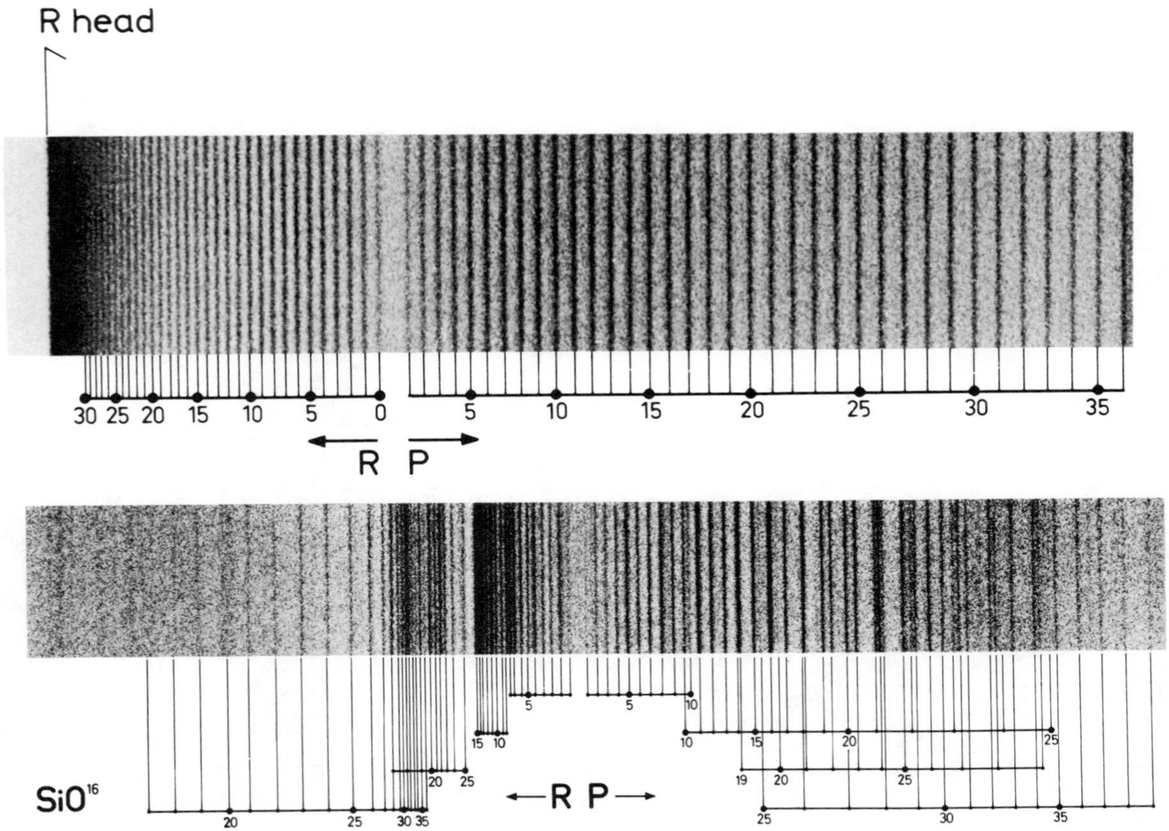

Fig. 1.1 A comparison of the SiO $H^1\Sigma - X^1\Sigma$ 0,0 (top) and 1,0 (bottom) bands. The 0,0 band at 1435 Å is perturbation-free, but perturbations in the $v = 1$ level of the $H^1\Sigma$ state cause the 1,0 band at 1413 Å to be shattered. (Courtesy I. Renhorn.)

behavior. In fact, irregular behavior is an excellent operational definition of perturbation phenomena.

One can define perturbations either pictorially, as deviations from a naive polynomial representation of the energy levels, or in terms of physical models of varying sophistication. Figure 1.1 shows spectra of two vibrational bands of the SiO $H^1\Sigma$–$X^1\Sigma$ electronic system. The rotational structure of an isolated, well-resolved $^1\Sigma$–$^1\Sigma$ band is so simple that the difference between the perturbed and perturbation-free bands is obvious. Typically, the pattern of a band is not so easy to recognize, but when that pattern is disrupted by a few lines being shifted from their expected positions or deviating from a smoothly varying intensity distribution, it is obvious that something in the model has gone wrong and that the band is perturbed. Effects are sometimes restricted to two or three spectral lines associated with a single rotation–vibration level (Fig. 1.2). Some perturbation effects can be dramatically dependent on the mode of observation of the spectrum, for example by a line being absent in emission versus apparently unperturbed in absorption or by being anomalously strong and weak under two different sets of excitation conditions (Fig. 1.3). It is even possible for all rotational lines in a given vibrational band to appear free of perturbation effects; yet, when an attempt is made to fit the band origin and rotational constants of that band into the regular pattern required for a vibrational progression, all lines are found to be strongly but smoothly affected by an unsuspected perturbation. Obviously, it is dangerous to rely on visual pattern recognition alone for detection or definition of perturbations.

The next step beyond pictorial patterns is a simple algebraic representation, which is based on the observed regularity of the pattern rather than any physical model. One expects that the energy levels E_{vJ} sampled in a band spectrum should be well represented by a simple, rapidly convergent, polynomial function of the rotational and vibrational quantum numbers, J and v. The Dunham expansion (1932),

$$E_{vJ} = \sum_{l,m} Y_{lm}(v + \tfrac{1}{2})^l [J(J+1)]^m \qquad (1.1.1)$$

is an example of such a polynomial representation. Questions of the physical meaning of the Y_{lm} coefficients and the efficiency of the parametrization aside, one can regard a Dunham expansion as a convenient, empirical, model-free way of organizing a large quantity of spectral data. A polynomial expression is simply a translation of the expectation of smooth v,J-dependence into an algebraic form. Perturbations may be viewed, still in a model-independent sense, as a failure of a subset of observed v,J levels to be accommodated by the empirical energy level expression defined by the majority of sampled levels. Thus a band that appears unperturbed in isolation from other vibrational

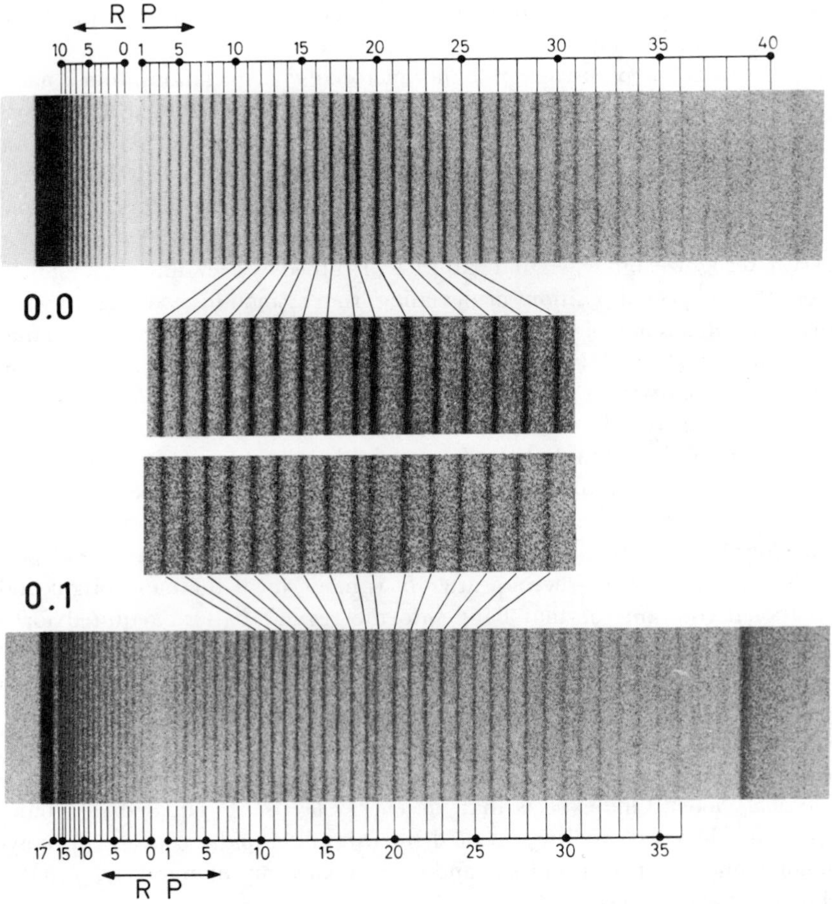

Fig. 1.2 Pictorial evidence that a perturbation affects the upper rather than lower electronic state. The SiO $F^1\Sigma - X^1\Sigma$ 0,0 (1460 Å) and 0,1 (1487 Å) bands are weakly perturbed at $J' = 18$ and 19. The enlargements clearly show identical level shifts for the P(19) line in the 0,0 and 0,1 bands. (Courtesy A. Lagerqvist.)

bands can be recognized as perturbed if its rotational constant or vibrational energy disagrees with an interpolation of

$$B(v) = \sum_l Y_{l1}(v + \tfrac{1}{2})^l \tag{1.1.2}$$

or

$$G(v) = \sum_l Y_{l0}(v + \tfrac{1}{2})^l. \tag{1.1.3}$$

However, a polynomial-based empirical definition of perturbations is limited in two important ways: there are no rules governing the permissible values of polynomial constants, and deviations from the polynomial expression have no meaning or predictive consequence.

Indeed, Dunham's energy-level formula [Eq. (1.1.1)] is based both on the concept of a potential energy curve, which rests on the separability of electronic and nuclear motions, and on the neglect of certain couplings between the angular momenta associated with nuclear rotation, electron spin, and electron orbital motion. The utility of the potential curve concept is related to the validity of the Born–Oppenheimer approximation, which will be discussed in Section 2.1.

The potential energy curve concept provides many self-consistency checks between molecular constants, for example, the Kratzer relation,

$$Y_{02} = -4Y_{01}^3/Y_{10}^2, \tag{1.1.4}$$

which is valid for harmonic or Morse potentials, and the Pekeris relation,

$$Y_{11} = 6Y_{01}^2[1 - (-Y_{20}/Y_{01})^{1/2}]/Y_{10}, \tag{1.1.5}$$

which is valid for a Morse potential. When fitted constants fail such consistency checks, this can indicate the presence of an unsuspected perturbation. Conversely, if interrelationships between molecular constants derived from eigenvalues of a molecular potential are satisfied, then one has confidence in the deperturbed nature of the molecular constants from which the potential is derived. Thus, the most useful definition of perturbed behavior is dependent on the potential curve concept. It is no longer sufficient that the energy levels be represented by an arbitrary polynomial in v or J; the coefficients in the polynomial expression must have values consistent with a simple mechanical oscillator in a well-behaved, isotopically invariant potential.

The study of perturbations involves going beyond the Born–Oppenheimer approximation. One starts with a zero-order model $\mathbf{H}^0$, in which each electronic state is associated with a potential energy curve $V(R)$, an electronic eigenfunction (which depends on internuclear distance) $\phi(r, R)$, and a set of rotation–vibration eigenvalues and eigenfunctions, E_{vJ} and $\chi_{vJ}(R)$. These zero-order $V(R)$, $\phi(r, R)$, and $\chi_{vJ}(R)$ functions can only be defined by ignoring certain terms in the exact molecular Hamiltonian, $\mathbf{H}$. Foremost among these neglected terms, $\mathbf{H}'$, are those that couple the zero-order functions associated with different electronic states.

The explicit choice of which terms to neglect is not unique. For example diabatic and adiabatic potentials are bases for two quite different but equally useful zero-order pictures. These two models and the reasons for selecting one over the other are discussed in Section 2.3. What appears as a large deviation

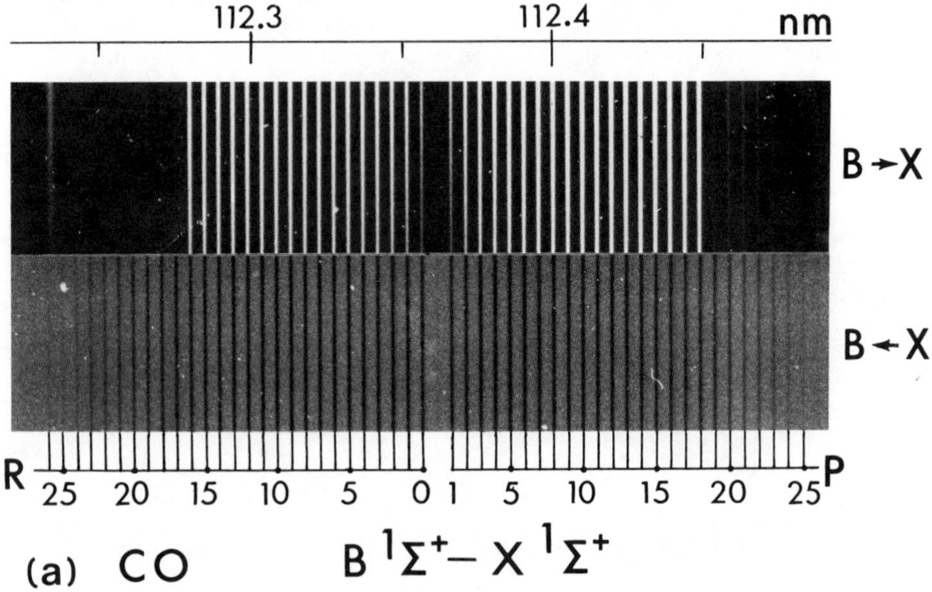

Fig. 1.3 Perturbations and predissociations affect absorption and emission line intensities in quite different ways. Two pairs of absorption and emission spectra are shown. The first pair illustrates the disappearance of a weakly predissociated line in emission without any detectable intensity or lineshape alteration in absorption. The second pair shows that emission from upper levels with slow radiative decay rates can be selectively quenched by collision induced energy transfer. The opposite effect, selective collisional enhancement of emission from perturbed, longer-lived levels, is well known in CN $B^2\Sigma^+ - X^2\Sigma^+$ ($v' = 0, v''$) emission spectra (see Fig. 5.11 and Section 5.5.4). (*a*) The CO $B^1\Sigma^+ - X^1\Sigma^+$ (1,0) band in emission (top) and absorption (bottom). The last strong lines in emission are R(16) and P(18). Emission from levels with $J' > 17$ is weak because the predissociation rate is larger than the spontaneous emission rate. (Courtesy F. Launay and J. Y. Roncin.) (*b*) The CO $A^1\Pi - X^1\Sigma^+$ (0,0) band in emission (bottom) and absorption (top). The $a'^3\Sigma^+$–$X^1\Sigma^+$(8,0) band lines appear in absorption because the $A^1\Pi \sim a'^3\Sigma^+$ spin-orbit interaction causes a small amount of $A^1\Pi$ character to be admixed into the nominal $a'^3\Sigma^+$ levels. These $a' - X$ lines are absent from the emission spectrum because collisional quenching and radiative decay into $a^3\Pi$ compete more effectively with radiative decay into $X^1\Sigma^+$ from the long-lived $a'^3\Sigma^+$ state than from the short-lived $A^1\Pi$ state. In addition, collisions and radiative decay into $a^3\Pi$ cause the P(31) extra line (E) (arising from a perturbation by $d^3\Delta\, v = 4$) to be weakened in emission relative to the main line (M). (Courtesy F. Launay, A. Le Floch, and J. Rostas.)

from one zero-order model may be vanishingly small for another. Consequently, the definition of a perturbation depends critically on one's choice of zero-order model, which is often a matter of personal preference.

Once a zero-order model is selected, the effects of all terms ($\mathbf{H}'$) in the Hamiltonian ($\mathbf{H}$) neglected in the definition of the zero-order basis functions (ψ_i^0, which are eigenfunctions of $\mathbf{H}^0$) may be taken into account exactly. One constructs a deperturbation model in which perturbation parameters govern

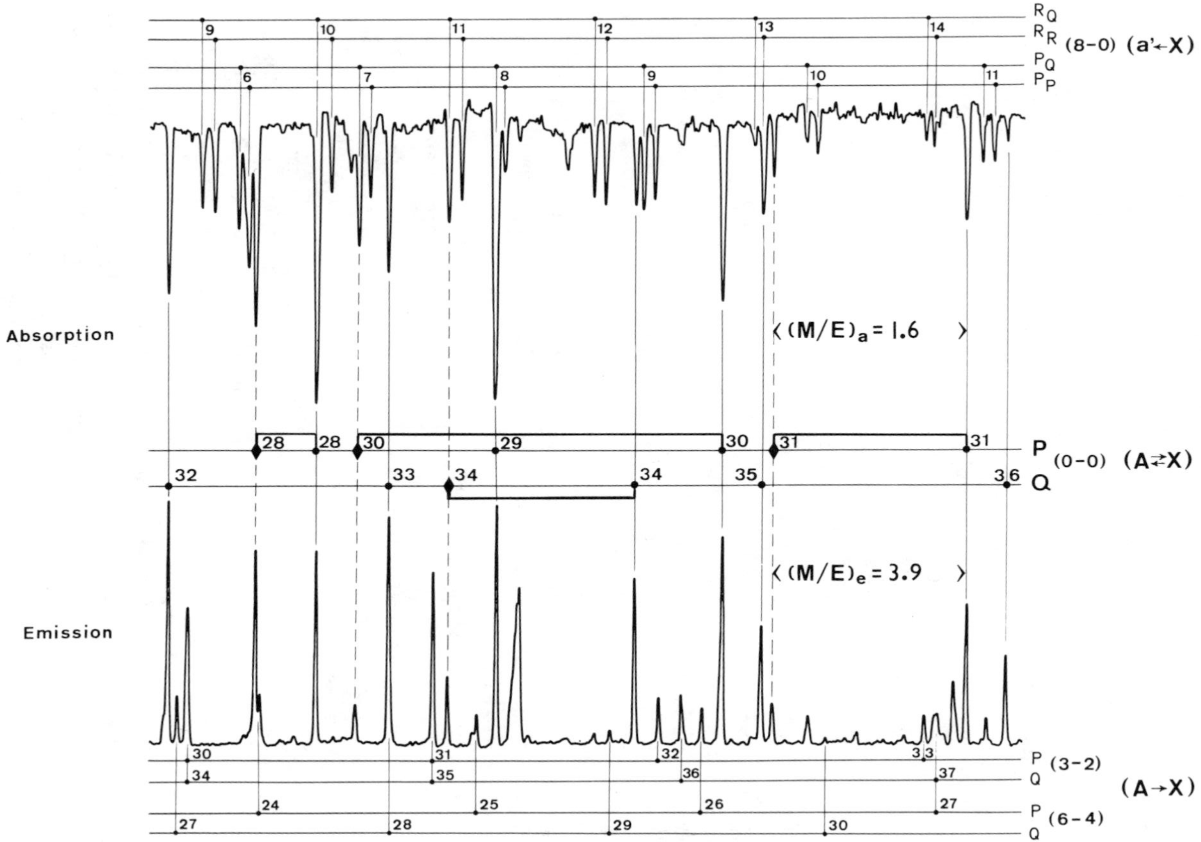

Absorption
Emission
(8-0) (a'←X)
P (0-0) (A⇄X)
Q
$\langle (M/E)_a = 1.6 \rangle$
$\langle (M/E)_e = 3.9 \rangle$
P (3-2)
Q
P (6-4)
Q
(A→X)

Fig. 1.3b

the strength of the perturbation or, in other words, the magnitudes of the differences between the (observable) eigenvalues of $\mathbf{H}$ (the perturbed levels) and the (unobservable but calculable) eigenvalues of $\mathbf{H}^0$ (the deperturbed levels). These perturbation parameters are matrix elements of $\mathbf{H}'$ between basis functions. By refining the molecular structural parameters that define both $\mathbf{H}^0$ and $\mathbf{H}'$, the deperturbation model is made to account for the properties of all observed levels. The deperturbation model should describe strongly perturbed and relatively perturbation free levels equally well.

The choice of a zero-order model determines the names used to label the observed energy levels. These names are a conventionally agreed upon (Jenkins, 1953) set of shorthand labels for the basis functions ψ_j^0, not for the eigenfunctions ψ_i. Since the eigenfunctions of the exact $\mathbf{H}$ are linear combinations of basis functions

$$\psi_i = \sum_j c_{ij}\psi_j^0,$$

it is not obvious how to compactly and unambiguously label the levels. It is customary to use the name of the basis function ψ_j^0, which has the largest coefficient $|c_{ij}|$ in the expansion of the ψ_i eigenfunction, as the name of the E_ith level. One must be careful not to be confused by the use of basis function names to specify both basis functions and the *nominal* (i.e., dominant) character of an eigenfunction.

It is important to emphasize that, although the label "perturbed" or "unperturbed" can depend on an arbitrary choice among several zero-order models (where each model in principle is equally capable of accounting for all molecular properties), there is no relativism in judging the quality of agreement between observed and calculated spectra. However, there are major differences in the ease of performing the calculations required by a particular deperturbation model or, more significantly, in the arduous process of refining a deperturbation model, starting from a fragmentary and incompletely assigned data base.

The primary purpose of a deperturbation should not be to obtain highly precise molecular constants. The most important objectives of a deperturbation are a perfect fit of the observed spectral lines, confirmation of line assignments, an assessment of the completeness of the deperturbation, proof of the mechanism of the perturbation, and determination of values of the perturbation parameters. The quality of the fit establishes the credibility of the model and reduces the possibility of misassigned lines. Assignments of perturbed lines are invariably difficult because of the absence of the usual spectral pattern and the low intensity of lines containing maximum information. Often, a deperturbation model can guide new assignments of a few levels at a time, testing the consequences of a dubious assignment by extrap-

Further, θ_{M_A} and $\theta_{M_A \pm 1}$ are arbitrary phase factors of the $|A\alpha M_A\rangle$ and $|A\alpha M_A \pm 1\rangle$ basis functions. If one requires that all matrix elements of $\mathbf{A}_X$ be real and positive, then the $2A + 1$ $|A\alpha M_A\rangle$ basis functions for $M_A = -A, -A + 1, \ldots, A$ must all have the same phase factor. This is generally referred to as the Condon and Shortley phase convention (Condon and Shortley, 1953, p. 48).

It is possible to derive all matrix elements of molecule-fixed components of $\vec{\mathbf{A}}$ in the $|A\alpha M_A\rangle$ basis from the commutation behavior of $\mathbf{A}_x$, $\mathbf{A}_y$, $\mathbf{A}_z$. Unfortunately, some angular momenta $(\mathbf{I}, \mathbf{L}, \mathbf{S}, \mathbf{J}_a)$ obey normal commutation rules,

$$[\mathbf{A}_i, \mathbf{A}_j] = i\hbar \sum_k \varepsilon_{ijk} \mathbf{A}_k, \tag{1.3.5}$$

while others $(\mathbf{R}, \mathbf{F}, \mathbf{J}, \mathbf{N}, \mathbf{O})$, *those describing rotations of the molecule-fixed coordinate system relative to the laboratory*, obey anomalous commutation rules,

$$[\mathbf{A}_i, \mathbf{A}_j] = -i\hbar \sum_k \varepsilon_{ijk} \mathbf{A}_k. \tag{1.3.6}$$

For normal commutation, the matrix elements for $\mathbf{A}_i$ are analogous to those for $\mathbf{A}_I$,

$$\langle A'\alpha' M'_A | \mathbf{A}_z | A\alpha M_A\rangle = +\hbar\alpha \delta_{A'A} \delta_{\alpha'\alpha} \delta_{M'_A M_A}, \tag{1.3.7}$$

$$\langle A'\alpha' M'_A | \mathbf{A}^{\pm} | A\alpha M_A\rangle = +\hbar e^{i(\phi_\alpha - \phi_{\alpha \pm 1})} [A(A+1) - \alpha(\alpha \pm 1)]^{1/2} \times \delta_{A'A} \delta_{\alpha'\alpha \pm 1} \delta_{M'_A M_A}, \tag{1.3.8}$$

where the *molecule-fixed raising and lowering operators are denoted here* $\mathbf{A}^{\pm}$,

$$\mathbf{A}^{\pm} \equiv \mathbf{A}_x \pm i\mathbf{A}_y,$$

to distinguish them from space-fixed $\mathbf{A}_{\pm}$ operators. Similarly, for molecule-fixed components of $\vec{\mathbf{A}}$ obeying anomalous commutation rules, the $\mathbf{A}_z$ matrix elements are identical to Eq. (1.3.7), but

$$\langle A'\alpha' M'_A | \mathbf{A}_x \pm i\mathbf{A}_y | A\alpha M_A\rangle = \langle A'\alpha' M'_A | \mathbf{A}^{\pm} | A\alpha M_A\rangle = \hbar e^{i(\phi_\alpha - \phi_{\alpha \mp 1})} [A(A+1) - \alpha(\alpha \mp 1)]^{1/2} \times \delta_{A'A} \delta_{\alpha'\alpha \mp 1} \delta_{M'_A M_A}. \tag{1.3.9}$$

The raising and lowering roles of $\mathbf{A}^+$ and $\mathbf{A}^-$ are interchanged when $\mathbf{A}$ behaves anomalously. It would be reasonable to require that the $\mathbf{A}_x$ matrix elements for both normal and anomalous $\mathbf{A}$ be real and positive, thereby requiring the ϕ_α phase factors for all $2A + 1$ basis functions to be equal. Such a phase choice is not in universal use. A standard phase convention has been proposed (Brown and Howard, 1976).

Brown and Howard (1976) have suggested that all molecule-fixed matrix elements be evaluated in terms of space-fixed operator components. The reasons for this are that the space-fixed components of all operators obey normal commutation rules, that it is natural to adopt the Condon and Shortley phase convention for the space-fixed part of the basis functions [$\theta_{M_A} = \theta_{M_A \pm 1}$ for all M_A, $\mathbf{A}_\pm$ matrix elements are real and positive, see Eq. (1.3.4)], and that all remaining ambiguities about the relative phases of the molecule-fixed part of the basis functions [ϕ_α; see (Eq. 1.3.9)] are settled by the choice of direction cosine matrix elements or rotation matrices used to transform between space- and molecule-fixed coordinate systems.

It is necessary to define precisely what is meant by a molecule-fixed component of any angular momentum that includes **R** (e.g., **R**, **N**, **O**, **J**, and **F**). Brown and Howard do this using spherical tensor methods; the following derivation is in terms of direction cosines. The direction cosine operator, $\boldsymbol{\alpha}$, is a 3×3 matrix, the components of which are defined by

$$\boldsymbol{\alpha}_I^j = \hat{\mathbf{I}} \cdot \hat{\mathbf{j}}, \tag{1.3.10}$$

where $\hat{\mathbf{I}}$, $\hat{\mathbf{J}}$, $\hat{\mathbf{K}}$ and $\hat{\mathbf{i}}$, $\hat{\mathbf{j}}$, $\hat{\mathbf{k}}$ are unit vectors defining right-handed Cartesian space- and molecule-fixed coordinate systems. Any operator which does not include **R** (e.g. **S**, **L**, **I**) will not cause relative rotation of the molecule- and space-fixed systems and will therefore commute with $\boldsymbol{\alpha}$. *For such operators only*, the transformation between molecule and space components is analogous to that for **S**,

$$\mathbf{S}_I = \sum_j \boldsymbol{\alpha}_I^j \mathbf{S}_j \qquad \text{for} \quad I = X, Y, Z, +, - \tag{1.3.11a}$$

$$\mathbf{S}_X = \tfrac{1}{2}[\boldsymbol{\alpha}_X^+ \mathbf{S}^- + \boldsymbol{\alpha}_X^- \mathbf{S}^+] + \boldsymbol{\alpha}_X^z \mathbf{S}_z \tag{1.3.11b}$$

$$\mathbf{S}_Y = \tfrac{1}{2}[\boldsymbol{\alpha}_Y^+ \mathbf{S}^- + \boldsymbol{\alpha}_Y^- \mathbf{S}^+] + \boldsymbol{\alpha}_Y^z \mathbf{S}_z \tag{1.3.11c}$$

$$\mathbf{S}_Z = \tfrac{1}{2}[\boldsymbol{\alpha}_Z^+ \mathbf{S}^- + \boldsymbol{\alpha}_Z^- \mathbf{S}^+] + \boldsymbol{\alpha}_Z^z \mathbf{S}_z, \tag{1.3.11d}$$

where

$$\boldsymbol{\alpha}_I^\pm = \boldsymbol{\alpha}_I^x \pm i\boldsymbol{\alpha}_I^y. \tag{1.3.12}$$

As a specific example, consider matrix elements of $\mathbf{J} \cdot \mathbf{S}$, which is a product of two operators, one having anomalous, the other regular molecule-fixed commutation rules. The value of a specific matrix element, $\langle nJ\Omega M | \mathbf{J} \cdot \mathbf{S} | n'J'\Omega' M' \rangle$, cannot depend on whether $\mathbf{J} \cdot \mathbf{S}$ is written in terms of space-fixed components,

$$\mathbf{J} \cdot \mathbf{S} = \mathbf{J}_Z \mathbf{S}_Z + \mathbf{J}_X \mathbf{S}_X + \mathbf{J}_Y \mathbf{S}_Y = \mathbf{J}_Z \mathbf{S}_Z + \tfrac{1}{2}(\mathbf{J}_+ \mathbf{S}_- + \mathbf{J}_- \mathbf{S}_+), \tag{1.3.13}$$

or molecule-fixed components,

$$\mathbf{J}\cdot\mathbf{S} = \mathbf{J}_z\mathbf{S}_z + \tfrac{1}{2}(\mathbf{J}^+\mathbf{S}^- + \mathbf{J}^-\mathbf{S}^+). \tag{1.3.14}$$

Inserting the Eq. (1.3.11) transformation from $\mathbf{S}_{\mathrm{SPACE}}$ to $\mathbf{S}_{\mathrm{molecule}}$ into Eq. (1.3.13) and comparing the coefficients of $\mathbf{S}_z$, $\mathbf{S}^-$, and $\mathbf{S}^+$ to those in Eq. (1.3.14), the following *definitions* of the molecule-fixed components of $\vec{\mathbf{J}}$ are obtained:

$$\mathbf{J}_z = \mathbf{J}_Z\boldsymbol{\alpha}_Z^z + \tfrac{1}{2}(\mathbf{J}_+\boldsymbol{\alpha}_-^z + \mathbf{J}_-\boldsymbol{\alpha}_+^z), \tag{1.3.15a}$$

$$\mathbf{J}^+ = \mathbf{J}_Z\boldsymbol{\alpha}_Z^+ + \tfrac{1}{2}(\mathbf{J}_+\boldsymbol{\alpha}_-^+ + \mathbf{J}_-\boldsymbol{\alpha}_+^+), \tag{1.3.15b}$$

$$\mathbf{J}^- = \mathbf{J}_Z\boldsymbol{\alpha}_Z^- + \tfrac{1}{2}(\mathbf{J}_+\boldsymbol{\alpha}_-^- + \mathbf{J}_-\boldsymbol{\alpha}_+^-), \tag{1.3.15c}$$

where

$$\boldsymbol{\alpha}_\pm^j = \boldsymbol{\alpha}_X^j \pm i\boldsymbol{\alpha}_Y^j. \tag{1.3.16}$$

More concisely,

$$\mathbf{J}_i = \sum_I \mathbf{J}_I\boldsymbol{\alpha}_I^i \qquad \text{for} \quad i = x, y, z, +, -. \tag{1.3.17}$$

The goal of this operator algebra is to obtain a foolproof prescription for evaluating the molecule-fixed matrix elements of $\mathbf{J}^\pm$. Consider the specific matrix element

$$\langle JM\Omega|\mathbf{J}^+|JM\Omega + 1\rangle.$$

Using Eq. (1.3.15b) and the completeness relationship,

$$\sum_{J',M',\Omega'} |J'M'\Omega'\rangle\langle J'M'\Omega'| = \mathbf{1} \qquad \text{(the unit matrix)},$$

to expand the products of $\mathbf{J}_I$ and $\boldsymbol{\alpha}_I^j$ operators, for example,

$$\langle JM\Omega|\mathbf{J}_+\boldsymbol{\alpha}_-^+|JM\Omega + 1\rangle = \sum_{J'M'\Omega'} \langle JM\Omega|\mathbf{J}_+|J'M'\Omega'\rangle \times \langle J'M'\Omega'|\boldsymbol{\alpha}_-^+|JM\Omega + 1\rangle, \tag{1.3.18}$$

noting that $\mathbf{J}_+$ selection rules require $J' = J$, $\Omega' = \Omega$, and $M' = M - 1$, the summation reduces to a single term,

$$\langle JM\Omega|\mathbf{J}_+\boldsymbol{\alpha}_-^+|JM\Omega + 1\rangle = \langle JM\Omega|\mathbf{J}_+|JM - 1\Omega\rangle \times \langle JM - 1\Omega|\boldsymbol{\alpha}_-^+|JM\Omega + 1\rangle. \tag{1.3.19}$$

The space-fixed $\mathbf{J}_+$ matrix element is

$$\langle JM\Omega|\mathbf{J}_+|JM - 1\Omega\rangle = \hbar[J(J+1) - M(M-1)]^{1/2}$$

and, using the $\Delta J = 0$ direction cosine matrix elements from Table 1.1 (see Hougen, 1970, Table 6),†

$$\langle JM - 1\Omega | \boldsymbol{\alpha}_X^+ - i\boldsymbol{\alpha}_Y^+) | JM\Omega + 1\rangle = \frac{[J(J+1) - M(M-1)]^{1/2}[J(J+1) - \Omega(\Omega+1)]^{1/2}}{J(J+1)}.$$

Thus

$$\langle JM\Omega | \mathbf{J}_+ \boldsymbol{\alpha}_-^+ | JM\Omega + 1\rangle = \frac{\hbar[J(J+1) - M(M-1)][J(J+1) - \Omega(\Omega+1)]^{1/2}}{J(J+1)}. \tag{1.3.20}$$

Similarly,

$$\langle JM\Omega | \mathbf{J}_- \boldsymbol{\alpha}_+^+ | JM\Omega + 1\rangle = \frac{\hbar[J(J+1) - M(M+1)][J(J+1) - \Omega(\Omega+1)]^{1/2}}{J(J+1)}, \tag{1.3.21}$$

and

$$\langle JM\Omega | \mathbf{J}_Z \boldsymbol{\alpha}_Z^+ | JM\Omega + 1\rangle = \frac{\hbar M^2[J(J+1) - \Omega(\Omega+1)]^{1/2}}{J(J+1)}, \tag{1.3.22}$$

thus

$$\begin{aligned}\langle JM\Omega | \mathbf{J}^+ | JM\Omega + 1\rangle &= \frac{\hbar J(J+1)[J(J+1) - \Omega(\Omega+1)]^{1/2}}{J(J+1)} \\ &= +\hbar[J(J+1) - \Omega(\Omega+1)]^{1/2}.\end{aligned} \tag{1.3.23}$$

The matrix elements for $\mathbf{J}_z$ and $\mathbf{J}^-$ are

$$\langle JM\Omega | \mathbf{J}_z | JM\Omega\rangle = +\hbar M, \tag{1.3.24}$$

$$\langle JM\Omega | \mathbf{J}^- | JM\Omega - 1\rangle = +\hbar[J(J+1) - \Omega(\Omega-1)]^{1/2}. \tag{1.3.25}$$

The reversal of raising/lowering roles of $J^{\pm}$ is made obvious by

$$\langle JM\Omega \pm 1 | \mathbf{J}^{\mp} | JM\Omega\rangle = \hbar[J(J+1) - \Omega(\Omega \pm 1)]^{1/2}.$$

† The phase choices implicit in the definitions of $|JM\Omega\rangle$ basis functions and tabulated $\boldsymbol{\alpha}_I^i$ direction cosine matrix elements are summarized in Section 1.3.3.

Table 1.1
Direction Cosine Matrix Elements for $\Delta J = 0$[a]

$\langle JM'\Omega'|\boldsymbol{\alpha}|JM\Omega\rangle$

$\Delta M = M' - M$	$\Delta\Omega = \Omega' - \Omega$: $+1$	0	-1
$+1$	$\langle\alpha_+^-\rangle = \langle\alpha_X^- + i\alpha_Y^-\rangle = \langle\alpha_X^x - i\alpha_X^y + i(\alpha_Y^x - i\alpha_Y^y)\rangle = \langle\alpha_+^x - i\alpha_+^y\rangle$ $= \dfrac{[J(J+1) - M(M+1)]^{1/2}[J(J+1) - \Omega(\Omega+1)]^{1/2}}{J(J+1)}$	$\langle\alpha_+^z\rangle = \langle\alpha_X^z + i\alpha_Y^z\rangle = \dfrac{\Omega[J(J+1) - M(M+1)]^{1/2}}{J(J+1)}$	$\langle\alpha_+^+\rangle = \langle\alpha_X^+ + i\alpha_Y^+\rangle = \langle\alpha_+^x + i\alpha_+^y\rangle$ $= \dfrac{[J(J+1) - M(M+1)]^{1/2}[J(J+1) - \Omega(\Omega-1)]^{1/2}}{J(J+1)}$
0	$\langle\alpha_Z^-\rangle = \langle\alpha_Z^x - i\alpha_Z^y\rangle = \dfrac{M[J(J+1) - \Omega(\Omega+1)]^{1/2}}{J(J+1)}$	$\langle\alpha_Z^z\rangle = \dfrac{\Omega M}{J(J+1)}$	$\langle\alpha_Z^+\rangle = \langle\alpha_Z^x + i\alpha_Z^y\rangle = \dfrac{M[J(J+1) - \Omega(\Omega-1)]^{1/2}}{J(J+1)}$
-1	$\langle\alpha_-^-\rangle = \langle\alpha_X^- - i\alpha_Y^-\rangle = \langle\alpha_-^x - i\alpha_-^y\rangle$ $= \dfrac{[J(J+1) - M(M-1)]^{1/2}[J(J+1) - \Omega(\Omega+1)]^{1/2}}{J(J+1)}$	$\langle\alpha_-^z\rangle = \langle\alpha_X^z - i\alpha_Y^z\rangle = \dfrac{\Omega[J(J+1) - M(M-1)]^{1/2}}{J(J+1)}$	$\langle\alpha_-^+\rangle = \langle\alpha_X^+ - i\alpha_Y^+\rangle = \langle\alpha_-^x + i\alpha_-^y\rangle$ $= \dfrac{[J(J+1) - M(M-1)]^{1/2}[J(J+1) - \Omega(\Omega-1)]^{1/2}}{J(J+1)}$

[a] From Hougen (1970), Table 6.

This justifies the general statement of Eq. (1.3.9) that, for all operators **A** obeying anomalous molecule-fixed commutation rules, Eq. (1.3.6) (any operator that includes **R**), the matrix elements of $\mathbf{A}^{\pm}$ are real and positive but that $\mathbf{A}^{+}$ and $\mathbf{A}^{-}$ act exactly as lowering and raising operators, respectively. This reversal of the lowering/raising roles of $\mathbf{A}^{\pm}$ may be viewed as arising from the direction cosine matrix elements [Eq. (1.3.17)] rather than any unusual property of the operator **A**.

1.3.2 Recipes for Evaluation of Molecule-Fixed Angular Momentum Matrix Elements

Consider the basis set $|nJLS\Omega\Lambda\Sigma M_J\rangle$, where n is a shorthand for all remaining quantum numbers (e.g., electronic configuration and vibrational level) needed to uniquely specify the basis function. Although, for reasons discussed in Section 2.2, L cannot usually be specified, this is the Hund's case (a) basis. Inclusion of L here simplifies the following discussion and illustrates an important point, namely, that when a basis set is defined in terms of a sufficient number of magnitude and molecule-fixed projection quantum numbers, then all conceivable magnitude and molecule-fixed angular momentum matrix elements are evaluable by elementary techniques. Matrix elements of $\mathbf{J}_a^2$ and $\mathbf{R}^2$ are evaluated below as illustration of this point.

The matrix elements of the $\mathbf{J}_a$ operator will be calculated as a first step toward those of **R**. Eigenvalues of the operators $\mathbf{J}_a^2$ do not appear in the list of quantum numbers which define the $|nJLS\Omega\Lambda\Sigma M_J\rangle$ basis. However,

$$\mathbf{J}_a = \mathbf{L} + \mathbf{S}, \tag{1.3.26}$$

$$\mathbf{J}_a^2 = \mathbf{L}^2 + \mathbf{S}^2 + \mathbf{L}\cdot\mathbf{S} + \mathbf{S}\cdot\mathbf{L}, \tag{1.3.27}$$

and, since **L** and **S** commute (because they act on different coordinates),

$$\mathbf{J}_a^2 = \mathbf{L}^2 + \mathbf{S}^2 + 2\mathbf{L}_z\mathbf{S}_z + (\mathbf{L}^{+}\mathbf{S}^{-} + \mathbf{L}^{-}\mathbf{S}^{+}). \tag{1.3.28}$$

Thus, $\mathbf{J}_a^2$ is expressed in terms of operators for which the matrix elements are easily evaluable. Further, $\mathbf{L}^2$, $\mathbf{S}^2$, and $\mathbf{L}_z\mathbf{S}_z$ have only diagonal matrix elements in the $|nJLS\Omega\Lambda\Sigma M_J\rangle$ basis, whereas the $\mathbf{L}^{\pm}\mathbf{S}^{\mp}$ terms follow $\Delta\Lambda = -\Delta\Sigma = \pm 1$, $\Delta\Omega = 0$, $\Delta J = \Delta L = \Delta S = 0$ selection rules:

1. Diagonal matrix elements of $\mathbf{J}_a^2$:

$$\langle n'JLS\Omega\Lambda\Sigma M_J|\mathbf{J}_a^2|nJLS\Omega\Lambda\Sigma M_J\rangle$$
$$= \hbar^2[L(L+1) + S(S+1) + 2\Lambda\Sigma] \tag{1.3.29}$$

2. $\Delta\Lambda = -\Delta\Sigma = \pm 1$ off-diagonal matrix elements of $\mathbf{J}_a^2$: The only term in

$\mathbf{J}_a^2$ that gives rise to $\Delta\Lambda = -\Delta\Sigma = \pm 1$ matrix elements is $\mathbf{L}^{\pm}\mathbf{S}^{\mp}$; thus the relevant matrix element is

$$\langle n'JLS\Omega, \Lambda \pm 1, \Sigma \mp 1, M_J | \mathbf{L}^{\pm}\mathbf{S}^{\mp} | nJLS\Omega\Lambda\Sigma M_J \rangle$$
$$= \hbar^2[L(L+1) - \Lambda(\Lambda \pm 1)]^{1/2}[S(S+1) - \Sigma(\Sigma \mp 1)]^{1/2}. \quad (1.3.30)$$

If L cannot be specified in the basis set, then the selection rules remain unchanged and $L(L+1)$, wherever it appears, must be replaced by an unknown constant.

The operator $\mathbf{R}^2$ is important because it appears in the rotational Hamiltonian operator,

$$\mathbf{R}^2 = (\mathbf{J} - \mathbf{L} - \mathbf{S})^2, \quad (1.3.31)$$

or, taking advantage of the fact that matrix elements of $\mathbf{J}_a^2$ have just been worked out,

$$\mathbf{R}^2 = (\mathbf{J} - \mathbf{J}_a)^2 = \mathbf{J}^2 + \mathbf{J}_a^2 - \mathbf{J}\cdot\mathbf{J}_a - \mathbf{J}_a\cdot\mathbf{J}, \quad (1.3.32)$$

and, since $[\mathbf{J}_i, \mathbf{J}_{ai}] = 0$, then

$$\mathbf{R}^2 = \mathbf{J}^2 + \mathbf{J}_a^2 - 2\mathbf{J}_z\mathbf{J}_{az} - (\mathbf{J}^+\mathbf{J}_a^- + \mathbf{J}^-\mathbf{J}_a^+). \quad (1.3.33)$$

Since $\mathbf{J}$ (but not $\mathbf{J}_a$) is an operator that involves rotation of the molecule relative to the space-fixed coordinate system, $\mathbf{J}^+$ acts as a lowering operator on Ω, whereas $\mathbf{J}_a^+$ acts as a raising operator on Λ or Σ (and thus indirectly on $\Omega = \Lambda + \Sigma$). Arranged according to selection rules, the matrix elements of $\mathbf{R}^2$ are:

1. Diagonal matrix elements of $\mathbf{R}^2$:

$$\langle n'JLS\Omega\Lambda\Sigma M_J | \mathbf{R}^2 | nJLS\Omega\Lambda\Sigma M_J \rangle$$
$$= \hbar^2[J(J+1) + L(L+1) + S(S+1) + 2\Lambda\Sigma - 2\Omega^2], \quad (1.3.34)$$

or, rearranged into a more symmetric form,

$$\hbar^2[J(J+1) - \Omega^2 + L(L+1) - \Lambda^2 + S(S+1) - \Sigma^2].$$

2. $\Delta\Omega = 0$, $\Delta\Lambda = -\Delta\Sigma = \pm 1$ off-diagonal matrix elements of $\mathbf{R}^2$:

$$\langle n'JLS\Omega\Lambda \pm 1\Sigma \mp 1 M_J | \mathbf{R}^2 | nJLS\Omega\Lambda\Sigma M_J \rangle$$
$$= -\hbar^2[L(L+1) - \Lambda(\Lambda \pm 1)]^{1/2}[S(S+1) - \Sigma(\Sigma \mp 1)]^{1/2}. \quad (1.3.35)$$

3. $\Delta\Omega = \Delta\Lambda = \pm 1$, $\Delta\Sigma = 0$ off-diagonal matrix elements of $\mathbf{R}^2$:

$$\langle n'JLS\Omega \pm 1\Lambda \pm 1\Sigma M_J | \mathbf{R}^2 | nJLS\Omega\Lambda\Sigma M_J \rangle$$
$$= -\hbar^2[J(J+1) - \Omega(\Omega \pm 1)]^{1/2}[L(L+1) - \Lambda(\Lambda \pm 1)]^{1/2}. \quad (1.3.36)$$

4. $\Delta\Omega = \Delta\Sigma = \pm 1$, $\Delta\Lambda = 0$, off-diagonal matrix elements of $\mathbf{R}^2$:

$$\langle n'JLS\Omega \pm 1\Lambda\Sigma \pm 1 M_J | \mathbf{R}^2 | nJLS\Omega\Lambda\Sigma M_J \rangle$$
$$= -\hbar^2[J(J+1) - \Omega(\Omega \pm 1)]^{1/2}[S(S+1) - \Sigma(\Sigma \pm 1)]^{1/2}. \quad (1.3.37)$$

These results are summarized in Table 2.2.

In order to illustrate the unique convenience of the case (a) basis, consider matrix elements of $\mathbf{J} \cdot \mathbf{S}$ and $\mathbf{L} \cdot \mathbf{S}$ in the case (b) $|nJLSN\Lambda M_J\rangle$ basis. The matrix elements of $\mathbf{J} \cdot \mathbf{S}$ (also $\mathbf{N} \cdot \mathbf{S}$ and $\mathbf{J} \cdot \mathbf{N}$) are evaluated easily by making the operator substitution:

$$\begin{aligned} \mathbf{N} &= \mathbf{J} - \mathbf{S}, \\ \mathbf{N}^2 &= \mathbf{J}^2 + \mathbf{S}^2 - 2\mathbf{J} \cdot \mathbf{S}, \\ \mathbf{J} \cdot \mathbf{S} &= \tfrac{1}{2}(\mathbf{J}^2 + \mathbf{S}^2 - \mathbf{N}^2). \end{aligned} \tag{1.3.38}$$

Thus $\mathbf{J} \cdot \mathbf{S}$ has only diagonal matrix elements,

$$\langle n'JLSN\Lambda M_J|\mathbf{J} \cdot \mathbf{S}|nJLSN\Lambda M_J\rangle = \tfrac{1}{2}\hbar^2[J(J+1) + S(S+1) - N(N+1)]. \tag{1.3.39}$$

The story is quite different for $\mathbf{L} \cdot \mathbf{S}$, because no suitable operator replacement exists and, when attempting to apply standard raising and lowering operator techniques to

$$\mathbf{L}_z\mathbf{S}_z + \tfrac{1}{2}(\mathbf{L}^+\mathbf{S}^- + \mathbf{L}^-\mathbf{S}^+),$$

one finds that the matrix elements and selection rules for $\mathbf{S}_z$ and $\mathbf{S}^{\pm}$ are not obvious. The solution to this problem, discussed in Section 2.2, requires expression of case (b) basis functions in terms of case (a) functions for which evaluation of $\mathbf{L} \cdot \mathbf{S}$ matrix elements is easy.

1.3.3 Euler Angles, $|JM\Omega\rangle$ Basis Functions, Direction Cosines, and Phases

In dealing simultaneously with space- and molecule-fixed coordinate systems, one must explicitly define the transformations between coordinate systems and specify the relative phases of all basis functions. Larsson (1981) has reviewed the numerous phase conventions currently in use for rotating diatomic molecules. (See also Wolf, 1969.)

The normalized rotating molecule wavefunction, expressed in terms of the three Euler angles α, β, and γ (see Fig. 1.1 in Edmonds (1974), p. 7) is

$$\begin{aligned} \langle\alpha\beta\gamma\,|\,JM\Omega\rangle &= \left(\frac{2J+1}{8\pi^2}\right)^{1/2} \mathscr{D}^J_{\Omega M}(\alpha, \beta, \gamma) \\ &= \left(\frac{2J+1}{8\pi^2}\right)^{1/2} \exp(iM\gamma)\, d^J_{\Omega M}(\beta) \exp(i\Omega\alpha) \end{aligned} \tag{1.3.40}$$

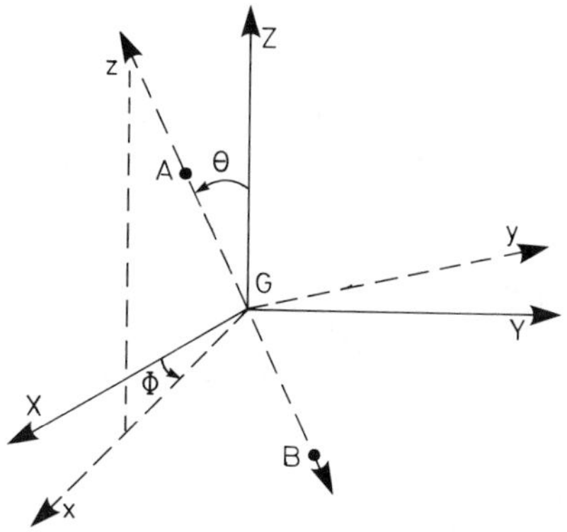

Fig. 1.4 Relationships between molecule-fixed (xyz) and space-fixed (XYZ) axis systems. The origin of both coordinate systems is located at the AB molecule center of mass, G. Transformation from XYZ to xyz ("active rotation"; Larsson, 1981) requires rotation of the molecule by $\phi = \gamma$ about $+Z$, followed by clockwise rotation by $\theta = \beta$ about $+Y$, and then rotation by $\alpha = \pi/2$ (not shown) about Z. Since diatomic molecules have cylindrical symmetry about the internuclear axis, the third angle required to specify the orientation of xyz relative to XYZ is fixed and implicitly specifies a phase choice for certain molecule-fixed $\mathbf{A}^{\pm}$ matrix elements.

where $d^J_{\Omega M}(\beta)$ is defined by Edmonds (1974, pp. 57–62).† The Euler angles relate the molecule-fixed axis system to the space-fixed system as follows (Edmonds, 1974, pp. 6–8): rotate the *molecule* first by γ about the space-fixed Z-axis, then by β about the space-fixed Y-axis, and finally by α about the space-fixed Z-axis. (Rotations are "active" and in the sense of advancing a right-handed screw along the specified axis.)

For a linear molecule, the position of the symmetry axis (the molecule-fixed z-axis) in space is specified by only two Euler angles, β and γ, which are respectively identical to the spherical polar coordinates θ and ϕ (see Fig. 1.4). The third Euler angle, α, which specifies the orientation of the molecule-fixed x- and y-axes, is unaffected by molecular rotation but appears explicitly as an Ω-dependent phase factor in the rotational basis functions [Eq. (1.3.40)]. Cartesian coordinates in space- and molecule-fixed systems are related by the geometrical transformation represented by the 3×3 direction cosine matrix (Wilson *et al.*, 1980, p. 286). The direction cosine matrix $\boldsymbol{\alpha}^i_I$ given by Hougen

† The notation $\langle \alpha\beta\gamma | JM\Omega \rangle$ is the translation, into Dirac notation, of the Schrodinger wavefunction

$$\psi_{JM\Omega}(\alpha, \beta, \gamma) = \langle \alpha\beta\gamma | JM\Omega \rangle.$$

One frequently sees $|JM\Omega\rangle$ defined as a function of α, β, γ, but strictly speaking $|JM\Omega\rangle$ is a vector, not a function of continuous variables.

(1970, p. 18) is obtained by setting $\alpha = \pi/2$ (notation of Wilson *et al.*, 1980: $\theta = \beta, \phi = \gamma, \chi = \alpha = \pi/2$). The direction cosine matrix is expressed in terms of sines and cosines of θ and ϕ. Matrix elements $\langle J'M'\Omega' |\alpha_I^i| JM\Omega\rangle$, evaluated in the $|JM\Omega\rangle$ basis, of the direction cosines, are expressed in terms of the J, M, and Ω quantum numbers. The direction cosine matrix elements of Hougen (1970, p. 31), Townes and Schawlow (1955, p. 96), and Table 1.1 assume the basis set definition derived from Eq. (1.3.40) and the phase choice $\alpha = \pi/2$:

$$\left\langle \frac{\pi}{2}\theta\phi \middle| JM\Omega \right\rangle = \left(\frac{2J+1}{4\pi}\right)^{1/2} \mathscr{D}^J_{\Omega M}\left(\frac{\pi}{2}, \theta, \phi\right)$$
$$= \left(\frac{2J+1}{4\pi}\right)^{1/2} \exp(iM\phi) d^J_{\Omega M}(\theta) \exp\left(i\Omega\frac{\pi}{2}\right). \quad (1.3.41)$$

The factor of $(2\pi)^{1/2}$ difference in normalization factor between Eqs. (1.3.40) and (1.3.41) occurs because Eq. (1.3.41) implies no integration over the angle α. The rotational basis function defined in Eq. (1.3.41) is identical to that used by Hougen (1970, p. 18) because

$$\mathscr{D}^J_{\Omega M}(\alpha\beta\gamma) = (-1)^{M-\Omega}\mathscr{D}^J_{M\Omega}(\alpha\beta\gamma).$$

Brown and Howard (1976) define $\langle(\pi/2)\theta\phi\,|\,JM\Omega\rangle$ in terms of Brink and Satchler's (1968) $\mathscr{D}^J_{M\Omega}{}^*(\alpha\beta\gamma)$. However, Brink and Satchler's $\mathscr{D}^J_{M\Omega}{}^*(\alpha\beta\gamma)$ corresponds to Edmonds' $\mathscr{D}^J_{M\Omega}{}^*(-\alpha, -\beta, -\gamma)$ (Brink and Satchler, 1968, p. 21) and, in Edmonds' notation,

$$\mathscr{D}^J_{M\Omega}{}^*(-\alpha, -\beta, -\gamma) = (-1)^{M-\Omega}\mathscr{D}^J_{M\Omega}(\alpha\beta\gamma) = \mathscr{D}^J_{\Omega M}(\alpha\beta\gamma);$$

thus the rotational basis functions used here are also identical to those of Brown and Howard (1976).

The result obtained in Section 1.3.1 that the matrix elements of the space-fixed $\mathbf{J}_\pm$ and molecule fixed $\mathbf{J}^\pm$ operator components are real and positive are consequences of the $|JM\Omega\rangle$ definition and α-phase choice.

1.4 Estimation of Parameters in a Model Hamiltonian

Whenever a perturbation, predissociation, or autoionization is observed, its strength is governed by an off-diagonal matrix element. Although these matrix elements can frequently be calculated from *ab initio* wavefunctions or estimated semiempirically from other perturbation-related information, experimentalists typically treat the perturbation matrix element as a purely empirical parameter of no interest other than its capacity to account for the spectrum. This book is based on the premise that perturbation matrix elements have intrinsic molecular structural significance, that their magnitudes are predictable, and that their measured values often provide unexpected clues to the global electronic structure of a molecule.

In Chapter 2, methods are described for using the electronic wavefunction to evaluate off-diagonal matrix elements of different terms that appear in the true microscopic Hamiltonian, **H**. Section 2.2.4 describes some simple procedures, based on the single-electronic-configuration approximation, for reducing many-electron matrix elements of **H** to simple molecular-orbital (Section 2.4.2) one- and two-electron matrix elements. Several examples where the single-configuration approximation fails but a two-configuration picture is adequate are discussed in Sections 4.7 and 6.9.1. Little insight is possible when it becomes necessary (as in large configuration-interaction *ab initio* computations) to go beyond two configurations.

In Sections 4.2, 4.3, and 4.4, the orders of magnitude of all types of perturbation parameters are related to known or easily estimable quantities in order to provide a basis for predicting their absolute magnitudes, relative sizes, and inter- or intramolecular trends.

The same electronic matrix elements that control the strength of perturbations also govern predissociation (Table 6.2) and, in "reduced" form, autoionization (Section 7.3).

The vibrational part of the **H** matrix elements is much easier to calculate than the electronic one. Methods for such calculations are described in Section 4.1 for bound$\sim$bound interactions and in Section 6.3 for bound–continuum vibrational interactions.

An understanding of observable properties is seldom trivial. Spectroscopic energy levels are, in principle, eigenvalues of an infinite matrix representation of **H**, which is expressed in terms of an infinite number of "true" deperturbed molecular constants. In practice, this matrix is truncated and the "observed" molecular constants are the "effective" parameters that appear in a finite-dimension "effective Hamiltonian." The Van Vleck transformation, so crucial for reducing **H** to a finite $\mathbf{H}^{\text{eff}}$, is described in Section 3.2.

Experimentalists often treat the electronic part of **H** matrix elements, H^{el}, as a phenomenological "black box." They regard H^{el} as an empirically determined variable parameter without structural significance and are overwhelmed by the difficulty of estimating H^{el} or comparing it with calculated values. This cuts in the opposite direction as well: often theorists are unaware of or unable to utilize a class of observable electronic properties simply because the relationship between the observable quantity and the relevant matrix element of **H** is neither traditionally exploited nor explicitly defined.

The assumption implicit throughout this book is that the parameters used to fit or represent molecular transition frequencies and intensities contain insights into molecular structure. These insights can be more useful than the multi-digit fit parameters themselves, especially when simplifying assumptions are made and tested. Comparisons of observable or effective parameters to those obtained from an exact calculation (true parameters) or a simplified electronic structure model (one-electron orbitals parameters) are seldom trivial or unique. The purpose of this book is to help experimentalists and theorists to go beyond molecular fit parameters to terms in the exact microscopic Hamiltonian on the one hand and to approximate electronic structure models on the other. Physical insight, not tables of spectral data and molecular constants, is the ultimate purpose of fundamental experimental and theoretical research.

Data Compilations

Barrow R. F., ed. (1973, 1975, 1979, 1982), "Molécules Diatomiques. Bibliographie Critique de Données Spectroscopiques," Tables Internationales de Constantes, CNRS, Paris.

Herzberg, G. (1950), "Molecular Spectra and Molecular Structure. I. Spectra of Diatomic Molecules," Van Nostrand-Reinhold, Princeton, New Jersey.

Huber, K. P., and Herzberg, G. (1979), "Molecular Spectra and Molecular Structure. IV. Constants of Diatomic Molecules," Van Nostrand-Reinhold, Princeton, New Jersey.

Kopp, I., Lindgren, R., and Rydh, B. (1974), "Table of Band Features of Diatomic Molecules in Wavelength Order," Institute of Physics, University of Stockholm, Tables Internationales de Constantes, Paris.

Kopp, I., Jansson, K., and Rydh, B. (1977), "Table of Band Features of Diatomic Molecules in Wavelength Order," Complement A1, Institute of Physics, University of Stockholm, Tables Internationales de Constantes, Paris.

Pearse, R. W. B., and Gaydon, A. G. (1976), "The Identification of Molecular Spectra," Chapman and Hall, London.

Rosen, B., ed. (1970), "Tables Internationales de Constantes Sélectionnées. 17. Données Spectroscopiques relatives aux Molécules Diatomiques," Pergamon, Oxford.

Suárez, C. B. (1972), "Bibliography of Spectra of Diatomic Molecules," 1960–1970, Universidad Nacional de La Plata.

References

Brink, D. M., and Satchler, G. R. (1968), "Angular Momentum," Oxford Univ. Press (Clarendon), London and New York.

Brown, J. M., and Howard, B. J. (1976), *Mol. Phys.* **31**, 1517; **32**, 1197.

Budó, A. (1937), *Z. Phys.* **105**, 579.

Budó, A., and Kovács, I. (1938), *Z. Phys.* **109**, 393.

Condon, E. U., and Shortley, G. H. (1953), "The Theory of Atomic Spectra," Cambridge Univ. Press, London and New York.

Dieke, G. H. (1935), *Phys. Rev.* **47**, 870.

Dunham, J. L. (1932), *Phys. Rev.* **41**, 713, 721.

Edmonds, A. R. (1974), "Angular Momentum in Quantum Mechanics," Princeton Univ. Press, Princeton, New Jersey.

Herzberg, G. (1950), "Molecular Spectra and Molecular Structure. I. Spectra of Diatomic Molecules", Van Nostrand-Reinhold Princeton, New Jersey.

Heurlinger, T. (1917), *Ark. Mat., Astron. Fys.* **12**, 1; (1918), *Phys. Z.* **19**, 316; (1919), *Phys. Z.* **20**, 188.

Hougen, J. T. (1970), "The Calculation of Rotational Energy Levels and Rotational Line Intensities in Diatomic Molecules," Nat. Bur. Stand. (U.S.), monograph 115.

Huber, K. P., and Herzberg, G. (1979), "Molecular Spectra and Molecular Structure. IV. Constants of Diatomic Molecules," Van Nostrand-Reinhold, Princeton, New Jersey.

Ittmann, G. P. (1931), *Z. Phys.* **71**, 616.

Jenkins, F. A. (1953), *J. Opt. Soc. Am.* **43**, 425.

Kovács, I. (1937), *Z. Phys.* **106**, 431.

Kovács, I. (1969), "Rotational Structure in the Spectra of Diatomic Molecules," Amer. Elsevier, New York.

Kronig, R. L. (1928), *Z. Phys.* **50**, 347.

Larsson, M. (1981), *Phys. Scr.* **23**, 835.

Mulliken, R. S. (1930), *Rev. Mod. Phys.* **2**, 60, 506.

Mulliken, R. S. (1931), *Rev. Mod. Phys.* **3**, 89.

Mulliken, R. S. (1932), *Rev. Mod. Phys.* **4**, 1.

Schmid, R., and Gerö, L. (1935), *Z. Phys.* **93**, 656.

Stepanov, B. I. (1940), *J. Phys. USSR* **2**, 81, 89, 197, 205, 381.

Stepanov, B. I. (1945), *J. Phys. USSR* **9**, 317.

Townes, C. H., and Schawlow A. L. (1955), "Microwave Spectroscopy," McGraw-Hill, New York.

Wilson, E. B., Decius, J. C., and Cross, P. C. (1980), "Molecular Vibrations," Dover, New York.

Wolf, A. A. (1969), *Am. J. Phys.* **37**, 531.

Van Vleck, J. H. (1929), *Phys. Rev.* **33**, 467.

Van Vleck, J. H. (1932), *Phys. Rev.* **40**, 544.

Van Vleck, J. H. (1936), *J. Chem. Phys.* **4**, 327.

Chapter 2

Terms Neglected in the Born–Oppenheimer Approximation

2.1 The Born–Oppenheimer Approximation

The exact Hamiltonian **H** for a diatomic molecule, with the electronic coordinates expressed in the molecule-fixed axis system, is rather difficult to derive. Bunker (1968) provides a detailed derivation as well as a review of the coordinate conventions, implicit approximations, and errors in previous discussions of the exact diatomic molecule Hamiltonian.

Our goal is to find the exact solutions, ψ_i^{T} (T = total), of the Schrödinger equation,

$$\mathbf{H}\psi_i^{\mathrm{T}} = E_i^{\mathrm{T}}\psi_i^{\mathrm{T}}, \tag{2.1.1}$$

which correspond to the observed (exact) E_i^{T} energy levels. **H** is the nonrelativistic Hamiltonian, which may be approximated by a sum of three operators,

$$\mathbf{H} = \mathbf{T}^N(R,\theta,\phi) + \mathbf{T}^e(r) + V(r,R), \tag{2.1.2}$$

where $\mathbf{T}^N$ is the nuclear kinetic energy, $\mathbf{T}^e$ is the electron kinetic energy, V is the electrostatic potential energy for the nuclei and electrons (including e–e, e–N and N–N interactions), R is the internuclear distance, θ and ϕ specify the orientation of the internuclear axis (molecule-fixed coordinate system) relative to the laboratory coordinate system (see Section 1.3.3 and Fig. 1.4), and r represents all electron coordinates in the molecule-fixed system.

The nuclear kinetic energy operator is given by

$$\mathbf{T}^N(R,\theta,\phi) = \frac{-\hbar^2}{2\mu R^2}\left[\frac{\partial}{\partial R}\left(R^2\frac{\partial}{\partial R}\right) + \frac{1}{\sin\theta}\frac{\partial}{\partial\theta}\left(\sin\theta\frac{\partial}{\partial\theta}\right) + \frac{1}{\sin^2\theta}\frac{\partial^2}{\partial\phi^2}\right], \tag{2.1.3a}$$

where

$$\mu = \frac{M_{\mathrm{A}}M_{\mathrm{B}}}{M_{\mathrm{A}} + M_{\mathrm{B}}}$$

is the nuclear reduced mass, with M_{A} and M_{B} the masses of atoms A and B. $\mathbf{T}^N$ can be divided into vibrational and rotational terms,

$$\mathbf{T}^N(R,\theta,\phi) = \mathbf{T}^N(R) + \mathbf{H}^{\mathrm{ROT}}(R,\theta,\phi). \tag{2.1.3b}$$

The electron kinetic energy operator is

$$\mathbf{T}^e(r) = \frac{-\hbar^2}{2m}\sum_i \nabla_i^2, \tag{2.1.3c}$$

where m is the electron mass and the summation is over all electrons.†

† The part of $\mathbf{T}^e$,

$$\frac{-\hbar^2}{2(M_{\mathrm{A}} + M_{\mathrm{B}})}\sum_{i,j}\nabla_i\nabla_j,$$

the diagonal matrix elements of which contribute, for example, 38 cm^{-1} for the ground state of H_2 and 27 cm^{-1} for the $B^1\Sigma_u^+$ excited state of H_2 (Bunker, 1968), is neglected here.

To solve Eq. (2.1.1), it would be useful to write the total energy as a sum of contributions from interactions between different particles. In decreasing order of importance, there are electronic energy, E^{el}; vibrational energy, $G(v)$; and rotational energy, $F(J)$. In fact, this separation is assumed whenever the expression

$$E^{\mathrm{T}}_{\mathbf{app}} = E^{\mathrm{el}} + G(v) + F(J) \tag{2.1.4}$$

is used to represent observed energy levels. However, Eq. (2.1.4) is always an approximation. It is never possible to express E^{T} exactly as in Eq. (2.1.4). This means that it is not possible to separate $\mathbf{H}$ rigorously into terms corresponding to the different motions of the particles. The approximate wavefunction suggested by the desired but approximate energy expression Eq. (2.1.4) is a product of two functions,

$$\psi^{\mathrm{BO}}_{i,v} = \Phi_{i,\Lambda,S,\Sigma}(r, R)\chi_v(R, \theta, \phi) \tag{2.1.5}$$

where the first factor is the electronic wavefunction and the second is the vibration–rotation function. Λ is the projection of the electronic orbital angular momentum on the internuclear axis, S the spin angular momentum, and Σ its projection on the internuclear axis. The approximate solution [Eq. (2.1.5)] is called a Born–Oppenheimer (BO) product function. It corresponds to a solution where the couplings in $\mathbf{H}$ between nuclear and electronic motions are ignored.

2.1.1 POTENTIAL ENERGY CURVES

Since $\mathbf{T}^N$ is smaller than $\mathbf{T}^e$ by the factor m/μ, it can be neglected initially. Φ is then the solution of the clamped nuclei electronic Schrödinger equation,

$$[\mathbf{T}^e(r) + V(r, R)]\Phi_i(r, R) = E^{\mathrm{el}}(R)\Phi^i(r, R), \tag{2.1.6}$$

or

$$\mathbf{H}^{\mathrm{el}}\Phi_i = E^{\mathrm{el}}\Phi_i.$$

If the approximate product function, $\Phi_i\chi$, is inserted into Eq. (2.1.1), after multiplying by Φ_i^* and integrating over the electronic coordinates r, one obtains

$$\langle\Phi_i|\mathbf{T}^N + \mathbf{T}^e + V|\Phi_i\rangle_r\chi = E^{\mathrm{T}}\langle\Phi_i|\Phi_i\rangle_r\chi.$$

The electronic wavefunction is normalized to unity because it is a probability distribution function, $\langle\Phi_i|\Phi_i\rangle_r = 1$. If the effect of $\partial/\partial R$ [contained in $\mathbf{T}^N(R)$] on the electronic wavefunction, which certainly depends on R, is neglected, then

$$\langle \Phi_i | \mathbf{T}^N | \Phi_i \rangle_r \chi \simeq \mathbf{T}^N \langle \Phi_i | \Phi_i \rangle_r \chi = \mathbf{T}^N \chi.$$

Thus, Eq. (2.1.6) becomes

$$\langle \Phi_i | \mathbf{T}^e + V | \Phi_i \rangle_r = E_i^{\mathrm{el}}(R),$$

and a nuclear Schrödinger equation,

$$[\mathbf{T}^N(R,\theta,\phi) + E_i^{\mathrm{el}}(R)]\chi(R,\theta,\phi) = E^{\mathrm{T}}\chi(R,\theta,\phi),$$

is obtained. Since $\mathbf{T}^N(R,\theta,\phi) = \mathbf{T}^N(R) + \mathbf{H}^{\mathrm{ROT}}$, the radial and angular variables can be separated, as in the case of the hydrogen atom (Pauling and Wilson, 1935), as

$$\chi(R,\theta,\phi) = \chi_{v,J}(R)\mathscr{D}^J_{\Omega M}(\alpha = \pi/2, \beta = \theta, \gamma = \phi) = \chi_{v,J}(R)\langle \alpha\beta\gamma | JM\Omega \rangle,$$

where $\langle \alpha\beta\gamma | JM\Omega \rangle$ is the symmetric rotor function defined by Eq. (1.3.41) and $\chi_{v,J}$ is a vibrational eigenfunction of

$$[\mathbf{T}^N(R) + (\hbar^2/2\mu R^2)[J(J+1) - \Omega^2] + E_i^{\mathrm{el}}(R)]\chi_{v,J}(R) = E^{\mathrm{T}}\chi_{v,J}(R), \quad (2.1.7)$$

where $\Omega = \Lambda + \Sigma$.

$E_i^{\mathrm{el}}(R)$ may be viewed as the potential energy curve in which the nuclei move, but it must be emphasized that potential energy curves do not correspond to any physical observable. They are a concept, derived from a specified set of assumptions for defining a particular type of approximate wavefunctions [Eq. (2.1.5)]. The observed levels are not exact energy eigenvalues of a given potential curve. In general, the separation between the electronic and nuclear motions, which constitutes the BO approximation, is convenient. But when the observed levels do not fit formulas such as Eq. (2.1.4), it is simply because the function [Eq. (2.1.5)] is a bad approximation in that particular case.

An exact solution, ψ_i^{T}, of the total Hamiltonian $\mathbf{H}$ must satisfy two conditions:

$$\begin{aligned} &\text{(i)} \qquad \langle \psi_i^{\mathrm{T}} | \mathbf{H} | \psi_i^{\mathrm{T}} \rangle = E_i^{\mathrm{T}} \\ &\text{(ii)} \qquad \langle \psi_i^{\mathrm{T}} | \mathbf{H} | \psi_j^{\mathrm{T}} \rangle = 0, \qquad \text{for any} \quad j \neq i. \end{aligned} \quad (2.1.8)$$

One then says that the Hamiltonian is diagonalized in the basis set $\{\psi^{\mathrm{T}}\}$.

In principle, it is possible to express any exact solution as an infinite expansion over the approximate BO products

$$\psi_i^{\mathrm{T}} = \sum_{j,v_j}^{\infty} c_{i,v_j} \Phi_j^{\mathrm{BO}} \chi_{v_j}. \quad (2.1.9)$$

The coefficients of this expansion are determined by diagonalizing a matrix representation of the total Hamiltonian, constructed by evaluating matrix elements between BO basis functions.

It is most useful to define a basis set of the type of Eq. (2.1.5) for which the off-diagonal elements of the total Hamiltonian are as small as possible. If one term of the expansion in Eq. (2.1.9) is sufficient, this means that the BO approximation is valid. Fortunately, when the BO approximation fails, often only two terms of the BO expansion are sufficient.

If large off-diagonal elements of the Hamiltonian exist that couple *many* vibrational wavefunctions belonging to two different electronic states, then it is much more convenient to write and solve the coupled differential equations that describe these two states (see Section 3.4). The BO approximation fails whenever off-diagonal elements of $\mathbf{H}$, H_{ij}, connecting different eigenstates of the Eq. (2.1.5) form, are large compared to the difference between the diagonal elements, $H_{ii} - H_{jj}$.

2.1.2 TERMS NEGLECTED IN THE BORN–OPPENHEIMER APPROXIMATION

This section deals with various types of nonzero off-diagonal matrix elements of $\mathbf{H}$ between approximate BO product basis functions. *In order to go beyond the BO approximation*, to try to obtain an exact solution, *it is necessary to use a BO representation.* In other words, the exact eigenvalues and eigenfunctions, which can be compared to observed energy levels, are expressed in terms of the BO representation. Presently, $\{\psi^{\mathrm{BO}}\}$ is the only available type of complete, rigorously definable basis set.

In the following, the off-diagonal matrix elements of

$$\mathbf{H} = \mathbf{H}^{\mathrm{el}} + \mathbf{T}^{N}(R) + \mathbf{H}^{\mathrm{ROT}}$$

of the form

$$\langle \Phi_{i,\Lambda,S,\Sigma}\chi_{v_i,J} | \mathbf{H} | \Phi_{j\Lambda',S',\Sigma'}\chi_{v'_j,J} \rangle$$

will be discussed. The off-diagonal matrix elements of $\mathbf{H}^{\mathrm{el}}$ give rise to *electrostatic perturbations*. The off-diagonal matrix elements of $\mathbf{T}^{N}(R)$ give rise to *nonadiabatic interactions*. The off-diagonal matrix elements of $\mathbf{H}^{\mathrm{ROT}}$ give rise to *rotational perturbations*. The total Hamiltonian discussed above does not include the relativistic part of the Hamiltonian. This contribution to $\mathbf{H}$ will be introduced as a phenomenological perturbation operator, $\mathbf{H}^{\mathrm{SO}}$, and will give rise to *spin–orbit perturbations*.

2.1.2.1 Electrostatic and Nonadiabatic Part

In Section 2.3 it will be shown that, for evaluating perturbations which result from neglected terms in the $\mathbf{H}^{el} + \mathbf{T}^{N}(R)$ part of the Hamiltonian, two different types of BO representations are useful. If a crossing (diabatic) potential curve representation is used, off-diagonal elements of $\mathbf{H}^{el}$ appear between the states of this representation. If a noncrossing (adiabatic) potential curve representation is the starting point, the $\mathbf{T}^{N}$ operator becomes responsible for perturbations.

2.1.2.1.1 Crossing or Diabatic Curves

If Φ_i and Φ_j are two different exact solutions of Eq. (2.1.6), then

$$\langle \Phi_i | \mathbf{H}^{el} | \Phi_j \rangle = 0.$$

It will be shown later that *exact* solutions of the electronic Schrödinger equation can give rise to double minimum potential curves. Such potentials can be inconvenient for treating some perturbation situations. It is often more convenient to start from *approximate* solutions of $\mathbf{H}^{el}$ where potential curves, which would have avoided crossing for the exact $\mathbf{H}^{el}$, actually cross. In such a case,

$$\langle \Phi_i^{app} | \mathbf{H}^{el} | \Phi_j^{app} \rangle_r = H_{i,j}^{e}(R) \neq 0. \tag{2.1.10}$$

The expression for $\mathbf{H}^{el}$ includes $\mathbf{T}^{e}(r)$ [Eq. (2.1.3c)] and $V(r, R)$, where

$$V(r, R) = V^{eN}(r, R) + V^{ee}(r) + V^{NN}(R);$$

$V^{eN}(r, R)$ is the Coulomb electron–nuclear attraction energy operator,

$$V^{eN}(r, R) = -\sum_{i=1}^{n} \left(\frac{Z_A e^2}{r_{Ai}} + \frac{Z_B e^2}{r_{Bi}} \right);$$

$V^{ee}(r)$ is the Coulomb interelectronic repulsion energy operator,

$$V^{ee}(r) = \sum_{\substack{i=1 \\ j>i}}^{n} \frac{e^2}{r_{ij}},$$

where $j > i$ ensures that each repulsion between the ith and jth electrons is considered only once; and $V^{NN}(R)$ is the Coulomb internuclear repulsion energy operator,

$$V^{NN}(R) = Z_A Z_B e^2 / R.$$

Note that the negative sign of V^{eN} implies that it contributes to energy stabilization. Crossing curves are obtained by excluding parts of the spin–orbit term, $\mathbf{H}^{SO}$, and the interelectronic term, V^{ee}, from the $\mathbf{H}^{el}$ operator in Eq. (2.1.10).† The effect of V^{ee}, discussed in Section 2.3.2, is extremely important as it compromises the validity of the *single electronic configuration* picture which is often taken as synonymous with the diabatic potential curve picture.

In the crossing and noncrossing curve approaches, perturbations between levels of the same symmetry can occur. In the diabatic picture, these are usually called "electrostatic perturbations" because they arise from V^{ee}. In the adiabatic picture, they arise from $\mathbf{T}^N$, the nuclear kinetic energy operator, but are often misleadingly called electrostatic perturbations.

2.1.2.1.2 *Noncrossing or Adiabatic Curves*

Equation (2.1.7) was obtained by assuming that $\mathbf{T}^N(R)$ does not act on the electronic wavefunction. Actually, the Φ_i are functions of the nuclear coordinate R. Adding to Eq. (2.1.7) the neglected R-dependent term

$$\langle \Phi_i | \mathbf{T}^N | \Phi_i \rangle_r,$$

where integration over all electronic coordinates and the θ,ϕ nuclear coordinates is implied, one obtains

$$[\mathbf{T}^N(R) + (\hbar^2/2\mu R^2)[J(J+1) - \Omega^2] + \langle \Phi_i | \mathbf{T}^N | \Phi_i \rangle_r + E_i^{el}(R)]\chi_{v,J}(R) = E^{\mathrm{T}}\chi_{v,J}(R). \tag{2.1.11}$$

The potential curves defined by

$$E_i^{ad}(R) = E_i^{el}(R) + \langle \Phi_i | \mathbf{T}^N | \Phi_i \rangle_r$$

are called adiabatic potential curves, but the second term makes a much smaller contribution to the energy than $E^{el}(R)$ (Bunker, 1968).

The off-diagonal element $\langle \Phi_i | \mathbf{T}^N | \Phi_j \rangle_r$, the nonadiabatic coupling term, is examined in Section 2.3.3. This type of matrix element appears between states of identical symmetry and gives rise to homogeneous perturbations.

2.1.2.2 The Spin Part of H

Equation (2.1.2) is the nonrelativistic Hamiltonian. This means that the spin-dependent part of the Hamiltonian ($\mathbf{H}^{SO}$ spin–orbit and $\mathbf{H}^{SS}$ spin–spin)

† It is not possible to give a unique definition of a diabatic potential curve without identifying the specific term in $\mathbf{H}^{el}$ that is excluded. The impossibility of identifying such a term and the consequent nonuniqueness of the *a priori* definition of diabatic curves is discussed by Lewis and Hougen (1968) and Smith (1969). Diabatic curves may be defined empirically (Section 2.3) by assuming a deperturbation model [e.g., $H^e_{i,j}(R)$ varies linearly with R].

has been neglected. The electronic angular momentum quantum numbers, which are well-defined for eigenfunctions of nonrelativistic adiabatic and diabatic potential curves, are Λ, Σ, and S (and, redundantly, $\Omega = \Lambda + \Sigma$).

Off-diagonal elements,

$$\langle \Phi_{i,\Lambda,\Sigma,\Omega} | \mathbf{H}^{\mathrm{SO}} + \mathbf{H}^{\mathrm{SS}} | \Phi_{j,\Lambda',\Sigma',\Omega} \rangle,$$

can be nonzero between states of different Λ and S (but identical Ω) quantum numbers, corresponding to different solutions [Eq. (2.1.5)] of the nonrelativistic Hamiltonian [Eq. (2.1.2)]. Matrix elements of this type are discussed in Section 2.4.

The spin–spin part, $\mathbf{H}^{\mathrm{SS}}$, of the spin Hamiltonian usually gives rise to matrix elements much smaller than $\mathbf{H}^{\mathrm{SO}}$ and $\mathbf{H}^{\mathrm{el}}$. The *relativistic* Hamiltonian may be defined by adding $\mathbf{H}^{\mathrm{SO}}$ to $\mathbf{H}^{\mathrm{el}}$. The eigenfunctions of this new Hamiltonian are the relativistic wave functions, $\Phi_{i,\Omega}$, which define the relativistic potential curves

$$\langle \Phi_{i,\Omega} | \{ \mathbf{H}^{\mathrm{el}}(R) + \mathbf{H}^{\mathrm{SO}}(R) \} | \Phi_{i,\Omega} \rangle_r = E_i^r(R), \tag{2.1.12}$$

where now the only good electronic angular momentum quantum number is $\Omega = \Lambda + \Sigma$.

Table 2.1 summarizes the different types of potential energy curves and the specific terms in $\mathbf{H}$ that are neglected in order to define the diabatic, adiabatic, relativistic–adiabatic, and relativistic–diabatic basis functions.

2.1.2.3 ROTATIONAL PART

An expression for $\mathbf{H}^{\mathrm{ROT}}$ is given by Hougen (1970):

$$\mathbf{H}^{\mathrm{ROT}} = (1/2\mu R^2)\mathbf{R}^2 = (1/2\mu R^2)(\mathbf{R}_x^2 + \mathbf{R}_y^2),$$

where $\mathbf{R}$ is the nuclear rotation angular momentum operator. The nuclear motion is necessarily in a plane that contains the internuclear axis: thus $\vec{\mathbf{R}}$ is perpendicular to the z direction and $\mathbf{R}_z = 0$.

The total angular momentum, $\mathbf{J}$, is defined by

$$\vec{\mathbf{J}} \equiv \vec{\mathbf{R}} + \vec{\mathbf{L}} + \vec{\mathbf{S}},$$

and this definition can be used to reexpress $\mathbf{H}^{\mathrm{ROT}}$ in a convenient form,

$$\begin{aligned} \mathbf{H}^{\mathrm{ROT}} &= (1/2\mu R^2)[(\mathbf{J}_x - \mathbf{L}_x - \mathbf{S}_x)^2 + (\mathbf{J}_y - \mathbf{L}_y - \mathbf{S}_y)^2] \\ &= (1/2\mu R^2)[(\mathbf{J}^2 - \mathbf{J}_z^2) + (\mathbf{L}^2 - \mathbf{L}_z^2) + (\mathbf{S}^2 - \mathbf{S}_z^2) \\ &\quad + (\mathbf{L}^+\mathbf{S}^- + \mathbf{L}^-\mathbf{S}^+) - (\mathbf{J}^+\mathbf{L}^- + \mathbf{J}^-\mathbf{L}^+) - (\mathbf{J}^+\mathbf{S}^- + \mathbf{J}^-\mathbf{S}^+)], \end{aligned} \tag{2.1.13}$$

Table 2.1
Definitions and Approximations Associated with Different Types of Potential Energy Curves

Function type	Definition of Φ	Definition of $E_i(R)$	Curve type
Adiabatic functions	$\langle \mathbf{\Phi}_i^{\mathrm{ad}} \vert \mathbf{H}^{\mathrm{el}} \vert \mathbf{\Phi}_j^{\mathrm{ad}} \rangle_r = 0$ for all $j \neq i$	$E_i^{\mathrm{BO}} = \langle \mathbf{\Phi}_i^{\mathrm{ad}} \vert \mathbf{H}^{\mathrm{el}} \vert \mathbf{\Phi}_i^{\mathrm{ad}} \rangle_r$	Born–Oppenheimer potential curve
	$\left\langle \mathbf{\Phi}_i^{\mathrm{ad}} \left\vert \frac{\partial}{\partial R} \right\vert \mathbf{\Phi}_j^{\mathrm{ad}} \right\rangle_r \neq 0$	$E_i^{\mathrm{ad}} = E_i^{\mathrm{BO}} + \langle \mathbf{\Phi}_i^{\mathrm{ad}} \vert \mathbf{T}^{\mathrm{N}} \vert \mathbf{\Phi}_i^{\mathrm{ad}} \rangle_r$	Adiabatic potential curve
Diabatic functions	$\langle \mathbf{\Phi}_i^{\mathrm{d}} \vert \mathbf{H}^{\mathrm{el}} \vert \mathbf{\Phi}_j^{\mathrm{d}} \rangle \neq 0$ for *specific* values of j		
	$\left\langle \mathbf{\Phi}_i^{\mathrm{d}} \left\vert \frac{\partial}{\partial R} \right\vert \mathbf{\Phi}_j^{\mathrm{d}} \right\rangle_r = 0$ for all $j \neq i$	$E_i^{\mathrm{d}} = \langle \mathbf{\Phi}_i^{\mathrm{d}} \vert \mathbf{H}^{\mathrm{el}} \vert \mathbf{\Phi}_i^{\mathrm{d}} \rangle_r$	Diabatic potential curve
Relativistic adiabatic functions	$\langle \mathbf{\Phi}_i^{\mathrm{ad\cdot r}} \vert \mathbf{H}^{\mathrm{el}} + \mathbf{H}^{\mathrm{SO}} \vert \mathbf{\Phi}_j^{\mathrm{ad\cdot r}} \rangle_r = 0$ for all $j \neq i$		
	$\left\langle \mathbf{\Phi}_i^{\mathrm{ad\cdot r}} \left\vert \frac{\partial}{\partial R} \right\vert \mathbf{\Phi}_j^{\mathrm{ad\cdot r}} \right\rangle_r \neq 0$	$E_i^{\mathrm{ad\cdot r}} = \langle \mathbf{\Phi}_i^{\mathrm{ad\cdot r}} \vert \mathbf{H}^{\mathrm{el}} + \mathbf{H}^{\mathrm{SO}} + \mathbf{T}^{\mathrm{N}} \vert \mathbf{\Phi}_i^{\mathrm{ad\cdot r}} \rangle_r$	Relativistic adiabatic potential curve
Relativistic diabatic functions	$\langle \mathbf{\Phi}_i^{\mathrm{d\cdot r}} \vert \mathbf{H}^{\mathrm{SO}} \vert \mathbf{\Phi}_j^{\mathrm{d\cdot r}} \rangle \neq 0$ for *one* value of j	$E_i^{\mathrm{d\cdot r}} = \langle \mathbf{\Phi}_i^{\mathrm{d\cdot r}} \vert \mathbf{H}^{\mathrm{el}} + \mathbf{H}^{\mathrm{SO}} \vert \mathbf{\Phi}_i^{\mathrm{d\cdot r}} \rangle_r$	Relativistic diabatic potential curve
	$\left\langle \mathbf{\Phi}_i^{\mathrm{d\cdot r}} \left\vert \frac{\partial}{\partial R} \right\vert \mathbf{\Phi}_j^{\mathrm{d\cdot r}} \right\rangle_r = 0$ for all $j \neq i$		

where

$$\mathbf{J}^{\pm} = \mathbf{J}_x \pm i\mathbf{J}_y \qquad \text{[cf. Eqs. (1.3.8 and 1.3.9)]},$$

$$\mathbf{L}^{\pm} = \mathbf{L}_x \pm i\mathbf{L}_y, \qquad \mathbf{S}^{\pm} = \mathbf{S}_x \pm i\mathbf{S}_y.$$

The first three terms of $\mathbf{H}^{\mathrm{ROT}}$ have diagonal matrix elements exclusively. This diagonal part of $\mathbf{H}^{\mathrm{ROT}}$ is the rotational energy of the $|JM\Omega\Lambda S\Sigma\rangle$ basis function. The eigenfunctions [Eq. (1.3.40)] of the rotational equation,

$$(1/2\mu R^2)[(\mathbf{J}^2 - \mathbf{J}_z^2) + (\mathbf{L}^2 - \mathbf{L}_z^2) + (\mathbf{S}^2 - \mathbf{S}_z^2)]\,|JM\Omega\rangle = E^{\mathrm{ROT}}(R)\,|JM\Omega\rangle, \tag{2.1.14}$$

correspond to the R-dependent eigenvalues

$$E^{\mathrm{ROT}}(R) = (\hbar^2/2\mu R^2)[J(J+1) - \Omega^2 + S(S+1) - \Sigma^2 + L(L+1) - \Lambda^2]. \tag{2.1.15}$$

In general, electronic wavefunctions are not eigenfunctions of $\mathbf{L}^2$, hence $L(L+1)$ is not quantized, but as the expectation value of $\mathbf{L}^2$ is a weakly R-dependent quantity for a given electronic state, the quantity

$$(\hbar^2/2\mu R^2)[L(L+1) - \Lambda^2]$$

may be incorporated into the electronic energy. Due to its μ-dependence, this term is responsible for part of the electronic isotope effect, the remainder arising from small terms neglected in the Hamiltonian given here (Bunker, 1968).

After integration of Eq. (2.1.14) over the vibrational coordinate,

$$E^{\mathrm{ROT}}(v, J) = B_v[J(J+1) - \Omega^2 + S(S+1) - \Sigma^2] \tag{2.1.16}$$

where

$$B_v \qquad (\mathrm{cm}^{-1}) = \langle v|(\hbar^2/2\mu R^2)|v\rangle(\hbar c)^{-1}. \tag{2.1.17}$$

The final three terms of the rotational operator in Eq. (2.1.13), which couple the orbital, spin, and total angular momenta, are responsible for perturbations between different electronic states:

1. $+(1/2\mu R^2)\mathbf{L}^{\pm}\mathbf{S}^{\mp}$ gives rise to *homogeneous* spin–electronic perturbations between basis functions of the same Ω and S, but different Λ and Σ.

2. $-(1/2\mu R^2)\mathbf{J}^{\pm}\mathbf{S}^{\mp}$ is responsible for heterogeneous electronic–rotational† perturbations between basis states of different Ω having identical values of S and Λ, but different values of Σ. This operator will be called the **S**-*uncoupling operator*.

† Perturbations resulting from the $\mathbf{J}\cdot\mathbf{L}$ and $\mathbf{J}\cdot\mathbf{S}$ terms are frequently called "Coriolis" perturbations. This use of the name Coriolis is misleading and should be reserved for vibration–rotation interactions in polyatomic molecules.

3. $-(1/2\mu R^2)\mathbf{J}^{\pm}\mathbf{L}^{\mp}$ causes heterogeneous electronic–rotational† perturbations between states of different Ω having identical values of S and Σ, but different values of Λ. This operator will be called the **L**-*uncoupling operator.*

These perturbation operators will be examined in Section 2.5.

Following the nomenclature suggested by Hougen (1970), perturbations are classified with respect to the *signed* Ω quantum number. The selection rule, $\Delta\Omega = 0$, implies a *homogeneous perturbation.* The important feature of such perturbations is that the *interaction matrix element does not depend on the rotational quantum number* J. The selection rule, $\Delta\Omega = \pm 1$, implies a *heterogeneous* perturbation. The important feature of such perturbations is that the *interaction matrix element depends on* J (usually approximately proportional to J). Table 2.2 summarizes the different perturbation selection rules. The matrix elements are written for the microscopic Hamiltonian, taking into account that L is not defined in a molecule, and in this sense differing slightly from those given in Chapter 1.

In all cases, the vibrational quantum number is no longer strictly defined because the perturbed level is a mixture of at least two vibrational levels belonging to different electronic states [see Eq. (2.1.9)]. The only quantum number that remains rigorously well defined is J, because only J is a conserved observable ($\mathbf{J}^2$ commutes with $\mathbf{H}$). The selection rule common to all perturbations is $\Delta J = 0$; this means that in the absence of nuclear spin and external perturbations (electric or magnetic fields, electromagnetic radiation fields, or collisions with other molecules), the total angular momentum of the molecule remains well defined. Even if the perturbation operator includes $\mathbf{J}^+$ or $\mathbf{J}^-$, this operator cannot change the value of J. Even in case (b), J (as well as N) remains well defined. Perturbations ($\sim$) correspond to an interaction between two levels, *not an electric dipole transition* (—) *between two levels.*

Mathematically, the conservation of the J quantum number on perturbation can be clearly shown, since no perturbation operator contains the θ and ϕ angular coordinates of the nuclei and consequently none can act on the rotational basis functions. As the rotational functions $|JM\Omega\rangle$ are orthogonal for different values of J, a matrix element of any perturbation operator, $\mathbf{H}'$, between two rotational functions is given by

$$\langle JM\Omega|\mathbf{H}'|J'M'\Omega\rangle = f(J)\langle JM\Omega|J'M'\Omega\rangle = f(J)\,\delta_{JJ'}\,\delta_{MM'}. \quad (2.1.18)$$

For homonuclear molecules, the g or u symmetry is always conserved. Only electric fields, hyperfine effects, and collisions can induce perturbations between g and u states. An additional symmetry will be discussed in Section 2.2.2:

† Perturbations resulting from the $\mathbf{J}\cdot\mathbf{L}$ and $\mathbf{J}\cdot\mathbf{S}$ terms are frequently called "Coriolis" perturbations. This use of the name Coriolis is misleading and should be reserved for vibration–rotation interactions in polyatomic molecules.

Table 2.2 Off-Diagonal Matrix Elements of Total Hamiltonian between Unsymmetrized Basis Functions

Neglected terms in the nonrelativistic Born–Oppenheimer approximation	Nature of the perturbation	Selection rules $\Delta J = 0$ and $g \not\leftrightarrow u$ always: $\Delta\Lambda$	$\Delta\Sigma$	ΔS	$\Delta\Omega$	Number of different spin-orbitals[d]
$\langle \Lambda, S, \Sigma, \Omega, v \vert \mathbf{H}^{\text{el}} \vert \Lambda, S, \Sigma, \Omega, v' \rangle \simeq H^e \langle v \vert v' \rangle$	Electrostatic homogeneous	0	0	0	0	1 or 2
$\langle \Lambda, S, \Sigma, \Omega, v \vert \mathbf{T}^{\text{N}} \vert \Lambda, S, \Sigma, \Omega, v' \rangle = \left\langle \Lambda, S, \Sigma, \Omega, v \left\vert -\frac{\hbar^2}{2\mu}\left(\frac{d^2}{dR^2} + \frac{2}{R}\frac{d}{dR}\right) \right\vert \Lambda, S, \Sigma, \Omega, v' \right\rangle$	Vibrational (or nonadiabatic) homogeneous	0	0	0	0	1
$\langle \Lambda, S, \Sigma, \Omega, v \vert \mathbf{H}^{\text{SO}} \vert \Lambda', S', \Sigma', \Omega, v' \rangle = \langle v \vert v' \rangle \left\langle \Lambda, S, \Sigma, \Omega \left\vert \sum_i \hat{a}_i \mathbf{l}_i \mathbf{s}_i \right\vert \Lambda', S', \Sigma', \Omega \right\rangle$	Spin–orbit homogeneous	0 or ± 1 $\Sigma^{\pm}$–$\Sigma^{\mp}$	0 or ∓ 1	0^a or 1 0 or 1	0	0 or 1 1
$\langle \Lambda, S, \Sigma, \Omega, v \vert \mathbf{H}^{\text{SS}} \vert \Lambda', S', \Sigma', \Omega, v' \rangle = \langle v \vert v' \rangle \langle \Lambda, S, \Sigma, \Omega \vert \mathbf{H}^{\text{SS}} \vert \Lambda', S', \Sigma', \Omega \rangle$	Spin–spin homogeneous	0 ± 1 ± 2 $\Sigma^{\pm}$–$\Sigma^{\pm}$	0 ∓ 1 ∓ 2	0^c, 1^b, or 2 0, 1, or 2 0, 1, or 2	0	0 1 2
$\left\langle \Lambda, S, \Sigma, \Omega, v \left\vert \frac{1}{2\mu R^2} \mathbf{L}^{\pm}\mathbf{S}^{\mp} \right\vert \Lambda \mp 1, S, \Sigma \pm 1, \Omega, v' \right\rangle \simeq B_{vv'} \left\langle \Lambda, S, \Sigma, \Omega \left\vert \sum_i \mathbf{l}_i^{\pm} \sum_j \mathbf{s}_j^{\mp} \right\vert \Lambda \mp 1, S, \Sigma \pm 1, \Omega \right\rangle^e$	Spin–electronic homogeneous	± 1	∓ 1	0	0	1 or 2
$\left\langle \Lambda, S, \Sigma, \Omega, v \left\vert -\frac{1}{2\mu R^2} \mathbf{J}^{\pm}\mathbf{L}^{\mp} \right\vert \Lambda \pm 1, S, \Sigma, \Omega \pm 1, v' \right\rangle \simeq -B_{vv'} \left\langle \Lambda, S, \Sigma, \Omega \left\vert \sum_i \mathbf{l}_i^{\mp} \right\vert \Lambda \pm 1, S, \Sigma, \Omega \pm 1 \right\rangle [J(J+1) - \Omega(\Omega \pm 1)]^{1/2\,e}$	Heterogeneous (**L**-uncoupling)	± 1	0	0	± 1	1
$\left\langle \Lambda, S, \Sigma, \Omega, v \left\vert -\frac{1}{2\mu R^2} \mathbf{J}^{\pm}\mathbf{S}^{\mp} \right\vert \Lambda, S, \Sigma \pm 1, \Omega \pm 1, v' \right\rangle = -B_{vv'}[S(S+1) - \Sigma(\Sigma \pm 1)]^{1/2}[J(J+1) - \Omega(\Omega \pm 1)]^{1/2\,e}$	Heterogeneous (**S**-uncoupling)	0	± 1	0	± 1	1

[a] Zero if $\Omega = 0$. [b] Zero if $\Sigma = \Sigma' = 0$. [c] Zero if $\Sigma = \Sigma' = 0$ *and* $S = 0$. [d] Each state must be well represented by a single configuration.
[e] $B_{vv'} \equiv \langle v \vert \hbar^2/(2\mu R^2) \vert v' \rangle$.

parity or, more usefully, the e and f symmetry character of the rotational levels remains well defined for both hetero- and homonuclear diatomic molecules. The matrix elements of Table 2.2 describe direct interactions between basis states. *Indirect* interactions can also occur and are discussed in Sections 3.2, 3.4.2, and 3.5.1. Even for indirect interactions the $\Delta J = 0$ and $e \nleftrightarrow f$ selection rules remain valid (see Section 2.2.2).

2.2 Basis Functions

2.2.1 HUND'S CASES

A choice of basis set implies a partitioning of the Hamiltonian into two parts: a part, $\mathbf{H}^0$, which is fully diagonal in the selected basis set, and a residual part, $\mathbf{H}'$. The basis sets associated with Hund's cases (a)–(d) reflect different choices of the parts of $\mathbf{H}$ that are included in $\mathbf{H}^0$. Although in principle the eigenvalues of $\mathbf{H}$ are unaffected by the choice of basis, as long as this basis set forms a complete set of functions, one basis set is usually more convenient or appropriate than the others for a particular problem. Convenience is a function of both the nature of the computational method and the relative sizes of electronic, spin–orbit, vibrational, and rotational energies. The angular momentum basis sets, from which Hund's cases (a)–(d) bases derive, are

$$\text{(a)} \quad |nJS\Omega\Lambda\Sigma\rangle \tag{2.2.1a}$$

$$\text{(b)} \quad |nJSN\Lambda\rangle \tag{2.2.1b}$$

$$\text{(c)} \quad |nJ\Omega\rangle \tag{2.2.1c}$$

$$\text{(d)} \quad |nJSNN^+\rangle \tag{2.2.1d†}$$

where the label n stands for all remaining state labels (e.g., vibration, electronic configuration). Since problems involving laboratory-fixed electromagnetic fields are beyond the scope of this book, the laboratory projection quantum number M_J is excluded from the basis set labels in Eqs. (2.2.1).

† Herzberg and Jungen (1982) introduce the quantum number N^+ which corresponds to the magnitude of the angular momentum $\mathbf{N}^+ = \mathbf{N} - \mathbf{l}$, where $\mathbf{l}$ is the orbital angular momentum of the electron in a Rydberg orbital and $\mathbf{N}^+$ is the total angular momentum, exclusive of spin, of the molecular ion core. The N^+ rather than R quantum number is a more appropriate label for the rotational levels of Rydberg states.

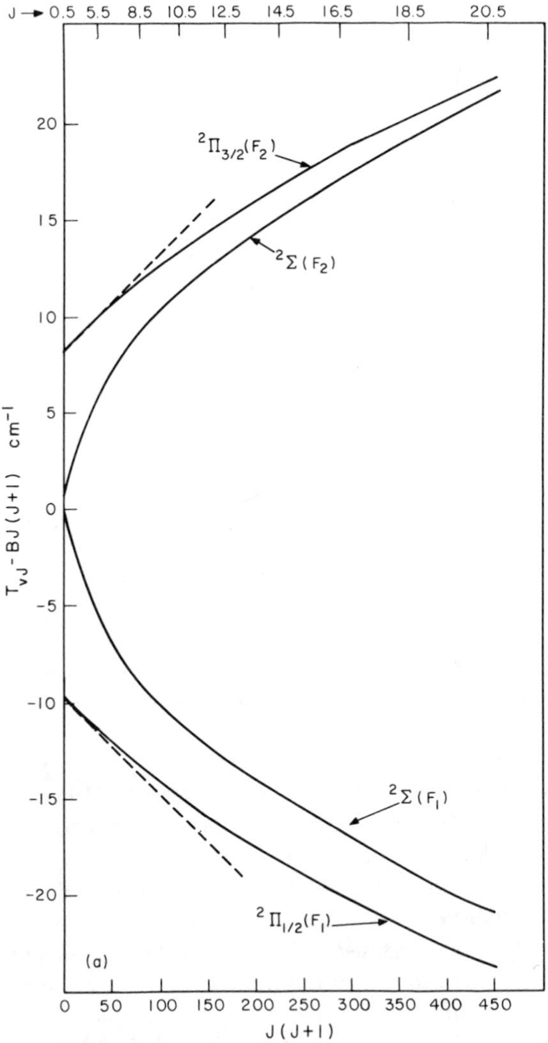

Fig. 2.1 Natural rotational quantum numbers for Hund's cases (a) and (b). Reduced term value plots for $^2\Sigma(B = 1.0\ \text{cm}^{-1})$ and $^2\Pi_r(B = 1.0\ \text{cm}^{-1},\ A = 20.0\ \text{cm}^{-1})$. (a) Plot of $T_{vJ} - BJ(J+1)$ versus $J(J+1)$ displays case (a) limiting behavior for the $^2\Pi$ state at very low J. The dotted lines illustrate the $\pm B^2/A$ corrections to the near case (a) effective B-values (see Section 2.5.4). The $^2\Sigma$ state does not exhibit case (a) behavior even at low J (at $J = 0$ the limiting slopes of the $^2\Sigma$ F_1 and F_2 curves are $-\infty$ and $+\infty$). (b) Plot of $T_{vJ} - BN(N+1)$ versus $N(N+1)$ displays case (b) limiting behavior for $^2\Sigma$ at all N and for $^2\Pi$ at very high N.

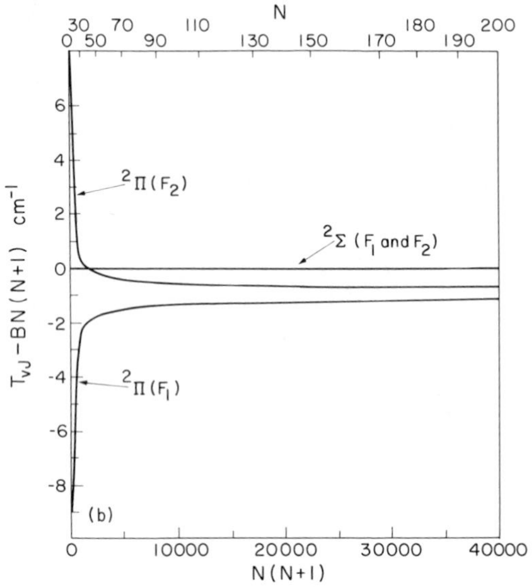

Fig. 2.1 *(Continued)*

l becomes a good quantum number and the molecular energy level pattern approaches that of case (d). This limiting case will be attained for small n if l is large but at large n if l is small.†

The good case (d) quantum numbers are J, S, l, N^+, and N (or $\mathscr{L} = N - N^+$, which is the projection of $\vec{\mathbf{l}}$ along the rotation axis, $\vec{\mathbf{N}}$). In replacing Λ of case (b) by N^+ of case (d), the dimension of the basis set does not change, provided that one keeps track of the quantum number l. In the case where the ion core has Σ symmetry, for each (J, S, l, N) there are $2l + 1$ N^+ values [case (d)] or Λ-values [case (b)]. The $2l + 1$ Λ-values correspond to the $2l + 1$ λ-values of an nl Rydberg orbital. If l is defined, Clebsch–Gordan coefficients allow transformation from case (b) to (d), just as for case (a) $\leftrightarrow$ (b) (see Section 7.4).

The case (d) basis functions, $|nJSlNN^+\rangle$, are eigenfunctions of $\mathbf{H}^{\mathrm{el}}$ provided that only the core-monopole part of $\mathbf{H}^{\mathrm{el}}$ is included in $\mathbf{H}^0$. The first term of $\mathbf{H}^{\mathrm{ROT}} = B\mathbf{N}^{+2} - B(\mathbf{J}^+\mathbf{l}^- + \mathbf{J}^-\mathbf{l}^+)$ is included in $\mathbf{H}^0$ and the resultant rotational energies are $E_{N^+} = BN^+(N^+ + 1)$, the energy of the rotating ion core. Each N^+-level is a $(2l + 1)(2S + 1)$-fold degenerate supermultiplet,

† Note that l can never be a rigorously good quantum number for a molecule. However, l can act as a good quantum number in one limiting case. If Rydberg complexes associated with different nl values are well separated, then it is possible to neglect $\Delta l \neq 0$ matrix elements of the electronic Hamiltonian that arise from the nonspherical terms in the multipole expansion of the electric field of the ion core.

consisting of $2l+1$ N-sublevels ($N^+ - l \leq N \leq N^+ + l$), each of which is ($2S+1$)-fold degenerate, corresponding to J-values $N-S \leq J \leq N+S$. The spin degeneracy ($2S+1$ J values) within each N-multiplet is lifted by $\mathbf{H}^{SR}$ (provided that $\mathbf{H}^{SR}$ is defined as $\gamma\mathbf{N}\cdot\mathbf{S}$ and not $\gamma\mathbf{N}^+\cdot\mathbf{S}$), which is fully diagonal in case (d) and has the same matrix elements [Eq. (2.2.7)] as in case (b).

The l-degeneracy ($2l+1$ N values) within each case (d) N^+-supermultiplet is lifted at low N^+ [transition to case (b)] by the difference between the electronic energies of the $nl\lambda$ components of the nl Rydberg orbital (see Section 7.4). $\mathbf{H}^{SO}$ can also distort the low-N^+ supermultiplet structure [transition to case (a)].

Case (c) is unusual in that its basis functions are specified by the fewest quantum numbers. In case (c), $\mathbf{H}^{el}$ and $\mathbf{H}^{SO}$ are included in $\mathbf{H}^0$. $\mathbf{H}^{el}+\mathbf{H}^{SO}$ is the relativistic electronic Hamiltonian. In case (c), separate Ω-components of a single case (a) multiplet state become separate electronic states, *each with its own potential curve.* In addition, if two case (a) states of different symmetry, say $^1\Sigma^+ \sim {}^3\Pi$, interact via $\mathbf{H}^{SO}$, then the two corresponding case (c) 0^+ potentials will exhibit an avoided crossing. The only remaining good quantum number in case (c) is Ω, and the basis functions are combinations of case (a) functions of the same Ω but different Λ, Σ, and S values. The rotational energies, as for case (a), are given by $BJ(J+1)$. The **L**-uncoupling and **S**-uncoupling operators that mix basis states of different Ω are combined into a single $\mathbf{J}_a$ uncoupling operator (Section 1.3.2), since neither Λ nor Σ are defined (see Veseth, 1973a). The sum of the **L**- and **S**-uncoupling perturbation operators can be written as

$$-2\,\mathbf{BJ}\cdot\mathbf{J}_a$$

with

$$\mathbf{J}_a \equiv \mathbf{L}+\mathbf{S}$$

and has $\Delta\Omega = 0, \pm 1$ matrix elements.

Case (c) arises when off-diagonal spin–orbit matrix elements are large relative to the energy separations of case (a) electronic basis states belonging to the same Ω value. Case (c) is often important for molecules that include one atom from at least the third row (K through Kr) of the periodic table. A good example is the I_2 molecule (Mulliken, 1971), for which perturbations among the $I_2X^1\Sigma_g^+$, $a1_g$, and $a'0_g^+$ states can be related to $B\mathbf{J}\cdot\mathbf{J}_a$ matrix elements (Martin *et al.*, 1983). Case (c) can be a good approximation also for light molecules in high-n Rydberg states (see Fig. 2.11 and Section 7.6).

Different groups of states of the same molecule can separately exhibit case (a), (b), or (c) behavior. Furthermore, within any given electronic state, it is possible for the coupling limit to change from case (a) to case (c) as R, the internuclear distance, increases, because, near a dissociation asymptote, the interval between electronic states can become small compared to the spin–orbit splittings (see Section 4.6).

Very few true relativistic *ab initio* calculations have been made except for hydride molecules such as TlH (Pyykkö and Desclaux, 1976). Other types of calculations involve treating the spin–orbit operator as a perturbation and adding its effects to the nonrelativistic potential curves (for example, Cohen and Schneider, 1974).

Case (e) coupling, although briefly mentioned by Herzberg (1950), has never been observed. It could appear in Rydberg complexes when the ion core has a spin–orbit splitting: for example, in the Rydberg series of N_2 that corresponds to a *d* Rydberg orbital with a $^2\Pi$ N_2^+ ion-core. In that case, the $\mathbf{H}^{SO}$ part of the Hamiltonian must be included in $\mathbf{H}^0$ along with $\mathbf{H}^{ROT}$, and the only good quantum numbers would be J, J^+, and the $\mathbf{\Omega}$-core value.

The electronic functions in Table 2.3 correspond to diagonalization of the total electronic Hamiltonian, $\mathbf{H}^{el}$. This means that the basis functions are adiabatic functions (Sections 2.1.2.1.2 and 2.3.3). Different Hund's subcases can be defined when only a part of the electronic Hamiltonian is diagonalized. These will be called *primitive Hund's cases* in this book. In the following, electronic case (a) wavefunctions that are not eigenfunctions of the entire $\mathbf{H}^{el}$ are considered by formally separating out a small residual part of the interelectronic repulsion as a perturbation term, $\mathbf{H}'$. The resultant basis functions are diabatic functions (Sections 2.1.2.1.1 and 2.3.1). Another example of primitive case (a) will be given in Section 7.5, where the zero-order electronic solutions correspond to those of one electron in the field of the ion core. In the study of the origin of *l*-mixing in Rydberg states [$s \sim d$ mixing, Jungen (1970)], the electronic potential is the long-range potential exclusive of exchange effects within the nonspherical core. For van der Waals potentials, the convenient electronic basis functions can be the separated atom wavefunctions; then $\mathbf{H}'$ contains the long-range forces between the permanent and induced multipoles of the individual atoms.

Similarly, a primitive case (c) can be defined by reducing the relativistic Hamiltonian, $\mathbf{H}^{el} + \mathbf{H}^{SO}$, to that of one of the constituent atoms, thus partitioning the field due to the other atom into $\mathbf{H}'$. For example, the low-lying states of CeO and PrO can be described in terms of primitive case (c) basis functions derived from the Ce^{2+} and Pr^{2+} free ion $f^N s$ configurations in $(J_c, j)_{J_a}$ coupling, where J_c, j, and J_a refer, respectively, to the total J of the f^N electrons, the s electron, and the resultant total atomic angular momentum (Dulick, 1982).

In conclusion, it is only in cases (a) or (c) that potential curves can be defined unambiguously and completely, since it is only in these cases that electronic motion can be completely separated from nuclear motion. However, it is the intent of this book to illustrate the point that observed molecular levels seldom correspond to any pure Hund's case. The real situation is generally intermediate between two limiting cases, because of nonzero matrix elements of the terms in $\mathbf{H}$ that were neglected in order to define the basis functions: $\mathbf{H}^{SO}$ is neglected for cases (a), (b), and (d); $\mathbf{BJ} \cdot \mathbf{L}$ for cases (a), (b), and (c); $\mathbf{BJ} \cdot \mathbf{S}$ for cases (a) and (c); and a part of $\mathbf{H}^{el}$ for case (d).

Consider an example intermediate between cases (a) and (d). At the case (d) limit, the electronic Hamiltonian matrix for an l-complex can be expressed in terms of $2l+1$ *degenerate* $-l \leq \lambda \leq l$ case (a) $|l\lambda S\Sigma\rangle$ basis functions. Real Rydberg complexes do not exhibit perfect λ-degeneracy. In the case (a) picture, this lack of degeneracy is attributed to a Λ-dependence of the *diagonal* matrix elements of the exact $\mathbf{H}^{el}$. In the case (d) picture, this effect must be attributed to *both nondiagonal and diagonal* matrix elements of a term in $\mathbf{H}^{el}$ that corresponds to higher-order terms in the multipole expansion of the ion-core field and that is neglected in defining the case (d) basis functions [see Jungen and Miescher (1969) for an example of an isolated $4f$ Rydberg complex]. In Section 7.4 a specific example of a p-complex is discussed, in which it is shown that the magnitude of the neglected term in $\mathbf{H}^{el}$ is proportional to the difference between the quantum defects for Π and Σ Rydberg series.

Similarly, if two same-Ω case (a) basis states belonging to different multiplicities of the same (nonrelativistic) electronic configuration are exactly degenerate, $\mathbf{H}^{SO}$ will cause these two states to be completely mixed. The resultant eigenfunctions of the relativistic $\mathbf{H}^{el}$ are perfect case (c) basis functions. However, when the case (a) basis functions are not exactly degenerate, then the term in $\mathbf{H}$ responsible for this splitting, although diagonal in the case (a) basis, must have both diagonal and nondiagonal matrix elements in the case (c) basis. In Section 7.6 a specific example of an interacting $^1\Pi \sim {}^3\Pi_1$ pair of $\Omega = 1$ basis functions is examined. An electrostatic exchange splitting in the case (a) picture is shown to be related in the case (c) picture to a difference in quantum defects for two $\Omega = 1$ Rydberg series converging to the $\Omega = \frac{1}{2}$ and $\frac{3}{2}$ components of a $^2\Pi$ ion-core state.

Throughout this book, matrix elements of various perturbation operators are evaluated in the case (a) basis set. To illustrate the irrelevance of the choice of basis, it is instructive to see that it is always possible to express the basis functions of one Hund's case in terms of those of another. Several methods exist whereby the basis functions of any Hund's case may be expanded in terms of those of case (a).

One especially simple procedure for defining transformations between cases (b), (c), or (d) and case (a) is to express the effective Hamiltonian in the case (a) basis, adjust the values of crucial parameters in order to create the degeneracies appropriate for the desired limiting Hund's case [$A = 0$ in $A\Lambda\Sigma$ for case (b),

$A = 0$ and $E_{\lambda=l} = E_{\lambda=l-1} = \cdots = E_{\lambda=0}$ for case (d), $E(^{2S+1}\Lambda_\Omega) = E(^{2S-1}\Lambda_\Omega)$ for case (c)†], and then diagonalize (see Sections 7.4 and 7.6). The transformations thus obtained may be used to transform any matrix expressed in a case (a) basis to the desired non-case (a) basis, regardless of whether the matrix to be transformed is near any limiting case.

Another systematic procedure makes use of Clebsch–Gordan (CG) coefficients. This use of CG coefficients is very familiar in atomic physics, for example, in expressing the $|LSJM_J\rangle$ basis functions in terms of $|LM_LSM_S\rangle$ or $|j_1 j_2 \omega_1 \omega_2\rangle$ basis functions (Condon and Shortley, 1953, pp. 73 and 284). Two angular momenta, **L** and **S**, or $\mathbf{j}_1$ and $\mathbf{j}_2$, are coupled to form the resultant total angular momentum, **J**. Similarly, for a molecule in case (b), **N** and **S** are coupled to give **J**, but because of the anomalous commutation rules for the molecule-fixed components of **J**, the coupling $\mathbf{N} + \mathbf{S} = \mathbf{J}$ must be replaced by $\mathbf{N} = \mathbf{J} - \mathbf{S}$ (Brown and Howard, 1976). Case (b) functions $|N, \Lambda, S, J\rangle$ may be expressed as linear combinations of case (a) functions $|\Lambda, S, \Sigma, J, \Omega\rangle$ using CG coefficients, as follows:

$$|N, \Lambda, S, J\rangle = \sum_{\Sigma=-S}^{+S} (JS\Omega - \Sigma|N\Lambda)\,|\Lambda, S, \Sigma, J, \Omega\rangle. \qquad (2.2.12)$$

Instead of CG coefficients it is often more convenient to use $3j$-coefficients. These are related to the CG coefficients as follows:

$$(JS\Omega - \Sigma|N\Lambda) = (-1)^{J-S+\Lambda}(2N+1)^{1/2}\begin{pmatrix} J & S & N \\ \Omega & -\Sigma & -\Lambda \end{pmatrix}. \qquad (2.2.13)$$

For example, consider the two case (a) $^2\Pi$ basis functions. If $A = 0$, the two case (a) basis functions $|^2\Pi_{1/2}\rangle$ and $|^2\Pi_{3/2}\rangle$ of the same J become almost completely mixed. Two $^2\Pi$ levels result: the F_1 component corresponds *by convention* to $J = N + \frac{1}{2}$, and the F_2 component to $J = N - \frac{1}{2}$. *Except at the lowest J values, the* F_i *indices label the energy order of the eigenvalues for a given J, with* $i = 1$ *the lowest and* $i = 2S + 1$ *the highest.* The general expression for both the F_1 and F_2 case (b) $^2\Pi$ basis functions is

$$\begin{aligned}|N, 1, \tfrac{1}{2}, J\rangle = {} & |1, \tfrac{1}{2}, -\tfrac{1}{2}, J, \tfrac{1}{2}\rangle\ (J\tfrac{1}{2}\tfrac{1}{2}\tfrac{1}{2}|N\,1) \\ & + |1, \tfrac{1}{2}, \tfrac{1}{2}, J, \tfrac{3}{2}\rangle\ (J\tfrac{1}{2}\tfrac{3}{2} - \tfrac{1}{2}|N\,1).\end{aligned} \qquad (2.2.14)$$

An extremely useful listing of $3j$-coefficients is found in Edmonds (1974, p. 125). Some of the properties of $3j$-coefficients are discussed in Section 2.4.5. For the special case of $|J - N| = \frac{1}{2}$, there is a convenient closed-form expression for the

† Strictly speaking, there is no unique way of specifying the case (c) limit. However, case (c) corresponds to a situation where $\mathbf{H}^{\mathrm{SO}}$ is more important than electrostatic terms. Thus, if one sets all isoconfigurational exchange splittings to zero, one has a computationally useful definition of the case (c) limit.

$3j$-coefficients,

$$\begin{pmatrix} J+\frac{1}{2} & J & \frac{1}{2} \\ M & -M-\frac{1}{2} & \frac{1}{2} \end{pmatrix} = (-1)^{J-M-\frac{1}{2}} \left[\frac{J-M+\frac{1}{2}}{(2J+2)(2J+1)} \right]^{1/2}. \quad (2.2.15)$$

For $N = J + \frac{1}{2}$ (the F_2 level),

$$|J+\tfrac{1}{2}, 1, \tfrac{1}{2}, J\rangle = (-1)^{J+\frac{1}{2}}(2J+2)^{1/2}\left[\begin{pmatrix} J & \frac{1}{2} & J+\frac{1}{2} \\ \frac{1}{2} & \frac{1}{2} & -1 \end{pmatrix}|^2\Pi_{1/2}\rangle + \begin{pmatrix} J & \frac{1}{2} & J+\frac{1}{2} \\ \frac{3}{2} & -\frac{1}{2} & -1 \end{pmatrix}|^2\Pi_{3/2}\rangle\right], \quad (2.2.16)$$

where the first $3j$-coefficient is [Eq. (2.2.15) with $M = -1$]

$$\begin{pmatrix} J & \frac{1}{2} & J+\frac{1}{2} \\ \frac{1}{2} & \frac{1}{2} & -1 \end{pmatrix} = (-1)^{J+\frac{1}{2}} \left[\frac{(J+\frac{3}{2})}{(2J+2)(2J+1)} \right]^{1/2} \quad (2.2.17)$$

and the second $3j$-coefficient is [Eq. (2.2.15) with $M = 1$]

$$\begin{pmatrix} J & \frac{1}{2} & J+\frac{1}{2} \\ \frac{3}{2} & -\frac{1}{2} & -1 \end{pmatrix} = \begin{pmatrix} J+\frac{1}{2} & J & \frac{1}{2} \\ -1 & \frac{3}{2} & -\frac{1}{2} \end{pmatrix}$$

$$= (-1)^{J-3/2} \left[\frac{(J-\frac{1}{2})}{(2J+2)(2J+1)} \right]^{1/2}, \quad (2.2.18)$$

thus

$$|F_2\rangle = \left[\frac{J+\frac{3}{2}}{2J+1}\right]^{1/2} |^2\Pi_{1/2}\rangle - \left[\frac{J-\frac{1}{2}}{2J+1}\right]^{1/2} |^2\Pi_{3/2}\rangle. \quad (2.2.19)$$

For $J = \frac{1}{2}$, only the F_2 component exists. The F_1 component for $J > \frac{1}{2}$ is derived by constructing a function which is orthogonal to the above F_2 function. The F_1 function ($N = J - \frac{1}{2}$) is

$$|F_1\rangle = \left[\frac{J-\frac{1}{2}}{2J+1}\right]^{1/2} |^2\Pi_{1/2}\rangle + \left[\frac{J+\frac{3}{2}}{2J+1}\right]^{1/2} |^2\Pi_{3/2}\rangle. \quad (2.2.20)$$

In Section 2.5.4, these expressions will be derived by diagonalizing the matrix that includes the spin-uncoupling interaction between the $^2\Pi$ case (a) basis functions.

2.2.2 SYMMETRY PROPERTIES

Up to this point, wavefunctions with signed values of Λ, Σ, or Ω have been discussed. However, the electronic basis functions associated with positive and negative signs of $(\Lambda, \Sigma, \Omega)$ are degenerate. In order to construct functions with a well-defined value of $|\Omega|$, it is necessary to combine functions with specific signed values of Λ and Σ. For example, a $^1\Pi_1$ state is doubly degenerate, the $\Lambda = +1$ and -1 components leading, respectively, to $\Omega = +1$ and -1 basis functions. These two Ω-components, degenerate in energy for the nonrotating molecule, are not properly symmetrized eigenfunctions of the total Hamiltonian, $\mathbf{H}$. Only linear combinations of these $\pm\Omega$ functions have well-defined symmetry. This symmetry is called *parity*, and $\boldsymbol{\sigma}_v$ is the operator used to classify basis functions as belonging to either even or odd parity. The total Hamiltonian commutes with the $\boldsymbol{\sigma}_v$ symmetry operator (see Hougen, 1970), which means that the only nonzero off-diagonal matrix elements of $\mathbf{H}$ are between basis functions that belong to the same eigenvalue of $\boldsymbol{\sigma}_v$. The eigenvalues of $\boldsymbol{\sigma}_v$ are $+1$ and -1 and correspond, respectively, to *even* and *odd total parity*.

Before using the $\boldsymbol{\sigma}_v$ operator to classify symmetrized basis functions as even versus odd, it is necessary to insert a note of caution about three distinct types of symmetry classifications, each frequently and misleadingly called "parity," and each associated with an effect of the $\boldsymbol{\sigma}_v$ operator on different parts of the basis function. The $\boldsymbol{\sigma}_v$ operator corresponds to a reflection through the molecule-fixed xz plane. If it operates on the spatial and spin coordinates of all electrons *and* on the nuclear displacement vectors (not the nuclei themselves because, for a diatomic molecule, their instantaneous positions define the z-axis, hence the molecule-fixed axis system), then the eigenvalues of $\boldsymbol{\sigma}_v$ label the *total parity*, $\pm$, of a rotating molecule basis function,

$$|J|\Omega|\Lambda S\Sigma\rangle = (2)^{-1/2}[|J\Omega\Lambda S\Sigma\rangle \pm |J -\Omega -\Lambda S -\Sigma\rangle]. \quad (2.2.21)$$

The total parity of a given class of levels (F_i for Σ-states, upper versus lower Λ-doublet component for Π-states) is found to alternate with J. The second type of label, often loosely called the e/f parity, factors out this $(-1)^J$ or $(-1)^{J-1/2}$ J-dependence (Brown *et al.*, 1975) and becomes a *rotation-independent label.* (Note that e/f is not the parity of the symmetrized nonrotating molecule $|\Lambda S\Sigma\rangle$ basis function. In fact, for half-integral S, it is not possible to construct functions of the form $[|\Lambda S\Sigma\rangle \pm |-\Lambda S -\Sigma\rangle]$ which belong to real eigenvalues of $\boldsymbol{\sigma}_v$.) The third type of parity label arises when $\boldsymbol{\sigma}_v$ is allowed to operate only on the spatial cordinates of all electrons, resulting in a classification of $\Lambda = 0$ states according to their intrinsic Σ^+ or Σ^- symmetry. Only $|\Lambda = 0\rangle$ basis functions have an intrinsic parity of this last type because, unlike $|\Lambda| > 0$ functions, they cannot be put into $[|\Lambda\rangle \pm |-\Lambda\rangle]$ symmetrized form. The peculiarity of this $\Sigma^\pm$ symmetry is underlined by the fact that the selection rule for spin-orbit perturbations (see Section 2.4.1) is $\Sigma^+ \leftrightarrow \Sigma^-$, whereas for all types of electronic states and all perturbation mechanisms (except hyperfine perturbations) the total parity and e/f selection rules are $+ \leftrightarrow +$, $- \leftrightarrow -$, $e \leftrightarrow e$, $f \leftrightarrow f$.

The effect of $\boldsymbol{\sigma}_v(xz)$ on electron spatial (molecule-fixed) coordinates is

$$\boldsymbol{\sigma}_v(xz)[x, y, z] = [x, -y, z], \tag{2.2.22}$$

which, for a right-handed spherical polar coordinate system (see Fig. 2.12), defined so that the $+z$-, $+y$-, and $+x$-axes correspond respectively to polar and azimuthal angles $[\theta, \phi]$ of $[0, \phi]$, $[\pi/2, \pi/2]$, and $[\pi/2, 0]$, becomes

$$\boldsymbol{\sigma}_v(xz)[r, \theta, \phi] = [r, \theta, -\phi]. \tag{2.2.23}$$

Note that here θ, ϕ are coordinates of the electron in the molecule-fixed coordinate system; θ, ϕ do not specify, nor are they affected by, the orientation of the molecule-fixed coordinate system relative to the laboratory-fixed system. Since the ϕ-dependence of an orbital angular momentum basis function of the one-electron ($|\lambda\rangle$) or many-electron ($|\Lambda\rangle$) type can be expressed in terms of a factor $e^{i\lambda(\phi+\phi_0)}$ or $e^{i\Lambda(\phi+\phi_0)}$, where ϕ_0 is an arbitrary phase factor, the effect of $\boldsymbol{\sigma}_v(xz)$ on a molecular orbital becomes

$$\begin{aligned}\boldsymbol{\sigma}_v(xz)\,|\lambda\rangle &= \boldsymbol{\sigma}_v[f(r, \theta)\, e^{i\lambda(\phi+\phi_0)}]\\ &= f(r, \theta)\, e^{-i\lambda\phi} e^{+i\lambda\phi_0}\\ &= e^{2i\lambda\phi_0}|-\lambda\rangle,\end{aligned}$$

which, for the specific phase choice $\phi_0 = n\pi$, becomes

$$\boldsymbol{\sigma}_v(xz)\,|\lambda\rangle = +|-\lambda\rangle. \tag{2.2.24}$$

This is the phase convention commonly used in *ab initio* molecular electronic structure calculations.

Unfortunately, diatomic molecule electronic spectroscopists most commonly use the phase convention

$$\boldsymbol{\sigma}_v|\lambda\rangle = (-1)^{\lambda}|-\lambda\rangle, \tag{2.2.25}$$

which corresponds either to the use of $\boldsymbol{\sigma}(yz)$ rather than $\boldsymbol{\sigma}(xz)$ or to the use of $\boldsymbol{\sigma}(xz)$ with $\phi_0 = (n + \frac{1}{2})\pi$. Then

$$\boldsymbol{\sigma}_v(yz)[r, \theta, \phi] = [r, \theta, \pi - \phi] \tag{2.2.26}$$

and thus

$$\boldsymbol{\sigma}_v e^{i\lambda\phi} = e^{i\lambda(\pi-\phi)} = (-1)^{\lambda} e^{-i\lambda\phi}. \tag{2.2.27}$$

This phase convention is similar to what is often called the *Condon and Shortley phase convention*, which specifies that

1. $\boldsymbol{\sigma}_v(xz)\,|A, \alpha, M_A\rangle = (-1)^{A-\alpha}|A, -\alpha, M_A\rangle$ (2.2.28a)

where **A** is any defined molecular angular momentum with magnitude

$[A(A+1)]^{1/2}$ and α and M_A are its molecule- (z) and space- (Z) fixed projections.

2. Matrix elements

$$\langle A, \alpha, M_A \pm 1 | \mathbf{A}_{\pm} | A, \alpha, M_A \rangle = +\hbar[A(A+1) - M_A(M_A \pm 1)]^{1/2} \quad (2.2.28b)$$

$$\langle A, \alpha \pm 1, M_A | \mathbf{A}^{\pm} | A, \alpha, M_A \rangle = +\hbar[A(A+1) - \alpha(\alpha \pm 1)]^{1/2} \quad (2.2.28c)$$

are real and positive where

$$\mathbf{A}_{\pm} = \mathbf{A}_X \pm i\mathbf{A}_Y, \qquad \mathbf{A}^{\pm} = \mathbf{A}_x \pm i\mathbf{A}_y,$$

are space-fixed (for any angular momentum) and molecule-fixed (only intrinsically molecule-fixed angular momenta, such as **l**, **s**, **L**, **S**, **I**) raising and lowering operators. See Section 1.3.1 and Brown and Howard (1976) for evaluation of matrix elements involving molecule- fixed components of the **R**, **N**, **J**, and **F** angular momenta.

A good reason for the widespread use of this phase convention is that, whenever the magnitude and molecule-fixed projection of an angular momentum are both specified, its transformation properties are automatically and uniformly specified. For example, if $\boldsymbol{\sigma}_v$ acts on spin or rotational basis functions,

$$\boldsymbol{\sigma}_v|S\Sigma\rangle = (-1)^{S-\Sigma}|S - \Sigma\rangle \quad (2.2.29)$$

$$\boldsymbol{\sigma}_v|J\Omega M\rangle = (-1)^{J-\Omega}|J - \Omega M\rangle. \quad (2.2.30)^{\dagger}$$

† Equation (2.2.30) is a direct consequence of the definition [Eq. (1.3.41)],

$$\left\langle \frac{\pi}{2}\theta\phi | J M \Omega \right\rangle = [(2J+1)/4\pi]^{1/2} \mathscr{D}^J_{\Omega M}\left(\frac{\pi}{2}, \theta, \phi\right)$$
$$= [(2J+1)/4\pi]^{1/2} \exp(iM\phi)\, d^J_{\Omega M}(\theta) \exp(i\Omega\pi/2);$$

and the equivalence, for diatomic molecules, of $\boldsymbol{\sigma}_v$ and **I** (inversion of all space-fixed coordinates). If θ and ϕ are the spaced-fixed spherical polar coordinates of the intermolecular axis (as distinct from the previous discussion of $|l\lambda\rangle$ where θ, ϕ were the molecule-fixed coordinates of the electron), then

$$\boldsymbol{\sigma}_v[\theta, \phi] = \mathbf{I}[\theta, \phi] = [\pi - \theta, \phi + \pi].$$

Now, to derive Eq. (2.2.30),

$$\boldsymbol{\sigma}_v \left\langle \frac{\pi}{2}\theta\phi | J M \Omega \right\rangle = \left[\frac{2J+1}{4\pi}\right]^{1/2} \exp[iM(\phi + \pi)]\, d^J_{\Omega M}(\pi - \theta) \exp\left(i\Omega\frac{\pi}{2}\right)$$
$$= \left[\frac{2J+1}{4\pi}\right]^{1/2} (-1)^M \exp(iM\phi)\, d^J_{\Omega M}(\pi - \theta)(-1)^{\Omega} \exp\left(-i\Omega\frac{\pi}{2}\right).$$

From Edmonds [1974, p. 60, Eq. (4.2.4)],

$$d^J_{\Omega M}(\pi - \theta) = (-1)^{J-\Omega} d^J_{M,-\Omega}(\theta),$$

Since the quantum numbers l or L are seldom well-defined in a one-electron orbital or a many-electron case (a) basis state, and since, for a linear molecule, there is no intrinsic distinction between the molecule-fixed x- and y-axes ($\boldsymbol{\sigma}_{xz}$ versus $\boldsymbol{\sigma}_{yz}$), nor reason to choose the phase of $|\lambda\rangle$ so that it is real and positive at any particular value of ϕ, the confusion over the transformation properties of the orbital angular momentum basis functions $|\lambda\rangle$ and $|\Lambda\rangle$ is understandable.

The choice of a phase convention is a matter of taste. However, the convention adopted must be internally consistent. See Brown and Howard (1976) for a discussion of the Condon and Shortley phase convention and molecule- versus space-fixed angular momentum components; see Larsson (1981) for a brief but comprehensive summary of all of the most frequently encountered phase conventions. Throughout this book, an attempt has been made to use the phase conventions of Eqs. (2.2.25), (2.2.28a), (2.2.29), (2.2.30), and

$$\boldsymbol{\sigma}_v(xz)\,|\lambda, s = \tfrac{1}{2}, \sigma\rangle = (-1)^{\lambda + 1/2 - \sigma}|-\lambda, \tfrac{1}{2}, -\sigma\rangle \qquad (2.2.31)^\dagger$$

$$\boldsymbol{\sigma}_v(xz)\,|\Lambda^\pm\, S\Sigma\rangle = (-1)^{\Lambda + S - \Sigma + s}|-\Lambda^\pm\, S - \Sigma\rangle \qquad (2.2.32)^\dagger$$

$$\boldsymbol{\sigma}_v(xz)\,|J\,\Omega\, M\,\Lambda^\pm\, S\Sigma\rangle = (-1)^{J - \Omega + \Lambda + S - \Sigma + s}|J\, -\Omega\, M\, -\Lambda^\pm\, S - \Sigma\rangle \qquad (2.2.33)$$

where $s = 1$ for a Σ^- state and $s = 0$ for all other states. Note that Eq. (2.2.33) is consistent with

$$\boldsymbol{\sigma}_v(xz)\,|J\,\Omega\, M\,\Lambda^\pm\, S\Sigma\rangle = (-1)^{J - S + s}|J\, -\Omega\, M\, -\Lambda^\pm\, S - \Sigma\rangle \qquad (2.2.34)$$

of Larsson (1981), because $(-1)^{S-\Sigma} = (-1)^{\Sigma - S + 2(S-\Sigma)} = (-1)^{\Sigma - S}$, since

---

and from Edmonds [1974, p. 60, Eq. (4.2.6)],

$$d^J_{M,-\Omega}(\theta) = (-1)^{M+\Omega} d^J_{-\Omega,M}(\theta),$$

one obtains

$$d^J_{\Omega M}(\pi - \theta) = (-1)^{J+M} d^J_{-\Omega,M}(\theta),$$

hence

$$\boldsymbol{\sigma}_v\left\langle\frac{\pi}{2}\theta\phi|JM\Omega\right\rangle = \left[\frac{2J+1}{4\pi}\right]^{1/2} \exp(iM\phi)\, d^J_{-\Omega,M}(\theta) \exp\left(-i\Omega\frac{\pi}{2}\right)(-1)^{J-\Omega}(-1)^{2(M+\Omega)}$$
$$= (-1)^{J-\Omega}\left\langle\frac{\pi}{2}\theta\phi|J, M, -\Omega\right\rangle.$$

† It is confusing but customary to use the same symbol for the one-electron spin quantum number [Eq. (2.2.31)] and the symmetry with respect to $\boldsymbol{\sigma}_v(xz)$ of the spatial part of the many-electron wavefunction for Σ-states.

$2(S-\Sigma)$ is an even integer. Replacing $S-\Sigma+\Lambda-\Omega$ in Eq. (2.2.33) by $\Sigma-S+\Lambda-\Lambda-\Sigma=-S$, one obtains Larsson's $(-1)^{J-S+s}$ factor.

One example of how one ensures that the one-electron phase convention of Eq. (2.2.31) is compatible with the many-electron phase convention of Eq. (2.2.32) will be discussed in Section 2.2.4.

Having specified the $\boldsymbol{\sigma}_v$ transformation properties of all case (a) basis functions, it becomes possible to return to the problem of constructing basis functions with well-defined parity. Two types of parity eigenfunctions are important, those labeled according to their total parity, $\pm$, and those labeled according to their e/f symmetry. Brown *et al.* (1975) define $\binom{e}{f}$ basis functions as those with total parity $(\pm)(-1)^J$ for molecules with an even number of electrons and $(\pm)(-1)^{J-1/2}$ for molecules with an odd number of electrons.

Two types of basis functions must be considered:

1. $\Lambda=0, \Omega=0$ (even number of electrons)

$$\boldsymbol{\sigma}_v|J\,\Omega=0\,\Lambda=0^{\pm}\,S\Sigma=0\rangle=(-1)^{J-S+s}|J\,0\,0^{\pm}S\,0\rangle.$$

Odd-multiplicity Σ-states always have exactly one $\Omega=0$ basis function. This function transforms into itself under $\boldsymbol{\sigma}_v$; that is, it has a well-defined parity. The relationships between $\pm$ and $\Sigma^{\pm}$ and between e/f and $\Sigma^{\pm}$ are as follows:

$$+\text{ total parity:}\quad |^{2S+1}\Sigma_0^s+\rangle\ \begin{cases}\xrightarrow{\text{even }J}\ {}^1\Sigma^+,\ {}^3\Sigma_0^-,\ {}^5\Sigma_0^+,\ \text{etc.}\\ \xrightarrow{\text{odd }J}\ {}^1\Sigma^-,\ {}^3\Sigma_0^+,\ {}^5\Sigma_0^-,\ \text{etc.}\end{cases}\tag{2.2.35a}$$

$$-\text{ total parity:}\quad |^{2S+1}\Sigma_0^s-\rangle\quad \text{Roles of even and odd } J \text{ reversed}$$

$$\begin{aligned}&e\text{ levels:}\quad |^{2S+1}\Sigma_0^s, Je\rangle\quad {}^1\Sigma^+,\ {}^3\Sigma_0^-,\ {}^5\Sigma_0^+,\ \text{etc.}\quad \text{all } J\\ &f\text{ levels:}\quad |^{2S+1}\Sigma_0^s, Jf\rangle\quad {}^1\Sigma^-,\ {}^3\Sigma_0^+,\ {}^5\Sigma_0^-,\ \text{etc.}\quad \text{all } J.\end{aligned}\tag{2.2.35b}$$

2. All other states

$$\begin{aligned}&\boldsymbol{\sigma}_v[|J\,\Omega\,\Lambda^{\pm}\,S\Sigma\rangle\pm|J\,-\Omega\,-\Lambda^{\pm}\,S\,-\Sigma\rangle]\\ &\quad=\pm(-1)^{J-S+s}[|J\,\Omega\,\Lambda^{\pm}\,S\Sigma\rangle\pm|J\,-\Omega\,-\Lambda^{\pm}\,S\,-\Sigma\rangle]\end{aligned}\tag{2.2.36a}$$

$$\begin{aligned}&\pm\text{ total parity:}\quad |^{2S+1}\Lambda_{\Omega}, J, \pm\rangle\\ &\quad=(2)^{-1/2}[|J\,\Omega\,\Lambda^{\pm}\,S\Sigma\rangle\pm(-1)^{J-S+s}|J\,-\Omega\,-\Lambda^{\pm}\,S-\Sigma\rangle]\end{aligned}\tag{2.2.36b}$$

$$\begin{aligned}&e/f\text{ symmetry (odd number of electrons):}\quad |^{2S+1}\Lambda_{\Omega}, J, {}^e_f\rangle\\ &\quad=(2)^{-1/2}[|J\,\Omega\,\Lambda^{\pm}\,S\Sigma\rangle\pm(-1)^{-S+s+\frac{1}{2}}|J\,-\Omega\,-\Lambda^{\pm}\,S\,-\Sigma\rangle]\end{aligned}\tag{2.2.37}$$

$$\begin{aligned}&e/f\text{ symmetry (even number of electrons)}\quad |^{2S+1}\Lambda_{\Omega}, J, {}^e_f\rangle\\ &\quad=(2)^{-1/2}[|J\,\Omega\,\Lambda^{\pm}\,S\Sigma\rangle\pm(-1)^{-S+s}|J\,-\Omega\,-\Lambda^{\pm}\,S\,-\Sigma\rangle].\end{aligned}\tag{2.2.38}$$

The $\pm$ and e/f labels are really two different bookkeeping devices for the same physical property, but the e/f labels are more convenient, mainly for optical transition and perturbation selection rules.

1. Selection rules for electric dipole transitions. In terms of e/f levels, the rules are

$$\left.\begin{array}{ll} Q \text{ branch:} & e \leftrightarrow f \\ P \text{ or } R \text{ branch:} & e \leftrightarrow e \quad \text{and} \quad f \leftrightarrow f \end{array}\right\} u \leftrightarrow g$$

2. Selection rules for perturbations:

$$\Delta J = 0: \quad e \leftrightarrow e \quad \text{and} \quad f \leftrightarrow f\} \quad u \leftrightarrow u \quad g \leftrightarrow g.$$

These selection rules are valid for all types of perturbations treated in this book. For perturbations due to the hyperfine Hamiltonian, which are beyond the scope of this book, the above selection rules are no longer rigorously valid (see Section 6.6.4). Figure 2.2 illustrates the utility of e/f compared to $\pm$ labels.

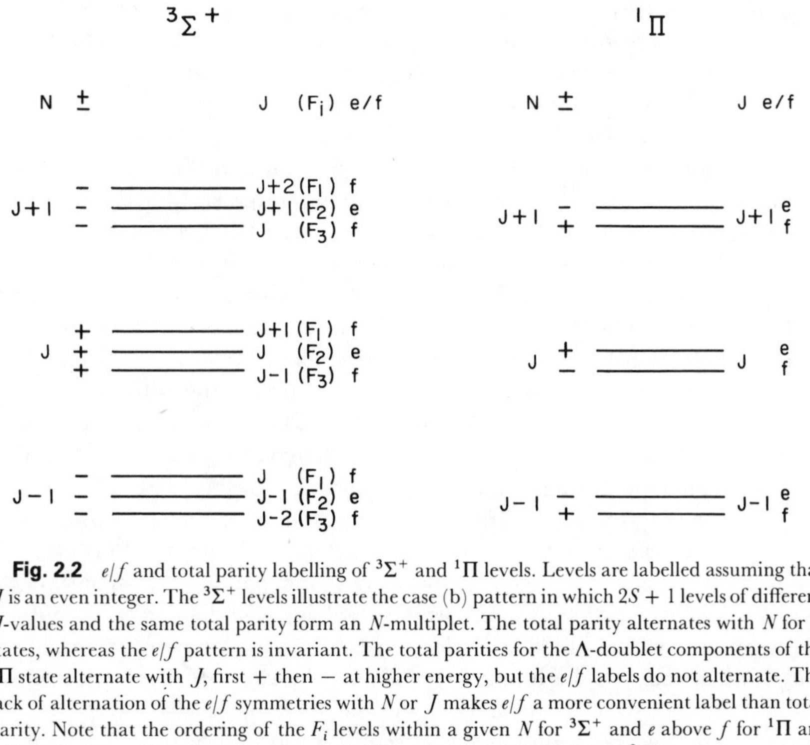

Fig. 2.2 e/f and total parity labelling of $^3\Sigma^+$ and $^1\Pi$ levels. Levels are labelled assuming that J is an even integer. The $^3\Sigma^+$ levels illustrate the case (b) pattern in which $2S + 1$ levels of different J-values and the same total parity form an N-multiplet. The total parity alternates with N for Σ states, whereas the e/f pattern is invariant. The total parities for the Λ-doublet components of the $^1\Pi$ state alternate with J, first $+$ then $-$ at higher energy, but the e/f labels do not alternate. The lack of alternation of the e/f symmetries with N or J makes e/f a more convenient label than total parity. Note that the ordering of the F_i levels within a given N for $^3\Sigma^+$ and e above f for $^1\Pi$ are arbitrarily chosen for this figure. The usual pattern of same-N J-levels for a $^3\Sigma$ state is that, at small N, the F_1 and F_3 components are close together $[F_1 - F_3 \simeq \gamma(2N + 1)]$ and removed by 2λ from the F_2 component [case (a)] but, at large N, they move apart and become separated from F_2 by an asymptotically N-independent splitting $[F_2 - (F_1 + F_3)/2 \simeq \lambda$, case (b)].

The value of *e*/*f* parity labels extends even to atom–molecule collisional processes. There is a strong propensity for conservation of *e*/*f* in *J*-changing collisions (Davis and Alexander, 1983) (see Section 5.5.4).

2.2.3 MOLECULAR ELECTRONIC WAVEFUNCTIONS

Hund's case (a) basis functions will be used here for estimating values of matrix elements, because these wavefunctions are conveniently calculated *ab initio*. In the examples given here, single-configuration representations of the electronic wavefunctions will be used frequently. Except for Rydberg states, the single-configuration approximation is often invalid, but it is usually a good starting point for estimates of perturbation magnitudes. In Section 4.7 one example is discussed where the single-configuration approximation is not sufficent, even for the ground state, to explain observed perturbations, but where a two-configuration model is adequate. Note also that the single-configuration representation, even if it is valid at small internuclear distances, can be invalid at large distances. As R increases the effect of contributions from other configurations becomes more and more important, and this can explain, for example, the variation with R of electronic quantities such as the spin-orbit coupling constant, A, or the spin–spin parameter, λ (see Section 4.6).

Several different notations for molecular orbitals are commonly used. The numbering $1\sigma, 2\sigma, \ldots$ adopted by most theoreticians corresponds to the energy order of the orbitals for each value of λ and not to any principal quantum number. Note that this notation is distinct from the λnl (e.g., $\sigma 1s$, $\sigma 2p$) notation, which specifies the composition of each molecular orbital in terms of the dominant atomic orbital ($1s, 2p$) contribution. The $nl\lambda$ notation (e.g., $3p\sigma$) is reserved for Rydberg molecular orbitals. The designations σ and σ^* or π and π^* are often used to distinguish between bonding (unstarred, excess electron density between atoms) and antibonding (starred, excess electron probability density behind rather than between atoms) orbitals. The term spin–orbital will be used when it is advantageous to specify both the spatial and spin parts of the one-electron wave function. Each orbital also has a well-defined value and sign for λ, the projection of the one-electron orbital angular momentum on the internuclear axis. In the notation used here, the right $\pm$ superscript on the Greek letter gives the sign of λ. For example, π^+ denotes $\lambda = +1$, π^- denotes $\lambda = -1$. Another notation has been used by Field *et al.* (1975) in which the main numeral specifies the signed value of λ and the superscript is the sign of σ ($+$ for α, $-$ for β). The correspondence between these two notations (on the left, Field's notation; on the right, the notation used in this book) is summarized as follows:

$$1^+ = \pi^+\alpha \qquad 1^- = \pi^+\beta$$
$$-1^+ = \pi^-\alpha \qquad -1^- = \pi^-\beta.$$

Still another notation designates β-spin by a bar over the wavefunction, α functions being unbarred; for example $\bar{\pi}^+ = \pi^+\beta$, or $-\bar{1} = \pi^-\beta$ (Field *et al.* 1972).

The electronic wavefunction, constructed from the spin-orbitals ϕ_i ($i = \mathrm{a}, \mathrm{b}, \ldots, n$), is written in the form of a determinant (or a sum of determinants),

$$\Phi(r_1, \ldots, r_n) = (n!)^{-1/2} \begin{vmatrix} \phi_\mathrm{a}(1) & \phi_\mathrm{a}(2) & \cdots & \phi_\mathrm{a}(n) \\ \phi_\mathrm{b}(1) & \phi_\mathrm{b}(2) & \cdots & \phi_\mathrm{b}(n) \\ \vdots & \vdots & \vdots & \vdots \\ \phi_n(1) & \phi_n(2) & \cdots & \phi_n(n) \end{vmatrix} \tag{2.2.39}$$

This determinantal form expresses the antisymmetry of the wavefunction with respect to interchange of two identical particles. The determinantal wavefunction will always be specified by an abbreviated notation that lists only the diagonal of the determinant,

$$\Phi = |\phi_\mathrm{a}(1)\,\phi_\mathrm{b}(2)\cdots\phi_n(n)| = |\phi_\mathrm{a}\,\phi_\mathrm{b}\cdots\phi_n|. \tag{2.2.40}$$

A molecular *electronic configuration* specifies only the total occupancies of molecular orbitals. It does not specify how the spins and angular momentum components of the individual electrons in these orbitals are coupled to form total molecular angular momenta, such as S, Σ, and Λ. When a given orbital has its maximum permissible occupancy, it is said to be *full* and to form a *closed shell*. It takes two electrons to fill a $\sigma(\lambda = 0)$ orbital and four to fill any $\lambda > 0$ orbital, two each in $\lambda > 0$ and $\lambda < 0$ *subshells*. In order to specify $\Lambda = \sum_i \lambda_i$, it is necessary to know the occupancy of each subshell. The Λ-values that can arise for a given electronic configuration may be determined by inspecting the possible subshell occupancies. Once the possible λ-values and the associated subshell occupancies are recognized, the eigenfunctions of $\mathbf{S}^2$ compatible with each Λ-value are obtained simply by counting the number of unpaired electrons, N_u (which is the same as the number of half-filled subshells). The possible values of S range from $S = N_u/2$ down to $S = 0$ (N_u even) or $S = \frac{1}{2}$ (N_u odd) in steps of 1.

When a configuration consists exclusively of closed shells, values of $\Lambda = 0$ and $S = 0$ are the only ones possible, giving rise to a single ${}^1\Sigma^+$ electronic state. For example,

$$(1\sigma_g)^2(1\sigma_u)^2(2\sigma_g)^2(2\sigma_u)^2(3\sigma_g)^2(1\pi_u)^4$$

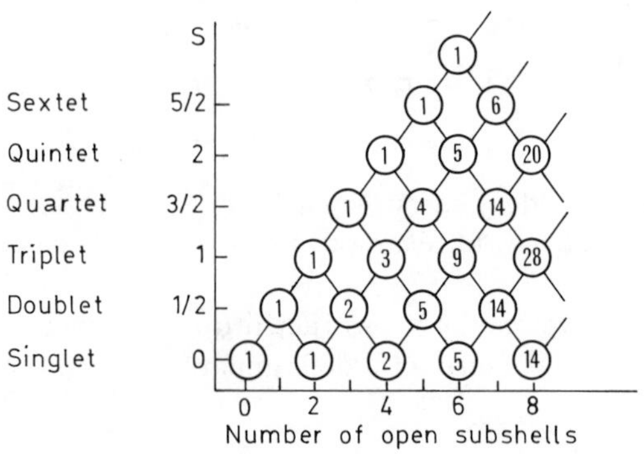

Fig. 2.3 Branching diagram shows the number of states of each spin multiplicity versus number of open (i.e., partly filled) subshells. It is frequently convenient to construct n-electron determinantal wavefunctions by adding a spin-orbital to an already known $(n-1)$-electron wavefunction. The explicit spin wavefunctions are summarized in Table 2.4.

is the configuration of the $N_2X^1\Sigma_g^+$ electronic ground state. If a configuration containing an odd total number of electrons has only one open (half-filled) subshell, an $S=\frac{1}{2}$ (doublet) state is the sole result. For example, removing one electron from the $N_2X^1\Sigma_g^+ 3\sigma_g$ or $1\pi_u$ orbital gives rise to the $N_2^+X^2\Sigma_g^+$ or $A^2\Pi_u$ state, respectively. A configuration with more than one open subshell gives rise to several electronic states.

The "branching diagram" (Fig. 2.3) shows how many eigenfunctions of $\mathbf{S}^2$ can be constructed from a given number of open subshells. The numbers in circles specify the number of states of a given multiplicity. For example, with two open subshells, one triplet and one singlet function result. If one electron is added to an empty orbital, yielding three unpaired electrons, the branching diagram shows that one doublet state is derived from the singlet parent, and one quartet and one doublet state are derived from the triplet parent.

Consider first the trivial case of the π^2 configuration. The following table gives all possible distributions of these two electrons among the two subshells: π^+ and π^- (the right $\pm$ superscript specifies whether $\lambda > 0$ or $\lambda < 0$; the spin projection is not yet specified).

π^+	π^-	$\Lambda = \sum_i \lambda_i$		Number of open subshells
2	0	2	Δ state	0
0	2	-2		
1	1	0	Σ state	2

Table 2.4
Spin Eigenfunctions[a] of $\boldsymbol{S}^2$ for $\Sigma = \boldsymbol{S}$

Open subshells	Eigenfunctions
One	
1 doublet: $S = \frac{1}{2}, \Sigma = \frac{1}{2}$	$[\alpha(1)]$
Two	
1 singlet: $S = 0, \Sigma = 0$	$(2)^{-1/2}[\alpha(1)\,\beta(2) - \beta(1)\,\alpha(2)]$
1 triplet: $S = 1, \Sigma = 1$	$[\alpha(1)\,\alpha(2)]$
Three	
2 doublets: $S = \frac{1}{2}, \Sigma = \frac{1}{2}$	$(2)^{-1/2}[\alpha(1)\,\alpha(2)\,\beta(3) - \alpha\beta\alpha]$
	$(6)^{-1/2}[2\beta(1)\,\alpha(2)\,\alpha(3) - \alpha\alpha\beta - \alpha\beta\alpha]$
1 quartet: $S = \frac{3}{2}, \Sigma = \frac{3}{2}$	$[\alpha(1)\,\alpha(2)\,\alpha(3)]$
Four	
2 singlets: $S = 0, \Sigma = 0$	$\frac{1}{2}[\alpha(1)\,\beta(2)\,\alpha(3)\,\beta(4) + \beta\alpha\beta\alpha - \beta\alpha\alpha\beta - \alpha\beta\beta\alpha]$
	$(12)^{-1/2}[2\alpha(1)\,\alpha(2)\,\beta(3)\,\beta(4) + 2\beta\beta\alpha\alpha - \beta\alpha\alpha\beta - \alpha\beta\alpha\beta - \beta\alpha\beta\alpha - \alpha\beta\beta\alpha]$
3 triplets: $S = 1, \Sigma = 1$	$(2)^{-1/2}[\alpha(1)\,\alpha(2)\,\alpha(3)\,\beta(4) - \alpha\alpha\beta\alpha]$
	$(6)^{-1/2}[-2\alpha(1)\,\beta(2)\,\alpha(3)\,\alpha(4) + \alpha\alpha\alpha\beta + \alpha\alpha\beta\alpha]$
	$(12)^{-1/2}[-3\beta(1)\,\alpha(2)\,\alpha(3)\,\alpha(4) + \alpha\alpha\beta\alpha + \alpha\alpha\alpha\beta + \alpha\beta\alpha\alpha]$
1 quintet: $S = 2, \Sigma = 2$	$[\alpha(1)\,\alpha(2)\,\alpha(3)\,\alpha(4)]$

[a] Eigenfunctions for up to six open subshells can be found in Yamazaki (1963a). Analytic formulas for constructing electronic interaction matrices between basis functions that belong to the same value of S (similar to example on $^2\Pi$ states) are given in Yamazaki (1963b).

The Δ state, which has no unpaired electron, can only be a singlet state. The Σ states, which are associated with two unpaired electrons, include one singlet and one triplet state.

A wavefunction, written as a sum of Slater determinants, can be deduced from Table 2.4. This table lists the spin functions, each of which appears along the diagonal of one determinant constructed from unpaired molecular spatial wavefunctions. For example, if there are two half-filled subshells, $\phi_a(1)$ and $\phi_b(2)$, using Table 2.4, the singlet wavefunction is written as a sum of two determinants:

$$(2)^{-1/2}\{|\phi_a(1)\alpha(1)\,\phi_b(2)\beta(2)| - |\phi_a(1)\beta(1)\,\phi_b(2)\alpha(2)|\}.$$

For a filled subshell, the spin functions are evidently $\alpha(1)\beta(2)$. Using the simplified notation of Eq. (2.2.40);

$$\begin{array}{llll} ^1\Delta & |\pi^+\alpha\,\pi^+\beta| & \Lambda = +2 & \\ & |\pi^-\alpha\,\pi^-\beta| & \Lambda = -2 & \end{array} \qquad \Sigma = 0 \quad (2.2.41a)$$

$$^1\Sigma^+ \qquad (2)\,(2)^{-1/2}\{|\pi^+\alpha\,\pi^-\beta| - |\pi^+\beta\,\pi^-\alpha|\} \qquad \Sigma = 0 \qquad (2.2.41b)^\dagger$$

$$|\pi^+\alpha\,\pi^-\alpha| \qquad \Sigma = +1$$

$$^3\Sigma^- \qquad (2)\,(2)^{-1/2}\{|\pi^+\alpha\,\pi^-\beta| + |\pi^+\beta\,\pi^-\alpha|\} \qquad \Sigma = 0 \qquad (2.2.41c)^\dagger$$

$$|\pi^+\beta\,\pi^-\beta| \qquad \Sigma = -1.$$

Table 2.4 only specifies the spin function for the maximum value of $\Sigma = S$, $|S = 1, \Sigma = 1\rangle = |\,|\alpha(1)\alpha(2)|\rangle$, but the wavefunctions for $-S \leq \Sigma < +S$ can be deduced by application of the $\mathbf{S}^-$ operator:

$$\mathbf{S}^-\,|S, \Sigma\rangle = +\hbar[S(S+1) - \Sigma(\Sigma - 1)]^{-1/2}\,|S, \Sigma - 1\rangle.$$

The $+$ sign is a consequence of the sign convention of Eq. (2.2.28c). The result of $\mathbf{S}^-$ operating on a determinantal wavefunction is

$$\mathbf{S}^-\,|\mathrm{det}| = \sum_i \mathbf{s}^-(i)\,|\cdots\phi_i(i)^{\alpha(i)}_{\beta(i)}\cdots| = \sum_i (1 - c_i)\,|\cdots\phi_i(i)\beta(i)\cdots|, \qquad (2.2.42)$$

where c_i is equal to 1 or 0 if the original spin function of $\phi_i(i)$ is $\beta(i)$ or $\alpha(i)$, respectively. The wavefunction must be renormalized after each application of $\mathbf{S}^-$. The orthogonality between wavefunctions of identical Σ but different S is often useful (but sometimes insufficient) to find the proper eigenfunctions of $\mathbf{S}^2$.

If another open shell is added to the π_g^2 configuration—for example, a singly occupied π_u orbital—the $\pi_g^2\pi_u$ configuration results. The possible states are shown in table form.

π_g^+	π_g^-	π_u^+	π_u^-	$\Lambda = \Sigma\lambda_i$	Number of open subshells	States
2	0	1	0	3		
0	2	0	1	-3	1	$^2\Phi_u$
$^1\Delta_g$						
2	0	0	1	1		
0	2	1	0	-1	1	$^2\Pi_u$
$^1\Delta_g$						
1	1	1	0	1		
1	1	0	1	-1	3	one $^4\Pi_u$, two $^2\Pi_u$
$^1\Sigma_g^+$,	$^3\Sigma_g^-$					

Evidently, the $\pi_g^2\pi_u$ configuration gives rise to three independent $^2\Pi_u$ states. One is formed from the $^1\Delta_g$ parent. The other two originate from the $(\pi_g^+)(\pi_g^-)$

† The two $\Sigma = 0$ Slater determinants appear with the same sign for $^3\Sigma^-$ and opposite signs for $^1\Sigma^+$. This may be demonstrated by applying $\mathbf{S}^-$ (Eq. 2.2.42) to the $\Sigma = +1$ $^3\Sigma^-$ determinant, $|\pi^+\alpha\,\pi^-\alpha|$.

subshell occupancy, one each associated with the ${}^3\Sigma_g^-$ and ${}^1\Sigma_g^+$ parent states. This is consistent with the branching diagram. The determinantal wavefunctions can be derived easily for these three ${}^2\Pi_u$ states, using this coupling scheme. When the singlet ${}^1\Sigma_g^+$ parent state ($\Lambda_1 = 0, \Sigma_1 = 0, S_1 = 0$) is coupled with a π_u electron ($\Lambda_2 = +1$, $\Sigma_2 = +\frac{1}{2}$), a doublet state results (because $S = \Sigma_{\max} = \Sigma_1 + \Sigma_2 = \frac{1}{2}$), which must be ${}^2\Pi_u$ (because $\Lambda = \Lambda_1 + \Lambda_2 = 1$). The wavefunction is

$$|{}^2\Pi(1), \pi_g^2\pi_u, \Omega = \tfrac{3}{2}\rangle$$
$$= (2)^{-1/2}\{|\cdots\pi_u^+\alpha\,\pi_g^+\alpha\,\pi_g^-\beta| - |\cdots\pi_u^+\alpha\,\pi_g^+\beta\,\pi_g^-\alpha|\}. \quad (2.2.43)$$

Another doublet state arises from the ${}^3\Sigma_g^-$ parent. There are two ways to construct a $\Sigma = \frac{1}{2}$, $\Lambda = 1$ state from ${}^3\Sigma_g^-$, one each from the ${}^3\Sigma_0^-$ and ${}^3\Sigma_1^-$ parent substates. Choosing the ${}^3\Sigma_0^-$ parent, $\Sigma_1 = 0$, $\Lambda_1 = 0$, $\{|\pi_g^+\alpha\,\pi_g^-\beta| + |\pi_g^+\beta\,\pi_g^-\alpha|\}$ and combining it with an electron in the $\pi_u^+\alpha(\Sigma_2 = \frac{1}{2}$, $\Lambda_2 = 1)$ spin-orbital, the result is

$$\Phi({}^3\Sigma_0^- \rightarrow \Pi_{3/2}) = \{|\pi_u^+\alpha\,\pi_g^+\alpha\,\pi_g^-\beta| + |\pi_u^+\alpha\,\pi_g^+\beta\,\pi_g^-\alpha|\}$$
$$\Sigma = \Sigma_1 + \Sigma_2 = \tfrac{1}{2}, \qquad \Lambda = \Lambda_1 + \Lambda_2 = 1, \qquad \Omega = \Sigma + \Lambda = \tfrac{3}{2}. \quad (2.2.43a)$$

When ${}^3\Sigma_1^-$, $\Sigma_1 = 1$, $\Lambda_1 = 0$ ($|\pi_g^+\,\alpha\,\pi_g^-\alpha|$) is combined with the $\pi_u^+\,\beta$ ($\Sigma_2 = -\frac{1}{2}$, $\Lambda_2 = 1$) spin-orbital, the result is

$$\Phi({}^3\Sigma_1^- \rightarrow \Pi_{3/2}) = |\pi_u^+\beta\,\pi_g^+\alpha\,\pi_g^-\alpha|$$
$$\Sigma = \Sigma_1 + \Sigma_2 = \tfrac{1}{2}, \qquad \Lambda = \Lambda_1 + \Lambda_2 = 1, \qquad \Omega = \Sigma + \Lambda = \tfrac{3}{2}, \quad (2.2.43b)$$

but neither of the states thus generated is an eigenstate of $\mathbf{S}^2$, being mixtures of ${}^2\Pi_{3/2}$ and ${}^4\Pi_{3/2}$. From these two $\Sigma = \frac{1}{2}, \Lambda = 1$ functions, two eigenfunctions of $\mathbf{S}^2$ can be constructed, a doublet,

$$|{}^2\Pi_u(2), \pi_g^2\pi_u, \Omega = \tfrac{3}{2}\rangle$$
$$= (6)^{-1/2}[2|\pi_u^+\beta\,\pi_g^+\alpha\,\pi_g^-\alpha| - |\pi_u^+\alpha\,\pi_g^+\alpha\,\pi_g^-\beta| - |\pi_u^+\alpha\,\pi_g^+\beta\,\pi_g^-\alpha|], \quad (2.2.44)$$

and a quartet,

$$|{}^4\Pi_u, \pi_g^2\pi_u, \Omega = \tfrac{3}{2}\rangle$$
$$= (3)^{-1/2}[|\pi_u^+\alpha\,\pi_g^+\alpha\,\pi_g^-\beta| + |\pi_u^+\alpha\,\pi_g^+\beta\,\pi_g^-\alpha| + |\pi_u^+\beta\,\pi_g^+\alpha\,\pi_g^-\alpha|]. \quad (2.2.45)$$

The spin parts of the ${}^2\Pi_u$ ($\Omega = \frac{3}{2}$) functions are taken from Table 2.4. Here the spatial part has been chosen as $\pi_u^+(1)\,\pi_g^+(2)\,\pi_g^-(3)$. Any permutation of the order of the orbitals defines the ${}^2\Pi(1)$ and ${}^2\Pi(2)$ states differently and with different energies. The spin part of ${}^4\Pi_u\,\Omega = \frac{3}{2}$ is derived either by applying $\mathbf{S}^-$ to the spin function for $|{}^4\Pi_u, \Omega = \frac{5}{2}\rangle$ from Table 2.4 or by requiring that $|{}^4\Pi_u, \Omega = \frac{3}{2}\rangle$ be orthogonal to $|{}^2\Pi_u, \Omega = \frac{3}{2}\rangle$. A third ${}^2\Pi_u$ state is derived when the $^1\Delta_g$ parent, $S_1 = 0, \Lambda_1 = 2$ ($|\pi_g^+ \alpha \pi_g^+ \beta|$), is combined with an electron in the $\pi_u^- \alpha$ ($\Sigma_2 = \frac{1}{2}, \Lambda_2 = -1$) spin-orbital,

$$|^2\Pi_u(3), \Omega = \tfrac{3}{2}\rangle = |\pi_u^- \alpha \pi_g^+ \alpha \pi_g^+ \beta|. \tag{2.2.46}$$

One might expect that the relative energies of the three $^2\Pi$ states would mirror the relative energies of the parent states, namely, that the $^2\Pi_u(2)$ and $^2\Pi_u(1)$ basis functions derived from the lowest ($^3\Sigma_g^-$) and highest ($^1\Sigma_g^+$) energy parent states would become, respectively, the lowest- and highest-lying $^2\Pi$ states. In fact, the observed $^2\Pi_u$ states are found to be mixtures of the three $^2\Pi_u$ functions:

$$|^2\Pi_{ui}(\text{obs})\rangle = c_{1i}|^2\Pi_u(1)\rangle + c_{2i}|^2\Pi_u(2)\rangle + c_{3i}|^2\Pi_u(3)\rangle$$
$$i = 1, 2, 3\,. \tag{2.2.47}$$

The c_{ji} coefficients are obtained by setting up and diagonalizing a 3×3 secular energy matrix for the three isoconfigurational $^2\Pi_u$ states. The diagonal matrix elements have the expected energy order, but the off-diagonal matrix elements that originate from the electrostatic part (e^2/r_{ij}) of the electronic Hamiltonian couple the Slater determinants of basis functions with identical $|\Lambda, S, \Sigma\rangle$ but that differ by two spin-orbitals. Using the rules given in the next section (Section 2.2.4), the matrix elements are

$$\begin{aligned}
\langle ^2\Pi(2)|\mathbf{H}^{\text{el}}|^2\Pi(2)\rangle &= E(^2\Pi(2)) = E(^3\Sigma_g^-, \pi_g^2) + a - \tfrac{1}{2}[K^{(0)}_{\pi_u\pi_g} + K^{(2)}_{\pi_u\pi_g}]\\
\langle ^2\Pi(3)|\mathbf{H}^{\text{el}}|^2\Pi(3)\rangle &= E(^2\Pi(3)) = E(^1\Delta g, \pi_g^2) + a - K^{(2)}_{\pi_u\pi_g}\\
\langle ^2\Pi(1)|\mathbf{H}^{\text{el}}|^2\Pi(1)\rangle &= E(^2\Pi(1)) = E(^1\Sigma_g^+, \pi_g^2) + a + \tfrac{1}{2}[K^{(0)}_{\pi_u\pi_g} + K^{(2)}_{\pi_u\pi_g}]\\
\langle ^2\Pi(1)|\mathbf{H}^{\text{el}}|^2\Pi(2)\rangle &= 3^{1/2}/2(K^{(0)}_{\pi_u\pi_g} - K^{(2)}_{\pi_u\pi_g})\\
\langle ^2\Pi(1)|\mathbf{H}^{\text{el}}|^2\Pi(3)\rangle &= (2)^{1/2} J^{(2)}_{\pi_u\pi_g}\\
\langle ^2\Pi(2)|\mathbf{H}^{\text{el}}|^2\Pi(3)\rangle &= -(\tfrac{2}{3})^{1/2} K^{(0)}_{\pi_u\pi_g}
\end{aligned} \tag{2.2.48}$$

where the quantity a represents the Coulombic interaction energy of the π_u electron with the field of the other electrons. The $K^{(0)}$ and $K^{(2)}$ exchange and $J^{(2)}$ Coulomb integrals are defined, using the abbreviated notation,

$$\langle \phi_a\phi_b|\phi_c\phi_d\rangle \equiv \int \phi_a^*(1)\phi_b(1)\frac{e^2}{r_{12}}\phi_c^*(2)\phi_d(2)\,d\tau_1\,d\tau_2 \tag{2.2.49}$$

$$K^{(0)}_{\pi_u\pi_g} = \langle \phi_{\pi_g^+}\phi_{\pi_u^+}|\phi_{\pi_u^+}\phi_{\pi_g^+}\rangle \tag{2.2.50a}$$

$$K^{(2)}_{\pi_u\pi_g} = \langle \phi_{\pi_g^+}\phi_{\pi_u^-}|\phi_{\pi_u^-}\phi_{\pi_g^+}\rangle \tag{2.2.50b}$$

$$J^{(2)}_{\pi_u\pi_g} = \langle \phi_{\pi_g^+}\phi_{\pi_g^-}|\phi_{\pi_u^-}\phi_{\pi_u^+}\rangle. \tag{2.2.50c}$$

The off-diagonal element between the $^2\Pi(2)$ and $^2\Pi(3)$ basis states associated with the $^3\Sigma_g^-$ and $^1\Delta_g$ parents is so large that the energy order of the roots of the secular equation is not that expected from the order of the diagonal elements. Consequently, the highest and lowest energy $^2\Pi$ states are predominantly mixtures of $^3\Sigma_g^-$ and $^1\Delta_g$ parents, even though the highest energy basis state is that associated with the $^1\Sigma_g^+$ parent.

An example has been discussed by Dixon and Hull (1969), who apply these ideas to the three states arising from the $O_2^+\ \pi_u^3\pi_g^2$ configuration. The eigenfunctions obtained are, starting with the lowest energy state, the $A^2\Pi_u$ state,

$$\begin{aligned}|A^2\Pi_u\rangle &= -0.20\,|^2\Pi(1)\rangle + 0.66\,|^2\Pi(2)\rangle + 0.72\,|^2\Pi(3)\rangle,\\ |2^2\Pi_u\rangle &= 0.91\,|^2\Pi(1)\rangle + 0.39\,|^2\Pi(2)\rangle - 0.15\,|^2\Pi(3)\rangle, \qquad (2.2.51)\\ |3^2\Pi_u\rangle &= 0.37\,|^2\Pi(1)\rangle + 0.74\,|^2\Pi(2)\rangle - 0.57\,|^2\Pi(3)\rangle.\end{aligned}$$

The values of the c_{ij} coefficients are very similar to those for the excited $^2\Pi$ states of NO (Field *et al.*, 1975) and of PO (Roche and Lefebvre-Brion, 1973), but the signs of the c_{ij} depend on the choice of phases for the electronic functions.

2.2.4 MATRIX ELEMENTS BETWEEN ELECTRONIC WAVEFUNCTIONS

The rules given by Slater must be used when evaluating matrix elements of any operator in a Slater determinantal representation (Slater, 1960, p. 291). These rules reduce the matrix elements of one- and two-electron operators between many-electron Slater determinantal functions to simple sums over spatial orbital integrals.†

Determinants have the property that they change sign when any two adjacent rows or columns are interchanged. Since electrons are Fermions, the total wavefunction must change sign upon all possible permutations of electrons. One finds that Slater determinantal basis functions are properly antisymmetrized with respect to electron interchange and that one need only keep track of the order of spin-orbitals along the diagonal in order to ascertain whether two determinants, constructed from the same spin-orbitals, are identical or differ by -1. Two determinants differing by an odd (even) number of adjacent spin-orbital permutations are identical except for a factor of -1 $(+1)$.

The procedure for evaluating determinantal matrix elements requires that all spin-orbitals common to both determinantal functions be permuted so that both the common spin-orbitals and the orbitals that differ in the space part but

† It is necessary to assume that the spin-orbitals are mutually orthogonal.

are identical in the spin part appear in the same locations. It is necessary to specify all common orbitals. For example, when a matrix element between states belonging to the two configurations $\sigma\pi^3$ and π^2 is to be calculated, it is important not to be confused by the abbreviated notation π^2, which suppresses the usually irrelevant closed σ shell, instead of the more complete designation, $\sigma^2\pi^2$.

If two determinantal functions, Φ_i and Φ_j, differ by only one spin-orbital, ϕ_i in Φ_i and ϕ_j in Φ_j, then a one-electron operator, $\mathbf{Op}_1$, can have a nonzero matrix element

$$\mathbf{Op}_1 \equiv \sum_{i=1}^{n} \mathbf{Op}_1(i)$$

$$\langle\Phi_i|\mathbf{Op}_1|\Phi_j\rangle = (-1)^P \int \phi_i^*(1)\mathbf{Op}_1(1)\phi_j(1)\, d\tau_1 \tag{2.2.52}$$

$$= (-1)^P \langle\phi_i|\mathbf{Op}_1|\phi_j\rangle,$$

where P is the number of permutations required to match the order of appearance of the two sets of spin-orbitals. Diagonal matrix elements of $\mathbf{Op}_1$ are given by

$$\langle\Phi_i|\mathbf{Op}_1|\Phi_i\rangle = \sum_{k=1}^{n} \langle\phi_k|\mathbf{Op}_1|\phi_k\rangle. \tag{2.2.53}$$

If two determinantal functions differ by two spin-orbitals, then their matrix elements for all one-electron operators will be zero. However, a two-electron operator,

$$\mathbf{Op}_{12} = \sum_{j>i=1}^{n} \mathbf{Op}_{12}(i,j),$$

can have nonzero matrix elements. Denoting the unique spin-orbitals as ϕ_i and ϕ_j in Φ_{ij}, and ϕ_k and ϕ_l in Φ_{kl}, then

$$\langle\Phi_{ij}|\mathbf{Op}_{12}|\Phi_{kl}\rangle = (-1)^P[\langle\phi_i\phi_k|\mathbf{Op}_{12}|\phi_j\phi_l\rangle - \langle\phi_i\phi_l|\mathbf{Op}_{12}|\phi_j\phi_k\rangle], \tag{2.2.54}$$

where, by convention [Eq. (2.2.49)],

$$\langle\phi_i\phi_k|\mathbf{Op}_{12}|\phi_j\phi_l\rangle \equiv \int \phi_i^*(1)\phi_k(1)\mathbf{Op}_{12}(1,2)\phi_j^*(2)\phi_l(2)\, d\tau_1\, d\tau_2. \tag{2.2.55}$$

In the frequently encountered situation when $\mathbf{Op}_{12}$ does not operate on electron spin wavefunctions, then one of the two integrals in Eq. (2.2.54) vanishes whenever the spin parts of ϕ_i and ϕ_j or ϕ_k and ϕ_l are not identical. When $\phi_i = \phi_k$, $\phi_j = \phi_l$, and $\mathbf{Op}_{12}(1,2) = e^2/r_{12}$, then

$$\langle\phi_i\phi_i|\mathbf{Op}_{12}|\phi_j\phi_j\rangle \equiv J_{ij}, \qquad \langle\phi_i\phi_j|\mathbf{Op}_{12}|\phi_j\phi_i\rangle \equiv K_{ij},$$

where J_{ij} and K_{ij} are direct (or Coulomb) and exchange intergrals. J_{ij} is the classical electrostatic repulsion between the charge densities associated with orbitals ϕ_i and ϕ_j. There is no classical analog for K_{ij}, but it is always true that $J_{ij} > K_{ij}$.

Diagonal matrix elements of $\mathbf{Op}_{12}$ are given by

$$\langle\Phi_i|\mathbf{Op}_{12}|\Phi_i\rangle = \sum_{j>i=1}^{n} [\langle\phi_i\phi_i|\mathbf{Op}_{12}|\phi_j\phi_j\rangle - \langle\phi_i\phi_j|\mathbf{Op}_{12}|\phi_j\phi_i\rangle], \quad (2.2.56)$$

and matrix elements of $\mathbf{Op}_{12}$ between functions differing by one spin-orbital are

$$\langle\Phi_i|\mathbf{Op}_{12}|\Phi_j\rangle = (-1)^P \sum_{k=1}^{n} [\langle\phi_i\phi_j|\mathbf{Op}_{12}|\phi_k\phi_k\rangle - \langle\phi_i\phi_k|\mathbf{Op}_{12}|\phi_k\phi_j\rangle], \quad (2.2.57)$$

where P is the number of permutations required to match the orbital orders in Φ_i and Φ_j. Matrix elements of $\mathbf{Op}_{12}$ between functions differing by two spin-orbitals are given by Eq. (2.2.54), and those between functions differing by more than two orbitals are necessarily zero.

The *absolute* sign of an off-diagonal matrix element cannot be determined, since it depends arbitrarily on the chosen phase for the determinantal wavefunction, namely, on the order of the spin-orbitals. However, the *relative* sign of *two* off-diagonal matrix elements can often be determined experimentally. Thus, care must be taken to define the phases of the wavefunctions consistently.

In Section 4.3 it will become clear why it is useful to reduce matrix elements between Slater determinantal functions to matrix elements between individual molecular orbitals.

Some operators that will be used here contain the orbital and spin angular momenta of the electrons. By definition,

$$\mathbf{l}_z(1)\,|p\lambda(1)\rangle = \hbar\lambda|p\lambda(1)\rangle$$

or

$$\mathbf{l}_z|\pi^+\rangle = +\hbar|\pi^+\rangle, \qquad \mathbf{l}_z|\pi^-\rangle = -\hbar|\pi^-\rangle. \quad (2.2.58)$$

For the spin,

$$\mathbf{s}_z(1)\,|s\sigma(1)\rangle = \hbar\sigma\,|s\sigma(1)\rangle$$

$$\mathbf{s}_z|\alpha\rangle = +\hbar/2\,|\alpha\rangle, \qquad \mathbf{s}_z|\beta\rangle = -\hbar/2\,|\beta\rangle. \quad (2.2.59)$$

Raising and lowering operators have the effect

$$\mathbf{s}^+\,|\beta\rangle = +\hbar|\alpha\rangle, \qquad \mathbf{s}^+\,|\alpha\rangle = 0,$$

and

$$\mathbf{s}^-\,|\alpha\rangle = +\hbar|\beta\rangle, \qquad \mathbf{s}^-\,|\beta\rangle = 0. \quad (2.2.60)$$

The effect of $\mathbf{S}^-$ on a determinantal function was illustrated by Eq. (2.2.42).

In general, the l value of a molecular orbital is not well defined. However, in some cases, such as for Rydberg orbitals, l has nearly an integer value. In such a case,

$$\mathbf{l}^{\pm}|l\lambda\rangle = \hbar[l(l+1) - \lambda(\lambda \pm 1)]^{1/2}|l, \lambda \pm 1\rangle, \tag{2.2.61}$$

and, for a p orbital,

$$\begin{aligned} &\mathbf{l}^+|p\sigma\rangle = (2)^{1/2}\hbar|p\pi^+\rangle \quad \text{and} \quad \mathbf{l}^-|p\pi^+\rangle = (2)^{1/2}\hbar|p\sigma\rangle \\ &\mathbf{l}^-|p\sigma\rangle = (2)^{1/2}\hbar|p\pi^-\rangle \quad \text{and} \quad \mathbf{l}^+|p\pi^-\rangle = (2)^{1/2}\hbar|p\sigma\rangle \\ &\mathbf{l}^+|p\pi^+\rangle = 0 \quad \text{and} \quad \mathbf{l}^-|p\pi^-\rangle = 0. \end{aligned} \tag{2.2.62}$$

Until now it has been assumed that the spatial part of all many-electron Σ-state basis functions is an eigenfunction of the $\boldsymbol{\sigma}_v$ operator. However, using the phase convention of Eq. (2.2.25),

$$\boldsymbol{\sigma}_v(xz)|\lambda\rangle = (-1)^{\lambda}|-\lambda\rangle,$$

it appears that there can be no one-electron wavefunction (i.e., orbital) of Σ^- symmetry. Evidently, $\Sigma^{\pm}$ symmetry is a property of many-electron wavefunctions. In order to understand the origin of states of Σ^- symmetry and to demonstrate the compatibility of Eq. (2.2.25) with

$$\boldsymbol{\sigma}_v(xz)|\Lambda^{\pm}\rangle = (-1)^{\Lambda+s}|-\Lambda^{\pm}\rangle$$

(where $s = 1$ for Σ^- and $s = 0$ for Σ^+ and all $\Lambda \neq 0$), it is necessary to examine the many-electron form of the $\boldsymbol{\sigma}_v$ operator. Let

$$\boldsymbol{\sigma}_v = \prod_{i=1}^{n} \boldsymbol{\sigma}_v^i,$$

where $\boldsymbol{\sigma}_v^i$ operates on the ith electron and, for the spatial part of the wavefunction only,

$$\begin{aligned} \boldsymbol{\sigma}_v(xz)|\lambda_1 \cdots \lambda_n| &= (-1)^{\Sigma_i^n \lambda_i}|-\lambda_1 \cdots -\lambda_n| \\ &= (-1)^{\Lambda}|-\lambda_1 \cdots -\lambda_n| \end{aligned} \tag{2.2.63}$$

or, for the complete wavefunction,

$$\begin{aligned} \boldsymbol{\sigma}_v(xz)|\lambda_1\sigma_1, \cdots, \lambda_n\sigma_n| &= (-1)^{\Sigma_i^n(\lambda_i + 1/2 - \sigma_i)}|-\lambda_1 - \sigma_1, \cdots, -\lambda_n - \sigma_n| \\ &= (-1)^{\Lambda - \Sigma + n/2}|-\lambda_1 - \sigma_1, \cdots, -\lambda_n - \sigma_n|. \end{aligned} \tag{2.2.64}$$

The correct $\Sigma^{\pm}$ symmetry appears when, for $\Lambda = 0$, the orbitals on the right side of Eq. (2.2.63) are rearranged to match the order of those on the left side, keeping track of the number of adjacent orbital permutations required.

Consider, as an example, the simplest configuration known to give rise to a Σ^- state, π^2. The determinantal wavefunctions corresponding to $^1\Delta$, $^1\Sigma^+$, and $^3\Sigma^-$ were given as Eqs. (2.2.41). Adopting a standard order for the spin-orbitals, $(\pi^+\alpha, \pi^+\beta, \pi^-\alpha, \pi^-\beta)$, and always rearranging the determinantal functions so that the spin-orbitals appear in this order, then

$$\begin{aligned}\boldsymbol{\sigma}_v|^3\Sigma_1^-\rangle &= \boldsymbol{\sigma}_v|\pi^+\alpha\,\pi^-\alpha| = (-1)^0|\pi^-\beta\,\pi^+\beta| \\ &= -|\pi^+\beta\,\pi^-\beta| = -|^3\Sigma_{-1}^-\rangle,\end{aligned}$$

which is consistent with Eq. (2.2.32), namely,

$$\begin{aligned}\boldsymbol{\sigma}_v|^3\Sigma_1^-\rangle &= \boldsymbol{\sigma}_v|\Lambda = 0^-, S = 1, \Sigma = 1\rangle \\ &= (-1)^{0+1-1+1}|0^-, 1, -1\rangle = (-1)|^3\Sigma_{-1}^-\rangle.\end{aligned}$$

Similarly,

$$\begin{aligned}\boldsymbol{\sigma}_v|^1\Sigma^+\rangle &= (2)^{-1/2}[\boldsymbol{\sigma}_v|\pi^+\alpha\,\pi^-\beta| - \boldsymbol{\sigma}_v|\pi^+\beta\,\pi^-\alpha|] \\ &= (2)^{-1/2}[-|\pi^-\beta\,\pi^+\alpha| + |\pi^-\alpha\,\pi^+\beta|] \\ &= (2)^{-1/2}[|\pi^+\alpha\,\pi^-\beta| - |\pi^+\beta\,\pi^-\alpha|] = +|^1\Sigma^+\rangle,\end{aligned}$$

which is consistent with

$$\boldsymbol{\sigma}_v|^1\Sigma^+\rangle = (-1)^{0+0+0+0}|^1\Sigma^+\rangle.$$

Finally,

$$\begin{aligned}\boldsymbol{\sigma}_v|^1\Delta_2\rangle &= -|\pi^-\beta\,\pi^-\alpha| = |\pi^-\alpha\,\pi^-\beta| = +|^1\Delta_{-2}\rangle \\ &= (-1)^{2+0+0+0}|^1\Delta_{-2}\rangle = +|^1\Delta_{-2}\rangle.\end{aligned}$$

Thus, at least for the π^2 configuration, Eqs. (2.2.25) and (2.2.32) are self-consistent.

The Σ^--states generally arise from configurations containing partially filled $\lambda > 0$ orbitals, for example,

$$|^4\Sigma_{3/2}^-\rangle = |\lambda^+\alpha\,\lambda^-\alpha\,\sigma\alpha|$$

and

$$|^2\Sigma_{1/2}^\pm\rangle = (2)^{-1/2}[|\lambda^+\alpha\,\lambda^+\beta\,(2\lambda)^-\alpha| \pm |\lambda^-\alpha\,\lambda^-\beta\,(2\lambda)^+\alpha|].$$

It is not possible to construct a Σ^- state from $\lambda = 0$ orbitals alone or from a combination of $\lambda = 0$ orbitals with only *one* $\lambda \neq 0$ orbital. When dealing with perturbations involving Σ^- states, it is useful to remember this configurational property.

2.3 Electrostatic Perturbations

Electrostatic perturbations occur between states of identical symmetry (Fig. 2.4), i.e., those with identical values of Λ, Σ, and S. In Section 2.1.2, two types of off-diagonal matrix elements were shown to connect these states: those of the $\mathbf{H}^{\mathrm{el}}$ and $\mathbf{T}^N$ operators,

$$\langle 1, \Lambda, \Sigma, S | \mathbf{H} | 2, \Lambda, \Sigma, S \rangle = \langle \Phi_1 | \mathbf{H}^{\mathrm{el}} | \Phi_2 \rangle + \langle \Phi_1 | \mathbf{T}^N | \Phi_2 \rangle. \quad (2.3.1)$$

Here the electronic wavefunctions can be taken as nonsymmetrized or symmetrized. The matrix elements are identical in either case. The ideal starting point would be a basis set, Φ_1 and Φ_2, that minimizes the values of the Eq. (2.3.1) off-diagonal matrix elements. Unfortunately, it is not possible to find solutions of the electronic Hamiltonian for which *both* terms of Eq. (2.3.1) are zero. Two possible types of deperturbed or zeroth-order electronic functions may be defined (see also Table 2.1):

1. *Diabatic* functions for which, by definition,

$$\langle \Phi_1^{\mathrm{d}} | \mathbf{T}^N | \Phi_2^{\mathrm{d}} \rangle_r = 0,$$

but then

$$\langle \Phi_1^{\mathrm{d}} | \mathbf{H}^{\mathrm{el}} | \Phi_2^{\mathrm{d}} \rangle_r = H_{12}^e(R) \neq 0. \quad (2.3.2)$$

This latter relation means that the diabatic potential curves associated with these Φ_i^{d} functions can cross. The noncrossing rule between states of identical symmetry applies only for exact solutions of the electronic Hamiltonian.

2. *Adiabatic* functions are precisely those that diagonalize the electronic Hamiltonian. By definition,

$$\langle \Phi_1^{\mathrm{ad}} | \mathbf{H}^{\mathrm{el}} | \Phi_2^{\mathrm{ad}} \rangle_r = 0$$

but then

$$\langle \Phi_1^{\mathrm{ad}} | \mathbf{T}^N | \Phi_2^{\mathrm{ad}} \rangle_r \neq 0. \quad (2.3.3)$$

In principle, whatever the initial model, after introducing the vibronic coupling terms corresponding to the chosen type of deperturbed potential curves, the experimental energy levels are obtained by diagonalizing one or the other type of interaction matrix. One example will be discussed later (Section 2.3.4).

If the deperturbed curves intersect and are characterized by very different molecular constants, then they are diabatic curves (see Fig. 2.5). If the crossing is avoided, adiabatic curves are involved, and one of these curves can have a double minimum. One frequently finds that, in the region of an avoided

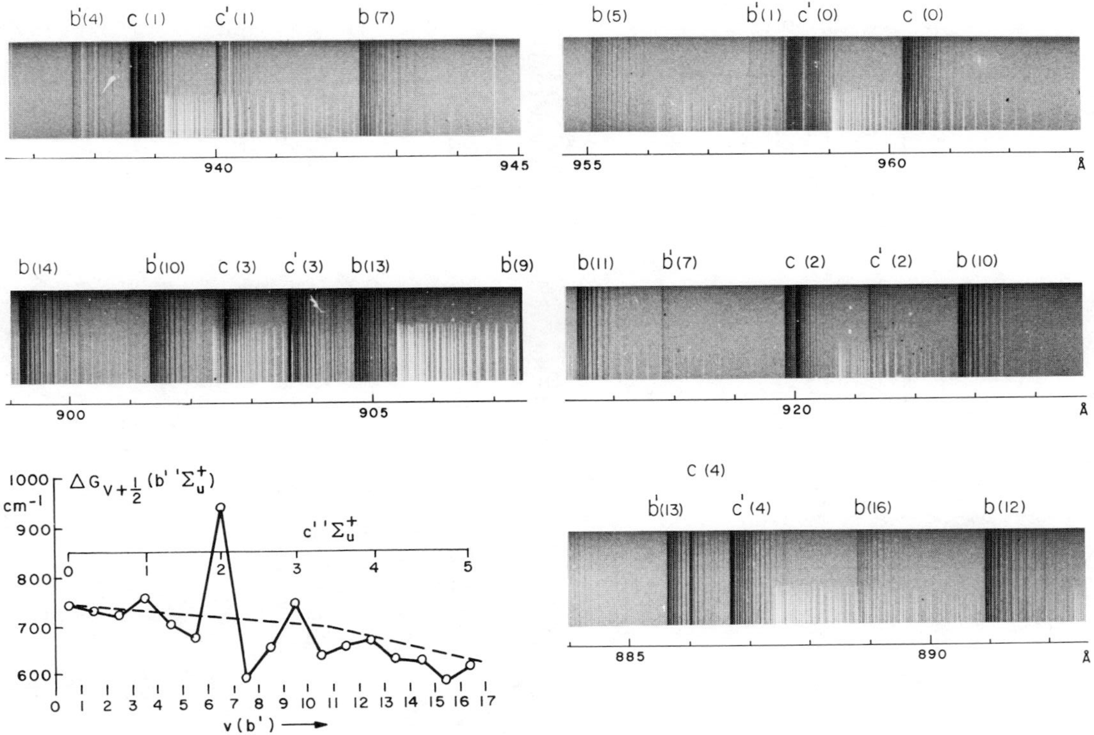

Fig. 2.4 Electrostatic valence–Rydberg N_2 $b'^1\Sigma_u^+(v_{b'}) \sim c'^1\Sigma_u^+(v_{c'})$ and $b^1\Pi_u(v_b) \sim c^1\Pi_u(v_c)$ perturbations. Each segment of the absorption spectrum (from Yoshino *et al.*, 1979) shows several perturbing states near a nominal $v_{c'} = 0$–4 $c'^1\Sigma_u^+$ level. The b and b′ valence states are perturbed by the c and c′ Rydberg states (nominally the $^1\Pi_u$ and $^1\Sigma_u^+$ components of a $3p$ complex) as well as by higher Rydberg states. The $\Delta G_{v+1/2}$ plot for the $b'^1\Sigma_u^+$ state (from Dressler, 1969; see also Fig. 2.6) in the lower left corner shows the massive level shifts that had made it difficult to recognize the electronic state parentage of the observed singlet vibronic levels of N_2. The largest positive deviations of the observed ΔG values from the smoothly varying deperturbed value (dashed line) occur when the $v_{c'} = 1$, 2, and 3 levels are sandwiched between the $v_c = 3$ and 4, 6 and 7, and 9 and 10 levels, respectively. These b ∼ c perturbations are discussed in Section 5.3.1 and further illustrated in Fig. 5.6.

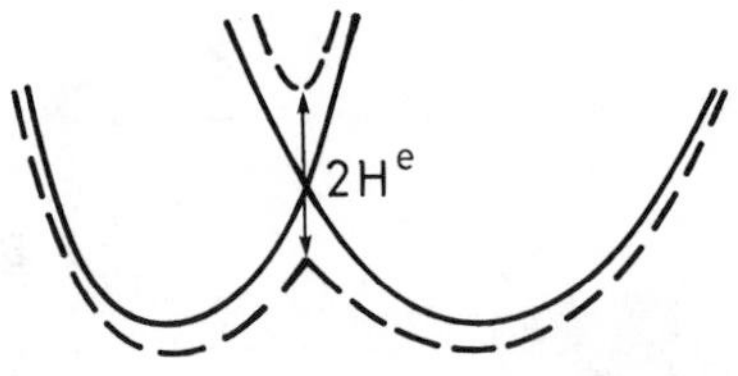

Fig. 2.5 Diabatic and adiabatic potential curves. The diabatic curves (solid lines) cross at R_C and are defined by neglecting the part of $\mathbf{H}^{el}$ that causes the adiabatic curves (dotted lines) to avoid crossing by $2H^e$ at R_C.

crossing, the adiabatic wavefunction changes electronic character abruptly and the derivative of the electronic wavefunction with respect to R can be large. In fact, it is this derivative that controls the size of the matrix elements of the $\mathbf{T}^N$ operator [Eq. (2.3.10)].

In the older literature (Dieke, 1935; Kovács, 1969), there is some confusion. Only the $\mathbf{T}^N$ operator has been assumed to connect states of the same symmetry, and the potential curves of these interacting states have been assumed to cross.

2.3.1 DIABATIC CURVES

The vibronic interaction between the level v_1 of the diabatic potential curve $V_1^d(R)$ and the level v_2 of another diabatic curve $V_2^d(R)$ is reduced to

$$H_{1,v_1;2,v_2} = \langle \Phi_1^d \chi_{1v_1}^d | \mathbf{H}^{el} | \Phi_2^d \chi_{2v_2}^d \rangle \tag{2.3.4}$$

since, by definition,

$$\langle \Phi_1^d | \mathbf{T}^N | \Phi_2^d \rangle = 0.$$

In the diabatic model, the electronic part of the matrix element $H_{1,v_1;2,v_2}$ is often assumed to be independent of R. Then the nuclear and electronic coordinates can be separated in the integration of Eq. (2.3.4). By integration over the electronic coordinates, one obtains

$$H_{1,v_1;2,v_2} = H^e \langle v_1^d | v_2^d \rangle, \tag{2.3.5}$$

where

$$H^e = \langle \Phi_1^d | \mathbf{H}^{el} | \Phi_2^d \rangle$$

and

$$\langle v_1^d | v_2^d \rangle = \int \chi_{v_1}^{d*}(R) \chi_{v_2}^d(R) \, dR.$$

Indeed, as for any electronic quantity, the value of H^e actually depends on R.

However, this dependence is usually weak. Equation (2.3.5) holds even if the R-dependence of H^e is a linear function of the R-centroid (Halevi, 1965) defined by

$$\bar{R}_{ij} = \frac{\langle v_i|R|v_j\rangle}{\langle v_i|v_j\rangle}. \tag{2.3.6a}$$

The significance of the R-centroid is illustrated as follows. The electronic matrix element, $H^e(R)$, may be expanded in a power series about an arbitrarily chosen internuclear distance, R' (most usefully, $R' = R_C$, the internuclear distance at which the two potential curves cross),

$$H^e(R) = H^e(R') + \left(\frac{dH^e}{dR}\right)_{R=R'}(R - R') + \frac{1}{2}\left(\frac{d^2H^e}{dR^2}\right)_{R=R'}(R - R')^2. \tag{2.3.7a}$$

Then the vibrational matrix elements of $H^e(R)$ are expressed in terms of R^n-centroids,

$$\overline{R^n} = \frac{\langle v_i|R^n|v_j\rangle}{\langle v_i|v_j\rangle} \tag{2.3.6b}$$

$$\langle v_i|H^e(R)|v_j\rangle = H^e(R')\langle v_i|v_j\rangle + \left(\frac{dH^e}{dR}\right)_{R=R'}[\langle v_i|R|v_j\rangle - R'\langle v_i|v_j\rangle]$$
$$+ \tfrac{1}{2}\left(\frac{d^2H^e}{dR^2}\right)_{R=R'}[\langle v_i|R^2|v_j\rangle - 2R'\langle v_i|R|v_j\rangle + R'^2\langle v_i|v_j\rangle]$$
$$\frac{\langle v_i|H^e(R)|v_j\rangle}{\langle v_i|v_j\rangle} = H^e(R') + \left(\frac{dH^e}{dR}\right)_{R=R'}[\bar{R} - R']$$
$$+ \tfrac{1}{2}\left(\frac{d^2H^e}{dR^2}\right)_{R=R'}[\overline{R^2} - 2R'\bar{R} + R'^2]. \tag{2.3.7b}$$

Now, making the R-centroid *approximation*, which is quite distinct from the R^n-centroid expansion, namely,

$$\bar{R} \simeq \frac{\overline{R^n}}{\overline{R^{n-1}}} = \frac{\langle v_i|R^n|v_j\rangle}{\langle v_i|R^{n-1}|v_j\rangle}$$

[or, in other words, $\overline{R^n} = (\bar{R})^n$], then Eq. (2.3.7b) becomes

$$\frac{\langle v_i|H^e(R)|v_j\rangle}{\langle v_i|v_j\rangle} = H^e(R') + \left(\frac{dH^e}{dR}\right)_{R=R'}(\bar{R} - R') + \tfrac{1}{2}\left(\frac{d^2H^e}{dR^2}\right)_{R=R'}(\bar{R} - R')^2,$$

which is identical to Eq. (2.3.7a) where R is set equal to $\bar{R}$.

For near-degenerate vibrational levels of any two crossing potential curves, the R-centroid has the convenient property of being nearly equal to R_C, the R-value where the two curves cross (Schamps, 1977). Thus,

$$\bar{R} = \overline{R^k}/\overline{R^{k-1}} = R_C,$$

and, setting $R' = R_C$,

$$\frac{\langle v_i|H^e|v_j\rangle}{\langle v_i|v_j\rangle} = H^e(R_C).$$

The R-centroid approximation has repeatedly been tested numerically. For perturbations between levels of potentials that intersect exactly once, the R-centroid approximation can be regarded as more accurate than experimentally measurable matrix elements.

Approximate deperturbed curves can be derived from unperturbed vibrational levels far from the energy of the curve crossing region. The overlap factor between vibrational wavefunctions is calculable numerically. (Note that a Franck–Condon factor is the absolute magnitude squared of the overlap factor.) From Eq. (2.3.5) and the experimental value of $H_{1,v_1;2,v_2}$, an initial trial value for H^e can be deduced. If the value of H^e is as large as the value of ω_e (see Table 4.4), then electrostatic interactions strongly perturb the entire set of vibrational levels. Figure 2.6 shows the irregular pattern of ΔG values for the perturbed $^1\Sigma_u^+$ states of N_2. Final deperturbed curves are obtained by diagonalization of the matrix, including all vibrational levels of both states, as described in Section 3.4.

2.3.2 APPROXIMATE REPRESENTATION OF THE DIABATIC ELECTRONIC WAVEFUNCTION

The single-configuration approximation is an approximate representation for the diabatic electronic function. Such an approximation is expected to result in smooth R-variation of the electronic wavefunction. In the case where the configurations of the two interacting states differ by two spin-orbitals, the electronic matrix element of the $\mathbf{T}^N$ operator between these two approximate functions is exactly zero, since it can be demonstrated that the $\mathbf{T}^N$ operator acts as a one-electron operator (Section 2.2.4) (Sidis and Lefebvre-Brion, 1971). However, as the electronic Hamiltonian, $\mathbf{H}^{el}$, contains a two-electron operator,

$$\sum_{i<j} \frac{e^2}{r_{ij}},$$

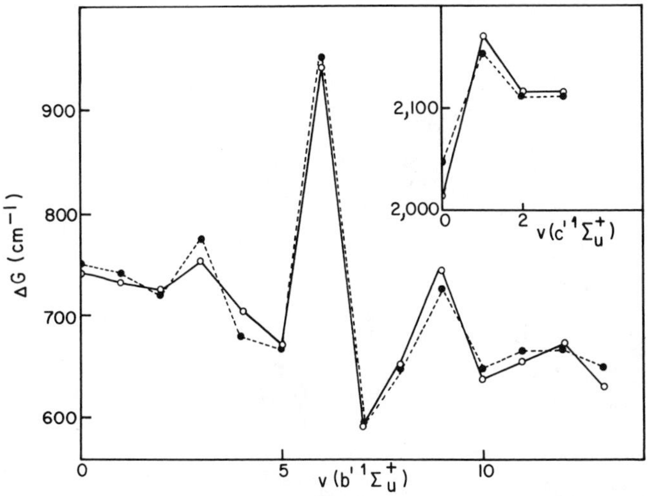

Fig. 2.6 Variation of ΔG_v for the mutually interacting b′ and c′ $^1\Sigma_u^+$ States of N_2. The solid and dashed lines correspond, respectively, to the observed and calculated (Lefebvre-Brion, 1969) values. The deperturbed c′ ($v = 2$) and b′ ($v = 7$) levels are nearly degenerate and interact strongly (see Table 4.4). This accounts for the largest ΔG anomalies.

$\mathbf{H}^{el}$ can have nonzero matrix elements. If the unique orbitals are ϕ_a, ϕ_b and ϕ_c, ϕ_d for the two configurations, then [Eqs. (2.2.54) and (2.2.55)]

$$H^e \propto \langle \phi_a \phi_c | 1/r_{12} | \phi_b \phi_d \rangle - \langle \phi_a \phi_d | 1/r_{12} | \phi_b \phi_c \rangle. \qquad (2.3.8)$$

If the spin parts of ϕ_a and ϕ_b are identical, both terms in Eq. (2.3.8) are nonzero. Otherwise, the single nonzero term is the one in which the spins of electron 1 and those of electron 2 are identical (i.e., $\langle \phi_a \phi_c | 1/r_{12} | \phi_b \phi_d \rangle = 0$ unless the spin parts of ϕ_a and ϕ_b, respectively, match those of ϕ_c and ϕ_d). In the case where the two configurations differ by only one *orbital*, but still by two *spin-orbitals*, more complicated formulas apply.† This situation would occur for two Rydberg states of the same symmetry that belong to series converging to ion-core states of different multiplicities (see Section 4.2.4 for an example of a $^3\Sigma_u^-$ Rydberg series converging to $^2\Sigma_g^-$ and $^4\Sigma_g^-$ ion cores).

It will be shown in Section 4.2 that electrostatic perturbations occur frequently between states whose configurations differ by two orbitals, especially

† The added complication arises from the necessity to express the wavefunctions for states with multiple open subshells as properly symmetrized sums of Slater determinantal functions. H^e will then include off-diagonal matrix elements between several pairs of Slater determinants.

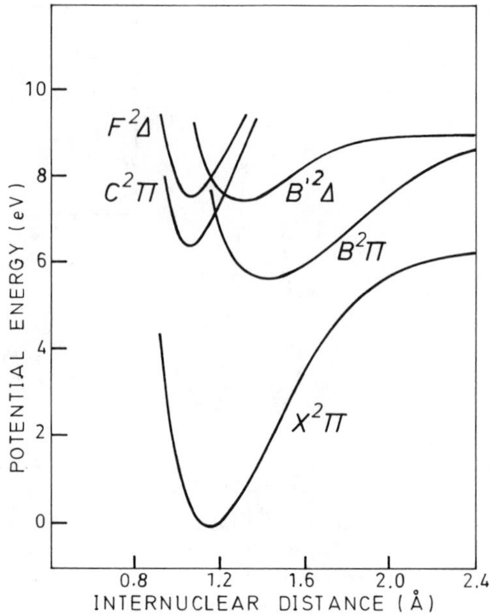

Fig. 2.7 $^2\Delta$ and $^2\Pi$ potential energy curves of NO. The $C^2\Pi$ Rydberg state is homogeneously perturbed by the $B^2\Pi$ valence state. Similarly, the $F^2\Delta$ Rydberg state is perturbed by the $B'^2\Delta$ valence state.

between Rydberg and valence states. An example from the NO spectrum is discussed here (Fig. 2.7). The $B'^2\Delta$ valence state of the NO molecule can be represented by the configuration $\sigma 2p\, \pi 2p^4\, \pi^* 2p^2$. The $F^2\Delta$ state belongs to a Rydberg series that converges to the ground state of the NO^+ ion and is represented by the $\sigma 2p^2\, \pi 2p^4\, 3d\delta$ configuration. The configurations of the $B'^2\Delta$ and $F^2\Delta$ states differ by two orbitals that have different spin functions. The electrostatic interaction is given by

$$H^e = \langle \sigma 2p\, \pi^* 2p | 1/r_{12} | 3d\delta\, \pi^* 2p \rangle.$$

This integral has been evaluated *ab initio* and found equal to 300 cm^{-1} (Felenbok and Lefebvre-Brion, 1966). (The one-center part of this integral is approximately the nonzero atomic integral $\langle sp | 1/r_{12} | dp \rangle$.) This calculated value is in fair agreement with the "semiexperimental" value of 450 cm^{-1} found by a deperturbation procedure (Jungen, 1966). Note that this electrostatic interaction is responsible not only for perturbations between states of identical symmetry but also for predissociation (Section 6.6.1) and autoionization (Section 7.7).

2.3.3 ADIABATIC CURVES

In the adiabatic model, the matrix element between the v_1 level of the first adiabatic curve $V_1^{\text{ad}}(R)$ and the v_2 level of the second adiabatic curve $V_2^{\text{ad}}(R)$ (see Fig. 2.5) is reduced to

$$H_{1,v_1;2,v_2} = \langle \Phi_1^{\text{ad}} \chi_{v_1}^{\text{ad}} | \mathbf{T}^N | \Phi_2^{\text{ad}} \chi_{v_2}^{\text{ad}} \rangle, \tag{2.3.9}$$

since

$$\langle \Phi_1^{\text{ad}} | \mathbf{H}^{\text{el}} | \Phi_2^{\text{ad}} \rangle = 0.$$

$\mathbf{T}^N$ is the nuclear kinetic energy operator, which is expressed in the molecular frame as

$$\mathbf{T}^N = -\frac{\hbar^2}{2\mu R^2}\left\{\frac{\partial}{\partial R}\left[R^2\frac{\partial}{\partial R}\right] - \mathbf{R}^2\right\} = -\frac{\hbar^2}{2\mu}\left[\frac{\partial^2}{\partial R^2} + \frac{2}{R}\frac{\partial}{\partial R}\right] + \frac{\hbar^2}{2\mu R^2}\mathbf{R}^2.^{\dagger}$$

Let us ignore the $\mathbf{R}^2$ rotational part ($\mathbf{R} = \mathbf{J} - \mathbf{L} - \mathbf{S}$) of this operator, which leads to off-diagonal matrix elements that are proportional to $J(J+1)$ but still very small compared to the matrix elements of the remaining radial term (Leoni, 1972). The effect of the derivatives with respect to R on the electronic and vibrational wavefunctions, both of which depend on R, is given by

$$\frac{\partial^2(\Phi\chi)}{\partial R^2} = \chi\frac{\partial^2\Phi}{\partial R^2} + \Phi\frac{\partial^2\chi}{\partial R^2} + 2\frac{\partial\Phi}{\partial R}\frac{\partial\chi}{\partial R}$$

and

$$\frac{2}{R}\frac{\partial}{\partial R}\Phi\chi = \frac{2}{R}\chi\frac{\partial\phi}{\partial R} + \frac{2}{R}\Phi\frac{\partial\chi}{\partial R}.$$

Combining this result with Eq. (2.3.9) yields

$$\begin{aligned} H_{1,v_1;\,2,v_2} = &-\frac{\hbar^2}{2\mu}\left\langle \chi_{v_1}^{\text{ad}}\left|\left\langle \Phi_1^{\text{ad}}\left|\frac{\partial^2}{\partial R^2} + \frac{2}{R}\frac{\partial}{\partial R}\right|\Phi_2^{\text{ad}}\right\rangle_r\right|\chi_{v_2}^{\text{ad}}\right\rangle_R \\ &-\frac{\hbar^2}{2\mu}\langle \Phi_1^{\text{ad}}|\Phi_2^{\text{ad}}\rangle_r\left\langle \chi_{v_1}^{\text{ad}}\left|\frac{\partial^2}{\partial R^2} + \frac{2}{R}\frac{\partial}{\partial R}\right|\chi_{v_2}^{\text{ad}}\right\rangle_R \\ &-\frac{\hbar^2}{\mu}\left\langle \chi_{v_1}^{\text{ad}}\left|\left\langle \Phi_1^{\text{ad}}\left|\frac{\partial}{\partial R}\right|\Phi_2^{\text{ad}}\right\rangle_r\right|\frac{\partial}{\partial R}\chi_{v_2}^{\text{ad}}\right\rangle_R. \end{aligned} \tag{2.3.10}$$

In Eq. (2.3.10), the second term is zero after integration over the electronic coordinates r, since Φ_1^{ad} and Φ_2^{ad} are two different solutions of the same equation

† Recall that R is the internuclear distance and $\mathbf{R}$ is the nuclear rotation angular momentum.

and must therefore be orthogonal. The off-diagonal matrix elements in Eq. (2.3.10) are often called nonadiabatic corrections to the energies.

The vibrational wavefunction is often written as $\chi = \xi/R$, where ξ is normalized with respect to dR (as opposed to $R^2\,dR$ as for χ). Then the derivative of the vibrational function with respect to R results in two terms. One of these terms exactly cancels the term in $(2/R)\,(\partial\Phi/\partial R)$ of Eq. (2.3.10), and the matrix element simplifies to

$$H_{1,v_1;\,2,v_2}\,(\mathrm{cm}^{-1}) = \frac{-16.856}{\mu}\left\langle \xi^{\mathrm{ad}}_{v_1}\left|\left\langle \Phi^{\mathrm{ad}}_1\left|\frac{\partial^2}{\partial R^2}\right|\Phi^{\mathrm{ad}}_2\right\rangle_r (\text{Å}^{-2})\right|\xi^{\mathrm{ad}}_{v_2}\right\rangle_R - \frac{33.712}{\mu}\left\langle \xi^{\mathrm{ad}}_{v_1}\left|\left\langle \Phi^{\mathrm{ad}}_1\left|\frac{\partial}{\partial R}\right|\Phi^{\mathrm{ad}}_2\right\rangle_r (\text{Å}^{-1})\right|\frac{\partial}{\partial R}\xi^{\mathrm{ad}}_{v_2}\right\rangle_R (\text{Å}^{-1}). \tag{2.3.11}$$

In the case where an avoided crossing is being represented by adiabatic curves, a relation between electronic matrix elements for basis functions belonging to adiabatic versus diabatic curves can be derived easily (Bandrauk and Child, 1970; Oppenheimer, 1972), as shown below.

Adiabatic potential curves can be obtained by diagonalizing, at a grid of R-values, the configuration-interaction matrix. This matrix is constructed in a particular diabatic (single-configuration) basis. The off-diagonal configuration-interaction matrix element is the familiar diabatic coupling term, $H^e_{12}(R)$, which involves integration over electronic coordinates at a fixed value of internuclear distance [Eq. (2.1.10)]. The configuration-interaction secular equation is

$$\begin{vmatrix} E^{\mathrm{d}}_1(R) - E & H^e_{12}(R) \\ H^e_{12}(R) & E^{\mathrm{d}}_2(R) - E \end{vmatrix} = 0, \tag{2.3.12}$$

where $E^{\mathrm{d}}_i(R)$ is defined by the fixed-R integral over electronic coordinates,

$$E^{\mathrm{d}}_i(R) = \langle \Phi^{\mathrm{d}}_i | \mathbf{H}^{\mathrm{el}} | \Phi^{\mathrm{d}}_i \rangle_r.$$

The resultant eigenstates are found to be linear combinations of diabatic electronic functions, for which the configuration interaction mixing coefficients are explicitly dependent on internuclear distance,

$$\begin{aligned} \Phi^{\mathrm{ad}}_1 &= \cos\theta(R)\Phi^{\mathrm{d}}_1 - \sin\theta(R)\Phi^{\mathrm{d}}_2, \\ \Phi^{\mathrm{ad}}_2 &= \sin\theta(R)\Phi^{\mathrm{d}}_1 + \cos\theta(R)\Phi^{\mathrm{d}}_2. \end{aligned} \tag{2.3.13}$$

These expressions imply that the functions Φ^{ad}_i are orthogonal. At the crossing point $R = R_{\mathrm{C}}$, $\theta = \pi/4$.

The vertical energy separation between two interacting diabatic potentials can be assumed to vary linearly with R in the crossing region,

$$E_1^{\rm d}(R) - E_2^{\rm d}(R) = a(R - R_{\rm C}).$$

Now, as the $\Phi_i^{\rm d}$ functions are diabatic, the derivative with respect to R acts only on the coefficients of the linear combinations in Eq. (2.3.13), thus

$$\left\langle \Phi_1^{\rm ad} \left| \frac{\partial}{\partial R} \right| \Phi_2^{\rm ad} \right\rangle_r = [\sin^2\theta + \cos^2\theta]\frac{\partial\theta}{\partial R} = \frac{\partial\theta}{\partial R}.$$

By definition, the adiabatic functions $\Phi_1^{\rm ad}$ and $\Phi_2^{\rm ad}$ diagonalize the electronic Hamiltonian. Using Eq. (2.3.13), one finds ($H^e \equiv H_{12}^e$)

$$\langle \Phi_1^{\rm ad} | \mathbf{H}^{\rm el} | \Phi_2^{\rm ad} \rangle_r$$
$$= + E_1^{\rm d} \sin\theta\cos\theta - H^e \sin^2\theta + H^e\cos^2\theta - E_2^{\rm d}\sin\theta\cos\theta = 0,$$

from which the R-dependence of θ near $R_{\rm C}$ may be determined,

$$\frac{\sin\theta\cos\theta}{\cos^2\theta - \sin^2\theta} = \tfrac{1}{2}\tan 2\theta = \frac{H^e}{E_2^{\rm d} - E_1^{\rm d}} = -\frac{H^e}{a(R - R_{\rm C})}.$$

Thus,

$$\theta = \frac{1}{2}\tan^{-1}\left[\frac{-2H^e}{a(R - R_{\rm C})}\right],$$

and since

$$\frac{d}{dx}\tan^{-1}\left(\frac{x}{c}\right) = \frac{c}{c^2 + x^2}$$

then

$$\left\langle \Phi_1^{\rm ad} \left| \frac{\partial}{\partial R} \right| \Phi_2^{\rm ad} \right\rangle_r = \frac{\partial\theta}{\partial R} = \frac{aH^e}{4(H^e)^2 + a^2(R - R_{\rm C})^2}.$$

Defining $b = H^e/a$, then

$$\left\langle \Phi_1^{\rm ad} \left| \frac{\partial}{\partial R} \right| \Phi_2^{\rm ad} \right\rangle_r = \frac{b}{4b^2 + (R - R_{\rm C})^2} = W^e(R). \qquad (2.3.14)$$

If the diabatic coupling matrix element, H^e, is R-independent, this $\partial/\partial R$ matrix element between two adiabatic states must have a Lorentzian R-dependence with a full width at half maximum (FWHM) of $4b$. Evidently, the adiabatic electronic matrix element $W^e(R)$ is not R-independent but is strongly peaked near $R_{\rm C}$. Its maximum value occurs at $R = R_{\rm C}$ and is equal to $1/4b = a/4H^e$.

Thus, if the diabatic matrix element H^e is large, the maximum value of the electronic matrix element between adiabatic curves is small. This is the situation where it is convenient to work with deperturbed adiabatic curves. On the contrary, if H^e is small, it becomes more convenient to start from diabatic curves. Table 2.5 compares the values of diabatic and adiabatic parameters. The deviation from the relation, $W^e(R)_{\max} \times \text{FWHM} = 1$, is due to a slight dependence of H^e on R and a nonlinear variation of the energy difference between diabatic potentials. When $W^e(R)$ is a relatively broad curve without a prominent maximum, the adiabatic approach is more convenient. When $W^e(R)$ is sharply peaked, the diabatic picture is preferable. The first two cases in Table 2.5 would be more convenient to treat from an adiabatic point of view. The description of the last two cases would be simplest in terms of diabatic curves. The third case is intermediate between the two extreme cases and will be examined later (see Table 2.6).

To obtain the second adiabatic electronic matrix element of Eq. (2.3.11), the ket, $(\partial/\partial R)\,|\Phi_2^{\text{ad}}\rangle$, is expanded using the complete set, $\{\Phi_j^{\text{ad}}\}$, of adiabatic functions:

$$\frac{\partial}{\partial R}\left|\Phi_2^{\text{ad}}\right\rangle = \sum_i \left|\Phi_i^{\text{ad}}\right\rangle \left\langle \Phi_i^{\text{ad}} \left| \frac{\partial}{\partial R} \right| \Phi_2^{\text{ad}} \right\rangle$$

$$\frac{\partial^2}{\partial R^2}\left|\Phi_2^{\text{ad}}\right\rangle = \frac{\partial}{\partial R}\frac{\partial}{\partial R}\left|\Phi_2^{\text{ad}}\right\rangle = \sum_i \left\{\frac{\partial}{\partial R}\left|\Phi_i^{\text{ad}}\right\rangle\right\} \left\langle \Phi_i^{\text{ad}} \left| \frac{\partial}{\partial R} \right| \Phi_2^{\text{ad}} \right\rangle$$
$$+ \sum_i \left|\Phi_i^{\text{ad}}\right\rangle \frac{\partial}{\partial R} \left\langle \Phi_i^{\text{ad}} \left| \frac{\partial}{\partial R} \right| \Phi_2^{\text{ad}} \right\rangle$$

$$\left\langle \Phi_1^{\text{ad}} \left| \frac{\partial^2}{\partial R^2} \right| \Phi_2^{\text{ad}} \right\rangle = \sum_{i \neq 1,2} \left[\left\langle \Phi_1^{\text{ad}} \left| \frac{\partial}{\partial R} \right| \Phi_i^{\text{ad}} \right\rangle \left\langle \Phi_i^{\text{ad}} \left| \frac{\partial}{\partial R} \right| \Phi_2^{\text{ad}} \right\rangle + \frac{\partial}{\partial R} \left\langle \Phi_1^{\text{ad}} \left| \frac{\partial}{\partial R} \right| \Phi_2^{\text{ad}} \right\rangle \right].$$

The second term in the summation reduces to a single term because the adiabatic functions are orthonormal (Hobey and McLachlan, 1960). In the simple case where only two electronic states interact (Eq. 2.3.13), one can assume that the matrix elements of $\partial/\partial R$ connecting either of these two states with other states are negligible and, from Eq. (2.3.14),

$$\left\langle \Phi_1^{\text{ad}} \left| \frac{\partial^2}{\partial R^2} \right| \Phi_2^{\text{ad}} \right\rangle_r = \frac{\partial}{\partial R} \left\langle \Phi_1^{\text{ad}} \left| \frac{\partial}{\partial R} \right| \Phi_2^{\text{ad}} \right\rangle_r = \frac{\partial}{\partial R} W^e(R).$$

This calculated matrix element of $\partial^2/\partial R^2$ acting on the electronic wavefunctions for the E, F and G, K states of H_2 (Fig. 2.8) is displayed in Fig. 2.9

Table 2.5
Comparison between Diabatic and Adiabatic Parameters

Interacting States	Molecule	H^e (cm^{-1})	Maximum value of $W^e(R)$ ($Å^{-1}$)	Width (Å) FWHM of $W^e(R)$
$B^2\Sigma_u^+ \sim C^2\Sigma_u^+$	N_2^{+a}	10,000	3	0.34
$E, F^1\Sigma_g^+ \sim G, K^1\Sigma_g^+$	$H_2{}^b$	~3,000	3.4	0.26
$G^1\Pi \sim I^1\Pi$	SiO^c	~400	6.4	0.16
$C^3\Pi_u \sim C'^3\Pi_u$	$N_2{}^d$	1,000	~20.4	0.05
$B^3\Sigma_u^- \sim B'^3\Sigma_u^-$	$O_2{}^e$	4,000	21.6	0.06

[a] A. L. Roche and H. Lefebvre-Brion, 1975.
[b] K. Dressler *et al.*, 1979.
[c] A. Lagerqvist and I. Renhorn, 1979 (semiexperimental value).
[d] J. M. Robbe, 1978.
[e] M. Yoshimine and Y. Tanaka, 1978.

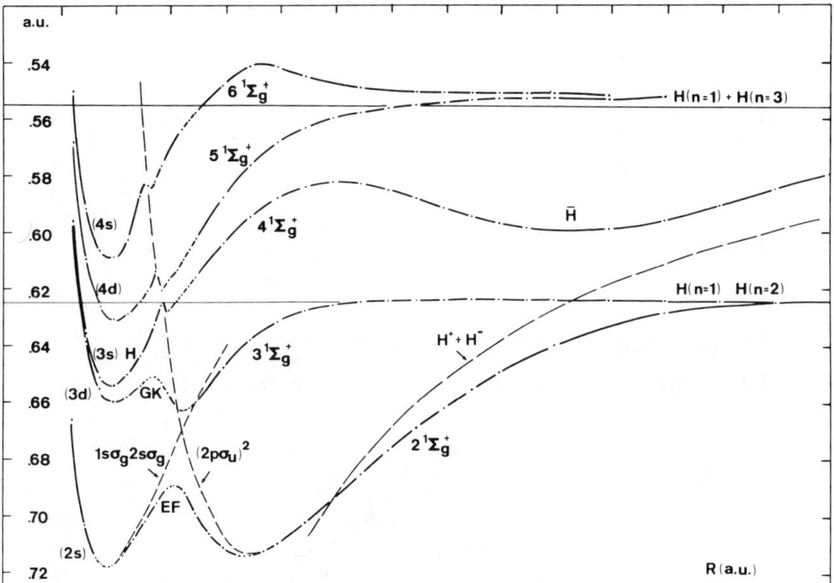

Fig. 2.8 *Ab initio* adiabatic potential curves for the ${}^1\Sigma_g^+$ states of H_2 (dash–dot line). The diabatic $1s\sigma_g 2s\sigma_g$ and $(2p\sigma_u)^2$ potentials are also plotted (dashed line). The (nl) labels on the small-R potential minima denote the dominant $1s\sigma_g nl\sigma_g$ configuration (Wolniewicz and Dressler, 1977). Note that more accurate results on the $4\,{}^1\Sigma_g^+$ $H\bar{H}$ state have been presented by Wolniewicz and Dressler (1979) and by Dressler and Wolniewicz (1981).

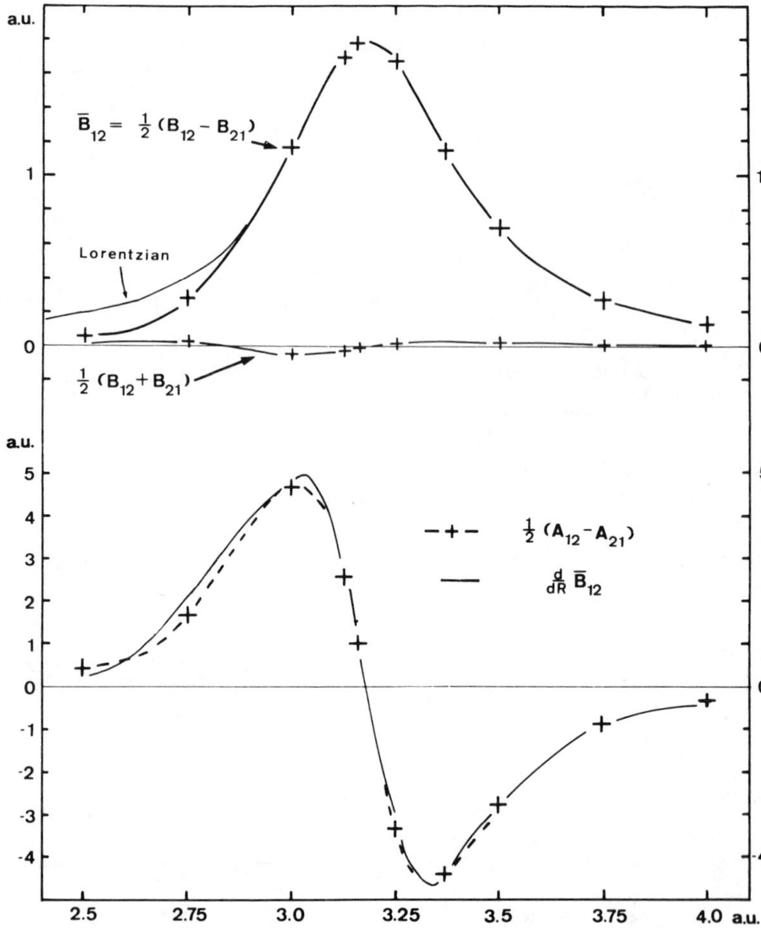

Fig. 2.9 *Ab initio* $\partial/\partial R$ and $\partial^2/\partial R^2$ matrix elements between the E, F and G, K adiabatic states of H_2:

$$B_{12} \equiv \left\langle \Phi_{1\mathrm{ad}} \left| \frac{\partial}{\partial R} \right| \Phi_{2\mathrm{ad}} \right\rangle_r$$

$$A_{12} \equiv \left\langle \Phi_{1\mathrm{ad}} \left| \frac{\partial^2}{\partial R^2} \right| \Phi_{2\mathrm{ad}} \right\rangle_r$$

where Φ_1 and Φ_2 are the adiabatic electronic wavefunctions for the E, F and G, K double-minimum states, respectively. Except for the smallest R values, $\bar{B}_{12}(R)$ is Lorentzian. The relationships $B_{12} = -B_{21}$ and

$$\tfrac{1}{2}(A_{12} - A_{21}) = \frac{d}{dR}\bar{B}_{12}$$

are not satisfied exactly because $\Phi_{1\mathrm{ad}}$ and $\Phi_{2\mathrm{ad}}$ are, in these calculations, not exactly orthogonal (Dressler *et al.*, 1979).

and is seen not to deviate appreciably from the derivative of a Lorentzian curve. Its contribution to the $H_{1,v_m;\,2,v_n}$ vibronic matrix element [Eq. (2.3.11)] is generally smaller than the contribution due to the $\partial/\partial R$ operator acting on the electronic functions, but it is in no case negligible.

2.3.4 CHOICE BETWEEN THE DIABATIC AND ADIABATIC MODELS

If the approximate deperturbed potential curves cross, they are diabatic curves. One can assume an interaction matrix element given by Eq. (2.3.5) and carry out a complete deperturbation.

The choice of an adiabatic picture leads to difficulties when one of the potentials has a double minimum (see Fig. 2.5). The vibrational level separations of such a curve do not vary smoothly with vibrational quantum number, as do the levels of a single minimum potential. In the separate potential wells (below the barrier), the levels approximately follow two different smooth curves. However, above the potential barrier the separation between consecutive energy levels oscillates. The same pattern of behavior is found for the rotational constants below and above the potential barrier. In addition, the rotational levels above the barrier do not vary as $B_v J(J+1)$. An adiabatic deperturbation of the (E, F + G, K) $^1\Sigma_g^+$ states of H_2 has been possible (Dressler *et al.*, 1979) only because the adiabatic curves were known from very precise *ab initio* calculations.

If the approximate deperturbed curves do not cross or have similar spectroscopic constants, the most convenient starting point is an adiabatic approach. Two situations must be considered:

1. The adiabatic curves result from an apparently avoided crossing. This means that the diabatic curves belong to very different electronic configurations. The coupling matrix element, $W^e(R)$, can be assumed to have a Lorentzian shape [Eq. (2.3.14)]. This is the situation for the G and I states of SiO.

2. The adiabatic curves correspond to configurations that differ by only one orbital. Rydberg states belonging to series which converge to the same state of the ion fall into this category. Then the matrix element of $\partial/\partial R$ is generally R-independent (or linear in R).† One example is the perturbations observed between identical-symmetry Rydberg states of H_2 that converge to the same

† If the unique orbital changes its character with R (e.g. "Rydbergization", Section 4.2.2), the coupling matrix element behaves similarly to case (1).

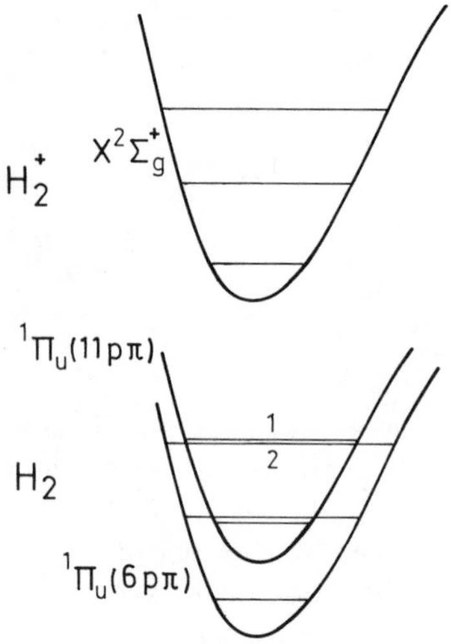

Fig. 2.10 Perturbations between the $6p\pi$ and $11p\pi$ $^1\Pi_u$ Rydberg states of H_2. Two Rydberg series converging to different vibrational levels of the H_2^+ $X^2\Sigma_g^+$ state interact via nonzero $\Delta v = \pm 1$ vibrational matrix elements of the $\partial/\partial R$ operator.

state of H_2^+ (the ground state) but to different vibrational levels of this state (Herzberg and Jungen, 1972). As the density of electronic levels increases near the ionization limit, the $v = 1$ level of the $^1\Pi_u$ $n = 11$ Rydberg state is nearly degenerate with $v = 2$ of the $^1\Pi_u$ $n = 6$ Rydberg state (Fig. 2.10). The electronic factor is found to be nearly independent of R, $\langle \Phi_1^{\text{ad}} | \partial/\partial R | \Phi_2^{\text{ad}} \rangle_r =$ constant. The vibrational wavefunctions of the two interacting states belong to virtually identical potentials; thus the vibrational factor is zero except for $\Delta v = 1$, which, in the harmonic approximation, is proportional to $(v + 1)^{1/2}$,

$$\left\langle \chi_{v_1} \left| \frac{\partial}{\partial R} \right| \chi_{v_2} \right\rangle (\text{Å}^{-1}) = \left(\frac{16.856}{\mu\omega\ (\text{cm}^{-1})} \right)^{1/2} (v_1 + 1)^{1/2} \delta_{v_1, v_2 - 1}. \quad (2.3.15)$$

Another approach to the treatment of these Rydberg ~ Rydberg interactions is given in Section 7.5, using, as zero-order electronic wavefunctions, those of a single electron in the field of the ion core (Jungen and Atabek, 1977).

The vibrational eigenstates of neither the diabatic nor the adiabatic potential curve exactly represent the observed levels. Interaction matrix elements between these zero-order levels (eigenstates of either diabatic or adiabatic potentials) must be added in order to reproduce the observed levels. Table 2.6 summarizes the results of two types of deperturbation approaches, diabatic and adiabatic, to the same problem, a pair of interacting $^1\Pi$ states of SiO. The purpose of a deperturbation calculation is to obtain a model, consisting of a pair of deperturbed potentials and an interaction matrix element (either R-dependent or R-independent),

Table 2.6

Comparison between Two Types of Deperturbation of the G and I $^1\Pi$ States of SiO (in cm^{-1})

	Diabatic deperturbation		Adiabatic deperturbation		
State	Diabatic levels	obs − dep	Adiabatic levels	obs − dep	Observed levels
G0	70180	− 78	70099	3	70102
G1	71056	− 91	70966	− 1	70965
G2	71919	− 87	71819	13	71832
G3	72769	−150	72658	−39	72619
G4	73606	−163	73482	−39	73443
G5	74431	−177	74292	−38	74254
G6	75243	−192	75087	−36	75051
G7	76042	−194	75866	−18	75848
I0	71689	89	71788	−10	71778
I1	72564	151	72675	40	72715
I2	73423	170	73546	47	73593
I3	74264	200	74401	63	74464
I4	75089	236	75241	84	75325
I5	75896	276	76066	106	76172

Spectroscopic constants for deperturbed curves (cm^{-1})

Diabatic curves	T_e	ω_e	$\omega_e x_e$	R_e (Å)
G	69,734.7	890.024	6.57	1.6139
I	71,245.02	892.014	8.66	1.6548

Interaction matrix element: $H^e = 365\ cm^{-1}$

Adiabatic curves	T_e	$\omega_e{}^a$	$\omega_e x_e{}^a$	R_e (Å)
G	69,633.0	881.07	7.02	1.6155
I	71,334.0	902.84	7.97	1.6526

Interaction matrix element[b]: $b = 0.1014$ Å
$R_C = 1.915$ Å

[a] Approximate values, since the adiabatic levels are obtained by direct numerical integration of the adiabatic curves constructed from the diabatic curves and H^e.

[b] The adiabatic interaction parameter, $W^e(R)$, is defined by Eq. (2.3.14) in terms of b and R_C.

that exactly reproduces the observed rotation–vibration energy levels. Whether this model is diabatic or adiabatic has no effect on the quality of the agreement between observed and calculated levels. Where the two approaches differ is in the complexity of the calculation (size of matrix to be diagonalized, number of fitting iterations required, etc.) and in the magnitudes of the differences between observed and zero-order deperturbed levels (the eigenvalues of the deperturbed potentials). Table 2.6 displays the differences (obs − dep) between the observed levels (far right column), and the *zero-order* diabatic (second column) and *zero-order* adiabatic (fourth column) levels. The energies of the zero-order adiabatic levels are closer to those of the observed levels; thus the adiabatic picture is a better starting point for an iterative deperturbation calculation.

The observed levels in Table 2.6 may be obtained from the diabatic potentials represented by the T_e, ω_e, $\omega_e x_e$, and R_e constants, which generate the deperturbed levels via $T_v = T_e + \omega_e(v + \frac{1}{2}) - \omega_e x_e(v + \frac{1}{2})^2$; an R-independent electronic matrix element, $H^e = 365\ \text{cm}^{-1}$; and vibrational overlap factors calculated using the vibrational eigenfunctions of the deperturbed diabatic potentials. Similarly, the observed levels may be computed from the adiabatic potentials, a Lorentzian interaction term [Eq. (2.3.14)] $W^e(R)$ with $b = 0.1014$ Å, $R_c = 1.915$ Å, and vibrational factors calculated using eigenfunctions of the deperturbed adiabatic potentials.

Finally, the eigenfunctions for the observed levels are obtained, in either representation, by diagonalizing the complete interaction matrix consisting of all vibrational levels of the two potentials in the diabatic picture,

$$\Psi = \sum_{v_1} a_{v_1} \Phi_1^{\mathrm{d}} \chi_{v_1}^{\mathrm{d}} + \sum_{v_2} b_{v_2} \Phi_2^{\mathrm{d}} \chi_{v_2}^{\mathrm{d}}, \tag{2.3.16}$$

or, in the adiabatic picture,

$$\Psi = \sum_{v_1} c_{v_1} \Phi_1^{\mathrm{ad}} \chi_{v_1}^{\mathrm{ad}} + \sum_{v_2} d_{v_2} \Phi_2^{\mathrm{ad}} \chi_{v_2}^{\mathrm{ad}}. \tag{2.3.17}$$

Equations (2.3.16) and (2.3.17) suggest one final indicator of whether the diabatic or adiabatic approach is preferable. The better approach is the one for which the sum over deperturbed functions involves fewer terms, especially if one term is dominant ($a_i \geq 2^{-1/2}$). An adiabaticity parameter

$$\gamma \equiv H^e / \Delta G^{\mathrm{ad}}$$

has been introduced recently by Dressler (1983), where ΔG^{ad} is the vibrational frequency of the higher-energy member of the pair of adiabatic electronic states. Near adiabatic behavior occurs for $\gamma \gg 1$; near diabatic behavior occurs for $\gamma \ll 1$. For $\gamma \simeq 1.0$, which corresponds to the case for the SiO G and I $^1\Pi$ states, the mixing of the vibrational basis functions is large in both diabatic and adiabatic descriptions.

2.4 Spin Part of the Hamiltonian

As in atoms, relativistic terms due to the interaction between the spin and orbital angular momenta of nuclei and electrons of the molecule must be added to the electronic Hamiltonian. There is also a magnetic interaction energy created by the orbital motion of the electrons and the rotational motion of the electrically charged nuclei. The relativistic effects consist mainly of three parts:

1. Interaction between the spin and orbital angular momenta of the electrons, $\mathbf{H}^{\mathrm{SO}}$ = spin–orbit operator.
2. Interaction between the electron spin and the rotational angular momenta of the nuclei, $\mathbf{H}^{\mathrm{SR}}$ = spin–rotation operator.
3. Interaction between the spins of different electrons, $\mathbf{H}^{\mathrm{SS}}$ = spin–spin operator.

These interactions remove the spin degeneracy of the levels of an electronic state and give rise to the multiplet splitting or the so-called zero-field splitting (i.e., the splitting that appears even in the absence of any external magnetic field). They are also responsible for mixing states of the same or different multiplicities, thus providing a mechanism for observation of nominally forbidden transitions (see Chapter 5), homogeneous ($\Delta\Omega = 0$, $\Delta S = 0, \pm 1$) perturbations (see Section 2.4.2), and predissociations (see Chapter 6).

2.4.1 THE SPIN–ORBIT OPERATOR

The form of the spin–orbit Hamiltonian has been given by Van Vleck (1951) and is an extension to diatomic molecules of the solution for the relativistic equation originally derived for a two-electron atom:

$$\begin{aligned}\mathbf{H}^{\mathrm{SO}} = \frac{g_S\mu_{\mathrm{B}}}{c}\sum_i \Bigg\{ & \frac{Z_{\mathrm{A}}e}{r_{i\mathrm{A}}^3}[(\mathbf{r}_i - \mathbf{r}_{\mathrm{A}}) \times \mathbf{v}_i/2]\cdot \mathbf{s}_i \\ & + \frac{Z_{\mathrm{B}}e}{r_{i\mathrm{B}}^3}[(\mathbf{r}_i - \mathbf{r}_{\mathrm{B}}) \times \mathbf{v}_i/2]\cdot \mathbf{s}_i \Bigg\} \qquad \text{direct spin–orbit interaction} \\ & - \frac{g_S\mu_{\mathrm{B}}}{c}\sum_{i\neq j}\frac{e}{r_{ij}^3}[(\mathbf{r}_i - \mathbf{r}_j) \times (\tfrac{1}{2}\mathbf{v}_i - \mathbf{v}_j)]\cdot \mathbf{s}_i \qquad \text{spin–other-orbit interaction}\end{aligned} \tag{2.4.1}$$

where $\mathbf{v}_i$ is the velocity of the ith electron measured in the molecule fixed coordinate frame, g_S is the Landé electronic factor, which is approximately 2, and μ_B is the Bohr magneton, $e\hbar/2mc$. If $\mathbf{v}_i$ is replaced by $\mathbf{p}_i/m$ where $\mathbf{p}_i$ is the momentum of the electron measured in the molecule fixed coordinate frame and if the angular momentum operator $\mathbf{l}_i$ is introduced in the form

$$\mathbf{l}_{i\mathrm{A}} = (1/\hbar)(\mathbf{r}_{i\mathrm{A}} \times \mathbf{p}_i), \qquad \mathbf{l}_{i\mathrm{B}} = (1/\hbar)(\mathbf{r}_{i\mathrm{B}} \times \mathbf{p}_i),$$

then the expression for the spin–orbit Hamiltonian becomes

$$\mathbf{H}^{\mathrm{SO}} = \frac{\alpha^2}{2}\sum_i \left\{ \frac{Z_{\mathrm{A}}}{r_{i\mathrm{A}}^3}\mathbf{l}_{i\mathrm{A}} \cdot \mathbf{s}_i + \frac{Z_{\mathrm{B}}}{r_{i\mathrm{B}}^3}\mathbf{l}_{i\mathrm{B}} \cdot \mathbf{s}_i \right\} - \frac{\alpha^2}{2}\sum_{i \neq j}\frac{1}{r_{ij}^3}(\mathbf{r}_{ij} \times \mathbf{p}_i)(\mathbf{s}_i + 2\mathbf{s}_j) \quad (2.4.2)$$

where α is the fine-structure constant, $\alpha = e^2/\hbar c$. If the matrix elements of the operators in the Eq. (2.4.2) summations are evaluated in atomic units, the final result must be multiplied by 5.8436 in order to obtain a value in cm^{-1}.

The first part of the Eq. (2.4.2) form of $\mathbf{H}^{\mathrm{SO}}$ represents the spin–orbit coupling of each electron in the field of the two bare nuclei with charges Z_{A} and Z_{B}. It is a single-electron operator.

The second part, the spin–other-orbit interaction, is due to interelectronic interactions and has the effect of partially counterbalancing the field of the bare nuclei. Its sign is opposite to that of the first part. This operator is a two-electron operator because of the r_{ij}^{-3} term.

It can be shown (Veseth, 1970) that all electron–nuclear distances, r_{iK}, can be referred to a common origin, and, neglecting only the contribution of spin–other-orbit interactions between unpaired electrons, the two-electron part of the spin-orbit Hamiltonian can be incorporated into the first one-electron part as a screening effect. The spin–orbit Hamiltonian of Eq. (2.4.2) can then be written as

$$\mathbf{H}^{\mathrm{SO}} = \sum_i \hat{a}_i \mathbf{l}_i \cdot \mathbf{s}_i \qquad \text{with} \quad \hat{a}_i \mathbf{l}_i = \sum_K \frac{\alpha^2}{2}\frac{Z_{\mathrm{eff},K}}{r_{iK}^3}\mathbf{l}_{ik} \quad (2.4.3)$$

where $\mathbf{l}_{iK}$ is the orbital angular momentum of electron i about nucleus K and $Z_{\mathrm{eff},K}$, the effective charge of the Kth nucleus, is less than Z_K because the spin–other-orbit part has the effect of screening the nuclear charge by typically 20–50%. [See the atomic calculations of Blume and Watson (1962, 1963) and Blume *et al.* (1964).] Note that $\hat{a}$ is an operator that acts only on the radial part of the wavefunction.

In the form of Eq. (2.4.3), $\mathbf{H}^{\mathrm{SO}}$ is a single-electron operator. It is prudent to avoid discussion of further simplified effective forms of this operator, such as $\mathbf{AL} \cdot \mathbf{S}$, because these forms of $\mathbf{H}^{\mathrm{SO}}$ have led to many errors in the literature. The selection rules for the microscopic $\mathbf{H}^{\mathrm{SO}}$ operator are based on its symmetry with respect to the operations of the point group of the diatomic molecule, $C_{\infty v}$(heteronuclear) or $D_{\infty h}$(homonuclear). The $\hat{a}$ factor of $\mathbf{H}^{\mathrm{SO}}$ is invariant under all symmetry operations. The $\mathbf{l}_i \cdot \mathbf{s}_i$ term may be expanded,

$$\mathbf{l}_i \cdot \mathbf{s}_i = \mathbf{l}_{iz} \cdot \mathbf{s}_{iz} + \tfrac{1}{2}(\mathbf{l}_i^+ \mathbf{s}_i^- + \mathbf{l}_i^- \mathbf{s}_i^+), \quad \text{where } \mathbf{l}_{iz} = \frac{\hbar}{i}\left(x\frac{\partial}{\partial y} - y\frac{\partial}{\partial x}\right) = \frac{\hbar}{i}\frac{\partial}{\partial \phi}, \quad (2.4.4)$$

and, as has been shown in Section 2.2.2, the $\partial/\partial\phi$ operator changes its sign (as does ϕ) upon reflection through *any* plane containing the two nuclei. Consequently, as expected, for $\Sigma^{\pm}$ states the spin–orbit operator must have zero diagonal matrix elements but, less obviously, nonzero off-diagonal matrix elements between $\Sigma^{\pm}$ and $\Sigma^{\mp}$ states. Ignorance of this $\Sigma^{+} \sim \Sigma^{-}$ spin–orbit matrix element selection rule has led to many errors in the literature.

The selection rules for matrix elements of $\mathbf{H}^{\mathrm{SO}}$ are summarized as follows (Kayama and Baird, 1967):

$$\begin{array}{lllll} \Delta J = 0 & \Delta\Omega = 0 & g \leftarrow\!\!\backslash\!\!\rightarrow u & e \leftarrow\!\!\backslash\!\!\rightarrow f & \Sigma^{+} \leftrightarrow \Sigma^{-} \\ \Delta S = 0 \quad \text{or} \quad \Delta S = \pm 1 & & \Delta\Lambda = \Delta\Sigma = 0 \quad \text{or} & \Delta\Lambda = -\Delta\Sigma = \pm 1. & \end{array} \tag{2.4.5}$$

In the single-configuration limit, if the two interacting states belong to the same configuration, then $\Delta\Lambda = \Delta\Sigma = 0$ or, if the two states differ by at most one spin-orbital, then $\Delta\Lambda = -\Delta\Sigma = \pm 1$. One example where the single-configuration approximation is not sufficient will be analyzed later (Section 4.7). If the more complex form of $\mathbf{H}^{\mathrm{SO}}$ [Eq. (2.4.2)] is used, the selection rules are identical but, because the spin–other-orbit interaction is a two-electron operator, additional nonzero matrix elements between configurations that differ by two spin-orbitals must be considered. Also, the spin–other-orbit interaction will contribute significantly to diagonal spin–orbit constants in cases where the direct spin–orbit interaction is zero [e.g., $^{2}\Delta$ states arising from a $\sigma\pi^{2}$ configuration (Lefebvre-Brion and Bessis, 1969), $^{3}\Delta_{g}$ derived from a $\pi_{u}^{2}\pi_{g}^{2}$ configuration (Veseth, 1973b)].

For diagonal matrix elements (diagonal in all quantum numbers including S), we can define A, the spin–orbit coupling constant, as

$$\langle \Lambda, \Sigma, S, \Omega, v | \mathbf{H}^{\mathrm{SO}} | \Lambda, \Sigma, S, \Omega, v \rangle = A_{\Lambda,v} \Lambda\Sigma. \tag{2.4.6}$$

This means that fine-structure levels are expected to be equally spaced (by $A\Lambda$); however, second-order spin–orbit effects can distort the equidistant fine-structure pattern.

For $\Lambda > 0$ states, second-order spin–orbit effects cause two types of J-independent level shifts: proportional to $\Lambda\Sigma$ (the form of the diagonal spin–orbit matrix element) or to $3\Sigma^{2} - S(S+1)$ (the form of the diagonal spin–spin matrix element). Second-order spin–orbit effects are unobservable for $^{1}\Sigma$ and $^{2}\Sigma$ states because they result in no new level splittings or changes in existing separations. For Σ states with $S > \frac{1}{2}$, second-order spin–orbit effects are observable only as shifts similar to the form of the spin–spin interaction. See Levy (1973) for a demonstration of this fact.

2.4.2 EXPRESSION OF SPIN–ORBIT MATRIX ELEMENTS IN TERMS OF ONE-ELECTRON MOLECULAR SPIN–ORBIT PARAMETERS

Using Eq. (2.4.3) for the spin–orbit operator and a single-configuration representation for the electronic states, it is possible to relate observable spin–orbit matrix elements to one-electron orbital integrals, which are called molecular spin–orbit parameters.

2.4.2.1 MATRIX ELEMENTS OF THE $\mathbf{l}_{zi}\cdot\mathbf{s}_{zi}$ TERM

When two states belong to the same values of the Λ and Σ quantum numbers, the only part of $\mathbf{H}^{\text{SO}}$ [Eq. (2.4.4)] by which they can interact is $\mathbf{l}_{zi}\cdot\mathbf{s}_{zi}$. The selection rules are $\Delta\Lambda = \Delta\Sigma = 0$ and $\Delta S = 0$ or ± 1.

If the two interacting states have the same value of S and $\Lambda \neq 0$, then the matrix element has the form of the diagonal matrix element [Eq. (2.4.6)]. From the effect of the $\mathbf{s}_z$ operator, which results in matrix elements with opposite signs for α and β spins, the effect of $\Sigma_i \mathbf{l}_{zi}\cdot\mathbf{s}_{zi}$ on closed shells is identically zero by inspection. Only open shells give a nonzero contribution; therefore the summation over i in Eq. (2.4.3) need only include open shells.

2.4.2.1.1 Diagonal Matrix Elements

Four examples are given for Π states: $\pi^1\,{}^2\Pi$, $\pi^3\,{}^2\Pi$, $\sigma\pi\,{}^3\Pi$, and $\delta\sigma\pi\,{}^4\Pi$.

${}^2\Pi$ States

Case of π^1 Configuration

For the substate ${}^2\Pi_{3/2}$, the wavefunction is

$$|\Lambda = 1, \Sigma = \tfrac{1}{2}, \Omega = \tfrac{3}{2}\rangle = |\pi^+\alpha\rangle.$$

For the substate ${}^2\Pi_{1/2}$, the wavefunction is

$$|\Lambda = 1, \Sigma = -\tfrac{1}{2}, \Omega = \tfrac{1}{2}\rangle = |\pi^+\beta\rangle.$$

From Eq. (2.4.6),

$$\langle {}^2\Pi_{3/2}|\mathbf{H}^{\text{SO}}|{}^2\Pi_{3/2}\rangle = A/2 \qquad \langle {}^2\Pi_{1/2}|\mathbf{H}^{\text{SO}}|{}^2\Pi_{1/2}\rangle = -A/2 \quad (2.4.7)$$

and

$$\langle {}^2\Pi_{3/2}|\mathbf{H}^{\text{SO}}|{}^2\Pi_{3/2}\rangle = \langle \pi^+\alpha|\hat{a}\mathbf{l}_z\mathbf{s}_z|\pi^+\alpha\rangle = \tfrac{1}{2}\langle\pi|\hat{a}|\pi\rangle = \tfrac{1}{2}a_\pi \quad (2.4.8)$$

$$\langle {}^2\Pi_{1/2}|\mathbf{H}^{\text{SO}}|{}^2\Pi_{1/2}\rangle = \langle \pi^+\beta|\hat{a}\mathbf{l}_z\mathbf{s}_z|\pi^+\beta\rangle = -\tfrac{1}{2}\langle\pi|\hat{a}|\pi\rangle = -\tfrac{1}{2}a_\pi. \quad (2.4.9)$$

From Eqs. (2.4.7) and (2.4.8), $A({}^2\Pi, \pi) = a_\pi$; a_π is a molecular spin–orbit parameter, analogous to the atomic $\zeta_K(nl)$ parameter defined by

$$\zeta_K(nl) = \left\langle nl \left| \frac{\alpha^2}{2} \frac{Z_{\text{eff},K}}{r_K^3} \right| nl \right\rangle.$$

As a_π is *always a positive quantity* (the orbital π appears twice in the matrix element), $A > 0$: the ${}^2\Pi$ state is *regular* for the π configuration, as is well known.

Case of π^3 Configuration

For the substate ${}^2\Pi_{3/2}$, the wavefunction is $|\pi^+\alpha\,\pi^+\beta\,\pi^-\alpha\rangle$.
For the substate ${}^2\Pi_{1/2}$, the wavefunction is $|\pi^+\alpha\,\pi^+\beta\,\pi^-\beta\rangle$.

The contribution of the $\pi^+\alpha\pi^+\beta$ subshell to the spin–orbit matrix element is zero; thus

$$\langle {}^2\Pi_{3/2}|\mathbf{H}^{\mathrm{SO}}|{}^2\Pi_{3/2}\rangle = \langle \pi^-\alpha|\hat{a}\mathbf{l}_z\mathbf{s}_z|\pi^-\alpha\rangle = -\tfrac{1}{2}a_\pi. \qquad (2.4.10a)$$

From Eq. (2.4.7),

$$A({}^2\Pi, \pi^3) = -a_\pi; \qquad (2.4.10b)$$

thus the state is *inverted.*

${}^3\Pi$ *States*

Case of $\sigma\pi$ *Configuration*

For ${}^3\Pi_2$ the wavefunction is $|\Lambda = 1, \Sigma = 1, \Omega = 2\rangle = |\sigma\alpha\pi^+\alpha\rangle$.

For ${}^3\Pi_1$ the wavefunction is $|\Lambda = 1, \Sigma = 0, \Omega = 1\rangle = 2^{-1/2}[|\sigma\alpha\pi^+\beta\rangle + |\sigma\beta\pi^+\alpha\rangle]$.

For ${}^3\Pi_0$ the wavefunction is $|\Lambda = 1, \Sigma = -1, \Omega = 0\rangle = |\sigma\beta\pi^+\beta\rangle$.

Only the π electron results in a nonzero contribution from the $\mathbf{l}_z$ operator. A and a_π are defined by Eqs. (2.4.6) and (2.4.8),

$$\langle {}^3\Pi_2|\mathbf{H}^{\mathrm{SO}}|{}^3\Pi_2\rangle = A({}^3\Pi)$$

and

$$\langle {}^3\Pi_2|\mathbf{H}^{\mathrm{SO}}|{}^3\Pi_2\rangle = \langle \pi^+\alpha|\hat{a}\mathbf{l}_z\mathbf{s}_z|\pi^+\alpha\rangle = \tfrac{1}{2}a_\pi \qquad (2.4.11)$$

thence

$$A({}^3\Pi, \sigma\pi) = \tfrac{1}{2}a_\pi. \qquad (2.4.12)$$

In the approximation where the orbital shapes are assumed to be identical for the molecule and positive ion, the following experimental values and Eqs. (2.4.10b) and (2.4.12),

$$A(B^3\Pi_g, \sigma_g\pi_g, \mathrm{N}_2) = 42.2\ \mathrm{cm}^{-1} = a_{\pi_g}/2$$
$$A(A^2\Pi_u, \pi_u^3, \mathrm{N}_2^+) = -80\ \mathrm{cm}^{-1} = -a_{\pi_u},$$

show that the ratio of the two observed magnitudes of A is not far from the predicted value of -2. In this homonuclear example, the equality of $a_{\pi_g} \approx a_{\pi_u}$ is expected to be approximately valid (see Section 4.3.1).

In general it is easiest to derive a relationship between the many-electron A_Λ parameter and the one-electron a_λ parameter from the diagonal matrix elements of the $|\Lambda S\Sigma = S\rangle$ basis function.† Consider, for example, the ${}^4\Pi$ state belonging to a $\delta\sigma\pi$ configuration (e.g., the VO molecule; Cheung *et al.*, 1982).

† The choice of $\Sigma = S$ is advantageous because the required linear combination of Slater determinants is simplest for this case (often a single Slater determinant).

For three open subshells (see Table 2.4);

$$|^4\Pi_{5/2}\rangle = |\Lambda = 1, S = \tfrac{3}{2}, \Sigma = \tfrac{3}{2}, \Omega = \tfrac{5}{2}\rangle$$
$$= |\delta^+\alpha\,\sigma\alpha\,\pi^-\alpha\rangle,$$
$$\langle ^4\Pi_{5/2}|\mathbf{H}^{\mathrm{SO}}|^4\Pi_{5/2}\rangle = A\Lambda\Sigma = A(1)(\tfrac{3}{2}) = \tfrac{3}{2}A,$$

and

$$\langle ^4\Pi_{5/2}|\mathbf{H}^{\mathrm{SO}}|^4\Pi_{5/2}\rangle = \left\langle \delta^+\alpha\,\sigma\alpha\,\pi^-\alpha \left| \sum_i \hat{a}_i \mathbf{l}_{zi}\mathbf{s}_{zi} \right| \delta^+\alpha\,\sigma\alpha\,\pi^-\alpha \right\rangle$$
$$= (2)(\tfrac{1}{2})\langle\delta|\hat{a}|\delta\rangle - (1)(\tfrac{1}{2})\langle\pi|\hat{a}|\pi\rangle;$$

thus

$$A(^4\Pi, \delta\sigma\pi) = \tfrac{1}{3}(2a_\delta - a_\pi).$$

Expressions for the more complicated cases appear in the literature. See, for the $\pi^3\pi^{*2}$ configuration of the NO molecule, the paper by Field *et al.* (1975).

2.4.2.1.2 Off-Diagonal Matrix Elements

Nonzero spin–orbit matrix elements for $\Delta\Lambda = \Delta\Sigma = 0$ but $\Delta S = 1$ occur mainly between states belonging to the same configuration.† When such states are far from each other, the effect causes only second-order effects called "isoconfigurational" spin–orbit effects. These isoconfigurational interactions are of greatest importance in understanding the strong spin–orbit transition to the case (c) limit. The first example is for the $\sigma\pi$ configuration, which gives rise to one $^3\Pi$ and one $^1\Pi$ state. As the selection rule for the spin–orbit operator is *always* $\Delta\Omega = 0$, only the states having $\Omega = 1$ $(\Lambda = 1, \Sigma = 0)$ can interact:

$$\langle ^3\Pi_1, v|\mathbf{H}^{\mathrm{SO}}|^1\Pi_1, v'\rangle$$
$$= \left\langle \pi^+\sigma 2^{-1/2}(\alpha\beta + \beta\alpha), v \left| \sum_i \hat{a}_i \mathbf{l}_{zi}\mathbf{s}_{zi} \right| \pi^+\sigma 2^{-1/2}(\alpha\beta - \beta\alpha), v' \right\rangle$$
$$= \tfrac{1}{2}[\tfrac{1}{2}\langle\pi|\hat{a}|\pi\rangle + \tfrac{1}{2}\langle\pi|\hat{a}|\pi\rangle]\langle v|v'\rangle$$
$$= \tfrac{1}{2}a_\pi\langle v|v'\rangle. \tag{2.4.13}$$

Since the two states belong to the same configuration, the following hypothesis can be adopted: the electronic wavefunctions for the $^3\Pi$ and $^1\Pi$ states are constructed from identical orbitals, and the two states have identical potential

† In Section 4.1.5 the notation $A_{12} = \langle\psi_1|\mathbf{H}^{\mathrm{SO}}|\psi_2\rangle$ is introduced for off-diagonal matrix elements of $\mathbf{H}^{\mathrm{SO}}$ (including the vibrational part of the total wavefunction).

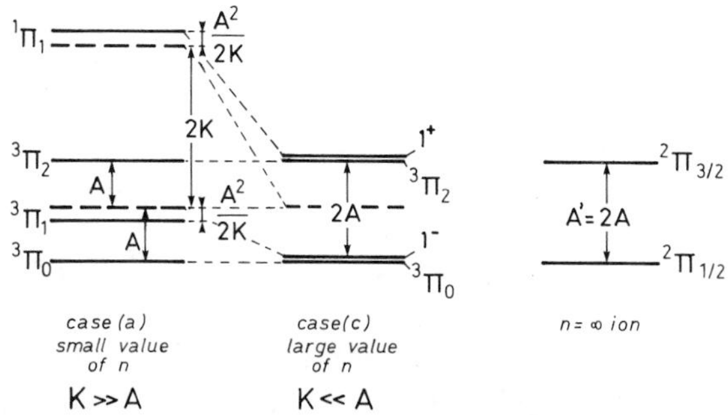

Fig. 2.11 Convergence of $^3\Pi$ and $^1\Pi$ Rydberg series to a $^2\Pi$ ion core. At small n, the exchange splitting (K) is large compared to the spin–orbit splitting of $^3\Pi$ (A) and the $^1\Pi_1 \sim {}^3\Pi_1$ ($-A$) interaction. The symmetric case (a) $^3\Pi$ spin splitting is slightly distorted. At large n, K has become much smaller than A (which arises entirely from the core π orbital) and the case (c) level pattern results. The level labels 1^+ and 1^- denote $|1^\pm\rangle = 2^{-1/2}[|^3\Pi_1\rangle \pm |^1\Pi_1\rangle]$.

curves; consequently,

$$\langle {}^3\Pi_1, v|\mathbf{H}^{\mathrm{SO}}|{}^1\Pi_1, v'\rangle \approx A({}^3\Pi)\,\delta_{vv'}. \tag{2.4.14}$$

When the two interacting states are relatively far apart, this spin–orbit mixing will result in second order energy shifts which destroy the equidistant splittings of the $^3\Pi$ substates (Fig. 2.11).

When the interacting states are close to each other relative to $A(^3\Pi)$, then the mixing is very strong and it is an example of what happens as the case (c) limit is approached. Even at the case (c) limit it is important to note that certain substates (in this case, $^3\Pi_0$ and $^3\Pi_2$) might remain well-described by a single case (a) basis function while the neighboring substates are at the case (c) limit where the quantum number S is undefined.

The series of N_2 $ns\sigma_g$ Rydberg states converging to the N_2^+ $A^2\Pi_u$ state (Ogawa and Tanaka, 1962) is an excellent example of the evolution from case (a) to case (c). The interval between the triplet and singlet Π_u states of the configuration $\pi_u p\, ns\sigma_g$ is given by $2K$, where K is the exchange integral:

$$K = \langle \pi_u p(1)\, ns\sigma_g(1)\,|1/r_{12}|\, ns\sigma_g(2)\, \pi_u p(2)\rangle. \tag{2.4.15}$$

At small n-values, only the $^1\Pi_u$–$X^1\Sigma_g$ series can be seen in absorption. The $^3\Pi_1$ and $^1\Pi_1$ states are mixed by the constant interaction $A = \frac{1}{2}a_{\pi_u p}$ [Eq. (2.4.14)]. As n increases, the interval between these two states, $2K$, which is proportional to n^{-3}, decreases while $a_{\pi_u p}$ remains constant. When n becomes infinite, the singlet and triplet basis functions become degenerate and the resulting eigenstates are completely singlet ~ triplet mixed and separated by $2A$, as shown in Fig. 2.11.

Consequently, in addition to the $^1\Pi_u$–$X^1\Sigma_g^+$ series (Worley's third Rydberg series), a $^3\Pi_{1_u}$–$X^1\Sigma_g^+$ Rydberg series appears in absorption at high n values. The observed separation between the two series reaches a constant value of 80 cm^{-1}, which is just the value of the spin–orbit coupling constant for the N_2^+ ion (which is inverted and thus opposite to the situation illustrated in Fig. 2.11).

A second example is given by the interaction between the $^3\Sigma^-$ and $^1\Sigma^+$ states that belong to a π^2 configuration. Note again that Eq. (2.4.6) is only valid for $\Delta S = 0$ diagonal matrix elements. For Σ states, a nonzero matrix element occurs between only those states of different multiplicity that behave differently under the $\boldsymbol{\sigma}_v$ operation (see selection rules given in Section 2.4.1). This has often been forgotten in the literature (see Tinkham and Strandberg, 1955; Veseth, 1970). The matrix element between $^3\Sigma_0^- e$ and $^1\Sigma_0^+ e$ basis functions from the $\sigma^2\pi^2$ configuration is given by

$$\langle {}^3\Sigma_0^- e, \Lambda = 0, \Sigma = 0|\mathbf{H}^{SO}|{}^1\Sigma_0^+ e, \Lambda = 0, \Sigma = 0\rangle$$

$$= \left\langle \pi^+\pi^-\ (2)^{-1/2}\,(\alpha\beta + \beta\alpha) \left| \sum_i \hat{a}_i \mathbf{l}_{iz}\mathbf{s}_{iz} \right| \pi^+\pi^-\ (2)^{-1/2}\,(\alpha\beta - \beta\alpha) \right\rangle$$

$$= \frac{1}{2}\left\{2\frac{a_\pi}{2} + 2\frac{a_\pi}{2}\right\} = a_\pi. \tag{2.4.16}$$

If one assumes that the orbitals in common for the $\sigma^2\pi^2\ {}^3\Sigma^-$ and $^1\Sigma^+$ and the $\sigma\pi^3\ {}^3\Pi$ states are identical, then

$$\langle {}^3\Sigma_0^- e|\mathbf{H}^{SO}|{}^1\Sigma_0^+\rangle = -2A({}^3\Pi, \sigma\pi^3).$$

Thus, an approximate value for the unknown $^3\Sigma^- \sim {}^1\Sigma^+$ interaction can be obtained from an observable diagonal spin–orbit constant. As will be discussed later (Sections 2.4.4 and 4.3.3), second-order spin–orbit effects of this type contribute significantly to the effective spin–spin interaction in $^3\Sigma^-$ states.

All matrix elements of the $\mathbf{l}_{zi}\mathbf{s}_{zi}$ part of the spin–orbit Hamiltonian can be calculated by using unsymmetrized wavefunctions. The $\mathbf{l}_z\mathbf{s}_z$ matrix elements between properly symmetrized states (eigenfunctions of the $\boldsymbol{\sigma}_v$ operator) are identical to those between unsymmetrized wavefunctions.

To summarize, matrix elements of the $\mathbf{l}_{zi}\mathbf{s}_{zi}$ part of $\mathbf{H}^{SO}$ are important mainly between states belonging to the same configuration. If these states lie far apart in energy, only minor perturbation effects (second-order shifts) result. If the interacting basis states are near degenerate, the result is a transition from case (a) toward case (c). Local $\Delta v \neq 0$ perturbations between electronic states belonging to the same configuration are rare and usually small, because their appearance requires differences between the potential curves of isoconfigurational states or an unusually rapid R-variation of the spin–orbit interaction. However, isoconfigurational $^1\Pi \sim {}^3\Pi$ perturbations with $\Delta v \neq 0$

have been observed for BeO (Lavendy *et al.* 1984), BeS (Pouilly *et al.*, 1982; Cheetham *et al.*, 1965), AsN (Perdigon and Femelat, 1982), and DCl (Huber and Alberti, 1983).

2.4.2.2 Matrix Elements for the $(\mathbf{l}_i^+\mathbf{s}_i^- + \mathbf{l}_i^-\mathbf{s}_i^+)$ Part of $\mathbf{H}^{\mathrm{SO}}$

Matrix elements of the total spin–orbit Hamiltonian between basis states differing by $\Delta\Lambda = \pm 1$, $\Delta\Sigma = \mp 1$, for a *given signed* value of Ω, may be calculated using only the $\frac{1}{2}\mathbf{l}_i^+\mathbf{s}_i^-$ *or* $\frac{1}{2}\mathbf{l}_i^-\mathbf{s}_i^+$ part of $\mathbf{H}^{\mathrm{SO}}$. Again, it is unnecessary to use symmetrized basis functions (except when one of the states involved is $\Sigma_0^{\pm}$):

$$\begin{aligned}
&\left\langle 2^{-1/2}[(\Lambda,\Sigma) \pm (-\Lambda,-\Sigma)]\left|\frac{1}{2}\sum_i \hat{a}_i(\mathbf{l}_i^+\mathbf{s}_i^- + \mathbf{l}_i^-\mathbf{s}_i^+)\right| 2^{-1/2}[(\Lambda+1,\Sigma-1)\right.\\
&\qquad \left. \pm (-\Lambda-1,-\Sigma+1)]\right\rangle\\
&= \langle \Lambda,\Sigma,+\Omega|\mathbf{H}^{\mathrm{SO}}|\Lambda+1,\Sigma-1,+\Omega\rangle = \left\langle +\Omega\left|\frac{1}{2}\sum_i \hat{a}_i\mathbf{l}_i^-\mathbf{s}_i^+\right|+\Omega\right\rangle\\
&= \langle -\Lambda,-\Sigma,-\Omega|\mathbf{H}^{\mathrm{SO}}|-\Lambda-1,-\Sigma+1,-\Omega\rangle\\
&= \left\langle -\Omega\left|\frac{1}{2}\sum_i \hat{a}_i\mathbf{l}_i^+\mathbf{s}_i^-\right|-\Omega\right\rangle\\
&= \left\langle \Omega_f^e\left|\frac{1}{2}\sum_i \hat{a}_i(\mathbf{l}_i^+\mathbf{s}_i^- + \mathbf{l}_i^-\mathbf{s}_i^+)\right|\Omega_f^e\right\rangle. \qquad (2.4.17)
\end{aligned}$$

When $2S+1$ is odd, all $\Sigma^{\pm}$ states have only one $\Omega = 0$ substate. Consequently, all $\mathbf{H}^{\mathrm{SO}}$ matrix elements between a $\Sigma_0^{\pm}$ basis function and a *symmetrized* Π_0 basis function must be obtained by multiplying the result from the third or fourth line of Eq. (2.4.17) by $2^{1/2}$. An example will be treated in Section 2.5.3.

The following example is for states where $\Omega \neq 0$, for the interaction between $^2\Pi_{1/2}$ and $^2\Sigma^+_{1/2}$ states.

The wavefunction for the $^2\Pi_{1/2}$ state from a π^1 configuration is $|\Lambda = 1, \Sigma = -\frac{1}{2}, \Omega = +\frac{1}{2}\rangle = |\pi^+\beta|$; the wavefunction for the $^2\Sigma^+_{1/2}$ state from a σ^1 configuration is $|\Lambda = 0, \Sigma = +\frac{1}{2}, \Omega = +\frac{1}{2}\rangle = |\sigma\alpha|$. Thus,

$$\begin{aligned}
\langle {}^2\Pi_{1/2}, v|\mathbf{H}^{\mathrm{SO}}|{}^2\Sigma^+_{1/2}, v'\rangle &= \langle \pi^+\beta|\tfrac{1}{2}\hat{a}\mathbf{l}^+\mathbf{s}^-|\sigma\alpha\rangle\langle v|v'\rangle\\
&= \tfrac{1}{2}\langle \pi^+|\hat{a}\mathbf{l}^+|\sigma\rangle\langle v|v'\rangle = \tfrac{1}{2}a_+\langle v|v'\rangle. \qquad (2.4.18)
\end{aligned}$$

The notation a_+ is used to avoid confusion with the a magnetic hyperfine constant. If one assumes that the σ and π molecular orbitals have the same well-defined value for l and the potential curves of the two states are identical,

$\langle v|v'\rangle = \delta_{vv'}$, then

$$a_+ = \langle \pi^+ |\hat{a}\mathbf{l}^+| \sigma\rangle = [l(l+1) - \lambda(\lambda+1)]^{1/2}\langle \pi|\hat{a}|\pi\rangle, \qquad (2.4.19)$$

or, from Eq. (2.4.9),

$$a_+ = [l(l+1) - \lambda(\lambda+1)]^{1/2} A(^2\Pi, \pi). \qquad (2.4.20)$$

If the value of l is 1, $a_+ = 2^{1/2}A$. This is known as part of the pure precession hypothesis and is approximately true only for hydrides, Rydberg states of an $l > 0$ complex, and certain highly ionic molecules (see Section 4.5).

These interactions between states of different symmetries arising from the $\frac{1}{2}(\mathbf{l}^+\mathbf{s}^- + \mathbf{l}^-\mathbf{s}^+)$ part of the spin–orbit Hamiltonian are very common. They often occur between states having $\Delta S = 1$. One example, dealing with the interactions between the excited states of the CO molecule, is given in Section 4.3.2. Another example shows that such an interaction explains the predissociation of the $A^2\Sigma^+$ state of OH (Section 6.9.1).

2.4.3 THE SPIN–ROTATION OPERATOR

This operator accounts for the interaction between the electron spins and the magnetic field created by nuclear motion. As the nuclei are heavy, their angular velocity is approximately m/M times smaller than the angular velocity of the electrons. Consequently, except for light molecules, the spin–rotation interaction is very small compared to the spin–orbit interaction. The microscopic Hamiltonian from Kayama and Baird (1967) and Green and Zare (1977) has the case (a) form,

$$\mathbf{H}^{\mathrm{SR}} = -2g_S\mu_B \frac{\mu_n M}{I}\left\{\sum_i\sum_K Z_K\left(\frac{R_{GK}\cos\theta_{iK}}{r_{iK}^2}\mathbf{s}_i\right)\right\}\cdot\mathbf{R}. \qquad (2.4.21)$$

The nuclear magneton, μ_n, is equal to $(m/M)\,\mu_B$, where m/M is the ratio of the mass of the electron to the proton, 1/1836, I is the molecular moment of inertia, Z_K is the nuclear charge, θ_{iK} is the angle between the electron coordinate r_{iK} and R_{GK} as defined in Fig. 2.12, and R_{GK} is the distance of nucleus K from the center of mass. If the moment of inertia, I, is related to the rotational constant, B_v, the result for a diatomic molecule is

$$\mathbf{H}^{\mathrm{SR}} = -1.05\times 10^{-4}B_v\left\{\sum_i\left(\frac{Z_A R_{GA}\cos\theta_{iA}}{r_{iA}^2} + \frac{Z_B R_{GB}\cos\theta_{iB}}{r_{iB}^2}\right)\mathbf{s}_i\right\}\cdot\mathbf{R} \qquad (2.4.22)$$

where B_v is in cm^{-1}, all distances are in atomic units, and the matrix elements of $\mathbf{H}^{\mathrm{SR}}$ will be in cm^{-1}. The angles are defined in Fig. 2.12.

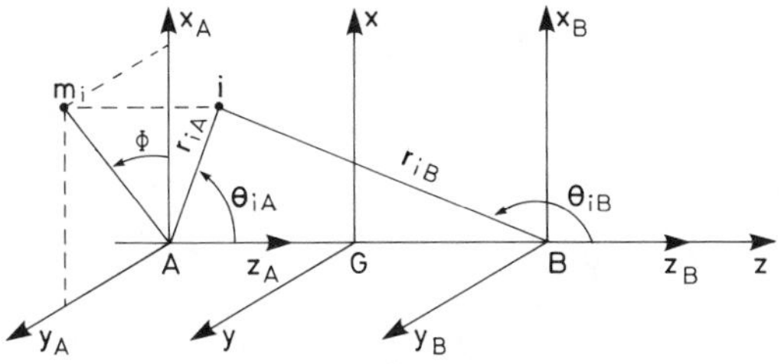

Fig. 2.12 Electronic coordinate system. The electron, at **i**, is located relative to three coordinate systems located respectively at atom A, at the center of mass G, and at atom B; m_i is the projection of **i** onto the x_A, y_A plane; Φ locates the electron relative to the xz plane ($0 \le \Phi \le 2\pi$; when $\Phi = 0$, electron is in xz plane; when $\Phi = \pi/2$, electron is in yz plane). The angles θ_{iA} and θ_{iB} ($0 \le \theta \le \pi$) locate the electron relative to the $+z_A$ and $+z_B$ axes.

The effective form for the spin–rotation Hamiltonian is given by Brown and Watson (1977) as

$$\mathbf{H}^{SR} = \gamma \mathbf{R} \cdot \mathbf{S} = \gamma(\mathbf{J} - \mathbf{L} - \mathbf{S}) \cdot \mathbf{S} = \gamma(\mathbf{N} - \mathbf{L}) \cdot \mathbf{S}, \tag{2.4.23}$$

which is convenient for the case (a) basis.

An alternative, commonly used form of $\mathbf{H}^{SR}$ [convenient for case (b)],

$$\mathbf{H}^{SR} = \gamma \mathbf{N} \cdot \mathbf{S}, \tag{2.4.24}$$

differs from Eq. (2.4.23) by $\gamma \mathbf{L} \cdot \mathbf{S}$. Since the matrix elements of $\mathbf{H}^{SO}$ within any multiplet state (i.e., $\Delta S = 0$) have the form $A\mathbf{L} \cdot \mathbf{S}$, the choice of whether to adopt Eq. (2.4.23) or Eq. (2.4.24) for $\mathbf{H}^{SR}$ affects the value of A by $\pm\gamma$, so that

$$A[\text{Eq. (2.4.23)}] = A[\text{Eq. (2.4.24)}] + \gamma.$$

However, since $\gamma \ll A$, this choice is immaterial for $\Lambda \neq 0$ states of all molecules except perhaps H_2. For a $^3\Pi_u$ state of H_2, Jette and Miller (1974) used the fact that the isotopic dependence of γ (proportional to μ^{-1}) is different from that of A (independent of μ) to separate $\gamma = -24$ MHz from $A = -3741$ MHz.

The diagonal matrix elements of $\mathbf{H}^{SR}$ in the case (a) basis are

$$\langle \Lambda S \Sigma \Omega | \mathbf{H}^{SR} | \Lambda S \Sigma \Omega \rangle = \gamma[\Sigma^2 - S(S+1)] \tag{2.4.25}$$

if Eq. (2.4.23) is used, or

$$\langle \Lambda S \Sigma \Omega | \mathbf{H}^{SR} | \Lambda S \Sigma \Omega \rangle = \gamma[\Omega \Sigma - S(S+1)] \tag{2.4.26a}$$

if Eq. (2.4.24) is used. Equation (2.4.24) is particularly useful for evaluating matrix elements of $\mathbf{H}^{SR}$ in the case (b) basis,

$$\langle NJS\Lambda | \mathbf{H}^{SR} | NJS\Lambda \rangle = (\gamma/2)[J(J+1) - N(N+1) - S(S+1)], \tag{2.4.26b}$$

because $\mathbf{N} \cdot \mathbf{S}$ has only diagonal matrix elements in the case (b) basis.

The spin–rotation Hamiltonian has $\Delta\Omega = \Delta\Sigma = \pm 1$, $\Delta\Lambda = 0$, $\Delta S = 0$ off-diagonal matrix elements,

$$\langle \Lambda, S, \Sigma, \Omega, v | \mathbf{H}^{\mathrm{SR}} | \Lambda, S, \Sigma \pm 1, \Omega \pm 1, v \rangle$$
$$= \left\langle \Sigma, \Omega, v \left| \frac{\gamma}{2} \mathbf{J}^{\pm} \mathbf{S}^{\mp} \right| \Sigma \pm 1, \Omega \pm 1, v \right\rangle$$
$$= (\gamma_v/2)[J(J+1) - \Omega(\Omega \pm 1)]^{1/2}[S(S+1) - \Sigma(\Sigma \pm 1)]^{1/2}. \quad (2.4.27)$$

This off-diagonal matrix element connects the same basis states as a term in $\mathbf{H}^{\mathrm{ROT}}$ ($\mathbf{S}$-uncoupling operator), but its sign is opposite to that of the B_v contribution. The off-diagonal B_v term appears with a negative sign while γ appears with a positive sign because of the phase convention [Eqs. (2.2.28c) and (2.2.29)],

$$\mathbf{S}^{\pm} |S, \Sigma\rangle = +\hbar[S(S+1) - \Sigma(\Sigma \pm 1)]^{1/2} |S, \Sigma \pm 1\rangle. \quad (2.4.28)$$

Zare *et al.* (1973) adopt a phase convention that requires a negative sign in Eq. (2.4.28). However, regardless of phase convention, the $\Delta\Lambda = 0$, $\Delta\Sigma = \Delta\Omega = \pm 1$ matrix elements always have the form

$$\pm(B - \gamma/2)[J(J+1) - \Omega(\Omega \pm 1)]^{1/2}[S(S+1) - \Sigma(\Sigma \pm 1)]^{1/2}.$$

Example. Consider the calculation of γ_{true} for ${}^3\Sigma^-$ of O_2. From Eq. (2.4.27) one obtains

$$\langle {}^3\Sigma^-_{+1} | \mathbf{H}^{\mathrm{SR}} | {}^3\Sigma^-_0 \rangle = (\gamma/2)[2J(J+1)]^{1/2}.$$

Following Eq. (2.4.22) and noting that for homonuclear molecules $R_{GA} = -R/2 = -R_{GB}$,

$$\langle {}^3\Sigma^-_{+1} | \mathbf{H}^{\mathrm{SR}} | {}^3\Sigma^-_0 \rangle = +1.05 \times 10^{-4}(B_{O_2})(Z_O)(R_{O_2})$$
$$\times \sum_i \left\langle {}^3\Sigma^-_{+1} \left| \frac{\cos\theta_{iA}}{r^2_{iA}} \mathbf{s}^+_i \right| {}^3\Sigma^-_0 \right\rangle (\tfrac{1}{2})[J(J+1)]^{1/2}.$$

The wavefunction for ${}^3\Sigma^-_{+1}$ is $|\Lambda = 0, \Sigma = 1, \Omega = 1\rangle = |\pi^+_g \alpha \pi^-_g \alpha|$, and the wavefunction for ${}^3\Sigma^-_0$ is $|\Lambda = 0, \Sigma = 0, \Omega = 0\rangle = 2^{-1/2}[|\pi^+_g \alpha \pi^-_g \beta| + |\pi^+_g \beta \pi^-_g \alpha|]$; thus,

$$\sum_i \left\langle {}^3\Sigma^-_{+1} \left| \frac{\cos\theta_{iA}\, \mathbf{s}^+_i}{r^2_{iA}} \right| {}^3\Sigma^-_0 \right\rangle = 2^{-1/2}\left[\left\langle \pi^-_g \alpha \left| \frac{\cos\theta_{1A}}{r^2_{1A}} \mathbf{s}^+_1 \right| \pi^-_g \beta \right\rangle \right.$$
$$\left. + \left\langle \pi^+_g \alpha \left| \frac{\cos\theta_{1A}}{r^2_{1A}} \mathbf{s}^+_1 \right| \pi^+_g \beta \right\rangle \right]$$
$$= 2^{1/2} \left\langle \pi_g \left| \frac{\cos\theta_{1A}}{r^2_{1A}} \right| \pi_g \right\rangle,$$

where the π_g^- and π_g^+ matrix elements are identical because the $\cos\theta_{1A}/r_{1A}^2$ operator does not operate on ϕ. The integral over the π_g molecular orbital may be simplified by expressing π_g in terms of the atomic orbitals, π_A and π_B,

$$\pi_g = [2(1 - S)]^{-1/2}[\pi_A - \pi_B],$$

where S is the atomic orbital overlap integral. The one-center integral $\langle\pi_A|\cos\theta_{1A}/r_{1A}^2|\pi_A\rangle$ is zero by symmetry (odd integrand). The two-center integral $\langle\pi_A|\cos\theta_{1A}/r_{1A}^2|\pi_B\rangle$ is negligible. The integral $\langle\pi_B|\cos\theta_{1A}/r_{1A}^2|\pi_B\rangle$ may be estimated by a point charge approximation to have the value $1/R^2$ (Kayama and Baird, 1967). With $R_{O_2} = 2.282$ a.u., $B_{O_2} = 1.44566\ \mathrm{cm}^{-1}$, and neglecting S (overlap);

$$\gamma = 2.66 \times 10^{-4}\ \mathrm{cm}^{-1} = 8\ \mathrm{MHz}.$$

A more exact calculation, which does not make the point charge approximation, gives 3.74 MHz. The experimental value of $\gamma_{\mathrm{eff}} = -8.4 \times 10^{-3}\ \mathrm{cm}^{-1}$ is much larger due to second-order effects.

In view of the approximate relationship

$$\frac{\langle \mathbf{H}^{\mathrm{SR}}\rangle}{\langle \mathbf{H}^{\mathrm{SO}}\rangle} \simeq -\frac{m}{M} \propto \frac{\gamma}{A}, \tag{2.4.29}$$

Van Vleck (1929) has suggested a scheme for determining γ from A. In general, Van Vleck's approximation is not justified. As pointed out by Green and Zare (1977), the relevant radial integrals, respectively

$$\langle\phi_A|1/r_{1A}^3|\phi_A\rangle$$

for the spin-orbit operator and

$$\langle\phi_A|\cos\theta_{1B}/r_{1B}^2|\phi_A\rangle$$

for the spin–rotation operator, act in different spatial regions. Furthermore, the integration often includes different orbitals for the two operators. For example, in the case of a $\sigma\pi\ {}^3\Pi$ state, the value of A is determined by the π orbital and the value of γ depends on both orbitals. This is particularly striking for the $1s\sigma_g np\pi_u\ {}^3\Pi_u$ state of H_2, where γ is determined nearly exclusively by the $1s\sigma_g$ orbital. The observed value of γ cannot be compared directly to the calculated value, because γ_{obs} is an effective constant that includes direct (or first-order) contributions from $\mathbf{H}^{\mathrm{SR}}$ and second-order effects from $\mathbf{H}^{\mathrm{SO}}$ and $\mathbf{H}^{\mathrm{ROT}}$.

Second-order effects arising from the product of matrix elements involving $\mathbf{J}^+\mathbf{L}^-$ and $\mathbf{L}^+\mathbf{S}^-$ operators have the same form as $\gamma\mathbf{J}^+\mathbf{S}^-$. In the case of H_2, the second-order effect seems to be smaller than the first-order effect, but in other molecules this second-order effect will be more important than the first-order contribution to the spin–rotation constant. These second-order contributions can be shown to increase in proportion with spin–orbit effects, namely roughly

Table 2.7
Comparison between Some Calculated and Observed Values for the First-Order Contribution to the Spin–Rotation Constant

	Molecule and state						
Parameter	H_2 $c^3\Pi_u$	OD $X^2\Pi$	HF^+ $X^2\Pi$	HCl^+ $X^2\Pi$	NO $X^2\Pi$	SiO $a^3\Pi$	CS $a^3\Pi\ v=0$
$\gamma(\text{calc}) \times 10^4\ \text{cm}^{-1}$	8.67[a]	3.9[b]	7.6[b]	2.9[b]	1.12[e]	0.09[e]	—
$\gamma(\text{depert}) \times 10^4\ \text{cm}^{-1}$	8.0[a]	16.4[c]	—	80.0[d]	−175.0[f]	—	−51[g]

[a] Jette and Miller (1974); [b] Green and Zare (1977); [c] Coxon (1975); [d] Brown *et al.* (1979); [e] Lefebvre-Brion (unpublished results); [f] Amiot *et al.* (1978); and [g] Cossart *et al.* (1977).

as Z^2, but the direct spin–rotation interaction is proportional to the rotational constant. For $^2\Pi$ states, γ is strongly correlated with A_D, the spin–orbit centrifugal distortion constant [see definition, Eq. (4.6.6)], and direct evaluation from experimental data is difficult. On the other hand, the main second-order contribution to γ is often due to a neighboring $^2\Sigma^+$ state. Table 2.7 compares calculated with deperturbed values of γ. γ^{eff} of $^2\Pi$ may be deperturbed with respect to $^2\Sigma^+$ by

$$\gamma^{\Pi}(\text{deperturbed}) = \gamma^{\text{eff}} + p^{\Pi}(^2\Sigma^+)/2,$$

where $p^{\Pi}(^2\Sigma^+)$ is defined by Eq. (2.5.32) (see also Coxon, 1975 and Brown *et al.* 1979). γ(deperturbed) includes not only the first-order $\mathbf{H}^{\text{SR}}$ contribution but second-order ($\mathbf{H}^{\text{SO}} \times \mathbf{H}^{\text{ROT}}$) contributions from all other states *except* the nearby $^2\Sigma^+$ (see Section 4.5).

For the $a^3\Pi$ state of CS, the second-order contribution to γ has been determined to be about 8×10^{-4} cm^{-1} and subtracted from the effective γ value. The resulting γ value is certainly not due to the first-order contribution (cf. the calculated value, $\gamma = 9 \times 10^{-6}$ cm^{-1}, for SiO $b^3\Pi$, Table 2.7) but to other high-order effects.

Perturbations between electronic states resulting from $\mathbf{H}^{\text{SR}}$ are usually so weak (comparable in magnitude to hyperfine perturbations for light molecules) as to be virtually undetectable.

2.4.4 THE SPIN–SPIN OPERATOR

The spin–spin operator represents the interaction energy between the magnetic dipoles associated with the spins of two different electrons. It is a rather complicated two-electron operator,

$$\mathbf{H}^{\mathrm{SS}} = -\alpha^2 \sum_{i<j} \frac{1}{r_{ij}^5} [3(\mathbf{r}_{ij} \cdot \mathbf{s}_i)(\mathbf{r}_{ij} \cdot \mathbf{s}_j) - (\mathbf{s}_i \cdot \mathbf{s}_j) r_{ij}^2]. \qquad (2.4.30)$$

The "contact" term, which gives an identical contribution for all components of a given multiplet, is neglected here. The selection rules for this operator are

$$\Delta S = 0, \pm 1, \pm 2$$

$$\Delta\Sigma = -\Delta\Lambda = 0 \quad \text{or} \quad \pm 1 \quad \text{or} \quad \pm 2$$

and always,

$$\Delta\Omega = 0.$$

The spin–spin interaction is zero for Σ states with $S \leq \frac{1}{2}$. The other selection rules for the $\mathbf{H}^{\mathrm{SS}}$ operator are $g \sim g$ or $u \sim u$, but the selection rule $\Sigma^\pm \sim \Sigma^\pm$ is opposite to that for the spin–orbit operator, which is $\Sigma^\pm \sim \Sigma^\pm$. Note, however, that the spin–spin interaction is zero between triplet and singlet states if both of them are Σ states (for example, a ${}^3\Sigma_0^+$ state has only f levels and the universal selection rule for perturbations is $e \nleftrightarrow f$; thus ${}^1\Sigma^\pm \sim {}^3\Sigma^\pm$ $\mathbf{H}^{\mathrm{SS}}$ perturbations are $e \nleftrightarrow f$ forbidden; see the end of Section 2.4.5).

Second-order spin–orbit effects result in matrix elements which have the same form as the spin–spin operator. Symbolically,

$$[(\mathbf{S} \cdot \mathbf{L})][(\mathbf{L} \cdot \mathbf{S})] \equiv (\mathbf{S})(\mathbf{S}).$$

Experimentally, this second-order spin–orbit effect is indistinguishable from the direct spin–spin interaction.

2.4.4.1 DIAGONAL MATRIX ELEMENTS OF $\mathbf{H}^{\mathrm{SS}}$: CALCULATION OF THE DIRECT SPIN–SPIN PARAMETER

The usual form of the effective spin–spin Hamiltonian, derived by application of the Wigner–Eckart theorem with $\Delta S = \Delta\Sigma = 0$ (see Section 2.4.5), is

$$\mathbf{H}^{\mathrm{SS}} = \tfrac{2}{3}\lambda(3\mathbf{S}_z^2 - \mathbf{S}^2), \qquad (2.4.31\mathrm{a})$$

which has nonzero case (a) matrix elements,

$$\langle S, \Sigma | \mathbf{H}^{\mathrm{SS}} | S, \Sigma \rangle = \tfrac{2}{3}\lambda[3\Sigma^2 - S(S+1)]. \qquad (2.4.31\mathrm{b})$$

Note that in the literature, two other symbols are also used:

$$\tfrac{2}{3}\lambda = -C_\Lambda = \varepsilon.$$

The splitting of Σ states with $S > \frac{1}{2}$ is due to the spin–spin effect. For example, for $^3\Sigma$ states,

$$E(^3\Sigma_{\pm1}) - E(^3\Sigma_0) = 2\lambda.$$

For states other than Σ states, the spin–spin interaction contributes only to the asymmetry of the spin–orbit splitting.

Expressed in the form of Eq. (2.4.30), the calculation of matrix elements of the spin–spin operator is not trivial. For $^3\Sigma$ states, only the following terms in $\mathbf{H}^{\mathrm{SS}}$ give rise to nonzero matrix elements:

$$\mathbf{H}^{\mathrm{SS}}(^3\Sigma) = -\alpha^2 \sum_{i<j} (3z_{ij}^2 - r_{ij}^2)/r_{ij}^5 \left[\mathbf{s}_{iz}\mathbf{s}_{jz} - \frac{1}{4}(\mathbf{s}_i^+\mathbf{s}_j^- + \mathbf{s}_i^-\mathbf{s}_j^+) \right]. \quad (2.4.32)$$

For example, the direct spin–spin interaction of a $\pi^2\ ^3\Sigma^-$ state may be derived as follows (see Kayama, 1965). From Eq. (2.4.31b),

$$\langle ^3\Sigma_0^- | \mathbf{H}^{\mathrm{SS}} | ^3\Sigma_0^- \rangle = -\tfrac{4}{3}\lambda.$$

Now, again evaluating in the one-electron spin–orbital basis for the π^2 configuration,

$$\begin{aligned}
&\langle (2)^{-1/2}\{|\pi^+\alpha\pi^-\beta| + |\pi^+\beta\pi^-\alpha|\} |\mathbf{H}^{\mathrm{SS}}| (2)^{-1/2}\{|\pi^+\alpha\pi^-\beta| + |\pi^+\beta\pi^-\alpha|\}\rangle \\
&= -\alpha^2 \left\{ \left\langle \pi^+\alpha(1)\,\pi^+\alpha(1) \left| \frac{3z_{12}^2 - r_{12}^2}{r_{12}^5} (\mathbf{s}_{1z}\mathbf{s}_{2z}) \right| \pi^-\beta(2)\,\pi^-\beta(2) \right\rangle \right. \\
&\quad \left. - \left\langle \pi^+\alpha(1)\,\pi^+\beta(1) \left| \frac{3z_{12}^2 - r_{12}^2}{r_{12}^5} \frac{1}{4} (\mathbf{s}_1^+\mathbf{s}_2^- + \mathbf{s}_1^-\mathbf{s}_2^+) \right| \pi^-\beta(2)\,\pi^-\alpha(2) \right\rangle \right\}.
\end{aligned} \quad (2.4.33)$$

After evaluating the effects of the spin operators on simple spin-orbitals, Eq. (2.4.33) reduces to

$$\frac{\alpha^2}{2} \langle \pi^+(1)\,\pi^+(1) | (3z_{12}^2 - r_{12}^2)/r_{12}^5 | \pi^-(2)\,\pi^-(2) \rangle.$$

The atomic spin–spin parameter, η_{A}, is defined in terms of the p_{A} orbitals by

$$\begin{aligned}
\eta_{\mathrm{A}} &= -\frac{\alpha^2}{16} \langle p_{\mathrm{A}}^+ p_{\mathrm{A}}^+ | (3z_{12}^2 - r_{12}^2)/r_{12}^5 | p_{\mathrm{A}}^- p_{\mathrm{A}}^- \rangle \\
&= (\alpha^2/20) \int_0^\infty p_{\mathrm{A}}^2(r_1)\, dr_1 \int_{r_1}^\infty \frac{p_{\mathrm{A}}^2(r_2)}{r_2^3}\, dr_2 \geq 0
\end{aligned}$$

where, if the integral is evaluated in atomic units and the factor $\alpha^2/20 = 0.58436$, the final result for η_{A} is in cm^{-1}. For hydride molecules, AH, it is a good approximation to reduce the π molecular orbital to the p_{A} atomic orbital.

Then

$$\langle {}^3\Sigma_0^- | \mathbf{H}^{\mathrm{SS}} | {}^3\Sigma_0^- \rangle = -8\eta$$

or

$$\lambda({}^3\Sigma^-, \mathrm{AH}) = 6\eta.$$

In the case of the homonuclear molecule, A_2, if two-center integrals are neglected, the semiempirical expression

$$\lambda({}^3\Sigma^-, \mathrm{A}_2) = 3\eta$$

is obtained.

For a configuration $\sigma\pi^3$, there can be an off-diagonal contribution from the spin–spin interaction, with $\Delta\Lambda = -\Delta\Sigma = \pm 2$ matrix elements between the two components of a ${}^3\Pi_0$ sublevel, which results in a nonzero Λ doubling at $J = 0$,

$$\begin{aligned} E({}^3\Pi_{0f}) - E({}^3\Pi_{0e}) &= -2\alpha^{\dagger} = +2C^{\delta} \\ &= 2[\langle \sigma\alpha\, \sigma\beta | \mathbf{H}^{\mathrm{SS}} | \pi^-\alpha\, \pi^+\beta \rangle - \langle \sigma\alpha\, \pi^+\beta | \mathbf{H}^{\mathrm{SS}} | \sigma\beta\, \pi^-\alpha \rangle]. \end{aligned} \quad (2.4.34)$$

The second integral, which is an exchange-type integral (see Section 2.2.4), is much smaller than the first, and consequently the Λ-doubling difference of Eq. (2.4.34) is always positive (Horani *et al.*, 1967).

Table 2.8 gives semiempirical expressions for the direct spin–spin part of several effective spin–spin constants and some numerical examples for valence states of homonuclear molecules where the π^* orbital is assumed to be an antibonding orbital. For Rydberg states, the contribution of the Rydberg π' orbital can be neglected.

In Table 4.9 some numerical values for the atomic spin–spin parameter are reported, calculated or deduced from fitting of atomic fine-structure spectra. These values are only a few hundredths of reciprocal centimeters for atoms heavier than first-row atoms.

† This spin–spin constant, α, should not be confused with the fine-structure constant $\alpha = e^2/\hbar c$. The direct spin–spin contribution to the Λ doubling, α, cannot be separated from a second-order spin–orbit term. All contributions are combined in the o_v parameter, which is the coefficient of the term in the effective Hamiltonian (Brown *et al.* 1979; Brown and Merer, 1979) with a $\Delta\Sigma = -\Delta\Lambda = \pm 2$ selection rule,

$$o_v(\mathbf{S}_x^2 - \mathbf{S}_y^2) = (o_v/2)(\mathbf{S}_+^2 + \mathbf{S}_-^2),$$

where

$$o_v = -\alpha + \text{second-order spin–orbit}.$$

Table 2.8
Semiempirical Expressions for Effective Spin–Spin Constants: Valence States of Homonuclear Molecules

Configuration	Order of states[a]	State	Direct $\mathbf{H}^{SS}$	Isoconfigurational second-order $\mathbf{H}^{SO}$		Numerical examples (cm^{-1})		
						$\lambda^{SS}_{calc.}$	$\lambda^{SO}_{calc.}$ (iso)	$\lambda^{eff}_{exp.}$
π^2	$^1\Sigma^+$, $^1\Delta$, $^3\Sigma^-$	$^3\Sigma^-$	3η	$\frac{2[A(^3\Pi, \sigma\pi^3)]^2}{E(^1\Sigma^+) - E(^3\Sigma^-)}$	$X^3\Sigma_g^-$ O_2[b]	0.75	1.34	1.98
$\pi^3\pi^*$	$^1\Sigma^+$, $^1\Delta$, $^1\Sigma^-$, $^3\Sigma^-$, $^3\Delta$, $^3\Sigma^+$	$^3\Sigma^+$	-3η	$\frac{1}{2}\frac{[2A(^3\Delta, \pi^3\pi^*) - 2A(^3\Pi, \sigma\pi^*)]^2}{E(^3\Sigma^+) - E(^3\Sigma^-)}$	$A^3\Sigma_u^+$ N_2[c]	−0.42	−0.22	−1.33
		$^3\Sigma^-$	3η	$\frac{1}{2}\frac{[2A(^3\Delta) - 2A(^3\Pi)]^2}{E(^3\Sigma^-) - E(^3\Sigma^+)}$	$B^3\Sigma_u^-$ N_2[c]	0.42	0.22	0.66
		$^3\Delta$	3η	$-\frac{1}{2}\frac{[2A(^3\Delta) - 2A(^3\Pi)]^2}{E(^3\Delta) - E(^1\Delta)}$	$W^3\Delta_u$ N_2	0.42	0.29	0.67[d]
$\pi\pi^*$ or $\pi^3\pi^{*3}$	$^1\Sigma^+$, $^1\Delta$, $^3\Sigma^-$, $^3\Sigma^+$, $^3\Delta$, $^1\Sigma^-$	$^3\Sigma^+$	0	$\frac{2[A(^3\Delta, \pi\pi^*)]^2}{E(^1\Sigma^-) - E(^3\Sigma^+)}$	$A^3\Sigma_u^+$ O_2[e]	0	−5.5	−5.1
		$^3\Sigma^-$	6η	$\frac{2[A(^3\Delta)]^2}{E(^1\Sigma^+) - E(^3\Sigma^-)}$	$B^3\Sigma_u^-$ O_2[e]	1.38	0	1.63
		$^3\Delta$	0	0				

$\sigma\pi^2$	$^2\Sigma^+$ $^2\Sigma^-$ $^4\Sigma^-$	$^4\Sigma^-$	0	$\frac{1}{6}\frac{[A(^2\Pi, \sigma^2\pi)]^2}{E(^2\Sigma^+) - E(^4\Sigma^-)}$	$b^4\Sigma_g^-\ O_2^+$	0	0.34[f]	0.21[g]
$\sigma\pi$ or $\sigma\pi^3$	$^1\Pi$ $^3\Pi$	$^3\Pi$	$-\frac{3}{2}\eta$	$+\frac{1}{2}\frac{[A(^3\Pi)]^2}{E(^1\Pi) - E(^3\Pi)}$	$B^3\Pi_g\ N_2$	−0.20[h]	0.09	0.21[i]
$\pi^3\pi^{*2}$[j]	$^2\Pi$ $^2\Pi$ $^4\Pi$	$^4\Pi$	2η	$\frac{1}{4}\frac{[\langle ^2\Pi_{3/2}\|\mathbf{H}^{SO}\|^4\Pi_{3/2}\rangle]^2}{E(^2\Pi) - E(^4\Pi)}$	$a^4\Pi_u\ O_2^+$[g]	0.62	0.17	0.95

[a] The ordering of the states within each configuration corresponds to the energy order in the single-configuration approximation (see Recknagel, 1934).
[b] See Table 4.10.
[c] Sink *et al.* (1975).
[d] Effantin *et al.* (1979a).
[e] Field and Lefebvre-Brion (1974).
[f] From calculated term values of Beebe *et al.* (1976).
[g] Brown *et al.* (1981).
[h] Value obtained using the pure precession approximation. (Sections 2.5.4 and 4.5). Calculated value is $-0.08\ cm^{-1}$ (Lefebvre-Brion unpublished calculation).
[i] Effantin *et al.* (1979b).
[j] In addition to the states listed, there is a third $^2\Pi$ state and a $^2\Phi$ state. The value of $\lambda^{SO}_{calc}(iso)$ for $O_2^+ a^4\Pi_u$ calculated by Brown *et al.* (1981) takes into account the spin–orbit interaction with only the $A^2\Pi_u$ state.

2.4.4.2 Calculation of Second-Order Spin–Orbit Effects

As has already been mentioned, second-order spin–orbit effects result in matrix elements that have the same form as the spin–spin operator. Consequently, $\lambda_{\text{eff}} = \lambda^{\text{SS}} + \lambda^{\text{SO}}$, where λ^{SS} is the direct spin–spin parameter and λ^{SO} is the second-order spin–orbit contribution. The main second-order contributions are due to the nearest states; often these are the states that belong to the same configuration as the state under consideration. Moreover, as these nearby states are in general spectroscopically well characterized, it is often relatively easy to semiempirically estimate the contribution of these nearby states to observed spin–spin constants. These contributions are called *isoconfigurational* second-order spin–orbit effects. Some selected examples are given in the following (see Fig. 2.13).

2.4.4.2.1 π^2 Configuration

The π^2 configuration gives rise to three states, $^3\Sigma^-$ (the lowest state), $^1\Delta$, and $^1\Sigma^+$ states (see Section 2.2.3). Spin–orbit interaction is possible only between the $^3\Sigma_0^-$ and $^1\Sigma_0^+$ basis functions that have the same value of Ω and the same e/f symmetry. The other selection rules, $\Delta S = 1$ and $\Sigma^+ \sim \Sigma^-$, are also satisfied. Since $\Delta\Lambda = \Delta\Sigma = 0$, only the $\mathbf{l}_{zi}\mathbf{s}_{zi}$ part of the spin–orbit operator is relevant

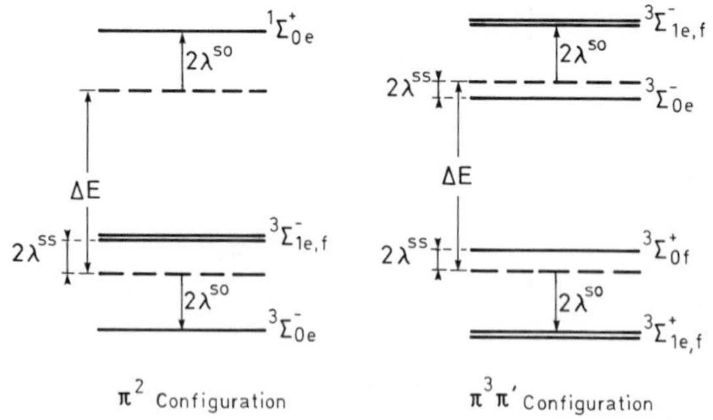

Fig. 2.13 Σ states derived from π^2 and $\pi^3\pi'$ configurations. For the π^2 states, the location of the $\Omega = 0$ levels, deperturbed with respect to the isoconfigurational spin–orbit interaction, is indicated by the dotted line. Since the deperturbed location of $^3\Sigma_0^-$ is below $^3\Sigma_1^-$, $\lambda^{\text{SS}} > 0$. For the π^2 configuration, the only Σ^+ state that can interact with $^3\Sigma^-$ always lies at higher energy, and the isoconfigurational spin–orbit contribution to λ is $\lambda^{\text{SO}} > 0$. For $\pi^3\pi'$, the dotted lines denote the deperturbed $\Omega = 1$ levels. On the figure, $\lambda^{\text{SS}}(^3\Sigma^+) = -\lambda^{\text{SS}}(^3\Sigma^-) < 0$. For $\pi^3\pi'$ there are two Σ states contributing, with opposite signs, to each λ^{SO}. The figure shows only the $^3\Sigma^- \sim {}^3\Sigma^+$ mutual contribution, which is negative for $\lambda(^3\Sigma^+)$ [and of equal magnitude and positive for $\lambda(^3\Sigma^-)$] if, as usual, $^3\Sigma^-$ lies above $^3\Sigma^+$.

and, as has been shown in Eq. (2.4.16),

$$\langle {}^3\Sigma_0^- |\mathbf{H}^{\mathrm{SO}}| {}^1\Sigma_0^+ \rangle = a_\pi \approx 2A({}^3\Pi, \sigma\pi^3).$$

Thus ${}^1\Sigma_0^+$ pushes ${}^3\Sigma_0^-$ down by

$$\frac{[\langle {}^3\Sigma_0^- |\mathbf{H}^{\mathrm{SO}}| {}^1\Sigma_0^+ \rangle]^2}{E({}^1\Sigma^+) - E({}^3\Sigma^-)} = \frac{4A^2}{E({}^1\Sigma^+) - E({}^3\Sigma^-)} = 2\lambda^{\mathrm{SO}}, \tag{2.4.35}$$

since, by convention, $E({}^3\Sigma_1) - E({}^3\Sigma_0) = 2\lambda$. In Table 4.10, values of effective spin–spin constants for ${}^3\Sigma^-$ states of π^2 configurations are collected. For molecules with one atom from beyond the first row, the second-order spin–orbit contribution always outweighs the direct spin–spin contribution.

2.4.4.2.2 $\pi^3\pi'$ (or $\pi^3\pi'^3$ and $\pi\pi'$) Configurations

The $\pi^3\pi'$, $\pi^3\pi'^3$, and $\pi\pi'$ configurations give rise to six states: ${}^3\Sigma^+$, ${}^3\Delta$, ${}^3\Sigma^-$, ${}^1\Sigma^-$, ${}^1\Delta$, and ${}^1\Sigma^+$. The ordering of these states is given in Table 2.8 in the single-configuration approximation (Recknagel, 1934). Let us consider the spin–orbit interaction between the ${}^3\Sigma^+$, which is the lowest state (of $\pi^3\pi'$), and the ${}^3\Sigma^-$ state, which lies above the ${}^3\Sigma^+$ state ($\Delta S = 0$ with $\Sigma^+ \sim \Sigma^-$). The component ${}^3\Sigma_{0f}^+$ has a different parity than the ${}^3\Sigma_{0e}^-$ component and consequently cannot interact with it, but the ${}^3\Sigma_{1e}^+$ component interacts with the ${}^3\Sigma_{1e}^-$ component and these two components repel each other (see Fig. 2.13). There is an identical interaction between the ${}^3\Sigma_{1f}^+$ and ${}^3\Sigma_{1f}^-$ components. Consequently, the λ^{SO} value of the ${}^3\Sigma^+$ state is negative, and the λ^{SO} value of the ${}^3\Sigma^-$ state is equal in magnitude but positive if only this ${}^3\Sigma^+ \sim {}^3\Sigma^-$ effect is considered. However, one must also take into account the effect of the higher-energy isoconfigurational ${}^1\Sigma_{0f}^-$ state, which repels the ${}^3\Sigma_{0f}^+$ component and thereby causes the λ_{eff} value of ${}^3\Sigma^+$ to become less negative. Similarly, the higher-energy isoconfigurational ${}^1\Sigma_{0e}^+$ state repels the ${}^3\Sigma_{0e}^-$ component and thereby increases the λ_{eff} value of the ${}^3\Sigma^-$ state. Putting the $\mathbf{H}^{\mathrm{SO}}$ interactions in terms of the a_π orbital parameter,

$$\begin{aligned} \langle {}^3\Sigma_1^+ |\mathbf{H}^{\mathrm{SO}}| {}^3\Sigma_1^- \rangle &= -\tfrac{1}{2}a_\pi - \tfrac{1}{2}a_{\pi'} \\ \langle {}^1\Sigma_0^- |\mathbf{H}^{\mathrm{SO}}| {}^3\Sigma_0^+ \rangle = \langle {}^1\Sigma_0^+ |\mathbf{H}^{\mathrm{SO}}| {}^3\Sigma_0^- \rangle &= +\tfrac{1}{2}a_\pi - \tfrac{1}{2}a_{\pi'} \end{aligned} \tag{2.4.36a}$$

where this result is valid for the $\pi^3\pi'$ configuration only. As the spin–orbit coupling constant of the isoconfigurational $\pi^3\pi'\,{}^3\Delta$ state is given by

$$2A({}^3\Delta, \pi^3\pi') = -\tfrac{1}{2}a_\pi + \tfrac{1}{2}a_{\pi'}$$

and that of the ${}^3\Pi$ state of the $\sigma\pi'$ configuration is given by

$$A({}^3\Pi, \sigma\pi') = +\tfrac{1}{2}a_{\pi'},$$

the semiempirical expressions,

$$\langle {}^3\Sigma_1^+ |\mathbf{H}^{\mathrm{SO}}| {}^3\Sigma_1^- \rangle = 2A({}^3\Delta) - 2A({}^3\Pi) = -\tfrac{1}{2}(a_\pi + a_{\pi'})$$
$$\langle {}^1\Sigma_0^\pm |\mathbf{H}^{\mathrm{SO}}| {}^3\Sigma_0^\mp \rangle = -2A({}^3\Delta) = \tfrac{1}{2}(a_\pi - a_{\pi'}),$$

may be used to estimate the isoconfigurational $\mathbf{H}^{\mathrm{SO}}$ contributions to λ_{eff} for the ${}^3\Sigma^+$ and ${}^3\Sigma^-$ states.

For valence states of homonuclear molecules, the π' orbital is the antibonding counterpart, π^*, of the bonding π orbital. One then finds that $a_{\pi_u} \simeq a_{\pi_g^*}(a_\pi \approx a_{\pi^*})$. Thus the interaction between the ${}^3\Sigma^+$ and ${}^3\Sigma^-$ states will dominate the λ^{SO} values for the $\pi^3\pi^*$ configuration and is given in Table 2.8. For example, in the P_2 molecule, it has been observed that $\lambda(a^3\Sigma_u^+) = -3.26\ \mathrm{cm}^{-1}$ and $\lambda(b^3\Sigma_u^-) = +3.20\ \mathrm{cm}^{-1}$ (Brion *et al.*, 1974, 1976). For this heavy (second-row) molecule, the direct spin–spin parameter is negligible.

For the $\pi\pi'$ (or $\pi^3\pi'^3$) configuration,

$$\begin{aligned} \langle {}^3\Sigma_1^+ |\mathbf{H}^{\mathrm{SO}}| {}^3\Sigma_1^- \rangle &= -\tfrac{1}{2}a_\pi + \tfrac{1}{2}a_{\pi'} \\ \langle {}^1\Sigma_0^\pm |\mathbf{H}^{\mathrm{SO}}| {}^3\Sigma_0^\mp \rangle &= -\tfrac{1}{2}a_\pi - \tfrac{1}{2}a_{\pi'}. \end{aligned} \qquad (2.4.36b)$$

Therefore, in valence states of homonuclear molecules belonging to the $\pi\pi^*$ (or $\pi^3\pi^{*3}$) configuration, the λ^{SO} effect from the ${}^1\Sigma^-$ state is dominant. The ${}^1\Sigma^+$ state of this configuration is generally very high-lying, and its effect has been neglected in Table 2.8. For more details, see Field and Lefebvre-Brion (1974). For Rydberg states, where π' is a Rydberg orbital, $a_{\pi'}$ is in general negligible relative to a_π for a valence π orbital.

2.4.4.3 Off-Diagonal Matrix Elements

Off-diagonal matrix elements, especially between states belonging to two different configurations, can give rise to perturbations. Such an example is the perturbation of the $B^2\Sigma^+$ state of CN by a ${}^4\Sigma^+$ state, which causes a barely detectable shift of about $0.02\ \mathrm{cm}^{-1}$ (Coxon *et al.*, 1975; Miller *et al.*, 1976). This may be due to a direct $\mathbf{H}^{\mathrm{SS}}$ effect or, more plausibly, to a second-order spin–orbit effect. For example, a second-order interaction via a ${}^2\Sigma^-$ state would have the form

$$\langle {}^2\Sigma^+ |\mathbf{H}^{\mathrm{SO}}| {}^4\Sigma^+ \rangle = \frac{\langle {}^2\Sigma^+ |\mathbf{H}^{\mathrm{SO}}| {}^2\Sigma^- \rangle \langle {}^2\Sigma^- |\mathbf{H}^{\mathrm{SO}}| {}^4\Sigma^+ \rangle}{E - E({}^2\Sigma^-)}$$

where E is the average energy of the interacting ${}^2\Sigma^+$ and ${}^4\Sigma^+$ levels. Recall that for this type of indirect or second-order interaction, the selection rule is $\Delta S = 0, \pm 1, \pm 2$. Another example of a very weak perturbation due to $\mathbf{H}^{\mathrm{SS}}$ has been detected between the $a'^3\Sigma^+$ and $a^3\Pi$ states of CO (Effantin *et al.*, 1982). This interaction can be distinguished from a direct $\mathbf{H}^{\mathrm{SO}}$ perturbation because the

$\Delta\Omega = 0$ and $\Omega = 1$ matrix elements have, respectively, opposite and equal signs for the spin–spin interaction and positive for the spin–orbit interaction (see Section 2.4.5).

Similarly to the $^2\Sigma^+$–$^4\Sigma^+$ interaction, a spin–spin interaction is possible between $^5\Sigma$ and $^1\Pi$ states as assumed, for example, in NH by Smith *et al.* (1976). For the spin–orbit operator, recall again the $\Delta S \leq 1$ selection rule; however, second-order spin–orbit effects can mix $^5\Sigma$ and $^1\Pi$ states as follows:

$$\langle {}^5\Sigma_1 |\mathbf{H}^{\mathrm{SO}}| {}^1\Pi_1\rangle = \frac{\langle {}^5\Sigma_1 |\mathbf{H}^{\mathrm{SO}}| {}^3\Pi_1\rangle \langle {}^3\Pi_1 |\mathbf{H}^{\mathrm{SO}}| {}^1\Pi_1\rangle}{E - E({}^3\Pi)},$$

where E is the average energy of the interacting $^5\Sigma$ and $^1\Pi$ levels.

2.4.5 TENSORIAL OPERATORS

Energy, which is the observable quantity associated with the Hamiltonian operator, is a pure number or, more precisely, a scalar quantity. Therefore, the Hamiltonian must be a scalar operator. In this section, the prescriptions for constructing a scalar operator from combinations of more complicated operators, such as vector angular momenta, and for evaluating matrix elements of these composite scalar operators are reviewed briefly.

Some operators, such as the interelectronic electrostatic interaction e^2/r_{ij}, are obviously scalar quantities. Others are scalar products of two tensorial operators. A tensor of rank zero is a scalar. A tensor of rank one is a vector. There are several ways of combining two vector operators: the scalar product

$$\vec{\mathbf{U}} \cdot \vec{\mathbf{V}} = \mathbf{U}_x\mathbf{V}_x + \mathbf{U}_y\mathbf{V}_y + \mathbf{U}_z\mathbf{V}_z \tag{2.4.37}$$

yields a scalar quantity, whereas the vector product

$$\vec{\mathbf{U}} \times \vec{\mathbf{V}} = (\mathbf{U}_y\mathbf{V}_z - \mathbf{U}_z\mathbf{V}_y)\hat{x} + (\mathbf{U}_z\mathbf{V}_x - \mathbf{U}_x\mathbf{V}_z)\hat{y} + (\mathbf{U}_x\mathbf{V}_y - \mathbf{U}_y\mathbf{V}_x)\hat{z} \tag{2.4.38}$$

yields a vector quantity. These schemes can be generalized to combinations of tensors of any rank yielding composite tensors of any rank. The reason for constructing operators of well-defined tensorial character is that the evaluation of matrix elements of such operators is tremendously simplified.

The general expression for the scalar product of two tensorial operators of rank k is

$$T_0^{(0)}(\vec{\mathbf{U}}, \vec{\mathbf{V}}) = \sum_{q=-1}^{+1} (-1)^q \mathbf{U}_q^{(k)} \mathbf{V}_{-q}^{(k)}, \tag{2.4.39}$$

where q is the *spherical* tensor (as opposed to Cartesian) component of the operator. For a vector operator, the relationship between spherical and Cartesian components is, as exemplified for $\vec{r}$, the position vector of the electron (e.g., the transition moment operator, $e\vec{r}$),

$$T_0^{(1)}(\vec{r}) = z = r\cos\theta \tag{2.4.40a}$$

$$T_{+1}^{(1)}(\vec{r}) = -(2)^{-1/2}(x + iy) = -r(\sin\theta)e^{i\phi} \tag{2.4.40b}$$

$$T_{-1}^{(1)}(\vec{r}) = (2)^{-1/2}(x - iy) = r(\sin\theta)e^{-i\phi}, \tag{2.4.40c}$$

or, for $\vec{\mathbf{S}}$

$$T_0^{(1)}(\vec{\mathbf{S}}) = \mathbf{S}_z \tag{2.4.41a}$$

$$T_{+1}^{(1)}(\vec{\mathbf{S}}) = -(2)^{-1/2}\mathbf{S}^+ \tag{2.4.41b}$$

$$T_{-1}^{(1)}(\vec{\mathbf{S}}) = +(2)^{-1/2}\mathbf{S}^-. \tag{2.4.41c}$$

The power of the Wigner–Eckart theorem (Messiah, 1960, p. 489; Edmonds, 1974, p. 75) is that it relates one nonzero matrix element to another, thereby vastly reducing the number of integrals that must either be explicitly evaluated or treated as a variable parameter in a least-squares fit to spectral data. For example, consider $\mathbf{S}^{(k)}$, a tensor operator of rank k that acts exclusively on spin variables. The Wigner–Eckart theorem requires

$$\langle S, \Sigma|\mathbf{S}_q^{(k)}|S', \Sigma'\rangle = (-1)^{S-\Sigma}\begin{pmatrix} S & k & S' \\ -\Sigma & q & \Sigma' \end{pmatrix}\langle S\|\mathbf{S}^{(k)}\|S'\rangle, \tag{2.4.42}$$

where $\begin{pmatrix} S & k & S' \\ -\Sigma & q & \Sigma' \end{pmatrix}$ is a $3j$-coefficient (introduced in Section 2.2.1) and $\langle S\|\mathbf{S}^{(k)}\|S'\rangle$ is a *reduced matrix element* that depends on neither Σ', Σ, nor q.

If a matrix element is evaluated for a particular value of Σ and Σ' by another method (see Section 2.4.1), for example for the maximum value of $\Sigma = S$ and $\Sigma' = S'$, the $\langle S\Sigma''|\mathbf{S}_q^{(k)}|S'\Sigma'''\rangle$ matrix elements for all values of Σ'' and Σ''' are related to the first matrix element by the ratio of two $3j$-coefficients,

$$\frac{\langle S, S|\mathbf{S}_q^{(k)}|S', S'\rangle}{\langle S, \Sigma''|\mathbf{S}_q^{(k)}|S', \Sigma'''\rangle} = (-1)^{(S-S)-(S-\Sigma'')}\begin{pmatrix} S & k & S' \\ -S & q & S' \end{pmatrix}\bigg/\begin{pmatrix} S & k & S' \\ -\Sigma'' & q & \Sigma''' \end{pmatrix}. \tag{2.4.43}$$

Equation (2.4.43) illustrates one use of the Wigner–Eckart theorem, the expression of matrix element ratios as simple ratios of $3j$-coefficients. Another important application is that selection rules for nonzero matrix elements of any operator are immediately evident. The $3j$-coefficient $\begin{pmatrix} J_1 & J & J_2 \\ M_1 & M & M_2 \end{pmatrix}$ can be nonzero when

1. $M_1 = M = M_2 = 0$, provided that $J_1 + J + J_2$ is even.

2. The triangle rule is satisfied,

$$|J_1 - J_2| \leq J \leq J_1 + J_2,$$

which means that it must be possible to construct a triangle with sides of length J_1, J, J_2.

3. $M_1 + M + M_2 = 0$. This rule can be understood by making the correspondences

$$J \to l, \qquad M \to m$$

$$|J, M\rangle \to |l, m\rangle = Y_{lm}$$

$$\langle lm| = Y^*_{lm} = (-1)^{-m} Y_{l-m}$$

$$Y_{lm}(\theta, \phi) = (-1)^m \left[\frac{(2l+1)(l-m)!}{4\pi(l+m)!}\right]^{1/2} P_l^m(\cos\theta)\, e^{im\phi}$$

$$T_m^{(l)}(r, \theta, \phi) = f(r)\, Y_{lm}(\theta, \phi).$$

Then the matrix element, $\langle l'm' | T_m^{(l)} | l''m''\rangle$, may be expressed as the $r^2 \sin\theta\, dr\, d\theta\, d\phi$ volume integral,

$$\langle l'm' | T_m^{(l)} | l''m''\rangle = K \int e^{i(m-m'+m'')\phi}\, d\phi \times \int r^2 f(r)\, dr \int P_{l'}^{-m'} P_l^m P_{l''}^{m''} \sin\theta\, d\theta,$$

where K is a constant. This form of the integral optimally displays the Δm selection rules, namely, that the $d\phi$ integral will vanish unless

$$m - m' + m'' = 0,$$

which is identical to the nonvanishing condition for the $3j$-coefficients of Eq. (2.4.42). Now, using Edmonds' result (1974, page 63) for the $\sin\theta\, d\theta\, d\phi$ integral over the product of three spherical harmonics, one obtains

$$\langle l'm' | T_m^{(l)} | l''m''\rangle = (-1)^{l'-m'} \begin{pmatrix} l' & l & l'' \\ -m' & m & m'' \end{pmatrix} \langle l' \| T^{(l)} \| l''\rangle, \quad (2.4.44a)$$

where the reduced matrix element

$$\langle l' \| T^{(l)} \| l''\rangle$$
$$= (-1)^{-l'} \left[\frac{(2l'+1)(2l+1)(2l''+1)}{4\pi}\right]^{1/2} \begin{pmatrix} l' & l & l'' \\ 0 & 0 & 0 \end{pmatrix} \int r^2 f(r)\, dr \quad (2.4.44b)$$

is a radial integral times some purely l, l', l''-dependent factors. Note that the electric dipole transition moment operator for atoms transforms as $T_m^{(1)}$; thus Eq. (2.4.44a) requires that $\Delta l = 0, \pm 1$ and $\Delta m = 0, \pm 1$, but Eq. (2.4.44b) forbids $\Delta l = 0$. Similarly, for electric quadrupole transitions, $T_m^{(2)}$ permits $\Delta l = 0, \pm 2$ and $\Delta m = 0, \pm 1, \pm 2$.

The $3j$-coefficients also identify groups of matrix elements that have identical magnitudes. Even or cyclic permutations,

$$\begin{pmatrix} J_1 & J & J_2 \\ M_1 & M & M_2 \end{pmatrix} \to \begin{pmatrix} J_2 & J_1 & J \\ M_2 & M_1 & M \end{pmatrix},$$

of the columns of the $3j$-symbol leave its value unchanged. Odd permutations of the columns or a sign reversal of all lower row entries are accounted for by the phase factor $(-1)^{J_1+J+J_2}$.

Consider the spin–orbit operator,

$$\mathbf{H}^{\mathrm{SO}} = \sum_i \hat{a}_i \mathbf{l}_i \cdot \mathbf{s}_i,$$

which is a sum of scalar operators, each of which is a scalar product of two vector operators. Thus,

$$\mathbf{H}^{\mathrm{SO}} = \sum_q (-1)^q \sum_{i=1}^{N} \hat{a}(i)\, \mathbf{l}_q(i)\, \mathbf{s}_q(i).$$

Matrix elements between two N-electron wavefunctions can be factored into spin and spatial (including all information about orbital angular momenta) parts. It has been shown frequently that the Wigner–Eckart theorem can be applied as follows (Langhoff and Kern, 1977; McWeeny, 1965; Cooper and Musher, 1972):

$$\langle S, \Sigma, \Lambda, \Omega | \mathbf{H}^{\mathrm{SO}} | S', \Sigma', \Lambda', \Omega \rangle$$
$$= (-1)^{S-\Sigma} \begin{pmatrix} S & 1 & S' \\ -\Sigma & q & \Sigma' \end{pmatrix} \langle S, \Lambda, \Omega \| \mathbf{H}^{\mathrm{SO}} \| S', \Lambda', \Omega \rangle, \quad (2.4.45)$$

where the reduced matrix element is reduced with respect to Σ but not with respect to Λ and Ω (which involve spatial rather than spin coordinates). From the $3j$-coefficient, the selection rules

$$q = \Sigma - \Sigma'$$

$$|S - 1| \leq S' \leq S + 1 \qquad \text{or} \quad \Delta S = 0, \pm 1,$$

stated without proof in Section 2.4.1, are verified. For $q = 0$ (diagonal and off-diagonal elements of the $\mathbf{l}_z \mathbf{s}_z$ part of $\mathbf{H}^{\mathrm{SO}}$),

$$\Sigma' = \Sigma \qquad \text{because} \quad \Sigma' - q - \Sigma = 0,$$

and for $q = \pm 1$,

$$\Sigma' = \Sigma \pm 1.$$

Note also that, for $S' = S$ and $\Sigma' = \Sigma$,

$$\langle \Lambda S \Sigma | \mathbf{H}^{\mathrm{SO}} | \Lambda S \Sigma \rangle = (-1)^{S-\Sigma} \begin{pmatrix} S & 1 & S \\ -\Sigma & 0 & \Sigma \end{pmatrix} \langle \Lambda S \| \mathbf{H}^{\mathrm{SO}} \| \Lambda S \rangle$$

where $\begin{pmatrix} S & 1 & S \\ -\Sigma & 0 & \Sigma \end{pmatrix}$ is proportional to Σ, which is consistent with Eq. (2.4.6).

It is now possible to examine a specific example, the spin–orbit part of the $^4\Sigma^- \sim {}^4\Pi$ interaction. The $\mathbf{H}^{\mathrm{SO}}$ operator has three types of $\Delta\Omega = 0$ matrix elements between $|\Lambda = 1\,{}^4\Pi\rangle$ and $|{}^4\Sigma^-\rangle$ basis functions:

$$|{}^4\Pi_{3/2}\rangle = |\Lambda = 1, \Sigma = \tfrac{1}{2}\rangle \sim |{}^4\Sigma^-_{3/2}\rangle = |\Lambda = 0, \Sigma = \tfrac{3}{2}\rangle$$

$$|{}^4\Pi_{1/2}\rangle = |\Lambda = 1, \Sigma = -\tfrac{1}{2}\rangle \sim |{}^4\Sigma^-_{1/2}\rangle = |\Lambda = 0, \Sigma = \tfrac{1}{2}\rangle$$

$$|{}^4\Pi_{-1/2}\rangle = |\Lambda = 1, \Sigma = -\tfrac{3}{2}\rangle \sim |{}^4\Sigma^-_{-1/2}\rangle = |\Lambda = 0, \Sigma = -\tfrac{1}{2}\rangle.$$

(There are also the same three types of $|\Lambda = -1\,{}^4\Pi\rangle \sim |{}^4\Sigma^-\rangle$ matrix elements.) In parameterizing the $^4\Pi \sim {}^4\Sigma^-$ matrix, it would be desirable to represent the three types of spin–orbit interactions by a single perturbation parameter. Applying Eq. (2.4.45) with $S = S' = \frac{3}{2}$,

$$\begin{aligned}
\langle {}^4\Sigma^-_{3/2} | \mathbf{H}^{\mathrm{SO}} | {}^4\Pi_{3/2} \rangle &= \langle S = \tfrac{3}{2} \Sigma = \tfrac{3}{2} | \mathbf{H}^{\mathrm{SO}} | S = \tfrac{3}{2} \Sigma = \tfrac{1}{2} \rangle \\
&= (-1)^{(3/2)-(3/2)} \begin{pmatrix} \frac{3}{2} & 1 & \frac{3}{2} \\ -\frac{3}{2} & 1 & \frac{1}{2} \end{pmatrix} \langle {}^4\Sigma^- \| \mathbf{H}^{\mathrm{SO}} \| {}^4\Pi \rangle \\
&= (-1)^0 \times [-(10)^{-1/2}] \langle {}^4\Sigma^- \| \mathbf{H}^{\mathrm{SO}} \| {}^4\Pi \rangle \\
\langle {}^4\Sigma^-_{1/2} | \mathbf{H}^{\mathrm{SO}} | {}^4\Pi_{1/2} \rangle &= \langle \tfrac{3}{2} \tfrac{1}{2} | \mathbf{H}^{\mathrm{SO}} | \tfrac{3}{2} - \tfrac{1}{2} \rangle \\
&= (-1)^{(3/2)-(1/2)} \begin{pmatrix} \frac{3}{2} & 1 & \frac{3}{2} \\ -\frac{1}{2} & 1 & -\frac{1}{2} \end{pmatrix} \langle {}^4\Sigma^- \| \mathbf{H}^{\mathrm{SO}} \| {}^4\Pi \rangle \\
&= (-1) \times (\tfrac{2}{15})^{1/2} \langle {}^4\Sigma^- \| \mathbf{H}^{\mathrm{SO}} \| {}^4\Pi \rangle \\
\langle {}^4\Sigma^-_{-1/2} | \mathbf{H}^{\mathrm{SO}} | {}^4\Pi_{-1/2} \rangle &= \langle \tfrac{3}{2} - \tfrac{1}{2} | \mathbf{H}^{\mathrm{SO}} | \tfrac{3}{2} - \tfrac{3}{2} \rangle \\
&= (-1)^{(3/2)+(1/2)} \begin{pmatrix} \frac{3}{2} & 1 & \frac{3}{2} \\ \frac{1}{2} & 1 & -\frac{3}{2} \end{pmatrix} \langle {}^4\Sigma^- \| \mathbf{H}^{\mathrm{SO}} \| {}^4\Pi \rangle \\
&= (-1)^2 [-(10)^{-1/2}] \langle {}^4\Sigma^- \| \mathbf{H}^{\mathrm{SO}} \| {}^4\Pi \rangle
\end{aligned}$$

$$\frac{\langle {}^4\Sigma^-_{3/2} | \mathbf{H}^{\mathrm{SO}} | {}^4\Pi_{3/2} \rangle}{\langle {}^4\Sigma^-_{1/2} | \mathbf{H}^{\mathrm{SO}} | {}^4\Pi_{1/2} \rangle} = (3)^{1/2}/2$$

$$\frac{\langle {}^4\Sigma^-_{3/2} | \mathbf{H}^{\mathrm{SO}} | {}^4\Pi_{3/2} \rangle}{\langle {}^4\Sigma^-_{-1/2} | \mathbf{H}^{\mathrm{SO}} | {}^4\Pi_{-1/2} \rangle} = 1.$$

The $^4\Pi \sim {}^4\Sigma^-$ example involved matrix elements evaluated using individual $|\Omega\Lambda S\Sigma\rangle$ basis functions, in which the sign of $\Omega = \Lambda + \Sigma$ is specified, rather than symmetrized,

$$(2)^{-1/2}[|\Omega\Lambda\Sigma\rangle \pm |-\Omega - \Lambda - \Sigma\rangle]$$

e/f linear combinations, in which the range of Ω is $S - |\Lambda| \leq \Omega \leq S + |\Lambda|$. Off-diagonal case (a) matrix elements in these two basis sets are identical, but with one important exception for odd-multiplicity systems. For odd-multiplicity systems (even number of electrons, integral S), each Ω-component is always doubly degenerate, except for $\Lambda = \Omega = 0$ which is nondegenerate. Matrix elements between nondegenerate $\Omega = 0$ and degenerate Ω basis functions in the signed-Ω basis, for example,

$$|^3\Pi_0\rangle = |\Lambda = 1, \Sigma = -1\rangle$$

$$|^3\Sigma_0^+\rangle = |\Lambda = 0, \Sigma = 0\rangle$$

$$\langle ^3\Pi_0 | \mathbf{H}^{\mathrm{SO}} | ^3\Sigma_0^+\rangle = (2)^{1/2}\alpha \tag{2.4.46}$$

(α is a reduced matrix element), are always a factor of $(2)^{1/2}$ smaller than the corresponding quantity in the symmetrized basis,

$$\langle ^3\Pi_{0f} | \mathbf{H}^{\mathrm{SO}} | ^3\Sigma_{0f}^+\rangle = 2\alpha. \tag{2.4.47}$$

The spin–spin operator is a scalar product of two tensorial operators of rank 2, one of which operates on spin coordinates only, the other (spherical harmonics of rank 2) of which affects only spatial coordinates. $\mathbf{H}^{\mathrm{SS}}$ can be factored into its spin and spatial parts, and the Wigner–Eckart theorem can be used to evaluate matrix elements of the spin part. Considering only diagonal matrix elements,

$$\langle \Lambda, S, \Sigma | \mathbf{H}^{\mathrm{SS}} | \Lambda, S, \Sigma\rangle = (-1)^{S-\Sigma}\begin{pmatrix} S & 2 & S \\ -\Sigma & 0 & \Sigma \end{pmatrix}\langle \Lambda, S \| \mathbf{H}^{\mathrm{SS}} \| \Lambda, S\rangle, \tag{2.4.48}$$

the $\begin{pmatrix} S & 2 & S \\ -\Sigma & 0 & \Sigma \end{pmatrix}$ $3j$-coefficient is proportional to $3\Sigma^2 - S(S+1)$, which is consistent with Eq. (2.4.31). ΔS and $\Delta\Sigma$ selection rules for nonzero matrix elements of $\mathbf{H}^{\mathrm{SS}}$ are obtained from the triangle rule,

$$S - 2 \leq S' \leq S + 2$$

$$\Delta S = 0, \pm 1, \pm 2,$$

and, for

$$q = 0 \qquad \Sigma' = \Sigma \qquad \Delta\Sigma = 0 \qquad (\text{and} \quad \Delta\Lambda = 0 \quad \text{because} \quad \Delta\Omega = 0)$$

$$q = \pm 1 \qquad \Sigma' = \Sigma \pm 1 \qquad \Delta\Sigma = \pm 1 = -\Delta\Lambda$$

$$q = \pm 2 \qquad \Sigma' = \Sigma \pm 2 \qquad \Delta\Sigma = \pm 2 = -\Delta\Lambda.$$

For the special case where $S' = S + 1$, the sum $S + 2 + S'$ is odd; consequently, the diagonal $\Sigma = \Sigma' = 0$ matrix element is zero. For example,

$$\langle {}^1\Sigma_0^+ | \mathbf{H}^{\mathrm{SS}} | {}^3\Sigma_0^+ \rangle = 0.$$

2.5 Rotational Perturbations

There are three terms in the rotational part of the Hamiltonian, $\mathbf{H}^{\mathrm{ROT}}$, that are neglected in the Born–Oppenheimer approximation:

1. $(1/2\mu R^2)(\mathbf{L}^+\mathbf{S}^- + \mathbf{L}^-\mathbf{S}^+)$, which causes spin–electronic homogeneous ($\Delta\Omega = 0$) perturbations.
2. $-(1/2\mu R^2)(\mathbf{J}^+\mathbf{S}^- + \mathbf{J}^-\mathbf{S}^+)$, which is the **S**-uncoupling operator.
3. $-(1/2\mu R^2)(\mathbf{J}^+\mathbf{L}^- + \mathbf{J}^-\mathbf{L}^+)$, which is the **L**-uncoupling operator.

The last two terms give rise to heterogeneous ($\Delta\Omega = \pm 1$) perturbations.

2.5.1 SPIN–ELECTRONIC HOMOGENEOUS PERTURBATIONS

For the off-diagonal elements of the $(1/2\mu R^2)(\mathbf{L}^+\mathbf{S}^- + \mathbf{L}^-\mathbf{S}^+)$ operator, $\mathbf{H}^{\mathrm{SE}}$, the selection rules are the same ($\Delta\Omega = 0$, $\Delta\Lambda = -\Delta\Sigma = \pm 1$) as for the part of the spin–orbit operator,

$$\frac{1}{2}\sum_i \hat{a}_i(\mathbf{l}_i^+\mathbf{s}_i^- + \mathbf{l}_i^-\mathbf{s}_i^+),$$

except that $\Delta S = 0$ only for $\mathbf{H}^{\mathrm{SE}}$. An important but subtle difference between these two operators is that

$$\mathbf{L}^+\mathbf{S}^- = \sum_i \mathbf{l}_i^+ \sum_j \mathbf{s}_j^- = \sum_i \mathbf{l}_i^+\mathbf{s}_i^- + \sum_{i<j} \mathbf{l}_i^+\mathbf{s}_j^-. \qquad (2.5.1)$$

In contrast, the spin–orbit operator only includes a summation $\sum_i \mathbf{l}_i^+\mathbf{s}_i^-$ involving the same electron for both $\mathbf{l}^+$ and $\mathbf{s}^-$. The terms in $\mathbf{H}^{\mathrm{SE}}$, $\sum_{i<j} \mathbf{l}_i^+\mathbf{s}_j^-$,

are two-electron operators and can cause nonzero matrix elements between wavefunctions that differ by two spin–orbitals.

Moreover, the operator $\mathbf{B} = \hbar^2/2\mu R^2$ acts on the vibrational part of the wavefunction,

$$\langle v|\mathbf{B}|v'\rangle = \langle v|\hbar^2/2\mu R^2|v'\rangle. \tag{2.5.2}$$

$\langle v|\hbar^2/2\mu R^2|v'\rangle$ can be approximated by the value $(\hbar^2/2\mu R_C^2)\langle v|v'\rangle$, where R_C is the crossing point of the two electronic potential curves. If the two interacting electronic states have identical potential curves (a prerequisite for pure precession), then

$$B_{vv'} = \langle v|\mathbf{B}|v'\rangle = B_v\,\delta_{vv'}.$$

When the two potential curves are well-characterized, the $B_{vv'}$ matrix elements may be calculated numerically. For *nearly* identical potentials, the $B_{vv'}$ matrix elements behave much less like $\delta_{vv'}$ functions than do $\langle v|v'\rangle$ vibrational overlaps.

As an example, consider the interaction between the $^3\Pi_1$ state ($\sigma\pi^4\pi'$ configuration) and the $^3\Sigma_1^+$ state ($\sigma^2\pi^3\pi'$ configuration). For $\mathbf{BL}\cdot\mathbf{S}$, as for the $\mathbf{H}^{\mathrm{SO}}$ operator, the interacting basis wavefunctions must have the same value of Ω. For this example, the $\Omega = \pm 1$ matrix elements of the $\mathbf{L}^+\mathbf{S}^-$ part of $\mathbf{H}^{\mathrm{SE}}$ will be evaluated.

Let $(\pi^+\alpha\,\pi^+\beta\,\pi^-\alpha\,\pi^-\beta\,\sigma\alpha\,\sigma\beta\,\pi'^+\alpha\,\pi'^+\beta\,\pi'^-\alpha\,\pi'^-\beta)$ be the standard reference order of spin-orbitals,

$$|^3\Pi_1\rangle = 2^{-1/2}\{|\pi^+\alpha\,\pi^+\beta\,\pi^-\alpha\,\pi^-\beta\,\sigma\alpha\,\pi'^+\beta| - |\pi^+\alpha\,\underline{\pi^-\beta}\,\pi^-\alpha\,\underline{\pi^+\beta}\,\sigma\beta\,\pi'^+\alpha|\}$$
$$= 2^{-1/2}[\det 1 - \det 2]$$
$$|^3\Sigma_1^+\rangle = 2^{-1/2}\{-|\pi^+\alpha\,\pi^+\beta\,\pi^-\alpha\,\underline{\sigma\beta}\,\underline{\sigma\alpha}\,\pi'^-\alpha| - |\pi^+\alpha\,\underline{\pi^-\beta}\,\underline{\pi^-\alpha}\,\sigma\alpha\,\sigma\beta\,\pi'^+\alpha|\}$$
$$= 2^{-1/2}[-\det 3 - \det 4]$$

where the underlined spin-orbitals have been permuted out of standard order. Note that determinants 1 and 3 differ by orbitals at the fourth $(\pi^-\beta \to \sigma\beta)$ and sixth $(\pi'^+\beta \to \pi'^-\alpha)$ positions, 2 and 4 differ only at the fourth $(\pi^+\beta \to \sigma\alpha)$ position, 1 and 4 differ at the sixth position $(\pi'^+\beta \to \pi'^+\alpha)$ and by $(\pi^+\beta \to \sigma\beta)$, which can be brought into registration by an even number of permutations, and 2 and 3 differ at the sixth position $(\pi'^+\alpha \to \pi'^-\alpha)$ and by $(\pi^-\beta \to \sigma\alpha)$. The matrix elements between determinants 1 and 3 and determinants 2 and 3 are zero because $\mathbf{l}^\pm$ cannot change π'^+ to π'^-. The remaining combinations are

$$\left\langle -\det 2\left|\sum_{i,j}\mathbf{l}_i^+\mathbf{s}_j^-\right|-\det 4\right\rangle$$
$$= \langle\pi^+\beta|\mathbf{l}_1^+\mathbf{s}_1^-|\sigma\alpha\rangle - \langle\pi^+\alpha\pi^+\beta|\mathbf{l}_1^+\mathbf{s}_2^-|\sigma\alpha\pi^+\alpha\rangle - \langle\sigma\beta\pi^+\beta|\mathbf{l}_2^+\mathbf{s}_1^-|\sigma\alpha\sigma\beta\rangle$$

$$= (1 - 1 - 1)\langle \pi^+ |\mathbf{l}^+| \sigma \rangle \equiv -b$$

$$\left\langle \det 1 \left| \sum_{i,j} \mathbf{l}_i^+ \mathbf{s}_j^- \right| -\det 4 \right\rangle = -\langle \pi'^+ \beta |\mathbf{s}^-| \pi'^+ \alpha \rangle \langle \pi^+ \beta |\mathbf{l}^+| \sigma \beta \rangle$$

$$= -\langle \pi^+ |\mathbf{l}^+| \sigma \rangle \equiv -b$$

thus

$$\langle {}^3\Pi_{1^e_f,v} |\mathbf{H}^{\mathrm{SE}}| {}^3\Sigma^+_{1^e_f,v'} \rangle = \langle {}^3\Pi_1, \Omega = +1, v |\mathbf{BL}^+\mathbf{S}^-| {}^3\Sigma_1^+, \Omega = +1, v' \rangle$$

$$= (2)^{-1/2}(2)^{-1/2}(-b - b)\langle v |\mathbf{B}| v' \rangle = -bB_{vv'}. \quad (2.5.3)$$

The ${}^3\Pi_1 \sim {}^3\Sigma_1^+$ matrix element of $\mathbf{H}^{\mathrm{SO}}$ is much easier to evaluate because $\mathbf{H}^{\mathrm{SO}}$ is a one-electron operator,

$$\langle {}^3\Pi_{1^e_f,v} |\mathbf{H}^{\mathrm{SO}}| {}^3\Sigma^+_{1^e_f,v'} \rangle = \tfrac{1}{4} a_+ \langle v | v' \rangle. \quad (2.5.4)$$

For light molecules, the $\mathbf{H}^{\mathrm{SE}}$ contribution from Eq. (2.5.3) cannot be separated from the much larger $\mathbf{H}^{\mathrm{SO}}$ contribution (Eq. 2.5.4) except by isotope effects and, much more importantly, by the fact that other fine-structure components of ${}^3\Sigma^+$ and ${}^3\Pi$ interact with different J-dependences for the b parameter (see Table 4.8). For example, in the CO molecule, the interaction matrix elements between $a\,{}^3\Pi_1 v = 4$ and $a'\,{}^3\Sigma_1^+ v' = 0$ can be separated into

$$\langle {}^3\Pi_1, v |\mathbf{H}^{\mathrm{SO}}| {}^3\Sigma_1^+, v' \rangle = 10.57 \text{ cm}^{-1}$$

and

$$\langle {}^3\Pi_1, v |\mathbf{H}^{\mathrm{SE}}| {}^3\Sigma_1^+, v' \rangle = -0.1158 \text{ cm}^{-1}$$

(Field *et al.*, 1972).

2.5.2 THE S-UNCOUPLING OPERATOR

The **S**-uncoupling operator, $-(1/2\mu R^2)(\mathbf{J}^+\mathbf{S}^- + \mathbf{J}^-\mathbf{S}^+)$, has the selection rules $\Delta S = 0$, $\Delta\Omega = \Delta\Sigma = \pm 1$, and $\Delta\Lambda = 0$. Generally, this operator mixes different components of the same multiplet electronic state. It is responsible for the transition, as J increases, from Hund's case (a) to case (b).

When the **S**-uncoupling operator acts between two components of a multiplet state that belong to the same vibrational quantum number, then the vibrational part of the $\mathbf{BJ} \cdot \mathbf{S}$ matrix element is

$$\langle v |\mathbf{B}| v \rangle \equiv B_v.$$

However, if by chance there is a near degeneracy between the $\boldsymbol{\Omega}$ spin-component of the vth level and the $\boldsymbol{\Omega}' = \boldsymbol{\Omega} \pm 1$ components of the $(v+1)$th level of the same electronic state, then the **S**-uncoupling operator can cause a perturbation between these levels. In the harmonic approximation and using the phase choice that all vibrational wavefunctions are positive at the inner turning point,

$$B_{vv'} = \langle v|\hbar^2/2\mu R^2|v+1\rangle = 2\left(\frac{B_e^3}{\omega_e}\right)^{1/2}(v+1)^{1/2}. \qquad (2.5.5)$$

Note that nonzero $\Delta v \neq 0$ matrix elements within an electronic state require an operator that is R-dependent. The total interaction from the $-2\mathbf{BJ}\cdot\mathbf{S}$ term is then given by

$$\begin{aligned}\langle {}^{2S+1}\Lambda_{\Omega,v}|&-\mathbf{B}(\mathbf{J}^+\mathbf{S}^- + \mathbf{J}^-\mathbf{S}^+)|{}^{2S+1}\Lambda_{\Omega+1,v+1}\rangle \\ &= -2\left(\frac{B_e^3}{\omega_e}\right)^{1/2}(v+1)^{1/2}[J(J+1)-(\Omega+1)\Omega]^{1/2} \\ &\quad\times [S(S+1)-(\Omega-\Lambda+1)(\Omega-\Lambda)]^{1/2} \qquad (2.5.6)\end{aligned}$$

with the same phase convention as given in Eq. (2.4.28) for the $\mathbf{S}^{\pm}$ operator. The rotational levels will cross if the spin splitting between the spin substates of one vibrational level is slightly smaller than the vibrational interval. This can occur for heavy molecules.

An example of such a perturbation due to this **S**-uncoupling operator has been observed by Jenouvrier *et al.*, (1973) in the ground electronic state of the NSe molecule, between the $v = 3$ level of the $X^2\Pi_{1/2}$ substate and the $v = 2$ level of the $X^2\Pi_{3/2}$ substate. The position of the levels is shown in Fig. 2.14. Figure 2.15 displays an anomaly in the $X^2\Pi_{3/2}$ substate Λ-doubling, which has borrowed its large value from the $X^2\Pi_{1/2}$ substate. Expression (2.5.6) accounts very well for all perturbation effects in this case. However, the actual potential curves are anharmonic, and matrix elements of **B** are small but nonzero for $|\Delta v| > 1$. A $\Delta v = 2$ perturbation of this type has been reported in the HgH molecule by Veseth (1972).

2.5.3 THE L–UNCOUPLING OPERATOR

The **L**-uncoupling operator, $-(1/2\mu R^2)(\mathbf{J}^+\mathbf{L}^- + \mathbf{J}^-\mathbf{L}^+)$, which is responsible for the transition as J increases from Hund's case (a) to case (d), causes numerous perturbations between states that differ by $\Delta\Omega = \Delta\Lambda = \pm 1$ and with $\Delta S = 0$. This specific type of rotational perturbation is often called a *gyroscopic perturbation.*

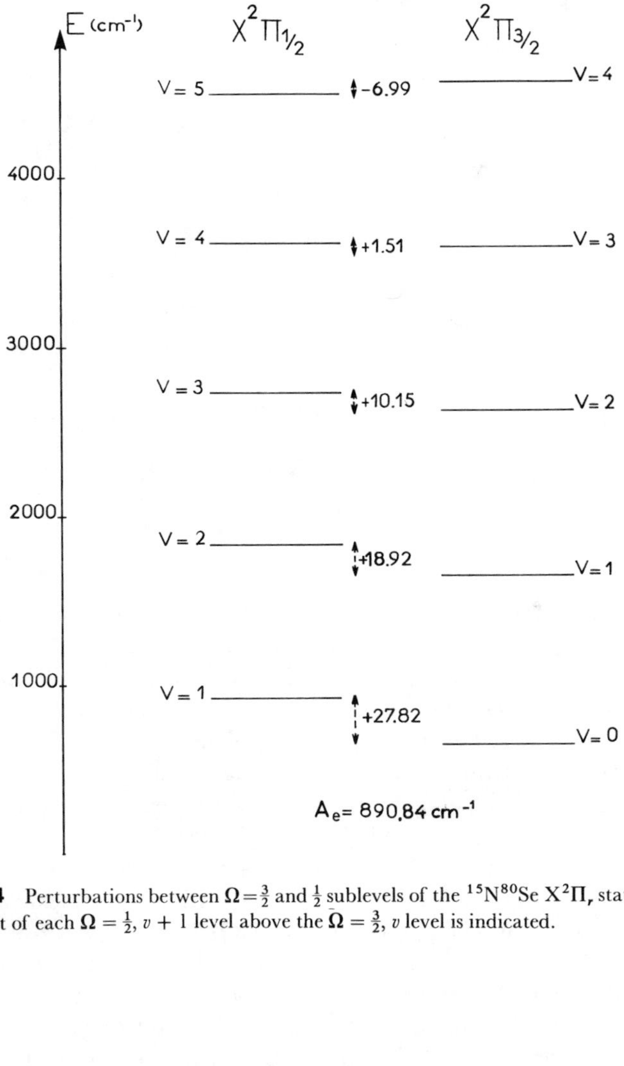

Fig. 2.14 Perturbations between $\Omega=\frac{3}{2}$ and $\frac{1}{2}$ sublevels of the $^{15}N^{80}Se$ $X^2\Pi_r$ state. The $J=\frac{1}{2}$ energy offset of each $\Omega=\frac{1}{2}$, $v+1$ level above the $\Omega=\frac{3}{2}$, v level is indicated.

Since $\mathbf{J}^+\mathbf{L}^- = \mathbf{J}^+\sum_i \mathbf{l}_i^-$, the **L**-uncoupling operator is a one-electron operator, and consequently, in the single-configuration approximation, the configurations describing the two interacting states can differ by no more than one spin-orbital. The electronic part of the perturbation matrix element is then proportional to the same $\langle \pi^+ |\mathbf{l}^+| \sigma\rangle$ or b parameter that appeared in the spin–electronic perturbation. However, owing to the presence of the $\mathbf{J}^+$ operator, the total matrix element of the $\mathbf{BJ}^+\mathbf{L}^-$ operator is proportional to $[J(J+1) - \Omega(\Omega-1)]^{1/2}$.

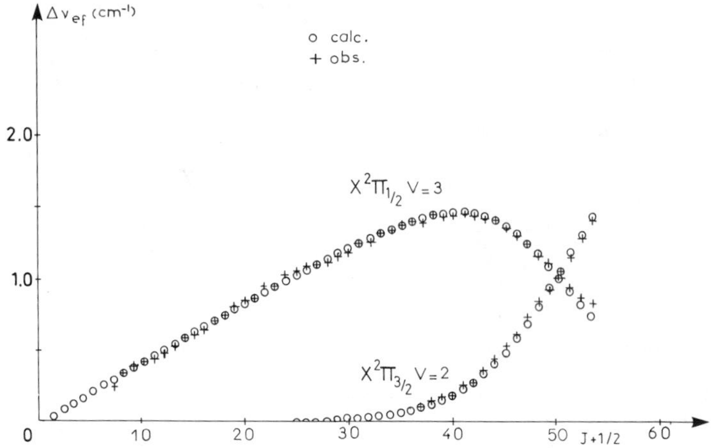

Fig. 2.15 Anomalous Λ-doubling in the $X^2\Pi_r$ state of $^{15}N^{80}Se$, due to **S**-uncoupling perturbations. (From Jenouvrier *et al.*, 1973.)

An example of this type of interaction between $^2\Pi_{3/2}$ and $^2\Sigma$ states of identical v is given in Section 2.5.4. If the two interacting states are sufficiently far apart in energy, this type of interaction gives rise to the splitting between the e and f components of a degenerate Λ state. This splitting is called Λ-doubling. The Λ-doubling of a $^1\Pi_1$ state, due to interaction with a $^1\Sigma_0$ state via the **L**-uncoupling operator, is treated by Hougen (1970).

Rydberg states belonging to the same l-complex are mixed by the **L**-uncoupling operator (see the end of Section 5.3.2 and Section 7.4). This most frequently involves states of the same value of v. At high n, very weak perturbations between states belonging to two Rydberg complexes of different nominal l values have been observed [in NO, $\Delta v = 0$, $8f \sim 9s$ and $6f \sim 6d$ perturbations were analyzed by Miescher (1976)]. These different l-complexes of identical v values converge to the same ion-core limit, form a supercomplex, and consequently an l-mixing results, due to the anisotropy of the molecular-ion field. Therefore, the **L**-uncoupling matrix element is slightly different from zero between states of different nominal l, since for a nonspherical system l is no longer a good quantum number (see Jungen and Atabek, 1977, p. 5607). However, note that the goodness of l is destroyed by an electrostatic term, not by the $\mathbf{l}^+$ or $\mathbf{L}^+$ operator.

If the energy interval between the Λ-components of a Rydberg l-complex is similar to the magnitude of the vibrational interval, degeneracy can occur, for example, between a $\Pi(np\pi)$ $v = 0$ level and a $\Sigma(np\sigma)$ $v = 1$ level. Equation (2.5.5) shows that a nonzero interaction can occur between these two levels.

L-uncoupling perturbations also occur between non-Rydberg states of two configurations that differ by a single orbital. Many examples of perturbations between nearly degenerate $^2\Pi_{3/2}$ and $^2\Sigma_{1/2}$ levels of different v are known in the literature. These perturbations result in a large and strongly J-dependent Λ-doubling, which changes sign at the perturbation culmination. A striking example is shown in Fig. 2.16, where strong perturbations of the CS^+ $A^2\Pi_{3/2}$

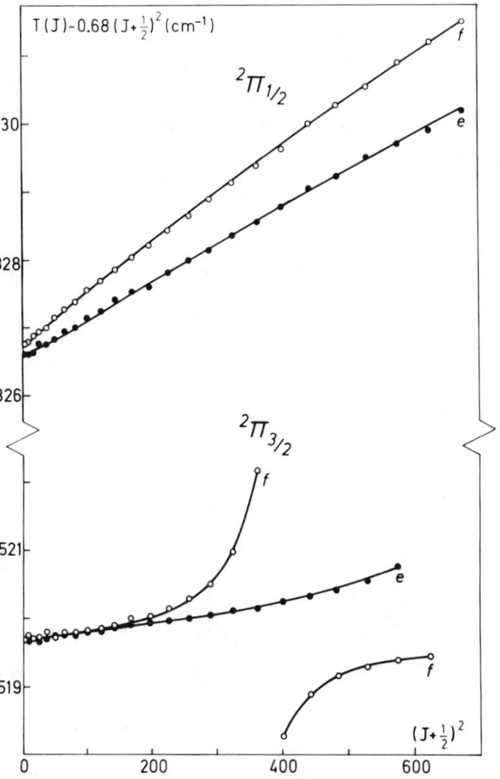

Fig. 2.16 Anomalous Λ-doubling in the CS^+ $A^2\Pi_i$ state due to **L**-uncoupling interaction with the $X^2\Sigma^+$ State. This is a reduced term-value plot for the CS^+ $A^2\Pi_i$ $v = 5$ level, which is perturbed by $X^2\Sigma^+$ $v = 13$. In order to expand the vertical scale, $0.68(J + \frac{1}{2})^2$ is subtracted from the observed term value (zero of energy at $v = 0$, $N = 0$ of $X^2\Sigma^+$). A plot of $T - B_{\text{eff}}(J + \frac{1}{2})^2$ versus $(J + \frac{1}{2})^2$ should show four nearly horizontal straight lines for a case (a) $^2\Pi$ state, with Λ-doubling in $^2\Pi_{1/2}$ and $^2\Pi_{3/2}$ proportional to J and J^3, respectively. The f-component of $X^2\Sigma^+$ (F_2, $N = J + \frac{1}{2}$) crosses $^2\Pi_{3/2f}$ near $J \simeq 19$ and disrupts the expected J^3 Λ-doubling pattern. The $^2\Pi_{3/2e}$ component is crossed at a J-value beyond the range of observed lines. The solid lines are calculated from the constants given by Gauyacq and Horani (1978, Table 8) and the circles represent observed spectral lines. (From Gauyacq and Horani, 1978).

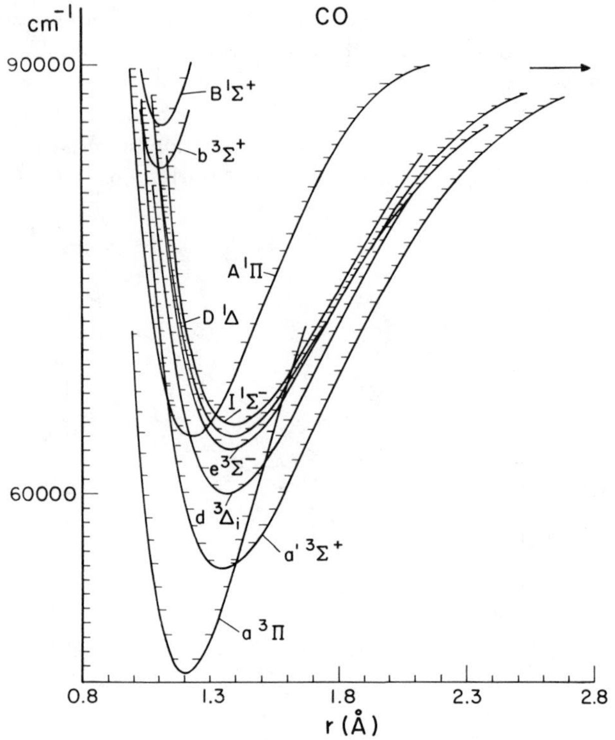

Fig. 2.17 Potential energy curves of the excited states of CO. (From Tilford and Simmons, 1972.)

$(v' = 5)$ component are shown. In this case, extra lines due to the perturbing states are not observed. However, it has been shown (Gauyacq and Horani, 1978) that the state responsible for the perturbation is a high vibrational level $(v'' = 13)$ of the $X^2\Sigma^+$ state.

Consider now a perturbation between a $\sigma\pi^4\pi'\ {}^1\Pi_1$ state and a $\sigma^2\pi^3\pi'\ {}^1\Sigma_0^-$ state. Such an example appears in the spectrum of CO (see Fig. 2.17; Field *et al.*, 1972). It is easy to see from the method presented in Section 2.2.3 that the $\pi^3\pi'$ configuration gives rise to two ${}^1\Sigma$ states from $\pi^+(\pi^-)^2\pi'^+$ and $(\pi^+)^2\pi^-\pi'^-$ subshell occupancies. The ${}^1\Sigma^-$ and ${}^1\Sigma^+$ wavefunctions, properly symmetrized with respect to the $\boldsymbol{\sigma}_v$ operator acting only on the spatial part of the wavefunction, must be constructed as linear combinations of the four possible $\Lambda = 0$, $\Sigma = 0$ Slater determinants:

$$|{}^1\Pi_1\rangle = (2)^{-1/2}\{|\pi^+\alpha\pi^+\beta\pi^-\alpha\pi^-\beta\sigma\alpha\pi'^+\beta| - |\pi^+\alpha\pi^+\beta\pi^-\alpha\pi^-\beta\sigma\beta\pi'^+\alpha|\}$$
$$= (2)^{-1/2}\{\det 1 - \det 2\}$$

$$|^1\Sigma_0^-\rangle = \tfrac{1}{2}\{|\pi^+\alpha\pi^+\beta\pi^-\alpha\sigma\alpha\sigma\beta\pi'^-\beta| - |\pi^+\alpha\pi^+\beta\pi^-\beta\sigma\alpha\sigma\beta\pi'^-\alpha|$$
$$- |\pi^+\alpha\pi^-\alpha\pi^-\beta\sigma\alpha\sigma\beta\pi'^+\beta| + |\pi^+\beta\pi^-\alpha\pi^-\beta\sigma\alpha\sigma\beta\pi'^+\alpha|\}$$
$$= \tfrac{1}{2}\{\det 3 - \det 4 - \det 5 + \det 6\}.$$

The only determinants of $|^1\Pi\rangle$ that differ from one of $|^1\Sigma^-\rangle$ by a single spin-orbital are det 1 $\sim$ det 5 ($\pi^+\beta \rightarrow \sigma\beta$) and det 2 $\sim$ det 6 ($\pi^+\alpha \rightarrow \sigma\alpha$). Thus, the nonzero matrix elements of $\mathbf{L}^+ = \sum_i \mathbf{l}_i^+$ are

$$\langle +\det 1\,|\mathbf{l}^+|\,-\det 5\rangle = -(-1)^P\langle \pi^+\beta\,|\mathbf{l}^+|\,\sigma\beta\rangle = +b \qquad (P \text{ is odd})$$
$$\langle -\det 2\,|\mathbf{l}^+|\,+\det 6\rangle = -(-1)^{P'}\langle \pi^+\alpha\,|\mathbf{l}^+|\,\sigma\alpha\rangle = +b \qquad (P' \text{ is odd});$$

thus

$$\langle ^1\Pi_1^+\,|\mathbf{L}^+|\,^1\Sigma_0^-\rangle = \tfrac{1}{2}(2)^{-1/2}(b + b) = (2)^{-1/2}b.$$

Transforming to the e/f parity basis, the above matrix element must be multiplied by $2^{1/2}$ since a Σ_0 function is involved. The final result is

$$\langle ^1\Pi_{1f}, v\,|\mathbf{H}^{\mathrm{ROT}}|\,^1\Sigma_{0f}^-, v'\rangle = -b\langle v|\mathbf{B}|v'\rangle[J(J+1)]^{1/2} \tag{2.5.7}$$

2.5.4 $^2\Pi \sim {}^2\Sigma^+$ INTERACTION

Matrix element expressions for spin–orbit and rotational interactions will be developed here in order to:

1. Treat the transition from case (a) to case (b) for $^2\Pi$ states and thereby provide an alternative derivation of the expressions for case (b) functions in terms of case (a) functions previously derived in Section 2.2.1 using Clebsch–Gordan coefficients.
2. Construct the interaction matrix between $^2\Pi$ and $^2\Sigma^+$ states.

Using Eqs. (2.1.16) and (2.4.6) to calculate the rotational and spin–orbit energies of the case (a) $^2\Pi_{3/2}$ state (without e/f symmetrization),

$$|^2\Pi_{3/2}\rangle = |\Lambda = 1, S = \tfrac{1}{2}, \Sigma = \tfrac{1}{2}, \Omega = +\tfrac{3}{2}\rangle,$$
$$E(^2\Pi_{3/2}, v, J) = T_\Pi^v + A_v/2 + B_v[J(J+1) - \tfrac{9}{4} + \tfrac{3}{4} - \tfrac{1}{4}]$$
$$= T_\Pi^v + A_v/2 + B_v[(J+\tfrac{1}{2})^2 - 2]. \tag{2.5.8}$$

For the $^2\Pi_{1/2}$ state, $|\Lambda = 1, S = \tfrac{1}{2}, \Sigma = -\tfrac{1}{2}, \Omega = +\tfrac{1}{2}\rangle$,

$$E(^2\Pi_{1/2}, v, J) = T_\Pi^v - A_v/2 + B_v[J(J+1) - \tfrac{1}{4} + \tfrac{3}{4} - \tfrac{1}{4}]$$
$$= T_\Pi^v - A_v/2 + B_v(J+\tfrac{1}{2})^2. \tag{2.5.9}$$

The only nonzero off-diagonal matrix element between these two substates, which differ by $\Delta\Omega = \Delta\Sigma = 1$, is given by the **S**-uncoupling part of the rotational Hamiltonian. Using Table 2.2, one obtains

$$\begin{aligned}
&\langle {}^2\Pi_{3/2}, v|\mathbf{H}^{\mathrm{ROT}}|{}^2\Pi_{1/2}, v\rangle \\
&\quad = \langle J, \Lambda, S, \Sigma + 1, \Omega + 1| -(1/2\mu R^2)\mathbf{J}^-\mathbf{S}^+|J, \Lambda, S, \Sigma, \Omega\rangle \\
&\quad = \langle v, J, 1, \tfrac{1}{2}, \tfrac{1}{2}, \tfrac{3}{2}| -(1/2\mu R^2)\mathbf{J}^-\mathbf{S}^+|v, J, 1, \tfrac{1}{2}, -\tfrac{1}{2}, \tfrac{1}{2}\rangle \\
&\quad = -(\hbar^2/2\mu)\langle v|R^{-2}|v\rangle[\tfrac{3}{4} - (-\tfrac{1}{2})(\tfrac{1}{2})]^{1/2}[J(J+1) - (\tfrac{1}{2})(\tfrac{3}{2})]^{1/2} \\
&\quad = -B_v[J(J+1) - \tfrac{3}{4}]^{1/2} = -B_v[(J+\tfrac{1}{2})^2 - 1]^{1/2}. \qquad (2.5.10)
\end{aligned}$$

It is necessary to assume that the two substates, ${}^2\Pi_{1/2}$ and ${}^2\Pi_{3/2}$, have identical potential curves, and thus the same vibrational wavefunctions. The other part of the **S**-uncoupling operator, $\mathbf{J}^+\mathbf{S}^-$, acts between ${}^2\Pi_{-3/2}$ ($\Omega = -\frac{3}{2}$) and ${}^2\Pi_{-1/2}$ ($\Omega = -\frac{1}{2}$), and gives the same matrix element. There are no off-diagonal matrix elements of the **S**-uncoupling operator between $\Omega > 0$ and $\Omega < 0$ ${}^2\Pi$ basis functions. Thus the e and f ${}^2\Pi$ basis functions, which are linear combinations of Ω and $-\Omega$ basis functions, have the same matrix elements of the rotational operator as the separate signed-Ω functions.

When the eigenvalues of the 2×2 ${}^2\Pi_{1/2}, {}^2\Pi_{3/2}$ matrix are found (see the matrix excluding ${}^2\Sigma$ in Table 2.9), the well-known result of Hill and Van Vleck (1928) is obtained. The energy levels, expressed as a function of $Y = A_v/B_v$, are

$$E = T_\Pi^v + B_v[(J+\tfrac{1}{2})^2 - 1] \pm (B_v/2)[Y(Y-4) + 4(J+\tfrac{1}{2})^2]^{1/2}, \qquad (2.5.11)$$

which can be rewritten as

$$E = T_\Pi^v + B_v[(J+\tfrac{1}{2})^2 - 1] \pm [-\tfrac{1}{2}(A_v - 2B_v)]\left(1 + \frac{B_v^2[(J+\tfrac{1}{2})^2 - 1]}{[-\tfrac{1}{2}(A_v - 2B_v)]^2}\right)^{1/2}. \qquad (2.5.12)$$

Table 2.9
${}^2\Pi(\pi)\sim{}^2\Sigma^+(\sigma)$ Interaction Matrix[a]

e/f[b]	${}^2\Pi_{1/2}$	${}^2\Pi_{3/2}$	${}^2\Sigma^+_{1/2}$
${}^2\Pi_{1/2}$	$T_v^\Pi - \frac{A_v}{2} + B_v(J+\frac{1}{2})^2$	$-B_v[(J-\frac{1}{2})(J+\frac{3}{2})]^{1/2}$	$\frac{a_+}{2} + [1 \mp (J+\frac{1}{2})]bB_v$
${}^2\Pi_{3/2}$		$T_v^\Pi + \frac{A_v}{2} + B_v[(J+\frac{1}{2})^2 - 2]$	$-bB_v[(J-\frac{1}{2})(J+\frac{3}{2})]^{1/2}$
${}^2\Sigma^+_{12}$			$T_v^\Sigma + B_v(J+\frac{1}{2})^2$

[a] Assuming identical potential curves for ${}^2\Pi$ and ${}^2\Sigma^+$ and $v_\Pi = v_\Sigma$.
[b] The *only* difference between the matrices for e and f levels is in the ${}^2\Pi_{1/2}\sim{}^2\Sigma^+_{1/2}$ element. The top sign is for e, bottom for f.

In the case where B/A is very small, the radical can be expanded according to

$$(1 + x)^{1/2} \simeq 1 + x/2,$$

resulting in an approximate expression for $^2\Pi$ levels intermediate between cases (a) and (b), but nearer the case (a) limit,

$$\begin{aligned} E(^2\Pi_{1/2}) = {} & T^v_\Pi - (1/2)(A_v - 2B_v) \\ & + B_v[1 - B_v/(A_v - 2B_v)][(J + \tfrac{1}{2})^2 - 1] + \cdots \\ E(^2\Pi_{3/2}) = {} & T^v_\Pi + (1/2)(A_v - 2B_v) \\ & + B_v(1 + B_v/(A_v - 2B_v)[(J + \tfrac{1}{2})^2 - 1] + \cdots \end{aligned}$$

These energy-level formulas may be expressed in a form similar to that in pure case (a) (i.e., without $^2\Pi_{1/2}$–$^3\Pi_{3/2}$ mixing) by introducing a B_{eff} constant for each Ω substate:

$$E(^2\Pi_{1/2}) = T^v_\Pi - \tfrac{1}{2}(A_v - 2B_v) + B_{1\text{eff}}[(J + \tfrac{1}{2})^2 - 1] \quad (2.5.13)$$

$$E(^2\Pi_{3/2}) = T^v_\Pi + \tfrac{1}{2}(A_v - 2B_v) + B_{2\text{eff}}[(J + \tfrac{1}{2})^2 - 1]. \quad (2.5.14)$$

The difference between $B_{1\text{eff}}$ and $B_{2\text{eff}}$, $2B_v^2/A_v$, is related to the value of the A constant. Each substate is doubly degenerate; the e and f levels of each substate have the same energy in the absence of interaction with $^2\Sigma^+$ or $^2\Sigma^-$ states.

Nearer the case (b) limit, if A is very small compared to BJ, the $^2\Pi$ eigenstates can be expressed in terms of case (b) functions, which can be obtained simply by setting $A = 0$ in the 2×2 $^2\Pi$ matrix. The energy eigenvalues for $BJ \gg A$ are, from Eq. (2.5.11),

$$E = T^v_\Pi + B_v[(J + \tfrac{1}{2})^2 - 1] \pm B_v(J + \tfrac{1}{2})\left[1 + \frac{Y(Y - 4)}{8(J + \frac{1}{2})^2}\right] + \cdots. \quad (2.5.15)$$

Equation (2.5.15) may be rearranged into case (b) form simply by replacing J with $N \mp \frac{1}{2}$ [$J = N - \frac{1}{2}$ for top sign in Eq. (2.5.15), $J = N + \frac{1}{2}$ for bottom sign]. The resulting energy level expressions are

$$\begin{aligned} F_2(N = J + \tfrac{1}{2}) \qquad E_2 &= T^v_\Pi + B_v(N^2 - 1) + B_vN + B_vY(Y - 4)/8N \\ &= T^v_\Pi + B_vN(N + 1) - B_v + B_vY(Y - 4)/8N, \end{aligned} \quad (2.5.16)$$

which, except for the last term (negligible when $BN \gg A$), is the F_2 energy level expression of Eq. (2.2.20), and

$$\begin{aligned} F_1(N = J - \tfrac{1}{2}) \qquad E_2 &= T^v_\Pi + B_v[(N + 1)^2 - 1] \\ &\quad - B_v(N + 1) - B_vY(Y - 4)/8(N + 1) \\ &= T^v_\Pi + B_vN(N + 1) - B_v - B_vY(Y - 4)/8(N + 1), \end{aligned} \quad (2.5.17)$$

which, except for the last term, is the F_1 energy level expression of Eq. (2.2.20).

Starting from a $^2\Pi$ matrix expressed in terms of case (a) basis functions, case (b) energy level expressions have been derived for $BJ \gg A$. This means that the eigenfunctions are almost exactly the case (b) basis functions. Alternatively, in the case (b) basis the **S**-uncoupling operator, $-2\mathbf{BJ}\cdot\mathbf{S}$, can be replaced by $-\mathrm{B}(\mathbf{N}^2 - \mathbf{J}^2 - \mathbf{S}^2)$ because

$$(\mathbf{N})^2 = (\mathbf{J} - \mathbf{S})^2 \qquad 2\mathbf{J}\cdot\mathbf{S} = \mathbf{N}^2 - \mathbf{J}^2 - \mathbf{S}^2.$$

Thus, it is evident that the only nonzero case (b) matrix elements of the **S**-uncoupling operator are $\Delta N = \Delta J = \Delta S = 0$ and that all off-diagonal matrix elements of this operator between F_1 and F_2 basis functions vanish. The departure from case (b) at low J is caused by $\Delta N = \pm 1$ matrix elements of the $\mathbf{H}^{\mathrm{SO}}$ and $\mathbf{H}^{\mathrm{SS}}$ operator.

Consider the levels of an isolated $^2\Sigma^+$ state. The case (a) basis function that corresponds to the $\Omega = +\frac{1}{2}$ component of this state is

$$|\Lambda = 0\,\Sigma = \tfrac{1}{2}\Omega = \tfrac{1}{2}\rangle$$

while the $\Omega = -\frac{1}{2}$ component is

$$|\Lambda = 0\,\Sigma = -\tfrac{1}{2}\Omega = -\tfrac{1}{2}\rangle.$$

The energies of both components are identical, given by

$$E_\Sigma = T_\Sigma^v + B_v[J(J+1) - \tfrac{1}{4} + \tfrac{3}{4} - \tfrac{1}{4}] = T_\Sigma^v + B_v(J + \tfrac{1}{2})^2. \quad (2.5.18)$$

These two components differ by $\Delta\Omega = +1$ and interact via a nonzero matrix element of the **S**-uncoupling operator. This interaction can *never* be neglected, since the two interacting components have the same energy. The matrix element,

$$\langle\Lambda = 0, \Sigma = \tfrac{1}{2}| -\frac{1}{2\mu R^2}\mathbf{J}^-\mathbf{S}^+|\Lambda = 0, \Sigma = -\tfrac{1}{2}\rangle = -B_v[J(J+1) + \tfrac{1}{4}]^{1/2}$$

$$= -B_v(J + \tfrac{1}{2}), \qquad (2.5.19)$$

causes the two case (a) components of $^2\Sigma^+$ belonging to the same J-value to become completely mixed. This gives rise to states that are e and f symmetry components,

$$|^2\Sigma_e^+\rangle = (2)^{-1/2}[|J, \Omega, \Lambda, \Sigma\rangle + |J, -\Omega, -\Lambda, -\Sigma\rangle]$$

$$= (2)^{-1/2}[|J, \tfrac{1}{2}, 0, \tfrac{1}{2}\rangle + |J, -\tfrac{1}{2}, 0, -\tfrac{1}{2}\rangle] \qquad (2.5.20\text{a})$$

$$|^2\Sigma_f^+\rangle = (2)^{-1/2}[|J, \Omega, \Lambda, \Sigma\rangle - |J, -\Omega, -\Lambda, -\Sigma\rangle]$$

$$= (2)^{-1/2}[|J, \tfrac{1}{2}, 0, \tfrac{1}{2}\rangle - |J, -\tfrac{1}{2}, 0, -\tfrac{1}{2}\rangle] \qquad (2.5.20\text{b})$$

Taking the matrix element of Eq. (2.5.19) into account, the energy eigenvalues are

$$E(^2\Sigma_e^+) = T_\Sigma^v + B_v[J(J+1) + \tfrac{1}{4} - (J + \tfrac{1}{2})],$$

or, if $N = J - \frac{1}{2}$ (F_1 levels),

$$E(^2\Sigma_e^+) = T_\Sigma^v + B_v[N(N+1)] \tag{2.5.21}$$

and

$$E(^2\Sigma_f^+) = T_\Sigma^v + B_v[J(J+1) + \tfrac{1}{4} + (J + \tfrac{1}{2})],$$

or, if $N = J + \frac{1}{2}$ (F_2 levels),

$$E(^2\Sigma_f^+) = T_\Sigma^v + B_v[N(N+1)]. \tag{2.5.22}$$

This is the usual case (b) energy level expression for $^2\Sigma^+$ states. The $e(F_1)$ and $f(F_2)$ levels of the same N are degenerate, but the $e(F_1)$ and $f(F_2)$ levels of the same J are separated by $2B(J + \frac{1}{2})$.

This e/f degeneracy between the two same-N components will be lifted by interaction with a $^2\Pi$ state. If the potential curves of the $^2\Sigma^+$ and $^2\Pi$ states are identical and the configurations of the $^2\Sigma^+$ and $^2\Pi$ states are σ^1 and π^1, respectively, then this interaction takes a particularly simple form. The discussion through the end of this chapter deals exclusively with these σ^1 and π^1 configurations. Perturbation selection rules for unsymmetrized basis functions require that the following interactions be considered. The $^2\Pi_{1/2}$ state experiences two types of $\Delta\Omega = 0$ interactions with $^2\Sigma_{1/2}^+$: *spin–orbit*,

$$\langle \Lambda = 1, \Sigma = -\tfrac{1}{2}, \Omega = \tfrac{1}{2}, v_\Pi | \mathbf{H}^{\mathrm{SO}} | \Lambda = 0, \Sigma = \tfrac{1}{2}, \Omega = \tfrac{1}{2}, v_\Sigma \rangle = \frac{a_+}{2} \langle v_\Pi | v_\Sigma \rangle \tag{2.5.23}$$

where $a_+ = \langle \pi | \hat{a} \mathbf{l}^+ | \sigma \rangle$ and $\langle v_\Pi | v_\Sigma \rangle = \delta_{v_\Pi v_\Sigma}$, and *spin–electronic*,

$$\langle 1, -\tfrac{1}{2}, \tfrac{1}{2}, v_\Pi | (1/2\mu R^2) \mathbf{L}^+ \mathbf{S}^- | 0, \tfrac{1}{2}, \tfrac{1}{2}, v_\Sigma \rangle = B_{v_\Pi v_\Sigma} b \tag{2.5.24}$$

where $b = \langle \pi | \mathbf{l}^+ | \sigma \rangle$ and $B_{v_\Pi v_\Sigma} = B_v \delta_{v_\Pi v_\Sigma}$.

The $^2\Pi_{1/2}$ state also experiences a $\Delta\Omega = +1$ interaction via the **L**-uncoupling operator with the $\Omega = -\frac{1}{2}$ component of the $^2\Sigma^+$ state,

$$\left\langle 1, -\tfrac{1}{2}, \tfrac{1}{2}, v_\Pi \left| -\frac{1}{2\mu R^2} \mathbf{J}^- \mathbf{L}^+ \right| 0, -\tfrac{1}{2}, -\tfrac{1}{2}, v_\Sigma \right\rangle = -B_{v_\Pi v_\Sigma} b(J + \tfrac{1}{2}). \tag{2.5.25}$$

In the following, it is assumed that $v_\Pi = v_\Sigma$. It is more convenient to construct the $^2\Pi \sim {}^2\Sigma^+$ interaction matrix between e/f-symmetrized basis functions. The e and f $^2\Sigma^+$ basis functions are given by Eqs. (2.5.20). The corresponding symmetrized $^2\Pi$ functions are

$$|^2\Pi^e_{1/2}\rangle = (2)^{-1/2}[|J, \tfrac{1}{2}, 1, -\tfrac{1}{2}\rangle + |J, -\tfrac{1}{2}, -1, \tfrac{1}{2}\rangle] \qquad (2.5.26a)$$

$$|^2\Pi^f_{1/2}\rangle = (2)^{-1/2}[|J, \tfrac{1}{2}, 1, -\tfrac{1}{2}\rangle - |J, -\tfrac{1}{2}, -1, \tfrac{1}{2}\rangle] \qquad (2.5.26b)$$

$$|^2\Pi^e_{3/2}\rangle = (2)^{-1/2}[|J, \tfrac{3}{2}, 1, \tfrac{1}{2}\rangle + |J, -\tfrac{3}{2}, -1, -\tfrac{1}{2}\rangle] \qquad (2.5.27a)$$

$$|^2\Pi^f_{3/2}\rangle = (2)^{-1/2}[|J, \tfrac{3}{2}, 1, \tfrac{1}{2}\rangle - |J, -\tfrac{3}{2}, -1, -\tfrac{1}{2}\rangle]. \qquad (2.5.27b)$$

The eigenvalues of the $^2\Pi_e$ and $^2\Pi_f$ matrices are identical with each other and with those of the $(^2\Pi_{1/2}, {}^2\Pi_{3/2})$ and $(^2\Pi_{-1/2}, {}^2\Pi_{-3/2})$ matrices [Eq. (2.5.11)]. The off-diagonal $^2\Pi \sim {}^2\Sigma^+$ elements are

$$\begin{aligned}\langle {}^2\Pi^e_{1/2}|\mathbf{H}^{SO} + \mathbf{H}^{ROT}|{}^2\Sigma^{+e}\rangle &= a_+/2 + bB_v - bB_v(J + \tfrac{1}{2}) \\ &= a_+/2 - bB_v(J - \tfrac{1}{2}) \\ \langle {}^2\Pi^f_{1/2}|\mathbf{H}^{SO} + \mathbf{H}^{ROT}|{}^2\Sigma^{+f}\rangle &= a_+/2 + bB_v + bB_v(J + \tfrac{1}{2}) \\ &= a_+/2 + bB_v(J + \tfrac{3}{2}).\end{aligned} \qquad (2.5.28)$$

Note that the crucial difference between the $e \sim e$ and $f \sim f$ $^2\Pi \sim {}^2\Sigma^+$ matrix elements arises from an interference effect between $\Delta\Omega = +1$ and $\Delta\Omega = -1$ interactions that is not evident from Eqs. (2.5.24) and (2.5.25).[†] The reader should verify that the $e \sim f$ matrix elements for the $^2\Pi$, $^2\Sigma^+$ basis functions are identically zero, consistent with the $e \not\leftrightarrow f$ perturbation selection rule.

[†] This interference effect, which occurs only for even multiplicity $\Pi \sim \Sigma$ and $\Sigma \sim \Sigma$ matrix elements of $\mathbf{H}^{ROT}$, is the second of two cases where extreme care must be taken in going from the unsymmetrized, signed-Ω basis to the e/f symmetrized basis.

For odd-multiplicity systems, all off-diagonal matrix elements involving a Σ_0 basis function increase by a factor of $2^{1/2}$ in going from the unsymmetrized to the symmetrized basis.

For even multiplicity systems, matrix elements between basis states of $\Omega > 0$ and $\Omega < 0$ cause problems. The $\mathbf{H}^{ROT}$ operator is responsible for both $\Delta\Omega = \Delta\Lambda = \pm 1$ and $\Delta\Omega = \Delta\Lambda = 0$ matrix elements. These two types of matrix elements are quite distinct in the signed basis, yet couple the same two basis states in the symmetrized basis. For example,

$$\langle {}^2\Sigma^+_{+1/2}|\mathbf{H}^{ROT}|{}^2\Pi_{-1/2}\rangle = -(J + \tfrac{1}{2})B_{v_\Sigma v_\Pi}b$$

yet

$$\langle {}^2\Sigma^+_{|1/2|{}^e_f}|\mathbf{H}^{ROT}|{}^2\Pi_{|1/2|{}^e_f}\rangle = [1 \mp (J + \tfrac{1}{2})]B_{v_\Sigma v_\Pi}b.$$

This sort of interference effect (resulting in an e/f-dependent matrix element in the symmetrized basis) can occur via $\mathbf{H}^{ROT}$ only between two $\Lambda \leq 1$ $|\Omega| = \frac{1}{2}$ levels.

It is interesting and significant that both of these differences between the signed-Ω and e/f basis sets are crucial to the existence of Λ-doubling. For odd-multiplicity, Λ-doubling is *caused* by the odd number of Σ_0 basis functions. For even multiplicity, Λ-doubling is *caused* by an interference effect unique to $\mathbf{H}^{ROT}$ matrix elements involving at least one $\Sigma_{1/2}$ basis function.

For $^2\Pi_{3/2}$, only the **L**-uncoupling interaction with $^2\Sigma$ is nonzero,

$$\langle 1, \tfrac{1}{2}, \tfrac{3}{2}, v_\Pi | -(1/2\mu R^2)\mathbf{J}^-\mathbf{L}^+ | 0, \tfrac{1}{2}, \tfrac{1}{2}, v_\Sigma \rangle = -bB_v[(J-\tfrac{1}{2})(J+\tfrac{3}{2})]^{1/2}$$

and

$$\langle {}^2\Pi^e_{3/2} | \mathbf{H}^{\mathrm{ROT}} | {}^2\Sigma^{+e} \rangle = \langle {}^2\Pi^f_{3/2} | \mathbf{H}^{\mathrm{ROT}} | {}^2\Sigma^{+f} \rangle = -bB_v[(J-\tfrac{1}{2})(J+\tfrac{3}{2})]^{1/2}. \tag{2.5.29}$$

These $^2\Pi$, $^2\Sigma^+$, and $^2\Pi \sim {}^2\Sigma^+$ matrix elements are collected in Table 2.9.

The $^2\Pi \sim {}^2\Sigma^+$ interaction is different for the e and f levels. This difference gives rise to Λ-doubling in $^2\Pi$ and a spin–rotation splitting of the $\Delta N = 0$ degeneracy of the $^2\Sigma^+$ state. If the $^2\Sigma^+$ state is sufficiently far from the $^2\Pi$ state, it is possible to use second-order perturbation theory to evaluate the Λ-doubling. The $^2\Pi^e_{1/2}$ level is repelled by the $^2\Sigma^+$ state with an energy shift of

$$\Delta T_{e_{1/2}} = [a_+/2 - bB_v(J-1)]^2/(E_\Pi - E_\Sigma). \tag{2.5.30a}$$

The $^2\Pi^e_{1/2}$ component is lowered in energy if the $^2\Sigma^+$ state lies above $^2\Pi$, as shown in Fig. 2.1.8. The $^2\Pi^f_{1/2}$ level is also repelled, but with a different energy shift:

$$\Delta T_{f_{1/2}} = [(a_+/2) + bB_v(J+\tfrac{3}{2})]^2/(E_\Pi - E_\Sigma). \tag{2.5.30b}$$

If a_+ and bB_v have the same signs, then $|\Delta T_f| > |\Delta T_e|$. The Λ-doubling contribution to $^2\Pi_{1/2}$ (more precisely, to the $\langle {}^2\Pi_{1/2} | \mathbf{H} | {}^2\Pi_{1/2} \rangle$ matrix element) is given by

$$\Delta \nu_{fe}({}^2\Pi_{1/2}) = \Delta T_{f_{1/2}} - \Delta T_{e_{1/2}} = \frac{(a_+ bB_v + 2b^2B_v^2)[(J+\tfrac{3}{2}) + (J-\tfrac{1}{2})]}{E_\Pi - E_\Sigma}$$

$$= \frac{(2J+1)(bB_v)(a_+ + 2bB_v)}{E_\Pi - E_\Sigma}. \tag{2.5.31}$$

If, as is typical, $2bB_v \ll a_+$, then

$$\Delta \nu_{fe}({}^2\Pi_{1/2}) = \frac{2a_+ bB_v}{E_\Pi - E_\Sigma}(J+\tfrac{1}{2}) = p(J+\tfrac{1}{2}). \tag{2.5.32}$$

The $^2\Sigma$ e and f levels will be shifted by an amount equal and opposite to that of the $^2\Pi$ e and f levels (Fig. 2.18), and

$$\Delta \nu_{fe}({}^2\Sigma^+, J) = E({}^2\Sigma^+_f, N = J+\tfrac{1}{2}) - E({}^2\Sigma^+_e, N = J-\tfrac{1}{2})$$

$$= 2(J+\tfrac{1}{2})B_v + [2a_+ bB_v/(E_\Sigma - E_\Pi)](J+\tfrac{1}{2})$$

or

$$\Delta \nu_{fe}({}^2\Sigma^+, N) = [2a_+ bB_v/(E_\Sigma - E_\Pi)](N+\tfrac{1}{2}) \equiv -\gamma(N+\tfrac{1}{2}). \tag{2.5.33}$$

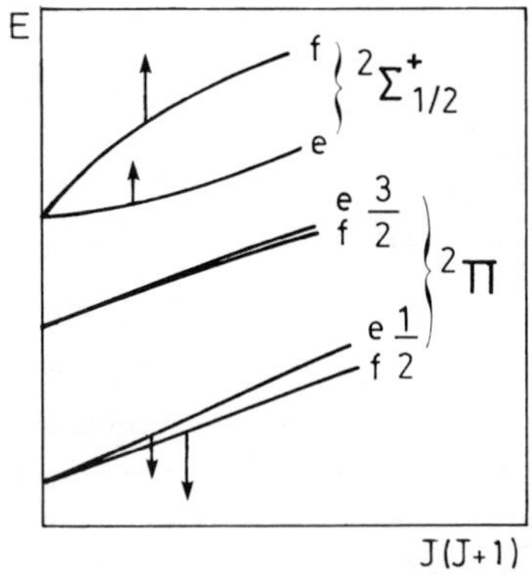

Fig. 2.18 Λ-doubling of $^2\Pi$ states and spin-doubling of $^2\Sigma^+$. The e/f-dependent interaction between $^2\Pi$ and $^2\Sigma$ states is shown schematically. The $^2\Sigma$ state, if plotted versus $N(N+1)$ rather than $J(J+1)$, would show spin–rotation splitting identical in magnitude but opposite in sign to the $^2\Pi_{1/2}$ Λ-doubling.

If the $^2\Pi$ and $^2\Sigma^+$ states (with identical potential curves) in question interact appreciably only with each other and not with other $^2\Sigma^\pm$ or $^2\Pi$ states, then $p = \gamma$. When the pair of interacting $^2\Pi$ and $^2\Sigma$ states are found to have $p = \gamma$, it is said that they are in a "unique perturber" relationship, which has *nothing* to do with the pure precession hypothesis, discussed below and in Section 4.5.

There is one other mechanism by which the $^2\Pi \sim {}^2\Sigma^+$ interaction results in Λ-doubling in the $^2\Pi$ state. The $\langle {}^2\Pi_{3/2}|H|{}^2\Pi_{3/2}\rangle$ matrix element is affected in a *parity-independent* sense by the $\Pi \sim \Sigma$ interaction,

$$\Delta T_{f_{3/2}} = \Delta T_{e_{3/2}} = \frac{b^2 B_v^2[(J-\tfrac{1}{2})(J+\tfrac{3}{2})]}{E_\Pi - E_\Sigma}, \tag{2.5.34}$$

but this makes no direct contribution to the $^2\Pi_{3/2}$ Λ-doubling. However (see Section 3.2 for a discussion of the Van Vleck transformation), the $\langle {}^2\Pi_{3/2}|\mathbf{H}|{}^2\Pi_{1/2}\rangle$ matrix element is affected in a parity-dependent manner,

$$\langle {}^2\Pi_{e_{3/2}}|\mathbf{H}|{}^2\Pi_{e_{1/2}}\rangle = -[(J-\tfrac{1}{2})(J+\tfrac{3}{2})]^{1/2}[B_v + p/4 - q(J-\tfrac{1}{2})/2] \tag{2.5.35a}$$

$$\langle {}^2\Pi_{f_{3/2}}|\mathbf{H}|{}^2\Pi_{f_{1/2}}\rangle = -[(J-\tfrac{1}{2})(J+\tfrac{3}{2})]^{1/2}[B_v + p/4 + q(J+\tfrac{3}{2})/2] \tag{2.5.35b}$$

where

$$q = 2b^2 B_v^2/(E_\Pi - E_\Sigma) \tag{2.5.36}$$

and p is defined by Eq. (2.5.32). Now, using second-order perturbation theory to estimate the parity-dependent effect of Eq. (2.5.35) on the Λ-doubling in $^2\Pi_{3/2}$,

$$\Delta\nu_{fe}(^2\Pi_{3/2}) \approx 2(J-\tfrac{1}{2})(J+\tfrac{3}{2})(J+\tfrac{1}{2})B_v q/A_v, \qquad (2.5.37)$$

then it is evident that Λ-doubling in $^2\Pi_{3/2}$ is proportional to $\sim qJ^3/Y$. (It can also be shown by second-order perturbation theory that the $^2\Pi_{1/2}$ Λ-doubling transferred to $^2\Pi_{3/2}$ by **S**-uncoupling is $\sim pJ^3/Y^2$.)

In Van Vleck's pure precession hypothesis (Van Vleck, 1929; Mulliken and Christy, 1931; see Section 4.5),

$$a_+ = [l(l+1)]^{1/2} A_v \qquad (2.5.38)$$

and

$$b = [l(l+1)]^{1/2}. \qquad (2.5.39)$$

Then

$$p = \gamma = \frac{2ABl(l+1)}{E_\Pi - E_\Sigma} \qquad (2.5.40)$$

and

$$q = \frac{2B^2 l(l+1)}{E_\Pi - E_\Sigma}. \qquad (2.5.41)$$

For example, the spin–rotation splitting (called "ρ-doubling" by Van Vleck) of the OH $A^2\Sigma^+$ state is given mainly by interaction with the inverted $X^2\Pi$ state. Consequently, for $A < 0$ and $E_\Pi - E_\Sigma < 0$, one predicts $\gamma > 0$. The *e*-levels of the $A^2\Sigma^+$ are well above the *f*-levels, in qualitative agreement with the pure precession prediction.

References

Amiot, C., Bacis, R., and Guelachvili, G. (1978), *Can. J. Phys.* **56**, 251.
Bandrauk, A. D., and Child, M. S. (1970), *Mol. Phys.* **19**, 95.
Beebe, N. H. F., Thulstrup, E. W., and Andersen, A. (1976), *J. Chem. Phys.* **64**, 2080.
Blume, M., and Watson, R. E. (1962), *Proc. R. Soc. London Ser. A* **270**, 127.
Blume, M., and Watson, R. E. (1963), *Proc. R. Soc. London Ser. A* **271**, 565.
Blume, M., Freeman, A. J., and Watson, R. E. (1964), *Phys. Rev. A* **134**, 320.
Brion, J., and Malicet, J. (1976), *J. Phys. B* **9**, 2097.
Brion, J., Malicet, J., and Guenebaut, H. (1974), *Can. J. Phys.* **52**, 2143.
Brown, J. M., and Howard, B. J. (1976), *Mol. Phys.* **31**, 1517.
Brown, J. M., and Merer, A. J. (1979), *J. Mol. Spectrosc.* **74**, 488.

Brown, J. M., and Watson, J. K. G. (1977), *J. Mol. Spectrosc.* **65**, 65.
Brown, J. M., Hougen, J. T., Huber, K. P., Johns, J. W. C., Kopp, I., Lefebvre-Brion, H., Merer, A. J., Ramsay, D. A., Rostas J., and Zare, R. N. (1975), *J. Mol. Spectrosc.* **55**, 500.
Brown, J. M., Colbourn, E. A., Watson, J. K. G., and Wayne, F. D. (1979), *J. Mol. Spectrosc.* **74**, 294.
Brown, J. M., Milton, D. J., Watson, J. K. G., Zare, R. N., Albritton, D. L., Horani, M., and Rostas, J. (1981), *J. Mol. Spectrosc.* **90**, 139.
Bunker, P. R. (1968), *J. Mol. Spectrosc.* **28**, 422.
Cheetham, C. J., Gissane, W. J. M., and Barrow, R. F. (1965), *Trans. Faraday Soc.* **61**, 1308.
Cheung, A. S. C., Taylor, A. W., and Merer, A. J. (1982), *J. Mol. Spectrosc.* **92**, 381.
Cohen, J. S., and Schneider, B. (1974), *J. Chem. Phys.* **61**, 3230.
Condon, E. U., and Shortley, G. H. (1953), "The Theory of Atomic Spectra," Cambridge Univ. Press, London and New York.
Cooper, I. L., and Musher, J. I. (1972), *J. Chem. Phys.* **57**, 1333.
Cossart, D., Horani, M., and Rostas, J. (1977), *J. Mol. Spectrosc.* **67**, 283.
Coxon, J. A. (1975), *J. Mol. Spectrosc.* **58**, 1.
Coxon, J. A., Ramsay, D. A., and Setser, D. W. (1975), *Can. J. Phys.* **53**, 1587.
Davis, S. L., and Alexander, M. H. (1983), *J. Chem. Phys.* **78**, 800.
Dieke, G. H. (1935), *Phys. Rev.* **47**, 870.
Dixon, R. N., and Hull, S. E. (1969), Chem. Phys. Lett. **3**, 367.
Dressler, K. (1969), *Can. J. Phys.* **47**, 547.
Dressler, K. (1983), *Ann. Isr. Phys. Soc.* **6**, 141.
Dressler, K., and Wolniewicz, L. (1981), *J. Mol. Spectrosc.* **86**, 534.
Dressler, K., Gallusser, R., Quadrelli, P., and Wolniewicz, L. (1979), *J. Mol. Spectrosc.* **75**, 205.
Dulick M. (1982), thesis, Massachusetts Institute of Technology.
Edmonds, A. R. (1974), "Angular Momentum in Quantum Mechanics," Princeton Univ. Press, Princeton, New Jersey.
Effantin, C., d'Incan, J., Bacis, R., and Vergès, J. (1979a), *J. Mol. Spectrosc.* **76**, 204.
Effantin, C., Amiot, C., and Vergès, J. (1979b), *J. Mol. Spectrosc.* **76**, 221.
Effantin, C., Michaud, F., Roux, F., d'Incan, J., and Vergès, J. (1982), *J. Mol. Spectrosc.* **92**, 349.
Felenbok, P., and Lefebvre-Brion, H. (1966), *Can. J. Phys.* **44**, 1677.
Field, R. W., and Lefebvre-Brion, H. (1974), *Acta Phys. Hung.* **35**, 51.
Field, R. W., Wicke, B. G., Simmons, J. D., and Tilford, S. G. (1972), *J. Mol. Spectrosc.* **44**, 383.
Field, R. W., Gottscho, R. A., and Miescher, E. (1975), *J. Mol. Spectrosc.* **58**, 394.
Gauyacq, D., and Horani, M. (1978), *Can. J. Phys.* **56**, 587.
Green, S., and Zare, R. N. (1977), *J. Mol. Spectrosc.* **64**, 217.
Halevi, P. (1965), *Proc. Phys. Soc.* **86**, 1051.
Herzberg, G. (1950), "Spectra of Diatomic Molecules," Van Nostrand-Reinhold, Princeton, New Jersey.
Herzberg, G., and Jungen, C. (1972), *J. Mol. Spectrosc.* **41**, 425.
Herzberg, G., and Jungen, C. (1982), *J. Chem. Phys.* **77**, 5876.
Hill, E. L., and Van Vleck, J. H., (1928), *Phys. Rev.* **32**, 250.
Hobey, W. D., and McLachlan, A. D. (1960), *J. Chem. Phys.* **33**, 1694.
Horani, M., Rostas, J., and Lefebvre-Brion, H. (1967), *Can. J. Phys.* **45**, 3319.
Hougen, J. T. (1970), *Monogr.* 115, Nat. Bur. Stand. (US) Washington D.C.
Huber, K. P., and Alberti, F. (1983), *J. Mol. Spectrosc.* **97**, 387.
Jenouvrier, A., Pascat, B., and Lefebvre-Brion, H. (1973), *J. Mol. Spectrosc.* **45**, 46.
Jette, A. N., and Miller, T. A. (1974), *Chem. Phys. Lett.* **29**, 547.
Jungen, C. (1966), *Can. J. Phys.* **44**, 3197.

Jungen, C. (1970), *J. Chem. Phys.* **53**, 4168.
Jungen, C. and Atabek, O. (1977), *J. Chem. Phys.* **66**, 5584.
Jungen, C, and Miescher, E. (1969), *Can. J. Phys.* **47**, 1769.
Kayama, K. (1965), *J. Chem. Phys.* **42**, 622.
Kayama, K., and Baird. J. C. (1967), *J. Chem. Phys.* **46**, 2604.
Kovács, I. (1969), "Rotational Structure in the Spectra of Diatomic Molecules," Am. Elsevier, New York.
Lagerqvist, A., and Renhorn, I. (1979), *Phys. Scr.* **19**, 289.
Langhoff, S. R., and Kern, C. W. (1977), *in* "Methods of Electronic Structure Theory" (H. F. Schaefer, III, ed.), Chap. 10, p. 381. Plenum, New York.
Larsson, M. (1981), *Phys. Scr.* **23**, 835.
Lavendy, H., Pouilly, B., and Robbe, J. M. (1984), *J. Mol. Spectrosc.* **103**, 379.
Lefebvre-Brion, H. (1969), *Can. J. Phys.* **47**, 541.
Lefebvre-Brion, H., and Bessis, N. (1969), *Can. J. Phys.* **47**, 2727.
Leoni, M. (1972), thesis, ETH, Zürich (unpublished).
Levy, D. H. (1973), *Adv. Magn. Reson.* **6**, 1.
Lewis, J. K., and Hougen, J. T. (1968), *J. Chem. Phys.* **48**, 5329.
Martin, F., Churassy, S., Bacis, R., Field, R. W., and Vergès, J. (1983), *J. Chem. Phys.* **79**, 3725.
McWeeny, R. (1965), *J. Chem. Phys.* **42**, 1717.
Messiah, A. (1960), "Mecanique Quantique," Dunod, Paris; (1961), "Quantum Mechanics," North-Holland Publ., Amsterdam.
Miescher, E. (1976), *Can. J. Phys.* **54**, 2074.
Miller, T. A., Freund, R. S., and Field, R. W. (1976), *J. Chem. Phys.* **65**, 3790.
Mulliken, R. S. (1971), *J. Chem. Phys.* **55**, 288.
Mulliken, R. S., and Christy, A. (1931), *Phys. Rev.* **38**, 87.
Ogawa, M., and Tanaka, Y. (1962), *Can. J. Phys.* **40**, 1593.
Oppenheimer, M. (1972), *J. Chem. Phys.* **57**, 3899.
Pauling, L., and Wilson, E. B. (1935), "Introduction to Quantum Mechanics," McGraw-Hill, New York.
Perdigon, P., and Femelat, B. (1982), *J. Phys. B* **15**, 2165.
Pouilly, B., Robbe, J. M., Schamps, J., Field, R. W., and Young, L. (1982), *J. Mol. Spectrosc.* **96**, 1.
Pyykkö, P., and Desclaux, J. P. (1976), *Chem. Phys. Lett.* **42**, 545.
Recknagel, A. (1934), *Z. Phys.* **87**, 375.
Roche, A. L., and Lefebvre-Brion, H. (1973), *J. Chem. Phys.* **59**, 1914.
Roche, A. L., and Lefebvre-Brion, H. (1975), *Chem. Phys. Lett.* **32**, 155.
Robbe, J. M. (1978), thesis, Université des Sciences et Techniques de Lille, Lille, France (unpublished).
Schamps, J. (1977), *J. Quant. Spectrosc. Radiat. Transfer* **17**, 685.
Sidis, V., and Lefebvre-Brion, H. (1971), *J. Phys. B* **4**, 1040.
Sink, M. L., Lefebvre-Brion, H., and Hall, J. A. (1975), *J. Chem. Phys.* **62**, 1802.
Slater, J. C. (1960), "Quantum Theory of Atomic Structure," Vol. 1, McGraw-Hill, New York.
Smith, F. T. (1969), *Phys. Rev.* **179**, 111.
Smith, W. H., Brzozowski, J., and Erman, P. (1976), *J. Chem. Phys.* **64**, 4628.
Tilford, S. G., and Simmons, J. D. (1972), *J. Phys. Chem. Ref. Data* **1**, 147.
Tinkham, M., and Strandberg, M. W. P. (1955), *Phys. Rev.* **97**, 937.
Van Vleck, J. H. (1929), *Phys. Rev.* **33**, 467.
Van Vleck, J. H. (1951), *Rev. Mod. Phys.* **23**, 213.
Veseth, L. (1970), *Theor. Chim. Acta* **18**, 368.
Veseth, L. (1972), *J. Mol. Spectrosc.* **44**, 251.

Veseth, L. (1973a), *J. Phys. B* **6**, 1473.
Veseth, L. (1973b), *Mol. Phys.* **26**, 101.
Wolniewicz, L., and Dressler, K. (1977), *J. Mol. Spectrosc.* **67**, 416.
Wolniewicz, L., and Dressler, K. (1979), *J. Mol. Spectrosc.* **77**, 286.
Yamazaki, M. (1963a), *Sci. Rep. Kanezawa Univ.* **8**, 371.
Yamazaki, M. (1963b), *Sci. Rep. Kanezawa Univ.* **8**, 397.
Yoshimine, M., and Tanaka, Y., (1978), private communication; (1976), *J. Chem. Phys.* **64**, 2254.
Yoshino, K., Freeman, D. E., and Tanaka, Y. (1979), *J. Mol. Spectrosc.* **76**, 153.
Zare, R. N., Schmeltekopf, A. L., Harrop, W. J., and Albritton, D. L. (1973), *J. Mol. Spectrosc.* **46**, 37.

Chapter 3

Methods of Deperturbation

This chapter deals with the process of obtaining molecular parameters from perturbed spectra. The terms in the Hamiltonian responsible for perturbations were discussed in Chapter 2, and the magnitudes of the off-diagonal matrix elements of these terms are interpreted in Chapter 4. The goal of a deperturbation calculation is not simply a phenomenological model that reproduces a specified set of spectral line frequencies and/or intensities; rather, the goal is a physical model, expressed in terms of potential energy curves and R-dependent interaction parameters, which not only is capable of reproducing all spectral data from which the model parameters were derived, but should also be capable of predicting additional properties of the observed levels and the characteristics of not yet observed levels.

There are two sides to the deperturbation process: the Hamiltonian model and the partially assigned and analyzed spectrum. The molecular Hamiltonian operator must be organized in a way that, of the infinite number of molecular bound states and continua, emphasis is placed on the finite subset of levels sampled in the spectra under analysis. A computational model is constructed in which the unknown information about the relevant states is represented by a set of adjustable parameters that may be systematically varied, until an acceptable match between calculated and observed properties is obtained.

The choice of a computational model depends in part on personal taste (matrix diagonalization versus numerical integration of partial differential equations, adiabatic versus diabatic potentials, Hund's coupling case), but the nature, quality, and completeness of the spectral data dictate the specific content of the model. This means that a great deal of work must be done on the raw spectra before the complete deperturbation model is set up and the final fitting process begun. Fragments of spectra must be partially assigned and reduced to phenomenological *effective molecular constants*. Traditional trial-and-error and graphical procedures are extremely useful at this initial stage.

Once lines are organized into branches and branches into subbands, computational models start to become useful by suggesting the identities of unassigned features and interrelationships between partially assigned features. The assembly of subbands into multiplet systems, vibrational levels into electronic states, and electronic states into configurations or Rydberg series often requires a succession of progressively more elaborate and extensive deperturbation models. At each stage, the parameters from one partial deperturbation model serve as initial parameter estimates for a more global and complete deperturbation.

The deperturbation methods discussed in this chapter range from non-adiabatic solutions of the complete Schrödinger equation (Section 3.1) to straight-line plots of quantities derived from observed energy levels (Section 3.3.1). Three computational techniques will be discussed: (1) variational calculations in which either the absolute energy (*ab initio* calculations, Section 3.1) or the sum of the squared deviations between observed and calculated energies (least-squares fitting, Section 3.4.1) is minimized in a finite basis set computational procedure; (2) perturbation theory in which the exact Hamiltonian is replaced by an approximate one, but where the Van Vleck (or "contact") transformation (Section 3.2) is used to replace an infinite basis set representation of $\mathbf{H}^{\text{approx}}$ by a finite-dimension effective Hamiltonian; (3) numerical solution of a system of coupled differential equations (Section 3.4.3).

3.1 Variational Calculations

The exact wavefunction ψ_i, associated with the energy level E_i, can, in principle, be expressed as a linear combination of the members of a complete set of basis functions. The mixing coefficients of each basis function are obtained by applying the variational principle [Eq. (2.1.9)]. As complete basis sets are infinite, the expansion of ψ_i must in practice be truncated and consequently the finite basis variational wavefunction obtained is only an approximation to ψ_i. The finite basis set can be the Born–Oppenheimer or, preferably, the adiabatic (i.e., solutions of the electronic Hamiltonian including the diagonal nuclear kinetic energy corrections, see Table 2.1) basis. The nonadiabatic vibronic wavefunctions (excluding rotation) have the form

$$\psi_i(r, R) = \sum_{n=1}^{q} \sum_{v_n=0}^{p_n} \Phi_n(r, R)\chi_{v_n}(R)c_{i,v_n}, \tag{3.1.1}$$

where the n-summation is over electronic states and the v_n-summation is over vibrational levels (for the basis states associated with electronic or vibrational continua, the summation is replaced by integration).

Application of the variational principle to the energy results in a system of linear equations of order $q^+ \sum_{n=1}^{q} p_n$:

$$0 = \sum_{n=1}^{q} \sum_{v_n=0}^{p_n} (H_{mv_m,nv_n} - E\delta_{mv_m,nv_n})c_{v_n} \tag{3.1.2}$$

for $m = 1, \ldots, q$ and $v_m = 0, \ldots, p_m$ where

$$H_{mv_m,nv_n} = \langle \Phi_m\chi_{v_m} |\mathbf{H}| \Phi_n\chi_{v_n}\rangle. \tag{3.1.3}$$

Equation (3.1.2) is written assuming that the members of the $\{\Phi\}$ and $\{\chi\}$ basis sets are orthogonal and normalized. The condition for the existence of a nontrivial set of c_{v_n} is that E has a value that satisfies the secular equation

$$0 = \left| \sum_{n=1}^{q} \sum_{v_n=0}^{p_n} (H_{mv_m,nv_n} - E\delta_{mv_m,nv_n}) \right|. \tag{3.1.4}$$

The number of E values satisfying Eq. (3.1.4) is equal to $q + \sum_{n=1}^{q} p_n$, although some eigenvalues may be degenerate. A different set of c_{v_n} is associated with each E_i value; these sets of mixing coefficients are distinguished by adding an index to specify the eigenvalue; hence c_{i,v_n} is the mixing coefficient for the $\Phi_n\chi_{v_n}$ basis function in the ψ_i eigenfunction.

When one does a variational calculation to find the eigenfunctions of the exact molecular **H**, the only quantum numbers available to label these eigenstates are symmetry labels (eigenvalues of an operator that commutes with **H**) and the energy order within a given set of symmetry labels. Only the index i appears in Eq. (3.1.1), not two indices to reflect the electronic *and* vibrational parentage of ψ. Although it is possible to label each ψ_i by its nominal character that corresponds to the largest $|c_{i,v_n}|^2$ in the Eq. (3.1.1) double summation, the only unambiguous (i.e., basis-set-independent) label is the energy order.

Very few nonadiabatic *ab initio* calculations have been performed. They require very accurate adiabatic calculations as a starting point. Variational nonadiabatic calculations have been completed for H_2^+ and H_2. In the H_2 $X^1\Sigma_g^+$ state, for which nonadiabatic effects are very small, the computed energy levels approach spectroscopic accuracy ($|\text{obs-calc}| \ll 1\ \text{cm}^{-1}$). For example, the nonadiabatic correction obtained by Bishop and Cheung (1978) with a 963 term wave function is only $0.42\ \text{cm}^{-1}$ for H_2 $X^1\Sigma_g^+$ $v = 0$.

An alternative method for obtaining the nonadiabatic wavefunctions [Eq. (3.1.1)], the coupled equations approach, will be discussed in Section 3.4.3.

The electronic–vibration levels of H_2^+ and the vibrational levels of the H_2 $X^1\Sigma_g^+$ state are of little relevance to the perturbation problems discussed in this book. However, the E, F and G, K double-minimum $^1\Sigma_g^+$ states of H_2, already discussed in Section 2.3.3, are typical examples of homogeneous perturbations. Dressler *et al.* (1979) have calculated, in a very accurate adiabatic basis set, the off-diagonal matrix elements necessary for a nonadiabatic variational calculation. The eigenvalues of the 42×42 secular determinant were obtained and found to differ from experimentally observed energy levels. This difference was only a few reciprocal centimeters for the lowest levels, but the error increased significantly for the highest levels. This is a consequence of the finite dimension of the variational calculation, in particular the exclusion of the higher Rydberg states.

At present, for molecules heavier than H_2, *ab initio* adiabatic or Born–Oppenheimer electronic wavefunctions are too inaccurate to be used as a starting point for a nonadiabatic calculation. Instead, a semiempirical variational procedure is used. The diagonal and off-diagonal matrix elements of Eq. (3.1.4) are taken as unknown parameters. The values of these parameters are then adjusted so as to minimize the squared deviation between the calculated eigenvalues and the observed energy levels. However, since there are $N(N+1)/2$ independent elements in a symmetric N by N matrix and only N eigenvalues, it is necessary to build in some constraints whereby the large number of matrix elements are expressed in terms of a small number of unknown molecular constants. For example, the diagonal matrix elements could be expressed as a Dunham expansion,

$$E_{vJ} = T_e + \sum_{lm} Y_{lm}(v + \tfrac{1}{2})^l[J(J+1)]^m, \tag{3.1.5}$$

many off-diagonal elements would be zero by symmetry, and the remaining off-diagonal elements would be interrelated by expressing them as a product of an explicitly calculable factor times an unknown variational parameter. In this way, a small number of variable parameter serves to define a model Hamiltonian.

It is not assured, just because the eigenvalues of a model Hamiltonian reproduce a group of spectroscopic energy levels, that the eigenfunctions of this Hamiltonian are identical to the eignefunctions of the exact **H** or that they would reproduce other properties of the observed levels. If a model Hamiltonian, least-squares fitting procedure had been used for the $^1\Sigma_g^+$ states of H_2, the obtained fitting parameters would not yield matrix elements identical to those accurately calculable from *ab initio* wavefunctions. The reason for this is that the fitting model is a finite-order Hamiltonian matrix, whereas an infinite-order matrix is required to represent the exact molecular **H**.

The errors associated with the replacement of an infinite **H** matrix by a finite *effective* $\mathbf{H}^{\text{eff}}$ must be, in principle, small and of estimable magnitude. The replacement of **H** by $\mathbf{H}^{\text{eff}}$ is equivalent to a partitioning of an infinite basis set into two classes: *class 1* includes those basis functions that belong to the proper symmetry species and have energy expectation values in the energy region sampled by the spectrum under analysis; *class 2* includes the infinite number of remaining basis functions that are energetically remote from the class 1 functions, and their effect on the class 1 functions may therefore be treated by nondegenerate perturbation theory. The procedure for truncating the infinite **H** into a finite $\mathbf{H}^{\text{eff}}$ is known as the *Van Vleck* or the *contact transformation.*

3.2 The Van Vleck Transformation and Effective Hamiltonians

The Van Vleck transformation is an approximate block diagonalization procedure. It allows one, in effect, to throw away an infinite number of "unimportant" (class 2) basis functions after taking into account, through second-order nondegenerate perturbation theory, the effect of these ignorable functions on the finite number of "important" (class 1) functions. The Van Vleck transformation, **T**, such that

$$\mathbf{T}^{\dagger}\mathbf{H}\mathbf{T} = \tilde{\mathbf{H}} = \left(\begin{array}{c|c} \tilde{\mathbf{H}}_2 & \tilde{\mathbf{H}}_{1\sim 2} \\ \hline \tilde{\mathbf{H}}_{2\sim 1} & \tilde{\mathbf{H}}_1 \end{array}\right), \tag{3.2.1}$$

has the effect of reducing the size of the class 1 $\sim$ class 2 matrix elements $\tilde{H}_{1\sim 2}$ relative to the corresponding $H_{1\sim 2}$ elements of the exact **H** so that the error in computing the energy of a class 1 level made by neglecting a nonzero $\tilde{H}_{1\sim 2}$

matrix element is significantly smaller than the error made by neglecting $H_{1\sim 2}$ itself. Thus it is a good approximation to define an effective Hamiltonian matrix as

$$\mathbf{H}^{\text{eff}} \equiv \tilde{\mathbf{H}}_1, \tag{3.2.2}$$

where only the class 1 portion of $\tilde{\mathbf{H}}$ is retained.

The approximate block diagonalization produced by the Van Vleck transformation may be understood by first illustrating how nondegenerate perturbation theory approximately transforms the two-level problem into two almost decoupled one-level blocks. Consider

$$\mathbf{H} = \mathbf{H}^0 + \mathbf{H}', \tag{3.2.3}$$

where the ψ_i^0 are eigenfunctions of $\mathbf{H}^0$ with eigenvalues E_i^0. The normalized eigenfunctions, correct through first-order perturbation theory, are

$$\begin{aligned} \psi_1 &\simeq c_1\psi_1^0 + \psi_2^0 H'_{12}/(E_1^0 - E_2^0) \\ \psi_2 &\simeq c_2\psi_2^0 + \psi_1^0 H'_{21}/(E_2^0 - E_1^0), \end{aligned} \tag{3.2.4}$$

where c_1 and c_2 are normalization factors,

$$c_1 = c_2 = [1 - |H'_{12}/(E_1^0 - E_2^0)|^2]^{1/2} \simeq 1 - |H'_{12}/(E_1^0 - E_2^0)|.$$

If ψ_1, ψ_2 define the transformation

$$\mathbf{T}^\dagger\mathbf{H}\mathbf{T} = \tilde{\mathbf{H}},$$

then the matrix elements of $\tilde{\mathbf{H}}$ are

$$\begin{aligned} \langle\psi_1|\tilde{\mathbf{H}}|\psi_1\rangle = {} & E_1^0 + H'_{11} + |H'_{12}|^2/(E_1^0 - E_2^0) \\ & + [H'_{12}/(E_1^0 - E_2^0)]^2[H'_{11} - H'_{22} - 2H'_{12}] \end{aligned} \tag{3.2.5a}$$

$$\begin{aligned} \langle\psi_2|\tilde{\mathbf{H}}|\psi_2\rangle = {} & E_2^0 + H'_{22} - |H'_{12}|^2/(E_1^0 - E_2^0) \\ & + [H'_{12}/(E_1^0 - E_2^0)]^2[H'_{11} - H'_{22} + 2H'_{12}] \end{aligned} \tag{3.2.5b}$$

$$\begin{aligned} \langle\psi_1|\tilde{\mathbf{H}}|\psi_2\rangle = {} & [H'_{12}/(E_1^0 - E_2^0)][(H'_{22} - H'_{11}) \\ & - |H'_{12}|^2/(E_1^0 - E_2^0)]. \end{aligned} \tag{3.2.6}$$

The first three terms on the right-hand side of Eqs. (3.2.5) are the zeroth-, first-, and second-order terms from nondegenerate perturbation theory. The lowest-order term in the off-diagonal matrix element, Eq. (3.2.6), is second-order in the perturbation term, $\mathbf{H}'$, and would give rise to a fourth-order correction to the energy. Thus neglect of $\tilde{H}_{12}$ would result in a less serious energy error than neglect of H_{12}.

An original derivation of the Van Vleck transformation was given by Löwdin (1951). The following derivation is adapted from Herschbach (1956) and Wollrab (1967). It is useful to express the $\mathbf{H}$ and $\tilde{\mathbf{H}}$ matrix representations of the Hamiltonian as

$$\mathbf{H} = \mathbf{H}^0 + \lambda\mathbf{H}' \tag{3.2.7}$$

$$\tilde{\mathbf{H}} = \tilde{\mathbf{H}}^0 + \lambda\tilde{\mathbf{H}}' + \lambda^2\tilde{\mathbf{H}}'' + \cdots \tag{3.2.8}$$

where λ is an order-sorting parameter. Any unitary matrix can be expressed as

$$\mathbf{T} = e^{i\lambda\mathbf{S}} = 1 + i\lambda\mathbf{S} - (\lambda^2/2)\mathbf{S}^2 + \cdots \tag{3.2.9}$$

where $\mathbf{S}$ is a Hermitian matrix. A specific unitary transformation, $\tilde{\mathbf{H}} = \mathbf{T}^\dagger\mathbf{H}\mathbf{T}$, called the contact transformation, is sought whereby all class 1 ~ class 2 matrix elements in $\tilde{\mathbf{H}}$ are smaller, by a factor of $[\mathrm{H}_{12}/(E_1^0 - E_2^0)]^2$, than the corresponding matrix element in $\mathbf{H}$. Writing out the effect of the general unitary transformation,

$$\tilde{\mathbf{H}} = \mathbf{H}^0 + \lambda\tilde{\mathbf{H}}' + \lambda^2\tilde{\mathbf{H}}'' = [1 - i\lambda\mathbf{S} - (\lambda^2/2)\mathbf{S}^2](\mathbf{H}^0 + \lambda\mathbf{H}')[1 + i\lambda\mathbf{S} - (\lambda^2/2)\mathbf{S}^2], \tag{3.2.10}$$

the terms with identical powers of λ can be collected,

$$\lambda^0: \quad \tilde{\mathbf{H}}^0 = \mathbf{H}^0 \tag{3.2.11}$$

$$\lambda^1: \quad \tilde{\mathbf{H}}' = \mathbf{H}' + i(\mathbf{H}^0\mathbf{S} - \mathbf{S}\mathbf{H}^0) \tag{3.2.12}$$

$$\lambda^2: \quad \tilde{\mathbf{H}}'' = i(\mathbf{H}'\mathbf{S} - \mathbf{S}\mathbf{H}') + \mathbf{S}\mathbf{H}^0\mathbf{S} - \tfrac{1}{2}(\mathbf{H}^0\mathbf{S}^2 - \mathbf{S}^2\mathbf{H}^0). \tag{3.2.13}$$

Equations (3.2.11)–(3.2.13) allow specific requirements to be built into $\mathbf{S}$ so that $\mathbf{T}$ transforms $\mathbf{H}$ into the approximately block diagonal form desired for $\mathbf{H}$. For clarity, class 1 and 2 basis functions will be denoted by Roman (a, b) and Greek (α, β) letters, respectively. The goal is to make all first-order class 1–class 2 matrix elements of $\tilde{\mathbf{H}}$ vanish,

$$\tilde{H}'_{a\alpha} = 0,$$

without causing any mixing among the class 1 (or class 2) basis functions. This is accomplished by requiring

$$S_{ab} = 0 \qquad \text{(no mixing among class 1 functions)} \tag{3.2.14}$$

$$S_{\alpha\beta} = 0 \qquad \text{(no mixing among class 2 functions)} \tag{3.2.15}$$

$$S_{a\alpha} = \frac{iH'_{a\alpha}}{E_a^0 - E_\alpha^0} \tag{3.2.16}$$

and inserting Eqs. (3.2.14)–(3.2.16) into Eq. (3.2.12). When Eq. (3.2.16) is inserted into Eq. (3.2.12),

$$\begin{aligned}\tilde{H}'_{a\alpha} &= H'_{a\alpha} + i[(\mathbf{H}^0\mathbf{S})_{a\alpha} - (\mathbf{S}\mathbf{H}^0)_{a\alpha}] \\ &= H'_{a\alpha} - \left[\frac{E_a^0 H'_{a\alpha}}{E_a^0 - E_\alpha^0} - \frac{H'_{a\alpha}E_\alpha^0}{E_a^0 - E_\alpha^0}\right] = 0,\end{aligned}$$

because

$$\mathbf{H}^0|a\rangle = E_a^0|a\rangle$$
$$\mathbf{H}^0|\alpha\rangle = E_\alpha^0|\alpha\rangle$$

Equations (3.2.14) and (3.2.15) cause the class 1 blocks of **H** and **S** (hence **T**) to commute (also the class 2 blocks), hence the only effect of **T** on the class 1 block of **H** is to fold into it some information from the off-diagonal $\mathbf{H}_{1\sim 2}$ block. Equation (3.2.16) forces the only nonzero elements of the second term in Eq. (3.2.12), $i(\mathbf{H}^0\mathbf{S} - \mathbf{S}\mathbf{H}^0)$, to exactly cancel all interblock elements of $\mathbf{H}'$. Thus the lowest-order interblock matrix elements of $\tilde{\mathbf{H}}$ occur in the second-order term, $\tilde{\mathbf{H}}''$, and may be neglected.

Equations (3.2.11)–(3.2.16) completely define the Van Vleck transformation. The matrix elements in the class 1 block of $\tilde{\mathbf{H}}$ are

$$\tilde{H}_{\mathrm{ab}} = E_{\mathrm{a}}^0\delta_{\mathrm{ab}} + \lambda H'_{\mathrm{ab}} + \frac{\lambda^2}{2}\sum_\alpha\left[\frac{H'_{\mathrm{a}\alpha}H'_{\alpha\mathrm{b}}}{E_{\mathrm{a}}^0 - E_\alpha^0} + \frac{H'_{\mathrm{a}\alpha}H'_{\alpha\mathrm{b}}}{E_{\mathrm{b}}^0 - E_\alpha^0}\right] + \cdots. \tag{3.2.17}$$

The first two terms in Eq. (3.2.17) are identical to the corresponding terms in the exact **H**. The correction terms introduced by the Van Vleck transformation appear both on and off diagonal in $\tilde{H}$, in contrast to the strictly on-diagonal location of energy correction terms from ordinary, nondegenerate second-order perturbation theory. Often, the off-diagonal correction term is replaced by a single summation over an averaged energy denominator,

$$\tilde{H}''_{\mathrm{ab}} = \sum_\alpha \frac{H'_{\mathrm{a}\alpha}H'_{\alpha\mathrm{b}}}{[(E_{\mathrm{a}}^0 + E_{\mathrm{b}}^0)/2 - E_\alpha^0]}, \tag{3.2.18}$$

which is negligibly different from the form of Eq. (3.2.17) if

$$|E_{\mathrm{a}}^0 - E_{\mathrm{b}}^0| \ll |(E_{\mathrm{a}}^0 + E_{\mathrm{b}}^0)/2 - E_\alpha^0|.$$

The third-order terms, $\tilde{H}'''$, may be obtained by retaining additional powers of λ in Eq. (3.2.10),

$$\begin{aligned}\tilde{H}'''_{\mathrm{ab}} &= \sum_{\alpha,\beta}\left[\frac{H'_{\mathrm{a}\alpha}H'_{\alpha\beta}H'_{\beta\mathrm{b}}}{(E_{\mathrm{a}}^0 - E_\alpha^0)(E_\beta^0 - E_{\mathrm{b}}^0)}\right] \\ &\quad - \frac{1}{2}\sum_{\alpha,\mathrm{c}}\left[\frac{H'_{\mathrm{ac}}H'_{\mathrm{c}\alpha}H'_{\alpha\mathrm{b}}}{(E_{\mathrm{c}}^0 - E_\alpha^0)(E_{\mathrm{b}}^0 - E_\alpha^0)} + \frac{H'_{\mathrm{a}\alpha}H'_{\alpha\mathrm{c}}H'_{\mathrm{cb}}}{(E_{\mathrm{a}}^0 - E_\alpha^0)(E_c^0 - E_\alpha^0)}\right].\end{aligned} \tag{3.2.19}$$

It is important to note that the summations implied in the Van Vleck transformation are over all electronic and all vibrational levels of class 2. The perturbation summations are restricted neither to only the nearest electronic state of a particular symmetry species nor to the bound vibrational levels. It is necessary, in principle, to integrate over the electronic (ionization) and nuclear (dissociation) continua.

The Van Vleck transformation specifies the functional form (Zare *et al.*, 1973; Hill and Van Vleck, 1928; Mulliken and Christy, 1930; Horani *et al.*, 1967; Freed, 1966; Miller, 1969; Brown *et al.*, 1979; Brown and Merer, 1979) of the effective Hamiltonian in terms of a small number of parameters from $\mathbf{H}^0$ and $\mathbf{H}'$ having explicit mechanical (rotation–vibration constants) or electromagnetic (T_e, A, λ, γ) significance and a much larger group of second-order parameters from $\tilde{\mathbf{H}}''$ (o, p, q, D, H, L, A_D, etc.) of significance ranging from explicit to phenomenological fitting parameters.

Values for many of the parameters in $\mathbf{H}^{\text{eff}}$ cannot be determined from a spectrum, regardless of the quality or quantity of the spectroscopic data, because of correlation effects. When two parameters enter into the effective Hamiltonian with identical functional forns, only their sum can be determined empirically. Sometimes it is possible to calculate, either *ab initio* or semiempirically, the value of one second-order parameter, thereby permitting the other correlated parameter to be evaluated from the spectrum. Often, although the parameter definition specifies a summation over an infinite number of states, the largest part or the explicitly vibration-dependent part of the parameter may be evaluated from an empirically determined electronic matrix element times a sum over calculable vibrational matrix elements and energy denominators (Wicke *et al.*, 1972).

Brown *et al.* (1979) and Brown and Merer (1979) have proposed that $\mathbf{H}^{\text{eff}}$ be expressed in terms of a minimal set of fitting parameters. In this way, the same $\mathbf{H}^{\text{eff}}$ would be used by most spectroscopists and the task of reproducing a spectrum from a set of published constants would be simplified. This proposal of a universal set of fitting parameters will not necessarily lead to the reduction of spectra to phenomenological rather than physical constants, because each of the fitting parameters is defined explicitly in terms of the parameters from $\mathbf{H}^0$, $\mathbf{H}'$, and $\tilde{\mathbf{H}}''$. There is, however, a danger that users of universal $\mathbf{H}^{\text{eff}}$ matrices will neither study the assumptions made in the derivation of $\mathbf{H}^{\text{eff}}$ nor be prepared to exploit the significance of or the implicit relationships among the fitting parameters.

Before concluding this section it will be useful to discuss two examples of the Van Vleck transformation: centrifugal distortion in a $^3\Pi$ state and Λ-doubling in a $^2\Pi$ state.

Although centrifugal distortion is not a perturbation effect, a derivation of the form of the centrifugal distortion terms in $\mathbf{H}^{\text{eff}}$ provides an excellent illustration of the Van Vleck transformation. If the vibrational eigenfunctions of the *nonrotating* molecular potential, $V(R)$ rather than $[V(R) + J(J+1)\hbar^2/2\mu R^2]$, are chosen as the vibrational basis set, then the rotational "constant" becomes an operator,

$$\langle v|B(R)|v'\rangle = (\hbar^2/2\mu)\langle v|R^{-2}|v'\rangle = B_{vv'} \neq 0,$$

which is not diagonal in the vibrational basis set.

The rotational Hamiltonian for a $^3\Pi$ state has matrix elements

$$\langle {}^3\Pi_0, v|\mathbf{H}^{\mathrm{ROT}}|{}^3\Pi_0, v'\rangle = B_{vv'}[J(J+1)+1] \tag{3.2.20a}$$

$$\langle {}^3\Pi_1, v|\mathbf{H}^{\mathrm{ROT}}|{}^3\Pi_1, v'\rangle = B_{vv'}[J(J+1)+1] \tag{3.2.20b}$$

$$\langle {}^3\Pi_2, v|\mathbf{H}^{\mathrm{ROT}}|{}^3\Pi_2, v'\rangle = B_{vv'}[J(J+1)-3] \tag{3.2.20c}$$

$$\langle {}^3\Pi_0, v|\mathbf{H}^{\mathrm{ROT}}|{}^3\Pi_1, v'\rangle = -B_{vv'}[2J(J+1)]^{1/2} \tag{3.2.20d}$$

$$\langle {}^3\Pi_0, v|\mathbf{H}^{\mathrm{ROT}}|{}^3\Pi_2, v'\rangle = 0 \tag{3.2.20e}$$

$$\langle {}^3\Pi_1, v|\mathbf{H}^{\mathrm{ROT}}|{}^3\Pi_2, v'\rangle = -B_{vv'}[2J(J+1)-4]^{1/2}. \tag{3.2.20f}$$

The Van Vleck transformation incorporates the effect of all $\Delta v \neq 0$ matrix elements into the $\Delta v = 0$ block of the $^3\Pi$ Hamiltonian. The perturbation summation

$$\sum_{v'} \frac{\langle v|B(R)|v'\rangle\langle v'|B(R)|v\rangle}{E_v^0 - E_{v'}^0} \equiv -D_v \tag{3.2.21}$$

defines D_v (Zare *et al.*, 1973; Albritton *et al.*, 1973), but this summation must be carried out by summing over the bound levels and integrating over the continuum. The matrix elements of the transformed Hamiltonian, $\tilde{\mathbf{H}}$, require summation over the $\Omega = 0$, 1, and 2 levels of the v' level:

$$\langle {}^3\Pi_0, v|\tilde{\mathbf{H}}|{}^3\Pi_0, v\rangle = -D_v[J(J+1)+1]^2 - D_v[2J(J+1)],$$

where the first term comes from the $\Omega = 0 \to 0 \to 0$ path and the second from $\Omega = 0 \to 1 \to 0$. The centrifugal distortion terms appear in $\mathbf{H}^{\mathrm{eff}}$ as follows:

$$\langle {}^3\Pi_0, v|\tilde{\mathbf{H}}|{}^3\Pi_0, v\rangle = -D_v[x^2 + 6x + 4] \tag{3.2.22a}$$

$$\langle {}^3\Pi_1, v|\tilde{\mathbf{H}}|{}^3\Pi_1, v\rangle = -D_v[x^2 + 8x] \tag{3.2.22b}$$

$$\langle {}^3\Pi_2, v|\tilde{\mathbf{H}}|{}^3\Pi_2, v\rangle = -D_v[x^2 - 2x] \tag{3.2.22c}$$

$$\langle {}^3\Pi_0, v|\tilde{\mathbf{H}}|{}^3\Pi_1, v\rangle = +D_v[2(x+2)(2x)^{1/2}] \tag{3.2.22d}$$

$$\langle {}^3\Pi_0, v|\tilde{\mathbf{H}}|{}^3\Pi_2, v\rangle = -D_v[4x(x-2)]^{1/2} \tag{3.2.22e}$$

$$\langle {}^3\Pi_1, v|\tilde{\mathbf{H}}|{}^3\Pi_2, v\rangle = +D_v[2x(2x-4)^{1/2}] \tag{3.2.22f}$$

$$x \equiv J(J+1).$$

It is evident from Eqs. (3.2.22) that the correct form of the centrifugal distortion terms is *not* given by the intuitive prescription: replace B_v everywhere it appears in Eqs. (3.2.20) by $B_v - D_vJ(J+1)$. Alternative methods for evaluating the Eq. (3.2.21) perturbation summation have been proposed by Tellinghuisen (1973) and Kirschner and Watson (1973).

As discussed in Sections 2.5.4 and 4.5, the Λ-doubling in a $^2\Pi$ state can result from interactions with remote $^2\Sigma^+$ and $^2\Sigma^-$ states via $\mathbf{H}^{\mathrm{ROT}}$ and $\mathbf{H}^{\mathrm{SO}}$. The Van Vleck transformation defines [Eqs. (4.5.1a)–(4.5.3a)] three second-order parameters (o, p, and q) that appear in the $^2\Pi$ block of $\tilde{\mathbf{H}}''$. These second-order $^2\Pi \sim {}^2\Sigma$ interaction parameters cause both e/f independent level shifts as well as Λ-doubling. The e/f-dependent terms all arise from the e/f-dependence of the

$$\langle {}^2\Pi_{1/2{}^e_f}, v_\Pi|\mathbf{H}^{\mathrm{ROT}}|{}^2\Sigma_{s{}^e_f}, v_\Sigma\rangle = [1 \mp (-1)^s(J+\tfrac{1}{2})]B_{v_\Pi v_\Sigma}\langle\Pi|\mathbf{L}^+|\Sigma\rangle$$

matrix element ($s = 0$ for Σ^+ and $s = 1$ for Σ^-). This means that e/f-dependent terms can arise in $\tilde{\mathbf{H}}''$ if (1) the $\mathbf{H}^{\mathrm{ROT}}$ operator appears once (p) or twice (q) in $H'_{a\alpha}H'_{\alpha b}$, and (2) either a or b $= |{}^2\Pi_{1/2}\rangle$. This explains why the o parameter cannot contribute to Λ-doubling in $^2\Pi$ states and why e/f-dependence appears in the $\langle {}^2\Pi_{1/2}|\tilde{\mathbf{H}}''|{}^2\Pi_{1/2}\rangle$ and $\langle {}^2\Pi_{1/2}|\tilde{\mathbf{H}}''|{}^2\Pi_{3/2}\rangle$ matrix elements.

3.3 Approximate Solutions

At an early stage in the analysis of a perturbed spectrum, graphical methods based on approximate algebraic solutions to the secular determinant [Eq. (3.1.4)] can be very useful, both as an aid in pattern recognition and in estimation of parameter values. At some point before completion of the analysis it is usually advantageous to switch from a graphical/algebraic approach to one based on exact diagonalization of $\mathbf{H}^{\mathrm{eff}}$ matrices. The labor in setting up an algebraic model for anything larger than a two-level interaction problem, even when the equations are presented in a convenient format (Kovács, 1969), is often greater than required to write a suitable $\mathbf{H}^{\mathrm{eff}}$ subroutine. In addition, a statistically rigorous least-squares fitting procedure is often better able to make use of fragmentary data than a graphical or algebraic procedure.

3.3.1 GRAPHICAL METHODS FOR DEPERTURBATION

There are many extremely useful graphical presentations of spectral data. Most are based on several of the following ideas: (1) straight-line plots, (2) scale expansion by subtracting away the expected, unperturbed behavior, (3) display of large-scale regularities, (4) the center-of-gravity rule.

Prior to the days when spectral line assignments could be checked by double-resonance or resolved-fluorescence techniques, assignments were based on both pattern recognition and redundancy. By constructing suitable combination differences, the spectroscopist is able to go directly from the spectrum to level separations in either the upper or lower electronic state without assuming any model (except that dipole selection rules restrict ΔJ to $\pm 1, 0$). This means that combination difference graphical presentations can serve the dual purposes of displaying a structural feature of one electronic state and verifying assignments by plotting corresponding combination differences from two bands thought to have an upper or lower level in common.

Figure 2.16 is a reduced term-value plot for the CS^+ $A^2\Pi_i$ ($v = 5$) state perturbed by $X^2\Sigma^+$ ($v = 13$). The form of the plot $[T - B(J + \frac{1}{2})^2$ versus $(J+\frac{1}{2})^2]$ is appropriate for a near case (a) $^2\Pi$ state. The scale expansion achieved by subtracting 0.68 $(J + \frac{1}{2})^2$ from all of the energy levels (this amounts to 460 cm^{-1} by $J = 25.5$) permits the $\sim$2-cm^{-1} perturbation-level shifts in $^2\Pi_{3/2}$ at $J = 18.5$ and 19.5 to remain visible and allows the values of the effective B_v and p_v $^2\Pi$ constants to be determined from the slopes and splitting of the nearly straight $^2\Pi_{1/2}$ e- and f-level plots. The pattern of the perturbation is unmistakably that of a Σ state (large splitting between e and f levels of the same J value) with $B_\Sigma > B_\Pi$ (level shift first positive, then negative). The maximum level shift provides a good estimate (lower bound) for the $B\mathbf{L}^+\mathbf{J}^-$ $^2\Pi_{3/2} \sim {}^2\Sigma^+$ matrix element [2.5 $cm^{-1}/(J + \frac{1}{2}) \simeq 0.13$].

Figure 4.7 is a crossing diagram [E versus $J(J+1)$] that shows the J-values at which the vibrational levels of the $Si^{16}O$ and $Si^{18}O$ $A^1\Pi$ state are crossed by various perturbers (Field *et al.*, 1976). The different classes of perturber ($^3\Sigma^-$ versus $^1\Sigma^-$ versus $^{1,3}\Delta$) were immediately evident from the patterns of level crossings in the e and f components of the $A^1\Pi$ state. The overall pattern of perturbations, including the expected locations of undetected weak perturbations or multiple crossings as shown by the nearly straight tie lines connecting the crossing points for each class of perturber, is very clear. This sort of diagram is very useful for arranging the perturbing levels into electronic states and establishing their relative vibrational numberings.

Several other kinds of information are available from the information in Fig. 4.7. Two consecutive vibrational levels of $A^1\Pi$ are crossed by the same vibrational level of $e^3\Sigma^-$. The Q branch (F_2) crossing occurs in $v_A = 1$ and 0 at $J = 48.9$ and 61.5. Since the deperturbed $A^1\Pi$ term energies at these two J-values are known by interpolation, accurate values for $B(^3\Sigma^-)$ and $E(^3\Sigma^-)$ can be determined. Alternatively, the J-values of all three $^1\Pi \sim {}^3\Sigma^-$ (F_1, F_2, F_3) crossings are determined accurately at many perturbations, for example, (56.5, 61.5, 66.0) in $Si^{16}O$ $A^1\Pi$ $v = 0$. Each $^3\Sigma^-$ N-level consists of three near-degenerate J-components. The perturbation selection rule is $\Delta J = 0$, thus the F_1 and F_3 crossings involve $N = J - 1$ and $N = J + 1$ levels, and the energy separation between these levels is

$$B(^3\Sigma^-)[(J_3+1)(J_3+2)-(J_1-1)J_1]$$
$$= B(^1\Pi)[J_3(J_3+1)-J_1(J_1+1)]. \qquad (3.3.1)$$

Thus,

$$B(^3\Sigma^-) = B(^1\Pi)\frac{J_3(J_3+1)-J_1(J_1+1)}{(J_3+1)(J_3+2)-(J_1-1)J_1}$$
$$= B(^1\Pi)\ 0.826. \qquad (3.3.2)$$

Finally, the absence of a detectable perturbation at a crossing point predicted by drawing nearly horizontal tie lines between corresponding perturbations on Fig. 4.7 (no unfilled Δ on $v = 1$ of $Si^{16}O$ near $J = 70$) implies that either the expected perturbing state would have "$v = -1$" or that the matrix element is less than

$$|H_{\text{min}}| < [\delta E_{\text{min}} \Delta B(J+1)]^{1/2}, \qquad (3.3.3)$$

where δE_{min} is the minimum detectable level shift, $\Delta B = |B(^1\Pi) - B(\Delta)|$, and J is the J-value just below the predicted noninteger crossing point.

Figure 3.1 is based on Gerö's (1935) extra-line method for determining the B-value of the perturbing state, B_p. It is based on the observation that a perturbation disrupts the regular series of lines in a branch. One series of lines includes the "main" lines below the crossing point and a few "extra lines" (see

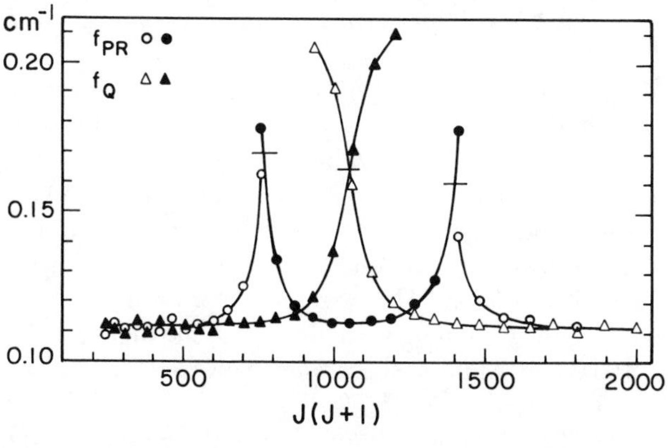

Fig. 3.1 Plot of effective ΔB-value versus $J(J+1)$ for the $Si^{16}O$ $e^3\Sigma^-$ $(v = 9) \sim A^1\Pi$ $(v = 2)$ perturbation. The quantities plotted [defined by Eqs. (3.3.1) and (3.3.2)] are related to the difference between the rotational constants of the lower and upper states, $B'' - B'$. When a perturbation occurs, the effective rotational constant of the perturbed state approaches the average of the B-values of the perturbed and perturbing states, $(B' + B_p)/2$. The horizontal marks locate the exact level crossing point, from which B_p may be determined. (From Field *et al.*, 1976.)

Section 5.2.1) above the crossing. This series of lines appears to break off abruptly at or above the crossing and is replaced by another, seemingly unrelated, series of lines starting just below or at the crossing. The discontinuity between these two series of lines is represented in tables of spectral lines by a column shift. Often there are two lines, one in each column, for each J-value at or near the crossing. Since the members of a given series appear to be related to each other, it is reasonable to use combination difference relationships to determine the effective B-values in the local J-region. Gerö (1935) derived

$$\frac{R(J-2) - R(J-1) + P(J) - P(J+1)}{4J} = B'' - B' - 2J^2(D'' - D') \tag{3.3.4a}$$

$$\frac{Q(J-1) - Q(J)}{2J} = B'' - B' - 2J^2(D'' - D') \tag{3.3.4b}$$

from the usual expressions for the frequencies of R, Q, and P branch lines, for example (neglecting centrifugal distortion),

$$R(J-2) = \nu_0 + B'(J-1)(J) - B''(J-2)(J-1) \tag{3.3.5}$$

$$\Delta R = R(J-2) - R(J-1) = -2JB' + 2(J-1)B'' \tag{3.3.6}$$

$$\Delta P = P(J) - P(J+1) = -2JB' + 2(J+1)B'' \tag{3.3.7}$$

$$\Delta R + \Delta P = -4JB' + 4JB''. \tag{3.3.8}$$

If $(\Delta R + \Delta P)/4J$ or $\Delta Q/2J$ is plotted versus J^2 for an unperturbed band, a straight line is obtained with intercept and slope, respectively, of $B'' - B'$ and $D'' - D'$.

Figure 3.1 shows how the $B'' - B'$ versus $J(J+1)$ plot is affected by a $^1\Pi \sim {}^3\Sigma^-$ perturbation in the upper state. Far above and below the crossings (at $J' = 26.6$ and 36.1 in the P and R branches for the e-levels, at $J' = 31.3$ in the Q branch for the f-levels), the $B'' - B'$ value is about 0.111 cm^{-1}. The limiting $B'' - B'$ value for the $\Delta Q/2J$ plot (filled triangles) from the $J' > 31$ continuation of the low-J series of lines is $\sim$0.22 cm^{-1}, implying that $B(^3\Sigma^-)$ is $\sim$0.11 cm^{-1} smaller than $B(^1\Pi)$. A more accurate measure of $B(^3\Sigma^-)$ is obtained from the intersection between the low-J and high-J $\Delta Q/2J$ curves at $B'' - B' = 0.163$ cm^{-1}. At the exact crossing between $^1\Pi_f$ and $^3\Sigma^-$ (F_2) levels, the effective B'-value is $[B(^1\Pi) + B(^3\Sigma^-)]/2$, thus $B(^1\Pi) - B(^3\Sigma^-) = 0.104$ cm^{-1}. The $B(^3\Sigma^-)$ values obtained from the plot at the $^1\Pi_e \sim {}^3\Sigma^-$ (F_1 and F_3) crossings are slightly lower and higher than at the F_2 crossing, indicating that $^1\Pi_e$ simultaneously feels the effect of both F_1 and F_3 $^3\Sigma^-$ e-components. Even though the F_1 and F_3 perturbations are not isolated from each other, the average of the two $B(^3\Sigma^-)$-values obtained graphically at these crossings is identical to the value from the isolated F_2 crossing.

The center of gravity rule can be useful for obtaining parameter estimates, locating extra lines, and detecting the presence of an unsuspected second perturber. When main and extra lines are observed for more than two J-values at a level crossing, then a plot of the average energy,

$$\overline{E(J)} = [E_{\mathbf{MAIN}}(J) + E_{\mathbf{EXTRA}}(J)]/2, \tag{3.3.9}$$

versus $J(J+1)$ has slope and intercept

$$\bar{B} = (B_{\mathbf{MAIN}} + B_{\mathbf{EXTRA}})/2 \tag{3.3.10}$$

$$\bar{E} = [E_{\mathbf{MAIN}}(0) + E_{\mathbf{EXTRA}}(0)]/2. \tag{3.3.11}$$

This is true regardless of the perturbation mechanism (J-dependent or J-independent matrix elements) because the trace of a matrix is representation invariant; the sum of the basis function (i.e., deperturbed) energies is equal to the sum of the eigenvalues. Since the approximately unperturbed $B_{\mathbf{MAIN}}$ and $E_{\mathbf{MAIN}}(0)$ values are usually known from a relatively perturbation-free portion of the band, the constants for the perturbing state can be inferred from Eqs. (3.3.10) and (3.3.11). If the average energy plot shows any deviation from linearity, this implies either an incorrect line identification or an additional perturber.

Kovács (1969) gives many examples of approximate graphical solutions to single- and multiple-state perturbation problems.

3.3.2 DIRECT DIAGONALIZATION VERSUS ALGEBRAIC APPROACHES

Perturbation theory is an extremely useful analytic tool. It is almost always possible to treat a narrow range of J-values in a multistate interaction problem by exactly diagonalizing a two-level problem after correcting, by nondegenerate perturbation theory or a Van Vleck transformation, for the effects of other nearby perturbers. Such a procedure can enable one to test for the sensitivity of the data set to the value of a specific unknown parameter.

There is a long history in molecular spectroscopy of elegant algebraic solutions for the eigenvalues and eigenfunctions of complicated secular equations. Whenever the secular determinant is larger than 3×3, algebraic solution requires the use of nondegenerate perturbation theory, which can only be valid when a critical off-diagonal matrix element H_{ab} is smaller than a zero-order energy difference, $E_a^0 - E_b^0$. Most algebraic solutions describe limiting behavior very well, but seldom describe all observable levels, from low to high J. It is not very satisfying when one must use different models to account for different portions of a single band. Perturbations make the situation even more difficult, necessitating frequent switches from one algebraic model to another.

There is a good historical reason for the algebraic tradition in spectroscopy. Without computers, exact matrix diagonalizations were impossibly time consuming. There was no choice but to resort to algebraic formulas. Now there is a choice.

3.4 Exact Solutions

3.4.1 LEAST-SQUARES FITTING

One starts with a model, either a set of equations or a matrix to be diagonalized. The least-squares fitting process involves comparison of the model predictions to a set of experimental data, followed by adjustment of the model parameters to minimize the squared deviations between the experimental data and model predictions. The experimental data may be of several types, either separately or in combination: spectral line frequencies, term values, combination differences, relative intensities, magnetic *g*-values, etc. Each input datum can be weighted according to its measurement precision and/or the probability that it resulted from a correct assignment; however, one often finds all input data given equal weights, regardless of quality or reliability.

The goal of a least-squares fitting process is to obtain a statistically rigorous and unbiased set of best possible parameters, with their uncertainties and correlations, for the specific combination of model and input data. If the model is inappropriate or the data set contains assignment or typographical errors, the least-squares procedure is only capable of providing a warning (systematic residuals, equality of fit poorer than measurement precision). The least-squares fitting process is often highly interactive; inflexible, few-parameter, diagnostic fits in the early stages followed by maximally flexible, multiparameter, model-refining fits in the final stages.

The treatment of least-squares fitting given here is superficial and nonrigorous. Albritton *et al.* (1976), Marquardt (1963), and Wentworth (1965) discuss the least-squares process more completely.

3.4.1.1 LINEAR LEAST-SQUARES FITTING

Consider a model

$$F(A_{1i}, \ldots, A_{ni}, X_1, \ldots, X_p) = y_i \tag{3.4.1}$$

defined by p initially unknown parameters X_k and n known coefficients that constitute the name (e.g., quantum numbers) of the data point whose measured value, y_i^{obs}, is to be compared to the computed value, y_i. For example, the rotation–vibration levels of a $^1\Sigma^+$ state are described by a model,

$$E(v, J) = T_e + \sum_{l=0}^{l_{max}} \sum_{m=0}^{m_{max}} Y_{lm}(v + \tfrac{1}{2})^l [J(J+1)]^m, \tag{3.4.2}$$

consisting of two A-values per level (J_i and v_i) and $(l_{max} + 1)$ $(m_{max} + 1) + 1$ unknown parameters. The correspondences between the quantities in the general model [Eq. (3.4.1)] and the specific example [Eq. (3.4.2)] are

$$y_i \leftrightarrow E(v_i, J_i)$$
$$A_{1i} \quad \text{and} \quad A_{2i} \leftrightarrow v_i \quad \text{and} \quad J_i$$
$$X_k \leftrightarrow Y_{lm}.$$

It is conceivable that the model would be asked to account for properties of the v_i, J_i level in addition to $E(v_i, J_i)$ (such as level population, zero-field g-value, etc.): in such case the data point would be specified by an additional index indicating which property is to be computed, but this is beyond the scope of the present discussion.

A model is said to be linear if

$$\frac{\partial y_i}{\partial X_k} = B_{ik}$$

for all i and k. This means that all higher derivatives, $\partial^n y_i / \partial X_k^n$, are zero. The model defined by Eq. (3.4.2) is linear where

$$\partial y_i / \partial X_k \equiv \partial E(v_i, J_i) / \partial Y_{lm} = (v_i + \tfrac{1}{2})^l [J(J+1)]^m.$$

The presence of $l, m > 1$ powers of v and J is irrelevant to whether the model is linear. Least-squares fits to a linear model have the convenient property of converging, in a single step, to a unique best set of parameter values.

The equations for a linear model with one dependent (y) and two independent (v, J) variables [for example, Eq. (3.4.2)] have the form

$$y_i = \sum_{m=1}^{p} B_{im} X_m, \tag{3.4.3}$$

where p is the number of parameters and each B_{im} is an explicit function of the two independent variables (v_i and J_i). A matrix formulation of Eq. (3.4.3) is useful,

$$\mathbf{BX} = \mathbf{y}, \tag{3.4.4}$$

where $\mathbf{y}$ and $\mathbf{X}$ are column matrices having respectively N (the number of input data points) and p rows and $\mathbf{B}$ is a matrix with N rows and p columns.

The sum of the squared residuals, Φ, is

$$\Phi \equiv \sum_{i=1}^{N} (y_i^{\text{obs}} - y_i^{\text{calc}})^2 = \sum_{i=1}^{N} \left(y_i^{\text{obs}} - \sum_{m=1}^{p} B_{im} X_m \right)^2. \tag{3.4.5}$$

The set of parameter values sought is that for which Φ is minimized,

$$\frac{\partial \Phi}{\partial X_m} = 0 \qquad \text{for} \quad m = 1, \ldots, p \tag{3.4.6}$$

or, taking the derivative of Eq. (3.4.5),

$$0 = \sum_{i=1}^{N} \frac{\partial}{\partial X_m} \left(y_i^{\text{obs}} - \sum_{m=1}^{p} B_{im} X_m \right)^2, \tag{3.4.7}$$

$$0 = \sum_{i=1}^{N} \sum_{l=1}^{p} (B_{il} B_{im} X_l - y_i^{\text{obs}} B_{im}), \tag{3.4.8}$$

or

$$\sum_{i=1}^{N} \sum_{l=1}^{p} B_{il} B_{im} X_l = \sum_{i=1}^{N} y_i^{\text{obs}} B_{im} \qquad \text{for} \qquad m = 1, \ldots, p. \tag{3.4.9}$$

This system of p equations in p unknowns can be rewritten in matrix notation as

$$\mathbf{B}^{\text{t}}\mathbf{B}\mathbf{X} = \mathbf{B}^{\text{t}}\mathbf{y}^{\text{obs}}, \tag{3.4.10}$$

where the right superscript t means transpose. Then

$$\mathbf{W} \equiv \mathbf{B}^{\text{t}}\mathbf{B} \tag{3.4.11}$$

defines the square $p \times p$ matrix $\mathbf{W}$, called the *normal matrix*, which has the property

$$\mathbf{X} = \mathbf{W}^{-1}\mathbf{B}^{\text{t}}\mathbf{y}^{\text{obs}}. \tag{3.4.12}$$

The left-hand side of Eq. (3.4.12) contains the desired optimized parameter values and the right-hand side consists entirely of measured ($\mathbf{y}^{\text{obs}}$) and calculable ($\mathbf{W}^{-1}$ and $\mathbf{B}$) quantities, provided that the columns of $\mathbf{B}$ are linearly independent (determinant of $\mathbf{W} \neq 0$) so that $\mathbf{W}^{-1}$ exists.

The quality of a nonweighted linear least-squares fit is given by the *standard deviation*,

$$\sigma \equiv \left(\frac{\Phi}{N - p} \right)^{1/2} \tag{3.4.13}$$

where Φ is the sum of the squared residuals and $N - p$ is the number of degrees of freedom (number of data points minus number of varied parameters). σ has the same units as the input data (cm^{-1}) and, if the fit is acceptable, should be comparable to the measurement uncertainty, δy. The standard error for each fitted parameter is given, in the same units as the parameter itself, by

$$\delta X_i = \sigma[(\mathbf{W}^{-1})_{ii}]^{1/2} \tag{3.4.14}$$

where $\sigma\mathbf{W}^{-1}$ is known as the *variance–covariance matrix*. In a fit with more than 10 degrees of freedom ($N - p > 10$), the true value of the parameter is expected to fall within the interval $X_i^{\text{fit}} - 2\delta X_i \leq X_i^{\text{true}} \leq X_i^{\text{fit}} + 2\,\delta X_i$ at the 92% confidence level.

The off-diagonal elements of $\sigma\mathbf{W}^{-1}$ reflect the correlations between fitted values of pairs of parameters. The *correlation matrix*, a matrix with ones on diagonal and numbers between $+1$ and -1 off-diagonal,

$$C_{ij} \equiv W_{ij}^{-1}/(W_{ii}^{-1}W_{jj}^{-1})^{1/2}, \tag{3.4.15}$$

reflects the sign and magnitude of the error in the jth parameter resulting from an error in the ith parameter. The true values of two parameters that have $C_{ij} \simeq 0$ are likely to fall anywhere in the rectangle defined by $X_i \pm 2\delta X_i$, $X_j \pm 2\delta X_j$, whereas the true values of two parameters with $C_{ij} \simeq 1$ are likely to fall along the $\delta X_i \simeq \delta X_j$ diagonal. Thus the joint confidence region for two strongly correlated constants is much smaller than the naive estimate from the uncorrelated standard errors.

In most real experiments, some measurements are made with greater precision than others (strong unblended lines versus weak or blended lines; a combined fit to microwave and optical data). A *weighted* least-squares procedure is appropriate. Typically, each measurement is weighted by the square of the reciprocal of its estimated uncertainty,

$$w_{ii} = (\delta y_i)^{-2}. \tag{3.4.16}$$

A weight matrix, $\mathbf{w}$, is defined by

$$w_{ij} = \delta_{ij}\sigma^2\,\delta y_i^{-2}, \tag{3.4.17}$$

where $\mathbf{w}$ is a diagonal matrix and σ^2 is the *a priori* undetermined [see Eq. (3.4.19)] variance of a measurement with unit weight. The weighted least-squares equations are very similar to the unweighted ones:

$$\mathbf{X} = (\mathbf{B}^{\text{t}}\mathbf{w}\mathbf{B})^{-1}\mathbf{B}^{\text{t}}\mathbf{w}\mathbf{y}^{\text{obs}} \tag{3.4.18}$$

$$\sigma^2 = \frac{1}{N-p}\sum_{i=1}^{N}(y_i^{\text{obs}} - y_i^{\text{calc}})^2/\delta y_i^2, \tag{3.4.19}$$

where σ^2, the *variance* of the fit, is a dimensionless number with value near 1 if the estimates of δy_i are reasonable and the fitting model is appropriate. The uncertainties in the parameters,

$$\delta X_i = \sigma[(\mathbf{B}^t\mathbf{wB})_{ii}^{-1}]^{1/2}, \tag{3.4.20}$$

do not depend on the initial estimate of σ^2 because the σ in Eq. (3.4.20) is cancelled by a σ^{-1} from $[(\mathbf{B}^t\mathbf{wB})_{ii}^{-1}]^{1/2}$.

3.4.1.2 Nonlinear Least-Squares Fitting

When the model [Eq. (3.4.1)] is nonlinear,

$$\frac{\partial^n y_i}{\partial X_k^n} \neq 0 \qquad \text{for} \quad n \geq 2,$$

the least-squares procedure is considerably more complicated because it is no longer possible to write a matrix equation analogous to Eq. (3.4.12) or Eq. (3.4.18). It is still possible to find a set of parameter values, $\mathbf{X}$, that minimizes the square of the residuals, $\mathbf{\Phi}$, but the process is iterative, and it is possible that a false convergence will occur to a local minimum of $\mathbf{\Phi}$.

A typical nonlinear model encountered in spectroscopy is one where the spectral line frequencies, y_i, are computed from differences between eigenvalues (specified by two groups of quantum numbers) of two effective Hamiltonian matrices, defined, in turn, by the molecular constants, X_j. For example, consider the 2 × 2 matrix of a $^2\Pi$ state:

$$\begin{aligned}
\mathbf{H} &= \begin{vmatrix} T_0 + A/2 + B(x^2-2) & -B(x^2-1)^{1/2} \\ -B(x^2-1)^{1/2} & T_0 - A/2 + Bx^2 \end{vmatrix} \\
&= T_0\begin{vmatrix} 1 & 0 \\ 0 & 1 \end{vmatrix} + A\begin{vmatrix} \frac{1}{2} & 0 \\ 0 & -\frac{1}{2} \end{vmatrix} + B\begin{vmatrix} x^2-2 & -(x^2-1)^{1/2} \\ -(x^2-1)^{1/2} & x^2 \end{vmatrix} \\
&= \sum_{m=1}^{p} X_m \mathbf{H}_m,
\end{aligned} \tag{3.4.21}$$

where $x = J + \frac{1}{2}$. This way of writing $\mathbf{H}$ as a pseudolinear form (Chedin and Cihla, 1967) takes advantage of the Hellmann–Feynman theorem (Hellmann, 1937; Feynman, 1939),

$$\left(\mathbf{U}^\dagger \frac{\partial \mathbf{H}}{\partial X_k}\mathbf{U}\right)_{ii} = \frac{\partial E_i}{\partial X_k} \tag{3.4.22}$$

where

$$\mathbf{H}|i\rangle = E_i|i\rangle$$

and $\mathbf{U}$ is the unitary matrix that diagonalizes $\mathbf{H}$. The $\mathbf{H}_m$ matrices in Eq. (3.4.21) are identical to the $\partial\mathbf{H}/\partial X_k$ in Eq. (3.4.22). The convenient point about the Hellmann–Feynman theorem is that it provides all of the $\partial E_i/\partial X_k$ after a single diagonalization of $\mathbf{H}$, rather than the $p + 1$ diagonalizations that would otherwise be needed to compute the derivatives by finite differences from

$$\frac{\partial E_i}{\partial X_k} \approx \frac{E_i(X_k^0) - E_i(X_k)}{X_k^0 - X_k}, \tag{3.4.23}$$

where X_k^0 are the initial estimates of the unknown parameter values. Matrix diagonalizations are usually the most time consuming portion of a nonlinear least-squares fit.

The value of Eq. (3.4.22) may be illustrated very simply. At very high J-values, a $^2\Pi$ state will approach the case (b) limit, at which point there is essentially no information in the spectrum from which the spin–orbit constant, A, may be determined. The $\mathbf{U}$ matrix for the case (a)$\rightarrow$(b) transformation at the high-J limit is

$$\mathbf{U} \approx \begin{vmatrix} 2^{-1/2} & 2^{-1/2} \\ -2^{-1/2} & 2^{-1/2} \end{vmatrix}.$$

Applying this transformation to the matrix $\mathbf{H}_m$ for $m = T_0$,

$$\mathbf{U}^\dagger \mathbf{H}_{T_0} \mathbf{U} = \mathbf{H}_{T_0} = \begin{vmatrix} 1 & 0 \\ 0 & 1 \end{vmatrix},$$

and to the $\mathbf{H}_A$ coefficient matrix,

$$\mathbf{U}^\dagger \mathbf{H}_A \mathbf{U} = \begin{vmatrix} 0 & \frac{1}{2} \\ \frac{1}{2} & 0 \end{vmatrix},$$

it is evident that $\partial E_i/\partial T_0 = (\mathbf{U}^\dagger \mathbf{H}_{T_0} \mathbf{U})_{ii} = 1$ at both the case (a) and (b) limits but $\partial E_i/\partial A = (\mathbf{U}^\dagger \mathbf{H}_A \mathbf{U})_{ii} = 0$ in the case (b) limit. The sensitivity of the eigenvalues to the value of the T_0 parameter is approximately J-independent, but the eigenvalues become insensitive to A as the case (b) limit is approached at high J.

It is now possible to derive a nonlinear equation analogous to Eq. (3.4.12):

$$y_i^{\text{calc}} = (\mathbf{U}^\dagger \mathbf{H} \mathbf{U})_{ii} = \left[\mathbf{U}^\dagger \left(\sum_{m=1}^{p} X_m \mathbf{H}_m\right) \mathbf{U}\right]_{ii} = \sum_{m=1}^{p} X_m (\mathbf{U}^\dagger \mathbf{H}_m \mathbf{U})_{ii} \equiv \sum_{m=1}^{p} X_m B_{im}. \tag{3.4.24a}$$

or

$$B_{im} = (\mathbf{U}^\dagger \mathbf{H}_m \mathbf{U})_{ii}. \tag{3.4.24b}$$

Equation (3.4.24) looks exactly like Eq. (3.4.3), but the hidden difference is that the B_{im} coefficients depend implicitly, through the $\mathbf{U}$ matrix, on the unknown parameters, X_m. It will be necessary to resort to an iterative procedure whereby the approximate $B_{im}^{(n)}$ matrix elements are recalculated after each adjustment of the parameter values,

$$\mathbf{X}^{(n)} = \mathbf{X}^{(n-1)} + \boldsymbol{\delta}^{(n-1)} \tag{3.4.25}$$

and the least-squares procedure is recast as a series of iterative calculations of $\boldsymbol{\delta}$ so that $\mathbf{X}^{(n)}$ converges to $\mathbf{X}$. Numerical convergence of the procedure is achieved when the $\mathbf{X}^{(n)}$ and $\mathbf{X}^{(n-1)}$ are found to differ by less than a preset convergence criterion.

The matrix formulation of the iterative process is derived by a series of steps identical to Eqs. (3.4.5)–(3.4.12),

$$\mathbf{X}^{(0)} + \boldsymbol{\delta}^{(0)} = \mathbf{X}^{(1)} = [\mathbf{W}^{(0)}]^{-1}\mathbf{B}^{(0)\mathrm{t}}\mathbf{y}^{\mathrm{obs}}, \qquad \text{where} \quad \mathbf{W}^{(0)} \equiv \mathbf{B}^{(0)\mathrm{t}}\mathbf{B}^{(0)}. \tag{3.4.26}$$

The right-hand side of Eq. (3.4.26) is expressed entirely in terms of known ($\mathbf{y}^{\mathrm{obs}}$) and calculable ($\mathbf{W}^{(0)}, \mathbf{B}(0)$) quantities, but since the model is nonlinear, the next iteration requires that the $\mathbf{B}^{(1)}$ matrix and y_i^{calc} be recalculated by Eq. (3.4.24) using the new parameter values, $\mathbf{X}^{(1)}$, and the new matrix $\mathbf{U}^{(1)}$, which diagonalizes the new $\mathbf{H}^{(1)} \equiv \Sigma_{m=1}^{p} \mathbf{H}_m X_m^{(1)}$; moreover, the true minimum of $\boldsymbol{\Phi}$ may not have been reached after one iteration. The iterative process described by Eqs. (3.4.24)–(3.4.26) is called the *Gauss–Newton method* (Marquardt *et al.*, 1961) and is implicitly based on a truncated Taylor series expansion,

$$\mathbf{y}^{\mathrm{calc}(1)} \simeq \mathbf{y}^{\mathrm{calc}(0)} + \sum_{m=1}^{p} \left(\frac{\partial \mathbf{y}^{\mathrm{calc}(0)}}{\partial X_m} \right)_{X_m = X_m^{(0)}} \delta_m^{(0)} \tag{3.4.27}$$

Its convergence properties are excellent near the minimum of the $\boldsymbol{\Phi}(\mathbf{X})$ hypersurface, but poor far from the minimum where the nonlinear dependence of $\boldsymbol{\Phi}$ on the X_i values is most important.

An alternative approach is based on a *method of steepest descents*. The Eq. (3.4.26) definition of $\boldsymbol{\delta}$ is replaced by

$$\boldsymbol{\delta}_g = -\left(\frac{\partial \Phi}{\partial X_1}, \ldots, \frac{\partial \Phi}{\partial X_p} \right)^{\mathrm{t}} \tag{3.4.28}$$

where the subscript g signifies that $\boldsymbol{\delta}_g$ points in the direction of the negative *gradient* of the $\boldsymbol{\Phi}(\mathbf{X})$ hypersurface and the superscript t means transpose ($\boldsymbol{\delta}$ is a $1 \times p$ column matrix). The $\partial\Phi/\partial X_m$ are computed from

$$\frac{\partial \Phi}{\partial X_m} = [\mathbf{W}^{(0)}\mathrm{X}^{(0)} - \mathbf{B}^{(0)\mathrm{t}}\mathbf{y}^{\mathrm{obs}}]_m. \tag{3.4.29}$$

The steepest descent method is very effective far from the minimum of $\boldsymbol{\Phi}$, but is always much less efficient than the Gauss–Newton method near the minimum of $\boldsymbol{\Phi}$. Marquardt (1963) has proposed a hybrid method that combines the advantages of both Gauss–Newton and steepest descent methods. Marquardt's method, combined with the Hellmann–Feynman pseudolinearization of the Hamiltonian energy level model, is the method of choice for most nonlinear molecular spectroscopic problems.

Birss (1983) has proposed a modification of the nonlinear fitting process that takes advantage of an extension of the Hellmann–Feynman theorem to computation of second derivatives,

$$\frac{\partial^2 E_i}{\partial X_k \partial X_l} = 2 \sum_{j \neq i} \frac{\langle i|\mathbf{U}^\dagger \mathbf{H}_k \mathbf{U}|j\rangle\langle j|\mathbf{U}^\dagger \mathbf{H}_l \mathbf{U}|i\rangle}{E_i - E_j}$$

$$= 2 \sum_{j \neq i} \frac{(\mathbf{U}^\dagger \mathbf{H}_k \mathbf{U})_{ij}(\mathbf{U}^\dagger \mathbf{H}_l \mathbf{U})_{ji}}{E_i - E_j}, \qquad (3.4.30)$$

where $\mathbf{U}$ defines the transformation that diagonalizes $\mathbf{H}$, and $\mathbf{H}_m$ is defined by Eq. (3.4.21). The availability of higher derivatives enables the most important nonlinearities in $\mathbf{\Phi}(\mathbf{X})$ to be dealt with explicitly by the fitting algorithm, yielding faster convergence.

Curl (1970) has proposed a *diagnostic least-squares* procedure that allows prior knowledge about the physically reasonable values (and their estimated 90% confidence intervals) for several X_i to be incorporated as constraints on the usual least-squares fitting procedure. When spectroscopic data are inadequate to determine all of the parameters in $\mathbf{H}^{\text{eff}}$, the normal matrix, $\mathbf{W} = \mathbf{B}^t\mathbf{B}$, will be nearly singular. Since it is impossible to compute the inverse, $\mathbf{W}^{-1}$, of a singular $\mathbf{W}$ matrix, an unmodified least-squares procedure must fail. Curl's procedure is sufficiently flexible that, if it turns out that the input data are capable of determining the value of a parameter more accurately than the prior estimate of its uncertainty, then that parameter is allowed to vary freely; otherwise the normal equation, $\mathbf{W}$, and parameter correction, $\boldsymbol{\delta}$, matrices are modified to remove the singularity by constraining the parameter. The variation ranges of more than one parameter can be constrained, and the standard errors and correlations between the unconstrained parameters are computed in a statistically rigorous manner.

Curl's method for recognizing an ill-determined parameter involves examination of the eigenvalues and eigenvectors of the normal matrix, $\mathbf{W}$. If $\mathbf{W}$ is nearly singular, then it must have at least one very small eigenvalue, λ_i. The eigenvector of $\mathbf{W}$ corresponding to this smallest λ_i is the specific linear combination of parameters X,

$$\chi_i = \sum_j a_{ji} X_j,$$

which is so poorly determined that roundoff error may prohibit or slow down convergence of the fit to the optimum value of χ_i. Lees (1970) has discussed the diagnostic significance of the eigenvalues and eigenvectors of $\mathbf{W}$ and has shown that $P + \log \lambda_i$ is the number of significant figures to which χ_i can be determined, where P is the number of significant figures stored by the computer.

3.4.1.3 PRACTICAL CONSIDERATIONS

The end results of a least-squares fit of the parameters in $\mathbf{H}^{\text{eff}}$ to a set of spectroscopic data are (1) a set of molecular constants, standard errors, and correlation coefficients for the constants, (2) a quality of fit indicator (σ), (3) a numerical model capable of reproducing the fitted data set without systematic error larger than the measurement precision, and (4) a model capable of extrapolating to unobserved levels or computing properties other than energies from the eigenvectors of $\mathbf{H}^{\text{eff}}$.

The fitted parameters may or may not be of intrinsic interest. It is important to remember that the parameter values obtained depend on the specific $\mathbf{H}^{\text{eff}}$ model, the parameters allowed to vary in the fit, the fixed values chosen for parameters not allowed to vary, and the specific subset of spectral data fitted. Generation of multidigit molecular constants should not normally be the primary objective or the stopping point of a spectroscopic investigation.

Even though molecular constants are model-dependent (see Section 3.4.2) and often strongly correlated, it is important to report these constants with one or two more digits than are statistically determined (Watson, 1977). These extra digits are required to reproduce the fitted data. The ability to reproduce a spectrum (typically $\sim$1000 lines reproduced by $\sim$20 constants), predict unobserved lines, and calculate seemingly unrelated properties is one of the most valuable and least model-dependent products of the spectrum-fitting process.

Before starting a multiparameter least-squares fitting process, it is important to perform diagnostic tests on the model Hamiltonian subroutine and on the input data set. An effective computer code testing scheme is to show that the $\mathbf{H}^{\text{eff}}$, molecular constants, and spectral lines in a published paper are self-consistent. The input data set should be examined for typographical and assignment errors by running the fitting program for at most one iteration, allowing it to vary only one obviously well determined parameter. The quality of the fit will be very poor, but the systematics of the residuals [discontinuities in (obs − calc) versus J] will identify most of the questionable lines.

It is of diagnostic value to perform a series of fits in which progressively larger numbers of parameters are simultaneously optimized. At each stage, the systematics of the residuals can suggest which additional parameter should be allowed to vary. If any fitted parameter is within two standard errors of zero or some prior estimate of its value, its value should be set to zero or to the prior estimate and that parameter should be temporarily removed from the fit. Sometimes, near convergence, a parameter that was poorly determined at an earlier stage (fewer parameters varied, larger σ) can be fitted.

When an additional parameter is allowed to vary, the quality of the fit should improve. If $\sigma(p)$ and $\sigma(p + 1)$ are the standard deviations of the fit when p and

$p+1$ parameters are varied, it is reasonable to vary the $(p+1)$th parameter when

$$\sigma(p+1) < \sigma(p)\left(\frac{N-p}{N-p-1}\right)^{1/2}. \tag{3.4.31}$$

Note that, if one $(p+1)$th parameter fails the Eq. (3.4.31) test, there may be a different $(p+1)$th parameter that will pass it. It is time to stop looking for new parameters when the variance of the fit is near one (the lines are being fit within their estimated measurement precision) and when there are no systematic residuals.

One problem unique to perturbed spectra concerns the correspondence between the spectroscopist's and a computer's names for a given energy level. An eigenstate of $\mathbf{H}^{\text{eff}}$ may be unambiguously labeled by specifying three quantities: its good quantum numbers, J and e/f, and its energy rank within the group of eigenstates belonging to the same values of the good quantum numbers. For example, an $\mathbf{H}^{\text{eff}}$ containing $^3\Delta$, $^1\Delta$, $^3\Sigma^+$, $^3\Sigma^-$, $^1\Sigma^+$, $^1\Sigma^-$, $^1\Pi$ basis functions has 18 eigenstates for each value of $J \geq 3$, nine each of e and f symmetry. It is possible to compute, from the eigenvectors of $\mathbf{H}^{\text{eff}}$, the fractional characters of each basis function present in each eigenfunction. If

$$(\mathbf{U}^{\dagger}\,\mathbf{HU})_{ij} = E_i\delta_{ij}, \tag{3.4.32}$$

then the fractional character of the kth basis function in the ith eigenfunction is U_{ki}^2. The nominal character of an eigenstate, the spectroscopist's name for that state, is the *leading character* determined by the largest of the U_{ki}^2 among all basis functions k contributing to the ith eigenstate. All J-levels of the same energy rank do not belong to the same nominal character. This is true for $J < 3$ because not all basis functions exist at low J ($J \geq \Omega$ is required). Rank changes also occur at higher J-values when level crossings occur. When the energy of the $^1\Pi$ basis function, $H_{^1\Pi,^1\Pi}$, overtakes the $^1\Sigma^-$ basis function energy, $H_{^1\Sigma^-,^1\Sigma^-}$, from below (because $B_{^1\Pi} > B_{^1\Sigma^-}$), the energy rank of the nominal $^1\Pi$ level must change. The least-squares parameter variation process can be improperly trapped by an incorrect energy ranking. What started as a correct ranking for the initial parameter estimates becomes incorrect after an iteration. The computer then thinks it has a level of nominal perturber ($^1\Sigma^-$) rather than main line ($^1\Pi$) character and adjusts the otherwise weakly determined constants of the perturber to accommodate the incorrectly ranked level. The possibility of computer-generated ranking errors can be eliminated if levels are labeled by their nominal character, which is almost always evident from a rotational analysis supplemented by level-shift and relative-intensity information. LeFloch and Legal (1986) have devised a character/rank dual-labeling scheme.

Fitted parameters are often strongly correlated. This can be a consequence of the structure of the data set or it can be an intrinsic property of the $\mathbf{H}^{\text{eff}}$ model (for example, γ and A_D in a $^2\Pi$ state, Brown *et al.*, 1979). The former effect can be minimized by supplementing the data set. A combined fit to optical and microwave data or to two electronic transitions sharing a common state (e.g., $^2\Pi$–$^2\Sigma^+$ and $^2\Sigma^+$–$^2\Sigma^+$ systems) can be very effective. Another approach is to replace the two correlated parameters by two new parameters that are the sum and difference of the original parameters (e.g., $B' + B''$ and $B' - B''$). The correlation matrix [Eq. (3.4.15)] and the magnitudes of the eigenvalues of the normal matrix [Eq. (3.4.11)] provide useful insights (Albritton *et al.*, 1976; Curl, 1970). Isotope relationships, computed D_v-values, and other semiempirical constraints are frequently used to minimize both data set and intrinsic correlation effects.

3.4.1.4 Types of Programs

Three types of spectrum-fitting programs exist that are suitable for reducing line spectra to molecular constants: the term-value method (Åslund, 1974), the direct approach (Zare *et al.*, 1973), and merging of separate band-by-band fits (Albritton *et al.*, 1977; Coxon, 1978).

The *term-value method* (Åslund, 1974) has the effect of focussing on a single electronic state. Transitions from many bands are converted, in a model-independent manner, to a set of term values. Many of the term values are multiply determined from several branches of many bands. The term-value approach reduces an enormous number of observed transitions to a much smaller number of preaveraged energy levels. In many cases, especially when one state is perturbed, the term-value approach can aid the assignment process and vastly simplify part of the least-squares parameter optimization process. The term value approach has been critiqued by Zare *et al.* (1973) and Albritton *et al.* (1977).

The *direct approach* had been used by spectroscopists (Birss *et al.*, 1970; Stern *et al.*, 1970; Merer and Allegretti, 1971; Field and Bergeman, 1971; Brand *et al.*, 1971; Meakin and Harris, 1972) for several years before Zare *et al.* (1973) presented a detailed description of both model Hamiltonians and numerical procedures. The direct approach is capable of accepting transitions (electronic, vibrational, or pure rotational), term values, combination differences, radiofrequency transitions between Λ-doublets, in short any sort of energy-level information. All information is treated in a statistically rigorous, properly weighted, uncorrelated, and unbiased manner. As many $\mathbf{H}^{\text{eff}}$ matrices are set up as are needed to calculate all observed energy levels and transition frequencies. A direct approach fit to many bands can require a lot of computer memory and time because of the need to store, invert, and diagonalize huge matrices.

The method of *merging* separate band-by-band fits can simplify the fitting process with no sacrifice in accuracy of parameter determination. One of the reasons for considering a simultaneous multiband fit is that parameter correlations and standard errors can be significantly reduced. The merge process takes the band-by-band fitted parameters and variance–covariance matrices and combines them to obtain global constants that are identical to those that would have been obtained from a global fit.

Several spectrum-fitting programs have been described in the literature (Lefebvre-Brion, 1969; Zare *et al.*, 1973; Johns and Lepard, 1975; LeFloch, 1984). Some of these programs are designed to handle transitions between virtually any pair of electronic states but are not set up for perturbation situations where actual level crossings occur within the sampled range of J-values. Other programs are designed for a specific problem and would require major modifications by a user wishing to study another class of state or fit a data set of different structure. Specific fitting models, such as the "unique perturber" approximation (Zare *et al.*, 1973), are often valuable in that they reduce the number of free parameters and, in some cases, prevent the mechanical significance of the major constants (E_v, B_v, D_v) from being contaminated by intrinsically correlated second-order constants (o, p, q). However, by attributing all perturbation effects to a single remote perturber and treating these effects by a Van Vleck transformation (Section 3.2) with constraints, the parameters obtained from the fit may not have the meaning implied by the microscopic definitions of the parameters (Brown *et al.*, 1979). It is important that every spectroscopic laboratory develop its own spectrum-fitting programs, usually by extensively modifying one of the widely distributed classes of program.

Most of the previous discussion has focussed on programs that fit rotational lines. Often, in the case of homogeneous perturbations, there are so many mutually interacting vibrational levels that a fit to individual rotational levels (rather than to G_v and B_v) could be contemplated only after developing a model which accounts for the nonrotating molecule vibrational structure. In such a model, the concept of potential-energy curves plays a crucial role. These curves may be either Morse (Lefebvre-Brion, 1969) or RKR curves (Stahel *et al.*, 1983). One assumes a perturbation interaction of the form

$$H_{1,v_1;2,v_2} = H^e_{12}\langle v_1|v_2\rangle, \qquad (3.4.33)$$

and the potential curves and vibrational overlap integrals are recalculated after each iteration of the vibrational constants. The rotational "constants" of the perturbed vibrational levels are also calculated after each iteration from

$$B_i^{\text{calc}} = \sum_n \sum_{v_n} |c_{i,v_n}|^2 B_{v_n}, \tag{3.4.34}$$

where the summations are over electronic states (n) and vibrational levels of the nth state (v_n), B_{v_n} is the deperturbed rotational constant of the v_n level,

$$B_{v_n} = (h^2/2\mu)\langle v_n|R^{-2}|v_n\rangle, \tag{3.4.35}$$

and $|c_{i,v}{}^n|^2$ is the mixing fraction of the v_n basis function in the ith eigenfunction as obtained from the transformation, $\mathbf{U}^\dagger\mathbf{H}\,\mathbf{U}$, that diagonalizes $\mathbf{H}$,

$$|c_{i,v_n}|^2 = U_{v_n,i}{}^2. \tag{3.4.36}$$

The B_i^{calc} are compared to the observed values, B^{obs}, obtained by extrapolating the observed levels to $J = 0$. The $B^{\text{obs}} - B^{\text{calc}}$ residuals allow the R_e-values of all interacting electronic states to be refined, whereas the vibrational data alone only determine differences in R_e-values.

This multivibrational-level homogeneous deperturbation approach, in which the potential curves and vibrational integrals ($\langle v_i|v_j\rangle$, B_{v_n}) are recalculated after each iteration of a least-squares fitting process, has one crucial feature in common with the way constants obtained from band-by-band local perturbation analyses are globally deperturbed. An initial series of fits is performed to obtain partially deperturbed G_v and B_v values. An initial set of RKR potentials are generated, and values for all quantities that are calculable from the potential curves are obtained (D_v, $\langle v_i|v_j\rangle$, $\langle v_i|R^{-2}|v_j\rangle$, A_D, etc.). These calculated values are then input into the fitting program as constraints on many weakly determined or highly correlated constants. The fit is repeated, new RKR curves and derivable constants are calculated, and the process is iterated to self-consistency. The multilevel and band-by-band deperturbation approaches both use potential-energy curves to place constraints on a myriad of seemingly unrelated molecular constants and perturbation parameters, thereby vastly reducing the number of degrees of freedom in the spectrum-fitting model. The use of such a tight model enables model inadequacies and unsuspected perturbations to be detected, and, most importantly, it provides a set of deperturbed molecular constants with clearly separate mechanical and magnetic meanings.

3.4.2 COMPARISON BETWEEN EFFECTIVE AND TRUE PARAMETERS

The parameters obtained by fitting the observed energy levels to the eigenvalues of an effective Hamiltonian are not unique. Their values depend on

(1) the precision and completeness of the input data, (2) the size of the matrix that is actually diagonalized, (3) the choice of model Hamiltonian, (4) choices such as which parameters to vary, which parameters will be held fixed at values different from zero, and the imposition of constraints on varied parameters. Spectroscopists seldom agree on the meaning of the term "deperturbed" or on what information belongs in $\mathbf{H}^{\text{eff}}$.

This book was written to help spectroscopists understand the relationship between the exact molecular Hamiltonian, effective Hamiltonians used in fitting spectral data, and the molecular parameters obtained from both spectra and *ab initio* calculations. Although the general ideas for constructing effective Hamiltonians (Section 3.2) and several examples appropriate to special cases (for example the $^2\Sigma^+ \sim {}^2\Pi$ interaction in Sections 2.5.4 and 4.5) are discussed, no attempt is made here to present a complete and universal effective Hamiltonian for diatomic molecules. Brown *et al.* (1979) derive an effective Hamiltonian that should be the starting point for the fitting of most non-$^1\Sigma$, perturbation-free, diatomic molecular spectra. Other, less general, effective Hamiltonians have been proposed, by DeSantis *et al.* (1973) for $^3\Sigma$ states, by Brown and Milton (1976) for $S \geq \frac{3}{2}$ Σ-states, and by Brown and Merer (1979) for $S \geq 1$ Π-states.

The idea of an effective Hamiltonian for diatomic molecules was first articulated by Tinkham and Strandberg (1955) and later developed by Miller (1969) and Brown *et al.* (1979). The crucial idea is that a spectrum-fitting model (for example eq. 18 of Brown *et al.*, 1979) be defined in terms of the minimum number of *linearly independent* fit parameters. These fit parameters have no physical significance. However, if they are defined in terms of sums of matrix elements of the exact Hamiltonian (see Tables I and II of Brown *et al.* 1979) or sums of parameters appropriate to a special limiting case (such as "unique perturber," see Table III of Brown *et al.* 1979, or pure precession, Section 4.5), then physically significant parameters suitable for comparison with the results of *ab initio* calculations are usually derivable from fit parameters.

The following example illustrates the critical dependence of the values of the fit parameters on the dimension of the effective Hamiltonian matrix.

Consider the 2 × 2 problem,

$$\begin{vmatrix} E_{v_1} - E & H_{1v_1,2v_2'} \\ H_{1v_1,2v_2'} & E_{v_2'} - E \end{vmatrix} = 0 \tag{3.4.37}$$

where $H_{1v_1,2v_2'}$ is independent of J, as in homogeneous electrostatic perturbations, and E_{v_1} and $E_{v_2'}$ are the zero-order energies (i.e., deperturbed) of the v_1 and v_2' vibrational levels belonging to electronic states 1 and 2. Of course, in the exact, infinite-dimension Hamiltonian matrix, the $1v_1$ and $2v_2'$ basis functions will have nonzero off-diagonal matrix elements with other, energetically remote, basis functions. If all but the $H_{1v_1,2v_2'}$ off-diagonal element are ignored, the E_{v_1}, $E_{v_2'}$, and $H_{1v_1,2v_2'}$ parameters in Eq. (3.4.37) become *effective parameters*, $E_{v_1}^{\text{eff}}$, $E_{v_2'}^{\text{eff}}$, $H_{v_1v_2'}^{\text{eff}}$. The effective parameters differ from the *true parameters* by perturbation summations defined by the Van Vleck transformation [Eqs. (3.2.17)–(3.2.19)],

$$E_{v_1}^{\text{eff}} = E_{v_1}^{\text{true}} + \sum_{n,v_n \neq 1v_1} \frac{(H_{1v_1,nv_n})^2}{E_{v_1} - E_{v_n}} \tag{3.4.38a}$$

$$E_{v_2'}^{\text{eff}} = E_{v_2'}^{\text{true}} + \sum_{n,v_n \neq 2v_2'} \frac{(H_{2v_2',nv_n})^2}{E_{v_2'} - E_{v_n}} \tag{3.4.38b}$$

$$H_{v_1,v_2'}^{\text{eff}} = H_{v_1,v_2'}^{\text{true}} + \sum_{n,v_n \neq 1v_1 \text{ or } 2v_2'} \frac{H_{1v_1,nv_n} H_{nv_n,2v_2'}}{(E_{v_1} + E_{v_2'})/2 - E_{v_n}}, \tag{3.4.39}$$

where corrections through second order are included. Thus, after a 2 × 2 *local deperturbation* is performed by fitting the observed levels to the E_{v_1}, E_{v_2}, $H_{1v_1,2v_2'}$ *effective parameters* in Eq. (3.4.37), the perturbation summations in Eqs. (3.4.38)–(3.4.39) can be evaluated [Eq. (3.4.33)] to yield the *more completely deperturbed* $E_{v_1}^{\text{true}}$, $E_{v_2'}^{\text{true}}$, $H_{v_1,v_2'}^{\text{true}}$ *parameters*. Lagerqvist and Miescher (1958) were the first to employ such a two-step deperturbation process to the G_v and B_v constants of the NO $B^2\Pi \sim C^2\Pi$ valence ~ Rydberg states.

However, Eqs. (3.4.38)–(3.4.39) can give erroneous results. Gallusser and Dressler (1982) have performed a simultaneous, multistate deperturbation on the NO $B^2\Pi$ ($v = 0$–37) and $L^2\Pi$ ($v = 0$–11) valence states and $C^2\Pi$ ($v = 0$–9), $K\,^2\Pi$ ($v = 0$–4), and $Q^2\Pi$ ($v = 0$–3) Rydberg states (see Section 5.2.2). They diagonalized 69 × 69 matrices rather than Van Vleck transform-corrected 2 × 2 matrices and found that in some cases the third-order correction to $H_{1v_1,2v_2'}^{\text{eff}}$,

$$H_{1v_1,2v_2'}^{\text{eff}} = H_{1v_1,2v_2'}^{\text{true}} + \sum_{n',v_n' \neq 1v_1} \sum_{n,v_n \neq 2v_2'} \frac{H_{1v_1,n'v_n'} H_{n'v_n',nv_n} H_{nv_n,2v_2'}}{(E_{v_1} - E_{v_n'})(E_{v_2'} - E_{v_n})}, \tag{3.4.40}$$

is larger than the first- and second-order contributions. In Eq. (3.4.40) the unprimed and primed state labels correspond to valence and Rydberg states, respectively. The second-order term is omitted from Eq. (3.4.40) because the deperturbed basis set is defined to consist of a group of prediagonalized $^2\Pi$ valence states (all $H_{1v_1,nv_n} = 0$) and a separately prediagonalized group of Rydberg states (all $H_{2v_2',n'v_n'} = 0$). The only nonzero off-diagonal matrix elements of $\mathbf{H}^{\text{true}}$ are valence ~ Rydberg ($H_{vv'} \neq 0$).

Gallusser and Dressler (1982) found that the first-order interaction between the $L^2\Pi$ ($v = 2$) and $K\,^2\Pi$ ($v = 0$) *basis states* calculated using Eq. (3.4.33) was $H_{\text{L2,K0}}^{\text{true}} = 0.07\ \text{cm}^{-1}$, although a matrix element of $H_{\text{L2,K0}}^{\text{eff}} = 5\ \text{cm}^{-1}$ was obtained from a 2 × 2 diagonalization (Dressler and Miescher, 1981) of the interaction between the *nominal* L2 and K0 levels. A single term was found to dominate the Eq. (3.4.40) summation,

$$H_{\mathrm{L2,C5}} = 40\ \mathrm{cm}^{-1}$$

$$H_{\mathrm{C5,B21}} = 180\ \mathrm{cm}^{-1}$$

$$H_{\mathrm{B21,K0}} = 39\ \mathrm{cm}^{-1}$$

$$E_{\mathrm{L2}} - E_{\mathrm{C5}} = 443\ \mathrm{cm}^{-1} \qquad (\Omega = \tfrac{3}{2})$$

$$E_{\mathrm{K0}} - E_{\mathrm{B21}} = 197\ \mathrm{cm}^{-1} \qquad (\Omega = \tfrac{3}{2})$$

$$\frac{H_{\mathrm{L2,C5}} H_{\mathrm{C5,B21}} H_{\mathrm{B21,K0}}}{(E_{\mathrm{L2}} - E_{\mathrm{C5}})(E_{\mathrm{K0}} - E_{\mathrm{B21}})} = 3.2\ \mathrm{cm}^{-1}.$$

It is a rather disturbing, yet not uncommon, situation where a third-order correction term is 70 times larger than the first-order term.

Another surprising result is found by Ngo *et al.* (1974) for some apparent homogeneous Rydberg$\sim$Rydberg perturbations in the PO molecule. Since Rydberg states have nearly identical potential energy curves and differ by a single electron orbital, both the H^e (see Section 2.3.4) and $\langle v|v' \neq v\rangle$ factors in the perturbation matrix element should be very small. The observation of $\mathrm{P^{16}O}$ $A^2\Sigma^+$ $(v = 12) \sim H^2\Sigma^+$ $(v = 0)$ $(H_{\mathrm{A12,H0}} = 11.5\ \mathrm{cm}^{-1})$ and $A^2\Sigma^+$ $(v = 9) \sim G^2\Sigma^+ (v = 0)$ $(H_{\mathrm{A9,G0}} = 30.0\ \mathrm{cm}^{-1})$ perturbations corresponds to a perturbation between two Rydberg states with $\Delta v \gg 1$ and Rydberg orbitals with different $l\lambda$ values. (The configurations of the PO Rydberg states are: $[\mathrm{PO}^+\ X^1\Sigma^+]nl\lambda$ with $A^2\Sigma^+$ $4s\sigma$, $G^2\Sigma^+$ $4p\sigma$, $H^2\Sigma^+$ $3d\sigma$.) A nonzero $H^{\mathrm{eff}}_{1v_1,2v_2'}$ matrix element could arise via second-order Van Vleck corrections

$$H^{\mathrm{eff}}_{1v_1,2v_2'} = 0 + \sum_{i,v_i \neq 1v_1 \text{ or } 2v_2'} \frac{H_{1v_1,iv_i} H_{iv_i,2v_2'}}{(E_{v_1} + E_{v_2'})/2 - E_{v_i}}. \tag{3.4.41}$$

For the PO examples, the leading terms in the perturbation summations are

$$H^{\mathrm{eff}}_{\mathrm{A12,H0}} \simeq \frac{H_{\mathrm{A12,F8}} H_{\mathrm{F8,H0}}}{(E_{\mathrm{A12}} + E_{\mathrm{H0}})/2 - E_{\mathrm{F8}}} = \frac{(36.4)(100.9)}{25{,}415 - 25{,}276} = 26.4\ \mathrm{cm}^{-1}$$

$$H^{\mathrm{eff}}_{\mathrm{A9,H0}} \simeq \frac{H_{\mathrm{A9,F3}} H_{\mathrm{F3,G0}}}{(E_{\mathrm{A9}} + E_{\mathrm{G0}})/2 - E_{\mathrm{F3}}} = \frac{(21.7)(172.0)}{21{,}669 - 21{,}509} = 23.3\ \mathrm{cm}^{-1}$$

where $F^2\Sigma^+$ is a valence state (see Table 4.4). These results for H^{eff} do not exactly reproduce the experimental values because only the largest term in the perturbation sum is considered here.

The next subsection shows how the solution of a system of coupled differential equations can yield molecular parameters that approach closer to the true parameter values than the usual matrix diagonalization approach.

3.4.3 COUPLED EQUATIONS

Instead of searching for an eigenfunction of $\mathbf{H}$ in the form of Eq. (3.1.2), where the unknown *coefficients* c_{i,v_n} describe a linear combination of members of a complete set of known vibrational wavefunctions, it is possible to obtain a solution of the type

$$\psi_i(r,R) = \sum_{n=1}^{q} \Phi_n(r,R)\chi'_{n,i}(R) \tag{3.4.42}$$

where the $\chi'_{n,i}$ are unknown *numerical* functions and the summation is over electronic states only. The standard approach [Eq. (3.1.1)] attempts to represent $\chi'_{n,i}$ as a superposition of the approximate χ_{v_n} vibrational wavefunctions associated with the nth potential curve,

$$\chi'_{n,i}(R) = \sum_{v_n=0}^{p_n} \chi_{v_n}(R)c_{i,v_n}. \tag{3.4.43}$$

The vibrational wavefunctions χ_{v_n} are analytic or numerical solutions of q *uncoupled* (independent) vibrational Schrödinger equations (see Section 4.1.3),

$$\left[\frac{-\hbar^2}{2\mu}\frac{d^2}{dR^2} + V_n(R) + \frac{\hbar^2}{2\mu R^2}J(J+1) - E\right]\chi_{v_n}(R) = 0. \tag{3.4.44}$$

The two cases where the Φ_n are the diabatic or the adiabatic electronic wavefunctions will be discussed here. In the case where the Φ_n are the diabatic basis functions, the *coupled* equations are written as

$$\left[\frac{-\hbar^2}{2\mu}\frac{d^2}{dR^2} + V_n^{\mathrm{d}}(R) + \frac{\hbar^2}{2\mu R^2}J(J+1) - E\right]\chi_n^{\prime\mathrm{d}}(R) = \sum_{m\neq n} H_{mn}^{e}(R)\chi_m^{\prime\mathrm{d}}(R) \tag{3.4.45}$$

where

$$H_{mn}^{e}(R) = \int \Phi_m^{\mathrm{d}*}(r,R)\mathbf{H}^{\mathrm{el}}\Phi_n^{\mathrm{d}}(r,R)\,dr. \tag{3.4.46}$$

The number of coupled equations is equal to the number, q, of different electronic state (or substate) potential curves.

Given a set of diabatic potential curves (V_n^{d}) and electronic interaction functions (H_{mn}^{e}), the coupled equations may be integrated numerically to find the energies of the bound levels. The numerical integration may be done by Fox's procedure (see references in Atabek and Lefebvre, 1980) or by the renormalized Numerov method of Johnson (1978) as employed by Stahel *et al.* (1983). If the V_n^{d} and H_{mn}^{e} were derived from a least-squares fit of the eigenvalues of a matrix $\mathbf{H}^{\mathrm{eff}}$ to the observed levels, then the energies obtained from the coupled equations method will all be lower than those obtained by the diagonalization method. This is easily explained because the coupled equations take into account *all* bound vibrational levels and a portion of the vibrational continuum associated with each coupled electronic state. The matrix method excludes some of the higher vibrational levels and all of the continuum. The net result is that, in the coupled-equations approach, the vibrational functions implicitly added to the basis set (all at energies above the observed and fitted levels) push all of the eigenvalues to lower energy. The only limitation on the coupled equations approach is in the limits of integration over R.

An attempt to combine the coupled-equations approach with a least-squares, matrix-diagonalization, parameter-iteration scheme was made by Stahel *et al.* (1983) for the multistate $N_2\,{}^1\Sigma_u^+ \sim {}^1\Sigma_u^+$ and ${}^1\Pi_u \sim {}^1\Pi_u$ valence $\sim$ Rydberg perturbations. The energy differences between the coupled equation solutions and experimentally observed energy levels were added to the experimental values. These fictitious observed levels were then refitted by the usual matrix diagonalization method. The new V_n^{d} and H_{mn}^e functions obtained from the matrix approach, when introduced into the coupled equations, were found to give energy levels in good agreement with experiment. Interestingly, the electronic interaction parameters were found to be slightly lower than previously. Since the basis set used in the coupled equations is more nearly complete than in the matrix diagonalization approach, the hybrid matrix/coupled-equations $V_n^{\mathrm{d}}(R)$ and $H_{mn}^e(R)$ functions are more completely deperturbed (closer to the true functions) than the ordinary $\mathbf{H}^{\mathrm{eff}}$ matrix solutions. This is true even though $\mathbf{H}^{\mathrm{eff}}$ included 30 levels for the ${}^1\Pi_u$ states and 41 levels for the ${}^1\Sigma_u^+$ states. The only feature lacking in the coupled equation solution by Stahel *et al.* (1983) is the influence of the higher electronic Rydberg states. This influence could be introduced using multichannel quantum defect theory (MQDT) (Raoult, 1985) (see Chapter 7).

A rigorous least-squares fitting procedure for the coupled-equations approach has been proposed by Dunker and Gordon (1976) for the case of a potential for a van der Waals complex.

The coupled equations have also been solved recently using the adiabatic electronic basis set for the $H_2\,{}^1\Sigma_g^+$ states (Glass-Maujean *et al.*, 1983). In this basis set, the d/dR derivative acting on the unknown $\chi_m'^{\mathrm{ad}}$ vibrational functions appears in the second term on the right-hand side of the coupled equations, namely,

$$\left[-\frac{\hbar^2}{2\mu}\frac{d^2}{dR^2} + V_n^{\mathrm{ad}}(R) + \frac{\hbar^2}{2\mu R^2}J(J+1) - E\right]\chi_n'^{\mathrm{ad}}(R)$$
$$= \sum_{m\neq n} + \frac{h^2}{2\mu}\left[H_{mn}^{(1)}(R)\chi_m'^{\mathrm{ad}}(R) + 2H_{mn}^{(2)}(R)\frac{d}{dR}\chi_m'^{\mathrm{ad}}(R)\right] \quad (3.4.47)$$

where

$$H^{(1)}_{mn} = \int \Phi^{\mathrm{ad}*}_m(r, R)\left(\frac{\partial^2}{\partial R^2} + \frac{2}{R}\frac{\partial}{\partial R}\right)\Phi^{\mathrm{ad}}_n(r, R)\, dr$$

and

$$H^{(2)}_{mn} = \int \Phi^{\mathrm{ad}*}_m(r, R)\frac{\partial}{\partial R}\Phi^{\mathrm{ad}}_n(r, R)\, dr.$$

Therefore, before solving the coupled equations, it was necessary to transform to the diabatic electronic basis in order to eliminate this d/dR derivative. To date there has been only one example for bound states, the H_2^+ three-particle system (Hunter and Pritchard, 1967), where the coupled equations were solved directly without such a transformation.

The coupled-equation method has also been used to calculate the widths and energy shifts arising from the interaction between a discrete and a repulsive state (predissociation, see Section 6.10) (Child and Lefebvre, 1978; Shapiro, 1982). An example of indirect predissociation, where two discrete states interact with the same repulsive state giving three coupled equations, has also been treated in the diabatic basis (see Section 6.11) (Lefebvre-Brion and Colin, 1977). Recently, the adiabatic basis coupled equations have been solved numerically without transformation to the diabatic basis. This has been done for the photopredissociation of the OH $^2\Pi$ states by van Dishoeck *et al.* (1984).

The coupled-equation numerical vibrational wavefunctions are difficult to visualize. A major difficulty is that their number of nodes is not simply related to a vibrational quantum number; hence prior expectations about the node count cannot be used to establish correspondences between observed and calculated levels. Johnson (1978) has presented a method for counting the number of nodes of the calculated wave functions to avoid inadvertently skipping over an eigenstate. The $\chi'^{\mathrm{ad}}_n(R)$ functions for the H_2 $^1\Sigma_g^+$ states have been expressed in the form of Eq. (3.4.43) as a linear combination of the known adiabatic vibrational functions. These $\chi'_n(R)$ functions were used to compute the nonadiabatic radiative lifetimes for all bound $J = 0$ vibrational levels of the EF, GK, and H$\bar{\text{H}}$ $^1\Sigma_g^+$ electronic states (Glass-Maujean *et al.* 1983).

The power of the coupled equation method is that, in principle, if the electronic interaction parameters are either calculated *ab initio* or taken as adjustable parameters, the microscopic vibrational Hamiltonian can be used without introducing a finite-basis effective Hamiltonian matrix. The coupled equations and MQDT methods are both rather new and recently borrowed from scattering theory. The scattering concepts are slowly beginning to penetrate into the field of molecular spectroscopy. It is not unlikely that, in the near future, perturbation problems will be treated largely using methods derived from scattering theory (Mies, 1980).

3.5 Typical Examples of Fitted Perturbations

3.5.1 AN INDIRECT HETEROGENEOUS PERTURBATION: NO $B^2\Pi \sim C^2\Pi \sim D^2\Sigma^+$

The NO $C^2\Pi$ and $D^2\Sigma^+$ states are the π and σ components of a $3p$ Rydberg complex belonging to a series which converges to the NO^+ $X^1\Sigma^+$ ion core,

$$5\sigma^2\ 1\pi^4\ 3p \begin{cases} \nearrow \pi\ C^2\Pi \\ \searrow \sigma\ D^2\Sigma^+. \end{cases}$$

The $B^2\Pi$ state is the lowest of three $^2\Pi$ valence states arising from the $5\sigma^2\,1\pi^3\,2\pi^2$ configuration. The configurations of the $B^2\Pi$ and $D^2\Sigma^+$ states differ by two orbitals, which means that the $B \sim D$ heterogeneous interaction is forbidden in the single configuration approximation. However, the electrostatic $B^2\Pi \sim C^2\Pi$ interaction is configurationally allowed (Section 2.3.2), and it is the admixture of $C^2\Pi$ character into both $B^2\Pi$ and $D^2\Sigma^+$ states that is responsible for the nominal $B \sim D$ perturbation. This is an indirect effect of the form

$$\langle B|\mathbf{H}^{\text{el}}|C\rangle\langle C|\mathbf{H}^{\text{ROT}}|D\rangle$$

but, owing to the near-degeneracy of the three interacting vibronic levels, this problem cannot be treated by a second-order Van Vleck correction to $H^{\text{eff}}_{\mathbf{BD}}$ [Eqs. (3.2.17), (3.4.39), and (3.4.41).]. A direct diagonalization of the two 5×5 matrices ($B^2\Pi_{1/2}$, $B^2\Pi_{3/2}$, $C^2\Pi_{1/2}$, $C^2\Pi_{3/2}$, $D^2\Sigma^+$; one matrix each for e and f levels) is required. One simplifying feature is that the pure precession approximation (Section 4.5) is expected to be valid for the $C \sim D$ interaction because these states differ by a single $3p$ Rydberg orbital. This means that (for $v_C = v_D$ interactions) reliable prior estimates are possible for the spin–orbit $[\langle v_C|v_D\rangle a_+ \simeq 2^{1/2}A(C^2\Pi)]$ and **l**-uncoupling $(\langle v_C|B|v_D\rangle b \simeq 2^{1/2}[B(C^2\Pi) + B(D^2\Sigma^+)]/2)$ [Eqs. (2.4.18), (2.4.19), (2.5.24), (2.5.28), (2.5.29)] perturbation parameters. The 5×5 $\mathbf{H}^{\text{eff}}$ matrices were constructed from a 3×3 $C \sim D$ interaction matrix (Table 2.9) augmented by a 2×2 $B^2\Pi$ matrix and a single homogeneous $H_{B\sim C}$ interaction parameter ($H_{B\sim D}$ is held fixed at zero because $B \sim D$ heterogeneous interactions are configurationally forbidden).

The $B \sim C \sim D$ perturbations were first successfully treated by Jungen and Miescher (1968) using a simple and pedagogical model. They used a fixed value of the $B \sim C$ perturbation parameter, obtained from previous analyses of isolated $B \sim C$ interactions (Lagerqvist and Miescher, 1958) combined with calculated vibrational overlap factors (Felenbok and Lefebvre-Brion, 1966). Rather than diagonalize the 5×5 matrices in a single step, a three-step diagonalization procedure was adopted. First, a transformation from case (a) to case (b) was performed in order to eliminate off-diagonal $^2\Pi_{1/2} \sim {}^2\Pi_{3/2}$ elements [Eqs. (2.5.16)–(2.5.17)]. The C–D matrix elements in the transformed representation are, for the e-levels,

$$H_{C(F_2)\sim D} = [(J + \tfrac{3}{2})/(2J+1)]^{1/2} a_+/2 \qquad (3.5.1a)$$
$$H_{C(F_1)\sim D} = -[(J - \tfrac{1}{2})/(2J+1)]^{1/2}[a_+/2 - 2bB_v(J+\tfrac{1}{2})],$$

and for the f-levels,

$$H_{C(F_2)\sim D} = [(J + \tfrac{3}{2})/(2J+1)]^{1/2}[a_+/2 + 2bB_v(J+\tfrac{1}{2})] \qquad (3.5.1b)$$
$$H_{C(F_1)\sim D} = -[(J - \tfrac{1}{2})/(2J+1)]^{1/2} a_+/2.$$

Next, the B$\sim$C interaction is diagonalized. The result is eight groups of B$\sim$C mixed $^2\Pi$ levels (four e, four f, the e and f levels having identical energies and mixing coefficients), which can be denoted $|^2\Pi, J, e/f, k\rangle$ where k ($k = 1, 2, 3, 4$) is the energy rank. The transformed basis functions are then

$$|^2\Pi, J, e/f, k\rangle = c_{k,C(F_1)}(J)|C^2\Pi, F_1\rangle + c_{k,C(F_2)}(J)|C^2\Pi, F_2\rangle$$
$$+ c_{k,B(F_1)}(J)|B^2\Pi, F_1\rangle + c_{k,B(F_2)}(J)|B^2\Pi, F_2\rangle \qquad (3.5.2)$$

where the mixing coefficients, c, are determined by the two transformations. The interaction matrix elements between the $|D^2\Sigma^+\rangle$ and the two-step transformed $|^2\Pi, J, e/f, k\rangle$ functions depend only on the $c_{k,C(F_1)}$ and $c_{k,C(F_2)}$ mixing coefficients and the Eq. (3.5.1) C$\sim$D matrix elements (the B$\sim$D interactions are assumed to be zero). The remaining 2×2 interactions between $D^2\Sigma^+$ and the eight $|^2\Pi, J, e/f, k\rangle$ functions are diagonalized in the final step of the deperturbation process. The matrix elements obtained from the separate 2×2 diagonalizations were found to have the relative magnitudes predicted by Eqs. (3.5.1) and (3.5.2).

The a_+ and b parameters obtained by Jungen and Miescher (1968) account very well for the observed level shifts and intensities. These parameters agree satisfactorily with those obtained by Amiot and Vergès (1982) for a larger and more precisely measured set of spectral lines and using a direct diagonalization fit to the full 5×5 $\mathbf{H}^{\text{eff}}$ matrices in which all parameters were simultaneously varied. The b-value obtained by Amiot and Vergès (1982) indicates that, as expected, the pure precession approximation, $b = 2^{1/2}$, is obeyed by members of a Rydberg complex,

$$b^{\text{obs}} = 1.36.$$

3.5.2 A STRONG MULTISTATE INTERACTION IN THE NO MOLECULE

The example treated in Section 3.5.1 dealt with the interaction between the $B^2\Pi$, $C^2\Pi$, and $D^2\Sigma^+$ states of NO. A two-step treatment yielded deperturbed molecular constants and a physical delineation of all processes affecting this three-state interaction. Here, a four-state interaction, also in NO, will be discussed. Strong interactions among three $^2\Pi$ states, the two valence states $B^2\Pi\,(v = 24)$ and $L^2\Pi\,(v = 5)$, and the $C^2\Pi\,(v = 6)$ Rydberg state, combined with the crossing of all three $^2\Pi$ states by the $D^2\Sigma^+\,(v = 6)$ Rydberg state result in an unusually complicated and confusing pattern of perturbed rotational levels. Figure 3.2 shows the rotational term values deduced from transitions observed in absorption (Lagerqvist and Miescher, 1966; Dressler and Miescher, 1981).

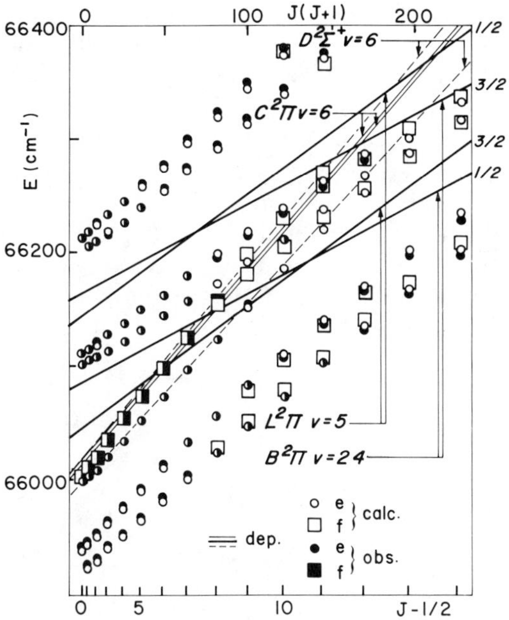

Fig. 3.2 The $^{14}N^{16}O$ $B^2\Pi$ $(v_B = 24) \sim C^2\Pi$ $(v_C = 6) \sim L^2\Pi$ $(v_L = 5) \sim D^2\Sigma$ $(v_D = 6)$ perturbation. The deperturbed $^2\Pi_{1/2}$ and $^2\Pi_{3/2}$ levels are plotted as solid lines (heavy lines for the valence $B^2\Pi$ and $L^2\Pi$ states, light lines for the Rydberg $C^2\Pi$ state). The $D^2\Sigma^+$ e and f levels are plotted as dashed lines. The observed and calculated e-levels are indicated by solid and open circles, respectively. The observed (solid squares) and calculated (open squares) f-levels are shown only when the Λ-doubling is larger than 2.5 cm^{-1}.

A numerical deperturbation of the NO $B^2\Pi$, $L^2\Pi$, and $C^2\Pi$ states has been performed by Gallusser and Dressler (1982) in which many vibrational levels were simultaneously deperturbed (see Section 5.2.2). Another deperturbation, which included the $D^2\Sigma^+$ state, has been performed (Lefebvre-Brion, unpublished calculation). In this four-state deperturbation, only one vibrational level from each state was treated. The deperturbed levels are represented by the straight lines on Fig. 3.2 (solid lines for $^2\Pi$ states, dashed lines for $D^2\Sigma$). The *effective* interaction matrix elements obtained from the four-state deperturbation are

$$H^{\mathrm{BC}}_{24,6} = 66.7 \text{ cm}^{-1}$$

$$H^{\mathrm{LC}}_{5,6} = 102.6 \text{ cm}^{-1},$$

which are very close to the *true* values of 66 cm^{-1} and 114 cm^{-1} obtained by Gallusser and Dressler (1982).

Each observed rotational eigenstate may be expressed as an explicit linear combination of deperturbed basis substates. The mixing coefficients are obtained from the eigenvectors of the four-state effective Hamiltonian (which are generated when the Hamiltonian is numerically diagonalized). For each value of J (except $J = \frac{1}{2}$) there are seven e-parity and seven f-parity basis substates (three $\Omega = \frac{1}{2}$ and three $\Omega = \frac{3}{2}$ $^2\Pi$ functions and one $^2\Sigma^+$ function). The eigenstates of the deperturbation Hamiltonian are specified by J and an integer between 1 and 14 (even $= f$, odd $= e$, 1 and 2 lowest energy, 13 and 14 highest energy). The mixing coefficients for the case (a) basis functions in the $J = 0.5$, 8.5, and 12.5 e-parity eigenfunctions are given in Table 3.1. For $J = 0.5$, only $\Omega = \frac{1}{2}$ basis states exist and the observed levels are exclusively of $\Omega = \frac{1}{2}$ character. At higher J-values, **S**-uncoupling within the $^2\Pi$ states causes Ω to become poorly defined. For example, the $J = 12.5e$ level number 7 has

$$\begin{aligned}\bar{\Omega} \equiv \langle J_z \rangle &= 3/2[(0.4325)^2 + (0.1779)^2 + (0.3361)^2] \\ &\quad + 1/2[(0.3920)^2 + (0.1810)^2 + (0.1943)^2 + (0.6664)^2] \\ &= 0.83.\end{aligned}$$

The electronic and vibrational character of this level is also poorly defined,

22.0%	$B^2\Pi$	$v_B = 24$
26.7%	$L^2\Pi$	$v_L = 5$
6.9%	$C^2\Pi$	$v_C = 6$
44.4%	$D^2\Sigma^+$	$v_D = 6$

as no basis state accounts for more than 50% of its character. Thus even the *nominal* vibronic character of this level is poorly defined. The only meaningful labels that can be attached to the observed levels are their good quantum

Table 3.1
Mixing Coefficients for *e*-Parity Levels of the ^{14}N ^{16}O $B^2\Pi \sim C^2\Pi \sim L^2\Pi \sim D^2\Sigma^+$ Perturbation[a]

Eigenstate index[b]	$B^2\Pi_{3/2}$	$L^2\Pi_{1/2}$	$B^2\Pi_{1/2}$	$C^2\Pi_{3/2}$	$C^2\Pi_{1/2}$	$L^2\Pi_{3/2}$	$D^2\Sigma^+$
$J = 0.5$							
1				does not exist			
3	0.0000	0.4074	0.3728	0.0000	−0.8335	0.0000	−0.0186
5	0.0000	0.0092	0.0101	0.0000	−0.0133	0.0000	0.9998
7	0.0000	−0.4211	0.8867	0.0000	0.1908	0.0000	−0.0026
9				does not exist			
11				does not exist			
13	0.0000	0.8103	0.2733	0.0000	0.5184	0.0000	−0.0033
$J = 8.5$							
1	−0.2078	−0.1929	−0.1905	0.6213	0.3414	−0.5933	0.1697
3	−0.0929	0.3627	0.4624	0.2899	−0.6622	−0.3384	−0.0944
5	0.0160	0.0727	0.1665	−0.0074	−0.0642	0.1499	0.9696
7	0.2045	−0.4858	0.7851	−0.0898	0.2828	−0.1159	−0.0658
9	0.6858	0.0741	−0.2203	−0.3511	−0.0818	−0.5790	0.1023
11	−0.6302	−0.2862	−0.0585	−0.5843	−0.1899	−0.3653	0.0813
13	0.1967	−0.7089	−0.2319	0.2392	−0.5640	0.1701	0.0279
$J = 12.5$							
1	0.2334	0.2298	0.3058	−0.5249	−0.3524	0.6087	−0.1703
3	0.0977	−0.2748	−0.6360	−0.2859	0.4865	0.4352	0.0510
5	−0.1799	0.4164	−0.5924	0.0305	−0.3254	−0.1119	−0.5692
7	−0.4325	0.3920	−0.1810	0.1779	−0.1943	0.3361	0.6664
9	−0.7095	−0.2687	0.2704	0.1668	0.1620	0.3611	−0.4084
11	0.4257	0.3189	0.0777	0.6789	0.2792	0.3743	−0.1789
13	−0.1786	0.6088	0.1978	−0.3484	0.6256	−0.2100	−0.0390

[a] From a calculation by Lefebvre-Brion, part of which has been published by Kimman *et al.* (1985).
[b] Odd-eigenstate indices correspond to the *e*-parity levels. The eigenvalues for each value of J are ranked in order of energy (index 1 for the lowest eigenstate, index 13 for the highest).

numbers: J, parity, and energy. Nevertheless, the observable properties of each level may be calculated from those of the basis states properly weighted by the mixing coefficients obtained from the eigenvectors of the deperturbation Hamiltonian.

References

Albritton, D. L., Harrop, W. J., Schmeltekopf, A. L., and Zare, R. N. (1973), *J. Mol. Spectrosc.* **46**, 25.

Albritton, D. L., Schmeltekopf, A. L., and Zare, R. N. (1976), *in* "Molecular Spectroscopy: Modern Research" Vol. 2, p. 1, (K. N. Rao, ed.), Academic Press, New York.

Albritton, D. L., Schmeltekopf, A. L., and Zare, R. N. (1977), *J. Mol. Spectrosc.* **67**, 132.
Amiot, C., and Vergès, J. (1982), *Phys. Scr.* **25**, 302.
Åslund, N. (1974), *J. Mol. Spectrosc.* **50**, 424.
Atabek, O., and Lefebvre, R. (1980), *Chem. Phys.* **52**, 199.
Birss, F. W. (1983), *J. Mol. Spectrosc.* **99**, 133.
Birss, F. W., Brown, J. M., Cole, A. R. H., Lofthus, A., Krishnamachari, S. L. N. G., Osborne, G. A., Paldus, J., Ramsay, D. A., and Watmann, L. (1970), *Can J. Phys.* **48**, 1230.
Bishop, D. M., and Cheung, L. M. (1978), *Phys. Rev. A* **18**, 1846.
Brand, J. C. D., Jones, V. T., and DiLauro, C. (1971), *J. Mol. Spectrosc.* **40**, 616.
Brown, J. M., and Merer, A. J. (1979), *J. Mol. Spectrosc.* **74**, 488.
Brown, J. M., Colbourn, E. A., Watson, J. K. G., and Wayne, R. D. (1979), *J. Mol. Spectrosc.* **74**, 294.
Brown, J. M., and Milton, D. J. (1976), *Mol. Phys.* **31**, 409.
Chedin, A., and Cihla, Z. (1967), *Cah. Phys.* **219**, 129.
Child, M. S., and Lefebvre, R. (1978), *Chem. Phys. Lett.* **55**, 213.
Coxon, J. A. (1978). *J. Mol. Spectrosc.* **72**, 252.
Curl, R. F. (1970), *J. Comput. Phys.* **6**, 367.
De Santis, D., Lurio, A., Miller, T. A., and Freund, R. S. (1973), *J. Chem. Phys.* **58**, 4625.
Dressler, K., and Miescher, E. (1981), *J. Chem. Phys.* **75**, 4310.
Dressler, K., Gallusser, R., Quadrelli, P., and Wolniewicz, L. (1979), *J. Mol. Spectrosc.* **75**, 205.
Dunker, A. M., and Gordon, R. G. (1976), *J. Chem. Phys.* **64**, 4984.
Felenbok, P., and Lefebvre-Brion, H. (1966), *Can. J. Phys.* **44**, 1677.
Feynman, R. P. (1939), *Phys. Rev.* **56**, 340.
Field, R. W., and Bergeman, T. H. (1971), *J. Chem. Phys.* **54**, 2936.
Field, R. W., Lagerqvist, A., and Renhorn, I. (1976), *Phys. Scr.* **14**, 298.
Freed, K. F. (1966), *J. Chem. Phys.* **45**. 4214.
Gallusser, R., and Dressler, K. (1982), *J. Chem. Phys.* **76**, 4311.
Gerö, L. (1935), *Z. Phys.* **93**, 669.
Glass-Maujean, M., Quadrelli, P., Dressler, K., and Wolniewicz, L. (1983), *Phys. Rev. A* **28**, 2868.
Hellmann, H. (1937), "Einführung in die Quantenchemie," Franz Deuticke, Vienna.
Herschbach, D. (1956), unpublished notes.
Hill, E. L., and Van Vleck, J. H. (1928), *Phys. Rev.* **32**, 250.
Horani, M., Rostas, J., and Lefebvre-Brion, H. (1967), *Can. J. Phys.* **45**, 3319.
Hunter, G., and Pritchard, H. O. (1967), *J. Chem. Phys.* **46**, 2153.
Johns, J. W. C., and Lepard, D. W. (1975), *J. Mol. Spectrosc.* **55**, 374.
Johnson, B. R. (1978), *J. Chem. Phys.* **69**, 4678.
Jungen, C., and Miescher, E. (1968), *Can. J. Phys.* **46**, 987.
Kimman, J., Lavollée, M., and Van der Wiel, J. (1985), *Chem. Phys.* **97**, 137.
Kirschner, S. M., and Watson, J. K. G. (1973), *J. Mol. Spectrosc.* **47**, 234.
Kovács, I. (1969), "Rotational Structure in the Spectra of Diatomic Molecules," Am. Elsevier, New York.
Lagerqvist, A., and Miescher, E. (1958), *Helv. Phys. Acta* **31**, 221.
Lagerqvist, A., and Miescher, E. (1966), *Can. J. Phys.* **44**, 1525.
Lees, R. M. (1970), *J. Mol. Spectrosc.* **33**, 124.
Lefebvre-Brion, H. (1969), *Can. J. Phys.* **47**, 541.
Lefebvre-Brion, H., and Colin, R. (1977), *J. Mol. Spectrosc.* **65**, 33.
LeFloch, A. and Legal, C. (1986), Computer Physics Communications. (To be published.)
Löwdin, P. O. (1951), *J. Chem. Phys.* **19**, 1396.
Marquardt, D. W. (1963), *J. Soc. Ind. Appl. Math.* **11**, 431.

Marquardt, D. W., Bennett, R. G., and Burrell, E.J. (1961), *J. Mol. Spectrosc.* **7**, 269.
Meakin, P., and Harris, D. O. (1972), *J. Mol. Spectrosc.* **44**, 219.
Merer, A. J., and Allegretti, J. M. (1971), *Can. J. Phys.* **49**, 2859.
Mies, F. M. (1980), *Mol. Phys.* **41**, 953, 973.
Miller, T. A. (1969), *Mol. Phys.* **16**, 105.
Mulliken , R. S., and Christy, A. (1930), *Phys. Rev.* **38**, 87.
Ngo, T. A., DaPaz, M., Coquart, B., and Couet, C. (1974), *Can. J. Phys.* **52**, 154.
Raoult, M. (1985), *in* "Photophysics and Photochemistry above 6 eV," p. 343. Elsevier, Amsterdam.
Shapiro, M. (1982), *Chem. Phys. Lett.* **91**, 12.
Stahel, D., Leoni, M., and Dressler, K. (1983), *J. Chem. Phys.* **79**, 2541.
Stern, R. C., Gammon, R. H., Lesk, M. E., Freund, R. S., and Klemperer, W. A. (1970), *J. Chem. Phys.* **52**, 3467.
Tellinghuisen, J. (1973), *Chem. Phys. Lett.* **18**, 544.
Tinkham, M., and Strandberg, M. W. P. (1955), *Phys. Rev.* **97**, 937.
van Dishoeck, E. F., van Hemert, M. C., Allison, A. C., and Dalgarno, A. (1984), *J. Chem. Phys.* **81**, 5709.
Watson, J. K. G. (1977), *J. Mol Spectrosc.* **66**, 502.
Wentworth, W. E. (1965), *J. Chem. Educ.* **42**, 96.
Wicke, B. G., Field, R. W., and Klemperer, W. (1972), *J. Chem. Phys.* **56**, 5758.
Wollrab, J. E. (1967), "Rotational Spectra and Molecular Structure," Appendix 7, Academic Press, New York.
Zare, R. N., Schmeltekopf, A. L., Harrop, W. J., and Albritton, D. L. (1973), *J. Mol. Spectrosc.* **46**, 37.

Chapter 4

Interpretation of the Perturbation Matrix Elements

The magnitudes of perturbation matrix elements are seldom tabulated in compilations of molecular constants. If deperturbed diagonal constants are listed, then the off-diagonal perturbation parameters should be listed as well, even though they cannot, without specialized narrative footnotes, be accommodated into the standard tabular format of such compilations. Without specification of at least the electronic part of the interaction parameters, it is impossible to reconstruct spectral line frequencies or intensities; thus the deperturbed diagonal constants by themselves have no meaning.

The purpose of this chapter is to show how to extract useful information from perturbation matrix elements. The uses of this information range from tactical (vibrational and electronic assignments) to insight into the global electronic structure of a molecule or family of molecules. It is tempting to suggest that perturbation matrix elements can contain at least as much structural information as the usual molecular constants.

The magnitude of the perturbation interaction, $H_{1,v_i,J;\,2,v_j,J}$, is determined by the product of two factors, one electronic and the other vibrational, both of which can depend on J. Recall (Section 2.3.1) that if the electronic factor varies linearly with internuclear distance in the neighborhood of the R-value of the crossing between the potential curves of states 1 and 2, R_C, then

$$H_{1,v_i,J;\,2,v_j,J} = H^e_{12}(\bar{R}_{v_i,v_j,J})\langle v_i, J|v_j, J\rangle,$$

where the R-centroid for each pair of near-degenerate perturbing levels has the value

$$\bar{R}_{v_iv_j} = \frac{\langle v_i|R|v_j\rangle}{\langle v_i|v_j\rangle} \simeq R_C$$

with a precision better than $\pm 5\%$.† Thus perturbations between states 1 and 2 sample $H^e_{12}(R, J)$ at $R = R_C$ and the magnitude of the electronic factor is nearly independent of v_i, v_j, provided that the curves cross and the interacting vibrational levels are near-degenerate.

The observability of a perturbation between levels $1, v_i, J \sim 2, v_j, J$ requires that *both* the electronic and vibrational factors be nonzero. For example, although the $J = 12.5$ levels of the $B^2\Pi_{1/2}$ $(v = 18)$ and $B'^2\Delta_{5/2}$ $(v = 1)$ states‡

† The vibrational part of H_{12} is not always simply a vibrational overlap integral. For example, matrix elements of $\mathbf{H}^{ROT}$ include the factor

$$\langle v_i, J|\mathbf{B}(R)|v_j, J\rangle = (\hbar/4\pi\mu c)\langle v_i, J|R^{-2}|v_j, J\rangle$$
$$\simeq \langle v_i, J|v_j, J\rangle R_C^{-2}\hbar/4\pi\mu c.$$

‡ Direct perturbations between *basis functions* differing in Ω by ± 2 are rigorously forbidden. However, at the $J = 12.5$ crossing between *nominal* $B_{1/2}$ and $B'_{5/2}$ substates, spin-uncoupling has introduced significant $\Omega = \frac{3}{2}$ basis function character into both nominal levels.

of $^{15}N^{18}O$ are separated by 0.3 cm^{-1} and $H^{e}_{B_{3/2} \sim B'_{3/2}}$ is estimated to be > 18 cm^{-1}, no perturbation was detectable within the 0.1 cm^{-1} measurement precision (Field *et al.*, 1975). This nonperturbation is explained by the computed $v_B = 18 \sim v_{B'} = 1$ vibrational overlap factor of 8×10^{-3}. Alternatively, if a perturbation is observed when a computed vibrational factor is prohibitively small, then the electronic or vibrational identification of either the perturbed or perturbing state is incorrect. For example, the $^{31}P^{16}O$ $D^2\Pi$ ($v = 0$) level is perturbed by what is now known (Coquart *et al.*, 1974; Ghosh *et al.*, 1975) to be a highly excited vibrational level of $B'^2\Pi$ (Verma and Dixit, 1968; Verma, 1971). Upon isotopic substitution, the $^{31}P^{18}O$ $D^2\Pi$ ($v = 0$) level is found to be perturbed by a level of $B'^2\Pi$ with T_{ve} nearly identical to that found for the $^{31}P^{16}O$ perturber. Verma initially assumed the same vibrational quantum number of the perturber for both isotopic species and concluded from the absence of a large isotope shift that $v_{\text{pert}} = 0$ (Verma and Dixit, 1968). However, if a potential curve for the perturber is derived from the incorrect vibrational numbering and molecular constants, the $v_D = 0 \sim v_{\text{pert}} = 0$ vibrational overlap is computed to be too small for any perturbation to have been detectable. It turns out that the $^{31}P^{16}O$ perturber of $D^2\Pi$ ($v = 0$) is $B'^2\Pi$ ($v = 24$) and, by chance, the isotope shift for the B′ state is nearly equal to the $\Delta G(24.5)$ value so that the perturber for $^{31}P^{18}O$ is actually $B'^2\Pi$ ($v = 25$) (Verma, 1971; Coquart *et al.*, 1974; Ghosh *et al.*, 1975). This situation is illustrated by Fig. 4.1.

The electronic factor can also provide an explanation for a nonperturbation and a test of a vibrational assignment (Robbe *et al.*, 1981). However, this factor can only be computed *ab initio* or, sometimes, estimated semiempirically. This dependence on theoreticians for the electronic factor is in contrast to the situation for the vibrational factor, which is routinely calculable by the experimentalist using widely available computer programs. Although *ab initio* computations of $H^{e}_{12}(R_C)$ are not always accessible to the spectroscopist, the fact that the electronic factor, for an important class of perturbations

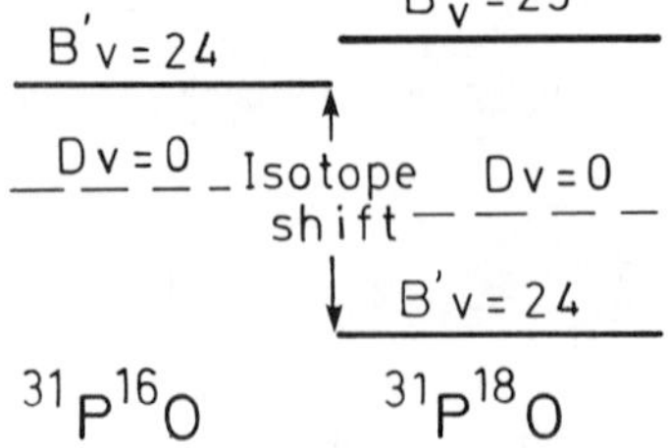

Fig. 4.1 Isotope effect on a vibrational level of the $B'^2\Pi$ state of the PO molecule.

considered here, is vibration-independent is the basis of a powerful method for determining vibrational and electronic assignments of perturbers. This "matrix element method" is described in Section 4.1.5.

The ultimate test of a deperturbation consists of (1) a demonstration that all observed perturbation matrix elements have the v, J-dependence required by the factorization into electronic and vibrational parts; (2) agreement between the observed and *ab initio* values of $H^e_{12}(R_C)$; (3) verification that the molecular constants for both electronic states are internally consistent [isotope shifts, D_v values calculable from $G(v)$ and B_v functions] (Gottscho *et al.*, 1979; Kotlar *et al.*, 1980).

Sections 4.1 and 4.2–4.4 deal respectively with the vibrational and electronic factors of the perturbation matrix element. The vibrational wavefunctions and matrix elements may be computed semiclassically (Section 4.1.1), derived analytically for special potentials, such as the Morse potential (Section 4.1.2), or obtained numerically via the RKR inversion followed by numerical integration (Section 4.1.3). In discussing the electronic factor, it is useful to distinguish between electrostatic perturbations (Section 4.2), which appear mainly between valence and Rydberg states, and spin or rotational perturbations, which appear mainly between different valence states (Sections 4.3 and 4.4). The distinction between valence and Rydberg states is discussed in Section 4.2.1. Electrostatic perturbations are divided into valence ~ Rydberg interactions (Sections 4.2.2 and 4.2.3) and Rydberg ~ Rydberg interactions (Section 4.2.4). Section 4.3 shows for the spin parameters, and Section 4.4 for the rotational parameters how comparisons of corresponding perturbations for isovalent molecules provide global insights into molecular electronic structure. The pure precession approximation is discussed in Section 4.5. After brief discussion of the origin of the R-variation of the electronic factor (Section 4.6) and the validity of the single configuration approximation (Section 4.7), it is shown how the study of perturbations can help to identify and locate metastable states (Section 4.8).

4.1 Calculation of the Vibrational Factor

Most spectroscopists are comfortable using RKR and Franck–Condon programs as black boxes. Since these programs are widely available, inexpensive to run, and generate more accurate vibrational factors than semiclassical or model potential approaches, the application-minded reader is advised to skip Sections 4.1.1 and 4.1.2.

4.1.1 SEMICLASSICAL APPROXIMATION

The power of this semiclassical approach is that it enables derivation of analytic expressions for the overlap integrals between vibrational eigenfunctions of two potentials. Since, for perturbations, one often knows one potential very well and the other potential poorly or not at all, these analytic relationships can provide a direct link between observed overlaps and some features of the poorly known potential, namely, the location of the crossing between the two potentials, E_C and R_C, and the slope of the unknown potential at R_C.

The physical idea behind this method is that, at R_C, if (E_{1,v_i}, χ_{v_i}) and (E_{2,v_j}, χ_{v_j}) are associated, respectively, with potentials 1 and 2, and $E_{1,v_i} \approx E_{2,v_j}$, then the vibrational wavefunctions χ_{v_i} and χ_{v_j} oscillate at the same frequency near R_C. The magnitude of the overlap integral is determined largely in the region where the two rapidly oscillatory functions oscillate at the same frequency (Tellinghuisen, 1984). The length of this region (how long it takes for the phase of χ_{v_i} to change by π relative to that of χ_{v_j}) is long or short depending on whether V_1 and V_2 have similar or different slopes at R_C. Three factors contribute to the magnitude of $\langle v_i|v_j\rangle$: the difference in slopes at R_C, the product of the maximum amplitudes of χ_{v_i} and χ_{v_j} near R_C, and the phase of χ_{v_i} relative to χ_{v_j} at R_C. The amplitude of χ at R_C is related to the classical probability density, which is inversely proportional to the velocity times the oscillation period, $1/\bar{\omega}$; thus the semiclassical amplitude will be proportional to

$$(\bar{\omega}/v)^{1/2} = \bar{\omega}^{1/2}(\mu/2)^{1/4}[E - V(R_C)]^{-1/4}$$

or

$$(\bar{\omega}/v)^{1/2} = |\bar{\omega}\mu/\hbar k(R_C)|^{1/2}$$

where μ is the reduced mass and k is the wavenumber, defined below.

Given a potential energy curve, it is possible to locate (iteratively) the vibrational energy levels using the semiclassical quantization condition

$$\int_a^b k(R)\, dR = (v + \tfrac{1}{2})\pi \tag{4.1.1}$$

$$k(R) = p(R)/\hbar = \{[2\mu]^{1/2}/\hbar\}[E - V(R)]^{1/2} \tag{4.1.2}$$

where p is the classical momentum associated with the kinetic energy, $E - V(R)$, and where a and b, respectively, are the left and right classical turning points [at which the total energy, E, is equal to the potential energy, $V(R)$] (Fig. 4.2). The eigenfunctions may be approximated by the semiclassical JWKB wavefunction (J=Jeffreys, W=Wentzel, K=Kramers, B=Brillouin),

$$\chi^{\text{JWKB}}(R) = N[k(R)]^{-1/2}\sin[\phi(R) + \pi/4] \tag{4.1.3}$$

$$\phi(R) = \int_a^R k(R')\,dR', \tag{4.1.4}$$

where N is a normalization factor. This χ^{JWKB} is a good approximation to the exact wavefunction between a and b but *not too close* to either turning point (see Merzbacher, 1970, p. 116).

The JWKB wavefunction defined by Eq. (4.1.3) is appropriate for either bound (two turning points) or unbound states (one turning point), provided that R is restricted to the region where $E > V(R)$. The normalization factor for the unbound χ^{JWKB} at energy E is

$$N_E = |2\mu/\pi\hbar^2|^{1/2}, \tag{4.1.5}$$

which corresponds to unity probability of finding the system with energy between E and $E + dE$. For a bound level, the normalization factor is

$$N_v = |2\mu\hbar\bar{\omega}/\pi\hbar^2|^{1/2}, \tag{4.1.6}$$

which corresponds to unity probability of finding the system in level v. N_E and N_v are related (discussed in Section 6.3) by

$$\hbar\bar{\omega} = \left(\frac{N_v}{N_E}\right)^2, \tag{4.1.7}$$

where $\hbar\bar{\omega}$ is the local energy-level spacing, which can be approximated by

$$\hbar\bar{\omega} \simeq \frac{E_{v+1} - E_{v-1}}{2}$$

or computed from the semiclassical quantization condition [Eq. (4.1.1)]

$$\hbar\bar{\omega} = \frac{\partial E}{\partial v} = \left(\frac{\partial v}{\partial E}\right)^{-1} = \left\{\frac{\partial}{\partial E}\left[\frac{1}{\pi}\int_a^b k(R)\,dR - \tfrac{1}{2}\right]\right\}^{-1}$$

$$\hbar\bar{\omega} = \left(\frac{1}{\pi}\int_a^b \frac{dk}{dE}\,dR\right)^{-1} = \frac{\pi\hbar^2}{\mu}\left[\int_a^b k^{-1}(R)\,dR\right]^{-1} \tag{4.1.8}$$

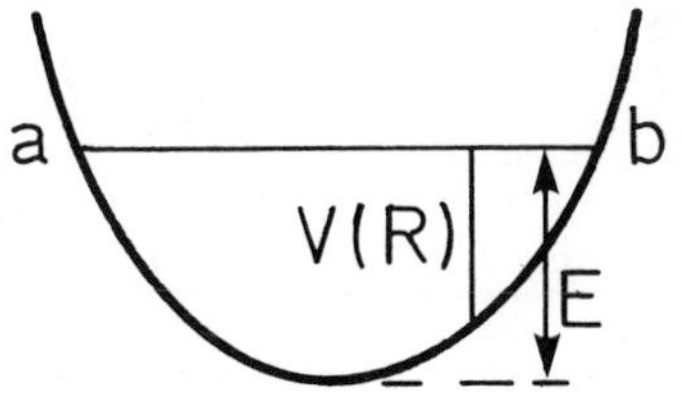

Fig. 4.2 Bound level in a potential.

since

$$\frac{dk}{dE} = \frac{\mu}{\hbar^2}\frac{1}{k}. \tag{4.1.9}$$

In order to take care of the problem that χ^{JWKB} is not a good approximation near the turning points, it is necessary to define $\chi^{\mathrm{USC}}(R)$, USC indicating uniform semiclassical wavefunction (Miller and Good, 1953; Langer, 1937; Miller, 1968), which is uniformly valid close to or far from the turning points

$$\chi^{\mathrm{USC}}(R) = \pi^{1/2} N \left[\frac{Z(R)}{k(R)^2}\right]^{1/4} \mathrm{Ai}[-Z(R)] \tag{4.1.10}$$

$$Z(R) = \left[\frac{3}{2}\int_a^R k(R')\,dR'\right]^{2/3} = \left[\frac{3}{2}\phi(R)\right]^{2/3}. \tag{4.1.11}$$

$\mathrm{Ai}[-Z(R)]$ is an Airy function, which is a decreasing exponential at large positive argument ($Z < 0$, k imaginary, nonclassical region) and oscillatory at large negative argument ($Z > 0$, k real, classical region). For $Z > 1.5$,

$$\mathrm{Ai}[-Z(R)] \longrightarrow \pi^{-1/2} Z(R)^{-1/4} \sin[\phi(R) + \pi/4].$$

Thus χ^{USC} is constructed so that when R is in the classical region but far from a turning point, $\chi^{\mathrm{USC}}(R) = \chi^{\mathrm{JWKB}}(R)$. Note also that for a linear potential,

$$V(R) = -FR$$

(F is an R-independent force between nuclei), the function [Eq. (4.1.10)] is the exact eigenfunction.

A typical Airy function, $\mathrm{Ai}(-Z)$, is plotted in Fig. 4.3; note, however, that

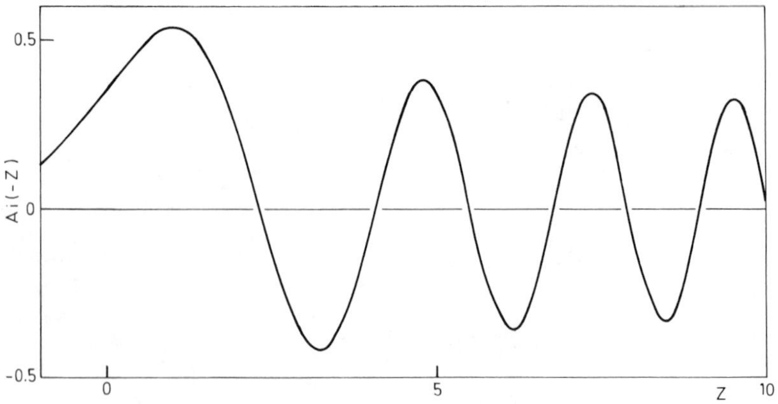

Fig. 4.3 The regular Airy function $\mathrm{Ai}(-Z)$.

the argument of the Airy function that appears in χ^{USC} is a function of R; thus Ai $[-Z(R)]$ does not behave exactly like Fig. 4.3.

The immediate objective is to obtain an analytic expression for vibrational overlap integrals, $\langle v_i|v_j\rangle$. Conveniently, the integral over the product of the Airy functions can be shown (Connor, 1981) to be approximately another Airy function (in fact, rigorously true for linear potentials). If

$$E = E_{1,v_i} = E_{2,v_j},$$

then

$$\langle v_i|v_j\rangle = \left|\frac{2(\hbar\bar{\omega}_1)(\hbar\bar{\omega}_2)\mu}{\hbar^2 k(R_C)|(dV_1/dR)-(dV_2/dR)|_{R=R_C}}\right|^{1/2} \xi(E)^{1/4}\mathrm{Ai}[-\xi(E)] \quad (4.1.12)$$

$$\hbar k(R_C) = (2\mu)^{1/2}(E-E_C)^{1/2} \quad (4.1.13)$$

$$\xi(E) = [(\tfrac{3}{2})\phi(E)]^{2/3} \quad (4.1.14)$$

$$\phi(E) = \int_{a_2}^{R_C} k_2(R)\,dR + \int_{R_C}^{b_1} k_1(R)\,dR, \quad (4.1.15)$$

where E_C and R_C are the energy and R value of the curve crossing. Equation (4.1.15) is valid for the specific case where the crossing point lies between the minima of the two potential curves (Fig. 4.4). The Airy function that appears in Eq. (4.1.12) is a function of E rather than R. Note that if $\xi(E) < 0$ (ϕ is imaginary), $\langle v_i|v_j\rangle$ is an exponentially decreasing function of $E_C - E$, whereas if $\xi(E) > 0$ (ϕ is real), the overlap is an oscillatory function of $E - E_C$. $\phi(E)$ is real if $E > E_C$; thus, as E approaches E_C from below, $|\langle v_i|v_j\rangle|$ increases monotonically to an absolute maximum just above $E = E_C$ and then passes

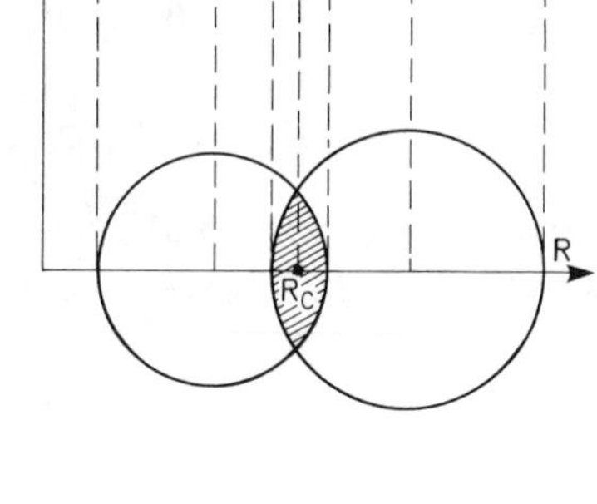

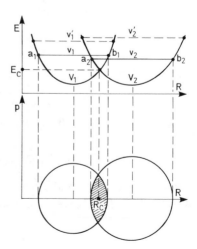

Fig. 4.4 Phase space mapping for the harmonic model. Two pairs of exactly degenerate interacting levels are shown. Since V_2 is less steep than V_1, $\omega_2 < \omega_1$ and $(v_1' - v_1) < (v_2' - v_2)$.

through a series of increasingly smaller relative maxima. For $\xi(E) > 1.5$ ($E > E_C$, ϕ real),

$$\langle v_i|v_j\rangle \to \left|\frac{2(\hbar\bar{\omega}_1)(\hbar\bar{\omega}_2)\mu}{\hbar(2\mu)^{1/2}|(dV_1/dR) - (dV_2/dR)|_{R=R_C}}\right|^{1/2} \times \pi^{-1/2}(E - E_C)^{-1/4}\sin[\phi(E) + \pi/4] \tag{4.1.16}$$

and for $\xi(E) < -1.5$ ($E_C > E$, $i\phi = \Phi < 0$, Φ is real),

$$\langle v_i|v_j\rangle \to \left|\frac{2(\hbar\bar{\omega}_1)(\hbar\bar{\omega}_2)\mu}{\hbar(2\mu)^{1/2}|(dV_1/dR) - (dV_2/dR)|_{R=R_C}}\right|^{1/2} \times \frac{\pi^{-1/2}}{2}(E_C - E)^{-1/4}e^{-\Phi(E)}. \tag{4.1.17}$$

Figure 4.4 is a pictorial description of $\phi(E)$ for $E \geq E_C$. In phase space, the classical motion of a system bound by a harmonic potential appears as a circle† with radius

$$p = (2\mu)^{1/2}[E - V(R_{\min})]^{1/2}$$

centered at $p = 0$, $R = R_{\min}$. The two circles in Fig. 4.4 intersect at $R = R_C$ and their overlap, shown shaded, is $2\phi(E)$. Obviously, a single experimentally measured perturbation matrix element will be insufficient to determine the unknown potential. This semiclassical method is based on the observed pattern of $\langle v_i|v_j\rangle$ versus v_i and upon isotopic substitution. The same sort of information forms the basis of a fully quantum mechanical, but trial-and-error, method for characterizing the unknown potential (Section 4.1.5).

Note that the three factors mentioned at the beginning of this section as determining the size of $\langle v_i|v_j\rangle$ appear explicitly in Eq. (4.1.12). The overlap integral contains

$$\left|\frac{dV_1}{dR} - \frac{dV_2}{dR}\right|_{R=R_C}^{-1/2}, \qquad \left[\frac{\bar{\omega}\mu}{\hbar k(R_C)}\right]^{1/2},$$

and a factor, $\phi(E)$, expressing the phase relationship at R_C.

The $(E - E_C)^{-1/4}\xi(E)^{1/4}\mathrm{Ai}[-\xi(E)]$ terms in Eq. (4.1.12) contain all information about the locations of the maxima and zeroes in the oscillatory variation of $\langle v_i|v_j\rangle$ with E, but the constant term,

$$\left|\frac{2(\hbar\bar{\omega}_1)(\hbar\bar{\omega}_2)\mu}{\hbar(2\mu)^{1/2}|(dV_1/dR) - (dV_2/dR)|_{R=R_C}}\right|^{1/2}$$

determines the magnitudes of $\langle v_i|v_j\rangle$ at its relative maxima. Absolute magnitudes of the vibrational overlap cannot be measured directly, but they

† The motion in phase space is a circle rather than an ellipse only when $\mu\omega = 1$.

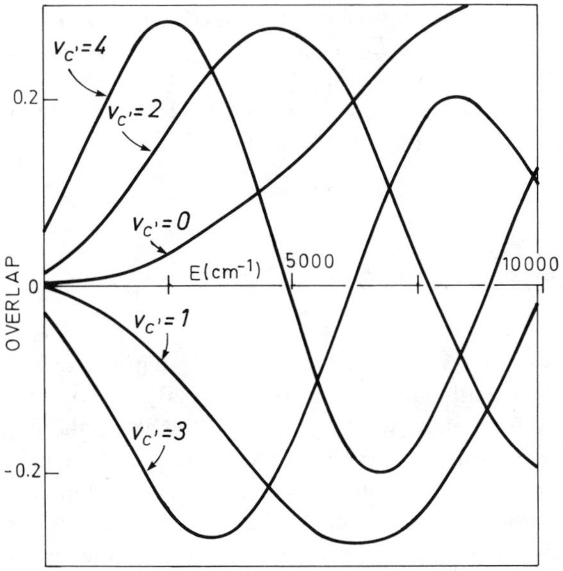

Fig. 4.5 Semiclassical calculation of vibrational overlap integrals between successive vibrational levels of the N_2 $c'^1\Sigma_u^+$ Rydberg state and the bound region of the $b'^1\Sigma_u^+$ valence state where the vibrational energy in the b′-state is treated as a continuous variable. Both states are represented by Morse potentials. The zero of energy is taken as $v=0$, $J=0$ of $b'^1\Sigma_u^+$. For the E values corresponding to the bound vibrational levels of the $b'^1\Sigma_u^+$ state, the values calculated using Eq. (4.1.12) reproduce the exact values (Table V of Lefebvre-Brion, 1969) to two decimal places. (Courtesy R. Lefebvre.)

can be inferred from the measured $H_{1,v_i;\,2,v_j}$ if an *ab initio* value of H_{12}^e is available. Alternatively, if sets of $H_{1,v_i;\,2,v_j}$ are available for two isotopic species, then the isotopic ratios of relative maxima in $\langle v_i|v_j\rangle$ versus E are obtainable from Eq. (4.1.12). Since the respective reduced mass-dependence of k, $\bar{\omega}$, $\phi(E)$, and $\xi(E)$ are $\mu^{+1/2}$, $\mu^{-1/2}$, $\mu^{+1/2}$, and $\mu^{+1/3}$, the first maximum value of $\langle v_i|v_j\rangle$ is proportional to $\mu^{-1/6}$,

$$\frac{\langle v_i|v_j\rangle_{E_{\max,l}}}{\langle v_i|v_j\rangle_{E_{\max,h}}} = \left(\frac{\mu_l}{\mu_h}\right)^{-1/6} \tag{4.1.18}$$

The accuracy of the semiclassical approach to $\langle v_i|v_j\rangle$ (typically better than $\pm 10\%$) is not limited to the case $E_{1,v_i} = E_{2,v_j}$.† This approach is particularly valuable for displaying the general qualitative form of $\langle v_i|v_j\rangle$ versus E (see Fig. 4.5 for an example) and for relating certain crucial features of the

† When $E_1, v_i \neq E_2, v_j$, it is necessary before applying Eq. (4.1.12) to vertically shift one of the potentials to ensure strict energy degeneracy. The new curve crossing point obtained in this way is called a pseudo-crossing point (Miller, 1968).

experimentally well-known oscillatory behavior of $H_{1,v_i;\,2,v_j}$ (Lagerqvist and Miescher, 1958, Fig. 16) to E_C, R_C, $(dV_2/dR)|_{R=R_C}$, and $a_2 - R_C$ or $b_2 - R_C$, where V_2 is the unknown potential. However, the semiclassical approach has not been widely applied to bound ~ bound interactions, its major use having been to bound ~ free interactions (Section 6.5).

4.1.2 MODEL POTENTIALS

When an electronic state is known only through its perturbations of a better known state, frequently only the energy and rotational constant of one or two vibrational levels of unknown absolute vibrational numbering can be determined. If the information available is insufficient to generate a realistic potential energy curve, then one has no choice but to adopt a model potential and exploit relationships between Dunham (Y_{lm}) and other derived constants (vibrational overlaps), which are rigorously valid for the model potential and approximately valid for general potentials.

The two most useful primitive model potentials are the harmonic and Morse oscillators,

$$V^{\text{harmonic}}(R)/hc = \tfrac{1}{2}(k/hc)(R - R_e)^2 \quad \text{cm}^{-1} \tag{4.1.19}$$

$$k/hc = (4\pi^2 c/h)\omega_e^2\mu = 2.966016 \times 10^{-2}\omega_e^2\mu \quad \text{cm}^{-1}\,\text{Å}^{-2} \tag{4.1.20}$$

where ω_e is in reciprocal centimeters, μ is in atomic mass units ($^{12}\text{C} = 12$), R is in angstroms, and V/hc is in reciprocal centimeters, and

$$V^{\text{Morse}}(R)/hc = D^e\{1 - \exp[-\beta(R - R_e)]\}^2 \tag{4.1.21}$$

where D^e is the dissociation energy

$$D^e = \omega_e^2/4\omega_e x_e \quad \text{cm}^{-1} \tag{4.1.22}$$

$$\beta = (8\pi^2 c\mu\omega_e x_e/h)^{1/2} = 2.435576 \times 10^{-1}(\mu\omega_e x_e)^{1/2} \quad \text{Å}^{-1} \tag{4.1.23a}$$

or

$$\beta = (2\pi^2 c\mu\omega_e^2/D^e h)^{1/2} = 1.2177881 \times 10^{-1}(\mu\omega_e^2/D^e)^{1/2} \quad \text{Å}^{-1} \tag{4.1.23b}$$

where D^e, ω_e, and $\omega_e x_e$ are in reciprocal centimeters, and μ is in atomic mass units. For both the harmonic and Morse oscillators (and general potentials),

$$R_e \quad (\text{Å}) = (8\pi^2 c\mu B_e/h)^{-1/2} = 4.105805(\mu B_e)^{-1/2} \quad \text{Å} \tag{4.1.24}$$

where B_e is in reciprocal centimeters.

In order to construct a harmonic or Morse potential from spectroscopic data, respectively, two (R_e and k) or three (D^e, R_e, and β) independent constants must be determined. The usual routes to these constants, through B_e, ω_e, and $\omega_e x_e$, are often impassable and it is necessary to exploit special relationships, two of the most useful of which are due to Kratzer (Kratzer, 1920),

$$D_e = 4B_e^3/\omega_e^2 \tag{4.1.25}$$

(D_e is the centrifugal distortion constant), and to Pekeris (Pekeris, 1934),

$$\omega_e x_e = \alpha_e^2\omega_e^2/36B_e^3 + \alpha_e\omega_e/3B_e + B_e, \tag{4.1.26}$$

where

$$\alpha_e \approx B_v - B_{v+1}.$$

Equation (4.1.25) is valid for both harmonic and Morse potentials; Eq. (4.1.26) is valid for the Morse potential only.

Isotope effects are another source of information. For a Morse oscillator,

$$\begin{aligned} T^l_{ve} - T^h_{(v+n)e} &= G_l(v) - G_{\text{h}}(v+n) \\ &\simeq (\Delta\mu/\mu_{\text{h}})\{n\omega_{\text{el}}/2 + (v+1/2) \\ &\quad \times [\omega_{\text{el}}/2 - \omega_e x_{\text{el}}(v + 1/2 - 2n)]\} - n\omega_{\text{el}} \end{aligned} \tag{4.1.27†}$$

$$B_{v,l} - B_{v+n,\text{h}} \simeq (\Delta\mu/\mu_h)[B_{\text{el}} - (3/2)(v + 1/2 + n)\alpha_{\text{el}}] + n\alpha_{\text{el}} \tag{4.1.28†}$$

$$\Delta\mu = \mu_{\text{h}} - \mu_l \tag{4.1.29}$$

where h and l refer to the heavy and light isotopic molecule. When the vibrational quantum number of the perturber is the same for both isotopic species ($n = 0$), Eqs. (4.1.27) and (4.1.28) reduce to simpler and more useful forms. Recall that, for the PO $B'^2\Pi \sim D^2\Pi$ perturbation, an apparent small isotope shift for the perturber of $D^2\Pi(v = 0)$ suggested that $n = v = 0$, but the correct assignment was $v = 24$, $n = 1$.

Once a model potential is derived, it is possible to verify and refine this potential, making use of observed perturbation matrix elements and calculated overlap integrals between the vibrational levels of the two interacting electronic states. Although it is usually more convenient to input the analytic form of $V(R)$ into a numerical integration program to calculate overlap integrals (Section 4.1.3), analytic expressions exist for harmonic and Morse $\langle v_i | v_j \rangle$ factors.

† These isotope shift equations are derived by expanding the ratio

$$Y^l_{mn}/Y^{\text{h}}_{mn} = (\mu_l/\mu_{\text{h}})^{-(m/2+n)} = (1 - \Delta\mu/\mu_{\text{h}})^{-(m/2+n)} \simeq 1 + (m/2 + n)\Delta\mu/\mu_{\text{h}}.$$

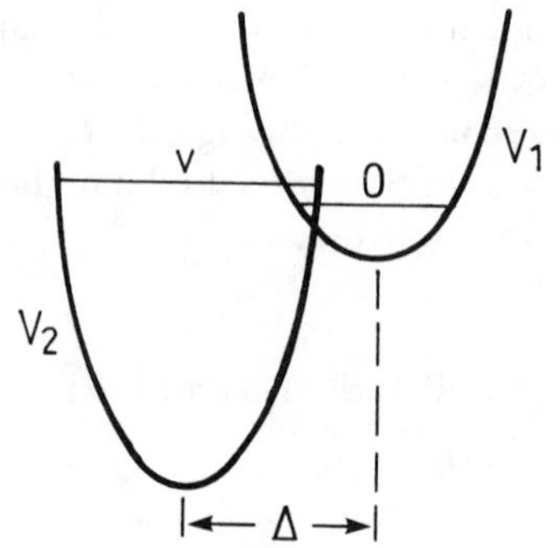

Fig. 4.6 Two displaced harmonic oscillators.

Even for two harmonic potentials, when $\omega_{e1} \neq \omega_{e2}$ and $R_{e1} \neq R_{e2}$, the analytic form for $\langle v_{1i}|v_{2j}\rangle$ is very complicated. For $\omega_{e1} = \omega_{e2}$ and $\Delta = R_{e1} - R_{e2}$ (Fig. 4.6),

$$\langle v_1 = 0|v_2 = v\rangle = \exp(-d^2/2)d^v(v!)^{-1/2} \tag{4.1.30}$$

$$d = \left(\frac{c\mu\omega_e}{2\hbar}\right)^{1/2}\Delta \tag{4.1.31}$$

where ω_e is in reciprocal centimeters. If $v_1 \neq 0$, formulas are given by Katriel (Katriel, 1970), but the sign of d is incorrect (Schamps, 1977).

The vibrational wavefunctions for a Morse oscillator can be expressed analytically in terms of Whittaker functions (Felenbok, 1963).

The harmonic potential is a model of last resort. Its behavior at $R = 0$ and $R = \infty$ is unphysical, as is the sign of α_e. Exact diatomic molecule vibrational wavefunctions for levels above $v = 0$, except for their number of nodes, differ from harmonic oscillator eigenfunctions (Hermite polynomials with an exponential factor), in that they are not symmetric about R_e and, increasingly so at high v, are skewed toward the outer turning point.

The Morse potential displays the qualitative features of realistic molecular potentials and serves adequately in vibrational overlap computations for levels up to about 10% of D^e (the Morse D^e, not the true D^e). The Morse curve rises sharply at small R, but not steeply enough. This deficiency is often minimized by splicing a short-R AR^{-12} segment onto V^{Morse}. The Morse potential goes asymptotically to a dissociation limit at large R, but with an R-dependence that is at variance with the required power law interaction of the separated atoms (LeRoy, 1973). One can splice a long range tail of the form $D^e - C_nR^{-n}$ onto V^{Morse}, where D^e is the measured dissociation energy (which can be quite different from the Morse D^e, determined from $\omega_e^2/4\omega_e x_e$).

One problem with splicing long and short R extensions onto V^{Morse} is that dV/dR and higher derivatives become discontinuous at R_{splice}. Such discontinuities cause difficulty with numerical integration of the vibrational Schrödinger equation. Lagrange polynomial interpolation or, even better, spline techniques exist by which an intermediate splicing function is used to ensure continuity of $V(R)$, dV/dR, and as many higher derivatives as the degrees of freedom of the splicing function permit.

The value of model potentials is not the availability of analytic formulas for vibrational overlap integrals; rather, it is the ability to pull together fragmentary information of diverse types into a coherent picture in order to test the observable consequences (location and strength of other perturbations) of alternative assignment schemes (Section 4.1.5). RKR and numerical integration comprise the most convenient way to obtain $\langle v_i|v_j\rangle$ factors and other quantities derivable from $V(R)$ [e.g., centrifugal distortion constants (Albritton *et al.* 1973)].

4.1.3 NUMERICAL POTENTIALS AND VIBRATIONAL WAVEFUNCTIONS

The Rydberg–Klein–Rees (RKR) procedure is the most widely used method of deriving $V(R)$ from the $G(v)$ and $B(v)$ functions of diatomic molecules. Countless RKR computer programs of independent origin exist, with differences primarily in the way a singularity at the upper limit of integration of Eqs. (4.1.37) and (4.1.39c) is handled, all of which give essentially identical results (see Mantz *et al.*, 1971 for the CO $X^1\Sigma^+$ potential), except for those based on the more accurate second-order JWKB quantization condition (Kirschner and Watson, 1974). The RKR potential is generated as pairs of turning points, $R_-(\mathrm{E})$ and $R_+(E)$, at specified energies [usually $E_n = G(n/4)$ for $-2 \leq n \leq 4v_{\max}$, where $v_{\max}$ is the highest observed v]. The RKR potential is not an analytic function, but can be compactly represented as such by least-squares fitting a suitable $V(R)$ function to the turning points.

RKR is based on Eq. (4.1.1), the semiclassical quantization condition; thus the $V^{\mathrm{RKR}}(R)$ it generates is not identical to $V^{\mathrm{BO}}(R)$, the "true" Born–Oppenheimer potential. The deperturbed experimental energies, E_{vJ}^{depert}, are, by definition, the eigenvalues of $V^{\mathrm{BO}}(R)$, E_{vJ}^{BO}. Thus, the small difference between V^{BO} and V^{RKR} is manifest in small (typically 0.1 cm^{-1}) and systematic discrepancies between E_{vJ}^{depert} and E_{vJ}^{RKR}. These discrepancies are typically 10–100 times smaller in magnitude than the unknown difference between partially and fully deperturbed experimental energies. The inverse perturbation approach (IPA) systematically distorts V^{RKR} to minimize the deviations between the partially deperturbed observed E_{vJ} and the exact eigenvalues of V^{IPA}, E_{vJ}^{IPA} (Kosman and Hinze, 1975; Vidal and Scheingraber, 1977). However, as long as the E_{vJ} are not fully deperturbed, V^{IPA} may be no better an approximation to V^{BO} than V^{RKR}.

Many detailed derivations of the RKR equations have been published (Zare, 1964; Miller, 1971; Elander *et al.*, 1979). The unknown potential energy for a rotating diatomic molecule,

$$V_J(R) = V_0(R) + J(J+1)\hbar^2/2\mu R^2, \tag{4.1.32}$$

is a function of both R and J. The function

$$A(E, J) = \int_{R_-(E)}^{R_+(E)} [E - V_J(R)]\, dR \tag{4.1.33}$$

is easily shown to have the convenient properties that

$$\frac{\partial A}{\partial E} = \int_{R_-(E)}^{R_+(E)} 1\, dR = R_+(E) - R_-(E) \tag{4.1.34}$$

and

$$\begin{aligned} \frac{\partial A}{\partial J} &= \frac{(2J+1)\hbar^2}{2\mu} \int_{R_-(E)}^{R_+(E)} \frac{1}{R^2}\, dR \\ &= -(2J+1)\frac{\hbar^2}{2\mu}\left[\frac{1}{R_+(E)} - \frac{1}{R_-(E)}\right] \end{aligned} \tag{4.1.35a}$$

and, for $J = 0$,

$$\frac{\partial A}{\partial J} = \frac{\hbar^2}{2\mu}\left[\frac{1}{R_-(E)} - \frac{1}{R_+(E)}\right]. \tag{4.1.35b}$$

Thus, if it is independently possible to define $A(E, J)$ in terms of $G(v)$ and $B(v)$ functions, Eqs. (4.1.34) and (4.1.35b) could be used to determine $V_0(R)$. The key to the procedure is that Eq. (4.1.33) has a form very similar to the semiclassical quantization condition [Eq. (4.1.1)],

$$\int_{R_-(E)}^{R_+(E)} [E - V_J(R)]^{1/2}\, dR = (v + \tfrac{1}{2})\pi. \tag{4.1.36}$$

Nontrivial manipulation of Eqs. (4.1.33) and (4.1.36) yields

$$A(E, J) = 2\pi \int_{v(0)=v_{\min}}^{v(E)} [E - E(v, J)]^{1/2}\, dv. \tag{4.1.37}$$

Taking $\partial/\partial E$ and $\partial/\partial J$ at $J = 0$,

$$\frac{\partial A}{\partial E} = \pi \int_{v_{\min}}^{v(E)} [E - G(v)]^{-1/2}\, dv \tag{4.1.38}$$

$$\frac{\partial A}{\partial J} = \pi \int_{v_{\min}}^{v(E)} [E - G(v)]^{-1/2} \frac{\partial E}{\partial J}\, dv \tag{4.1.39a}$$

but

$$\frac{\partial E}{\partial J} = (2J + 1)B(v) - (4J^3 + 6J^2 + 2J)D(v) \tag{4.1.39b}$$

so, for $J = 0$

$$\frac{\partial A}{\partial J} = \pi \int_{v_{\min}}^{v(E)} B(v)[E - G(v)]^{-1/2}\, dv. \tag{4.1.39c}$$

Note that the integrals in Eqs. (4.1.38) and (4.1.39c) become infinite at the upper limit of integration, but this singularity is integrable by various techniques [Mantz *et al.*, 1971]. Note also that, although $G(-1/2) = 0$,

$$E_{v0} = Y_{00} + G(v), \tag{4.1.40}$$

(Kaiser, 1970) where (Dunham, 1932)

$$Y_{00} = \frac{B_e - \omega_e x_e}{4} + \frac{\alpha_e \omega_e}{12 B_e} + \left(\frac{\alpha_e \omega_e}{12 B_e}\right)^2 \frac{1}{B_e} \tag{4.1.41}$$

(Y_{00} is zero for harmonic and Morse oscillators) and

$$v_{\min} \cong -1/2 - Y_{00}/\omega_e. \tag{4.1.42}$$

Thus, knowledge of $G(v)$ is sufficient to determine $R_+(E) - R_-(E)$ from Eqs. (4.1.34) and (4.1.38). Eqs. (4.1.35b) and (4.1.39c), based on *both* $G(v)$ and $B(v)$, then determine $R_+(E)$ and $R_-(E)$. There is nothing about the RKR equations which restricts the energies at which turning points are computed to those corresponding to integral values of v.

The RKR potential may be tested against the input $G(v)$ and $B_{(v)}$ values by numerical integration of the nuclear Schrödinger equation [Shore, 1973 reviews the various procedures, e.g. the Numerov method]. $G(v) + Y_{00}$ typically deviates from $E_{v,J=0}$ by $< 1\,\text{cm}^{-1}$ except near dissociation. B_v may be computed from $\chi_{v,J=0}(R)$ by

$$B_{(v)} = (h^2/2\mu)\langle v, J = 0 | R^{-2} | v, 0\rangle \tag{4.1.43}$$

and is typically found to agree with experimental values to 1 part in 10^5 except near dissociation. Alternatively, the completeness of the deperturbation of the experimental data may be tested by comparison of observed and computed

(Albritton *et al.*, 1973; Brown *et al.*, 1973; Tellinghuisen, 1973; Kirschner and Watson, 1973) centrifugal distortion constants or $G(v)$, $B_{(v)}$ values for other isotopic species.

4.1.4 SOME REMARKS ABOUT "BORROWED" COMPUTER PROGRAMS

The signs of off-diagonal matrix elements are often important (see Section 5.3), even though the phase of any given basis function is arbitrary. The sign of a computed vibrational overlap integral (or other vibrational matrix element) depends on the phase convention for $\chi_{v_i}(R)$ and $\chi_{v_j}(R)$ used in the computer program. A frequently used convention is that the vibrational wave functions with even or odd quantum numbers have opposite signs at the inner turning point. An opposite convention, for which all $\chi(R_{\min}) > 0$, can be more convenient when dealing with the vibrational continuum. The important point here is that the choice of one or the other convention has no physical consequence, but it is necessary to know which convention one has chosen and to use it self-consistently in all computations. It could be disastrous to calculate $\langle v_i|v_j\rangle$ integrals with a program following the $\chi(R_{\min}) > 0$ convention and then calculate $\langle v_i|R^{-2}|v_j\rangle$ integrals with another program that requires alternating signs.

The following example illustrates the necessity for self-consistent phases (as manifest in the sign of off-diagonal matrix elements) in order to predict correctly which of two transitions is more intense. Let basis functions ϕ_2 and ϕ_3 interact with each other (see Sections 5.2.1 and 5.3):

$$\begin{aligned} E_2 &= \frac{H_{22}+H_{33}}{2} + \frac{H_{23}^2}{H_{22}-H_{33}} \qquad \Psi_2 = \phi_2 + \frac{H_{23}}{H_{22}-H_{33}}\phi_3 \\ E_3 &= \frac{H_{22}+H_{33}}{2} - \frac{H_{23}^2}{H_{22}-H_{33}} \qquad \Psi_3 = \phi_3 - \frac{H_{23}}{H_{22}-H_{33}}\phi_2 \end{aligned} \tag{4.1.44}$$

Assume that ϕ_2 and ϕ_3 have identical transition probabilities to basis function ϕ_1, $P_{12}^0 = P_{13}^0$,

$$P_{12}^0 \propto |\langle\phi_1|\boldsymbol{\mu}|\phi_2\rangle|^2 \qquad P_{13}^0 \propto |\langle\phi_1|\boldsymbol{\mu}|\phi_3\rangle|^2, \tag{4.1.45}$$

then, even though $P_{12}^0 = P_{13}^0$, the perturbed intensities, P_{12} and P_{13}, will not be equal:

$$P_{12}=P_{12}^0+\left(\frac{H_{23}}{H_{22}-H_{33}}\right)^2 P_{13}^0+\frac{2H_{23}}{H_{22}-H_{33}}\langle\phi_1|\boldsymbol{\mu}|\phi_2\rangle\langle\phi_1|\boldsymbol{\mu}|\phi_3\rangle \tag{4.1.46a}$$

$$P_{13}=P_{13}^0+\left(\frac{-H_{23}}{H_{22}-H_{33}}\right)^2 P_{12}^0-\frac{2H_{23}}{H_{22}-H_{33}}\langle\phi_1|\boldsymbol{\mu}|\phi_2\rangle\langle\phi_1|\boldsymbol{\mu}|\phi_3\rangle. \tag{4.1.46b}$$

Note that whether the higher energy perturbed level has greater ϕ_2 or ϕ_3 character has nothing to do with any arbitrary choice of phase factors. It depends only on the sign of $(H_{22} - H_{33})$. However, the relative intensities of the transitions into the higher- versus lower-energy perturbed level depend on the product of six signed quantities [H^e, $\mu_{12}(R)$, and $\mu_{13}(R)$ can be determined *ab initio*]:

$$H_{23} = \langle 2|3 \rangle H^e \tag{4.1.47a}$$

$$\langle \phi_1 | \boldsymbol{\mu} | \phi_2 \rangle = \langle 1|2 \rangle \mu_{12}(\bar{R}_{12}) \tag{4.1.47b}$$

$$\langle \phi_1 | \boldsymbol{\mu} | \phi_3 \rangle = \langle 1|3 \rangle \mu_{13}(\bar{R}_{13}). \tag{4.1.47c}$$

Each vibrational wavefunction appears twice; thus the phase of the vibrational wave function is irrelevant to whether the upper or the lower eigenstate has the larger transition intensity to level 1. However, it is crucial to use internally consistent vibrational phases when computing the three sets of overlap integrals in order to correctly predict which of the two transitions is more intense (Section 5.3.1).

It is important to realize that most computer programs work with

$$\xi(R) = R\chi(R),$$

where $\chi(R)$ is the vibrational eigenfunction, which means that integration must be over dR, not $R^2\,dR$. This point is illustrated by noting the difference between Eqs. (2.3.10) and (2.3.11) (Section 2.3.3). It is common practice to combine subroutines from various sources, for example, eigenvalue search, numerical generation of radial eigenfunctions, matrix element evaluation, etc. No matter how widely a subroutine is used, it is essential to check for compatibility between subroutines.

4.1.5 VIBRATIONAL ASSIGNMENT BY THE MATRIX ELEMENT METHOD

Often, the energies and rotational constants of several vibrational levels of an otherwise unknown electronic state are obtained from analysis of perturbations of a better known electronic state. Figure 4.7 illustrates the ease with which the perturbing levels may usually be grouped into classes corresponding to separate electronic states and how, *within each class*, the *relative* vibrational numbering of the observed levels is established. There are two sources of information that enable determination of the *absolute* vibrational numbering of the perturbers; isotope shifts and the magnitudes of observed perturbation matrix elements. The use of isotope shifts is considerably better known (Herzberg, 1950, pp. 162–166) than that of matrix elements.

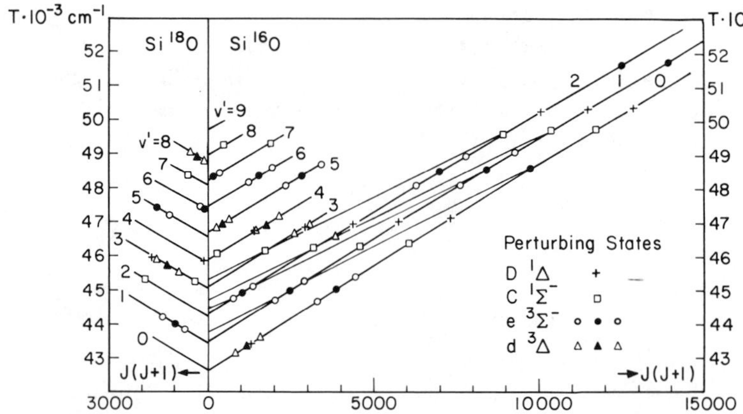

Fig. 4.7 Perturbations in the $^{28}Si^{18}O$ and $^{28}S^{16}O$ $A^1\Pi$ state. The J-values where each $A^1\Pi$ vibrational level crosses a perturber are marked with a different symbol for each type of perturbing state. The signs of the level shifts near each level crossing indicate that $A^1\Pi$ overtakes each perturber from below as J increases. This means that the same v-level of a perturbing state will cross the v_A and $v_A - 1$ levels of $A^1\Pi$ at low J and high J, respectively. Two consecutive vibrational levels of the $Si^{16}O$ $e^3\Sigma^-$ and $C^1\Sigma^-$ states are shown, each of which crosses two vibrational levels of $A^1\Pi$. [From Field *et al.*, (1976).]

The matrix element method is based on the fact that if two potential curves intersect exactly once (at $E = E_C$, $R = R_C$), then the matrix element connecting *near-degenerate* perturbing levels can be factored (see Section 2.3.1),

$$H_{1v_1;\,2v_2} = H^e_{12}(R_C)\langle v_1|v_2\rangle,$$

and the electronic matrix element is independent of vibrational level,

$$\frac{H_{1v_1;\,2v_2}}{\langle v_1|v_2\rangle} \simeq \frac{H_{1v_1+n;\,2v_2+m}}{\langle v_1+n|v_2+m\rangle} \simeq H^e_{12}(R_C). \qquad (4.1.48)$$

The correctness of a trial vibrational numbering of electronic state 2 is then indicated by the constancy of the ratio of the observed matrix element to the calculated vibrational overlap. This method is similar to the Franck–Condon method for establishing the absolute vibrational numbering. There, one compares observed transition intensities to sets of Franck–Condon factors calculated for a series of trial numberings. The key to the success of both methods is the existence of at least one relative minimum or maximum among the observed $H_{1v;\,2v}$ matrix elements or transition intensities.

Consider a well-characterized electronic state whose *absolutely* numbered vibrational levels are perturbed by a set of levels of another electronic state for which energies, rotational constants, and a *relative* vibrational numbering are known. A set of trial absolute numberings for the perturber is obtained, starting with the numbering that assigns the quantum number $v = 0$ to the lowest energy observed perturbing level. This first trial numbering corresponds to trial T_v and B_v functions,

$$T_v^0 = T_e^0 + \omega_e^0(v + \tfrac{1}{2}) - \omega_e x_e(v + \tfrac{1}{2})^2 \qquad (4.1.49)$$

$$B_v^0 = B_e^0 - \alpha_e(v + \tfrac{1}{2}). \qquad (4.1.50)$$

New T_v and B_v functions are obtained by adding n ($n = 1, 2, 3, \ldots$, some large integer) to each trial vibrational quantum number of the perturber

$$v + n = v_n \qquad (4.1.51a)$$

$$T_{v_n}^n = T_e^n + \omega_e^n(v + n + \tfrac{1}{2}) - \omega_e x_e(v + n + \tfrac{1}{2})^2 \qquad (4.1.51b)$$

$$B_{v_n}^n = B_e^n - \alpha_e(v + n + \tfrac{1}{2}) \qquad (4.1.51c)$$

and, adjusting the T_e^n, ω_e^n, and B_e^n constants,

$$T_e^n = T_e - n\omega_e - n^2\omega_e x_e \qquad (4.1.52a)$$

$$\omega_e^n = \omega_e + 2n\omega_e x_e \qquad (4.1.52b)$$

$$B_e^n = B_e + n\alpha_e \qquad (4.1.52c)$$

so that the transformed functions

$$T_{v_n}^n = T_{v+n}^n = T_v^0$$

$$B_{v_n}^n = B_{v+n}^n = B_v^0$$

change only the name of a level, not its observed T_v or B_v value. Each set of constants (T_e^n, ω_e^n, $\omega_e x_e$, B_e^n, α_e) defines an RKR potential energy curve and a set of vibrational wavefunctions. It is a routine and inexpensive matter to generate tens of trial RKR curves and, for each trial curve, calculate a set of vibrational overlap factors with the single, well-defined RKR curve of the better known, perturbed electronic state.

Each time the trial numbering is increased by 1, the trial potential is shifted to $\sim\omega_e$ lower energy and to slightly smaller internuclear distance,

$$\delta R = -\alpha_e(2B)^{-3/2}\hbar\mu^{-1/2}. \qquad (4.1.53)$$

Consequently, the point of intersection between the potentials of the two interacting states is shifted. It is not possible to derive a general expression for the direction and magnitude of the shift in E_C or R_C. However, in Section 4.1.1 it was shown that the vibrational overlap increases monotonically as the energy of the interacting pair of levels approaches E_C from below, reaches a maximum just above E_C, and then undergoes weakly damped oscillations [proportional to

$(E - E_C)^{-1/4}$], increasingly closely spaced. Thus, if E_C is above the highest observed perturbation pair, the observed matrix elements will vary monotonically and the matrix element method will almost certainly fail. Similarly, if the lowest observed perturbation pair is far above E_C, where the vibrational overlap oscillates so rapidly that successive maxima are separated by less than ω_e, the matrix element method will probably fail. Even when E_C is bracketed by observed perturbation pairs, if the absolute numbering suggested is such that the lowest observed perturber has $v > 15$, this assignment must be regarded as uncertain to about ± 1 quantum. The RKR curve is very sensitive to extrapolated ω_e and B_e values, and the overlaps calculated for observed high-v levels can be significantly affected by extrapolation errors.

When the matrix element method fails, two possibilities for establishing the vibrational numbering remain, *ab initio* $H^e(R)$ functions and isotope shifts. When E_C appears to lie above the highest observed perturbing level, isotope shifts are the method of choice. However, if $H^e(R)$ is available, then a modified matrix element method may prove successful. Each trial numbering determines R_C^{trial}, hence $H^e(R_C^{\text{trial}})$. The calculated vibrational overlap should be equal to the observed perturbation matrix element divided by $H^e(R_C^{\text{trial}})$. However, if the R-dependence of H^e matches the effect of a shift in E_C^{trial}, R_C^{trial} on the calculated $\langle v_1 | v_2 \rangle$ elements, the vibrational numbering will be ambiguous by ± 1 quantum. When the lowest sampled vibrational level of the perturber is > 10, isotope shifts are no more reliable than matrix elements.

The first rigorous test of the matrix element method was with the ($C^1\Sigma^-$, $D^1\Delta$, $d^3\Delta$, and $e^3\Sigma^-$)$\sim A^1\Pi$ perturbations of the $^{28}Si^{16}O$ and $^{28}Si^{18}O$ molecules (Field *et al.*, 1976). Vibrational assignments derived from isotope shifts were found to be identical to those deduced from the vibrational variation of observed matrix elements. Previously, the near-perfect vibration-independence of the electronic part of perturbation matrix elements observed between levels of well-characterized potentials had been experimentally demonstrated for ($a'^3\Sigma^+$, $d^3\Delta$, $e^3\Sigma^-$, and $I^1\Sigma^-$)$\sim$($A^1\Pi$ and $a^3\Pi$) perturbations of CO (Field *et al.*, 1972). The first use (Field, 1974) of the matrix element method was in identifying the electronic ($^3\Pi$ and $^1\Pi$) and vibrational perturbers of the $A^1\Sigma^+$ state of CaO (Hultin and Lagerqvist, 1951), SrO (Almkvist and Lagerqvist, 1950), and BaO (Lagerqvist *et al.*, 1950).

The perturbations of the SiO $A^1\Pi$ state are summarized in Fig. 4.7. The J-values where the $A^1\Pi$ and perturber basis function energies become equal are marked using a different symbol for each type of electronic perturber. This not necessarily integral J-value is called the culmination of the perturbation and is easily recognized because:

1. The energy shift of the main line (by definition, the main line corresponds to the eigenfunction that has the largest perturbed basis function fractional

character and is usually more intense than extra lines (see Section 5.2.1)) switches sign as one passes through this J-value.

2. The main and extra lines reach their minimum energy separation,

$$|E_{\text{main}} - E_{\text{extra}}| = 2|H_{12}|. \tag{4.1.54}$$

3. The intensities of the main and extra lines become equal.

Grouping the perturbers into classes is not as trivial as it may seem from the figure.

The qualitative pattern of the perturbers of a given vibrational level provide clues to the electronic identity of the perturber, particularly when the perturber is a Σ-state of any multiplicity. These clues consist of:

1. As J increases, does the perturber approach from lower energy (larger B-value) or higher energy (smaller B-value)? This is indicated by the sign of the level shifts above and below the culmination.

2. Is the perturbation matrix element J-dependent (heterogeneous perturbation or **S**-uncoupling in either the perturbed or perturbing state)? Figure 4.8 contrasts the level shifts resulting from J-independent perturbation matrix elements with those from matrix elements proportional to J. It is important to note that when perturbations by two or more states are not well separated, it is difficult to distinguish between the J-independent and J-dependent cases. In addition, the actual matrix element J-dependence is often more complicated than either of these two simple cases, particularly for weak perturbations ($H_{12} \lesssim 0.1$ cm^{-1}) and at high $J (J \approx A/B)$.

3. When the perturbed state has two near-degenerate parity components for each J ($\Lambda \geq 1$), then it is possible to distinguish between three classes of perturbers: single-parity perturbers ($^1\Sigma^\pm$ has e or f parity only, $^{2S+1}\Sigma$ has widely separated e and f components belonging to the same J and only one or the other total parity for the near degenerate J components of a given N, etc.), states with significant Λ-doubling [usually only Π states or $\Omega < 2$ states in case (c)], and states without Λ-doubling. Kovács (1969, pp. 227–284) illustrates the qualitative perturbation patterns for 42 combinations of states with $S \leq \frac{3}{2}$, $\Lambda \leq 2$.

4. On the E_{vJ} versus $J(J+1)$ plot, all crossings for a particular perturbation sequence (constant $\Delta v = v_{\text{main}} - v_{\text{perturber}}$) fall onto a family of approximately straight lines with slope

$$\left(\frac{dE_{\text{cross}}}{d[J(J+1)]}\right)_{\Delta v} = B_{\text{main}}(v) - \Delta G_{\text{main}}(v - \tfrac{1}{2}) \frac{\delta B_{\Delta v}}{\delta G_{\Delta v}} \tag{4.1.55}$$

where

$$\delta B_{\Delta v} = B_{\text{main}}(v) - B_{\text{perturber}}(v - \Delta v)$$

$$\delta G_{\Delta v} = \Delta G_{\text{main}}(v - \tfrac{1}{2}) - \Delta G_{\text{perturber}}(v - \tfrac{1}{2} - \Delta v).$$

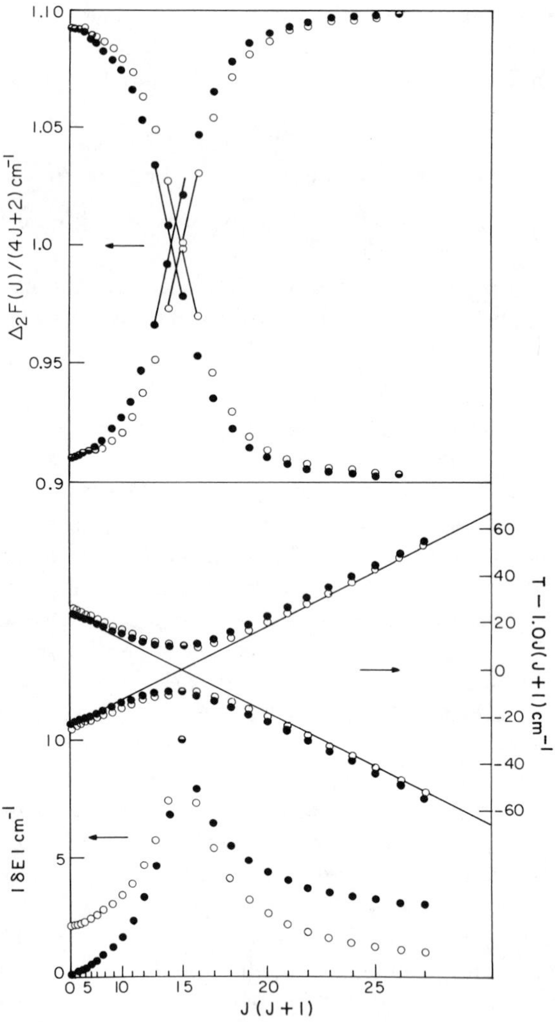

Fig. 4.8 Level shifts, reduced term values, and effective B-values $[\Delta_2F(J)]$ for homogeneous versus heterogeneous perturbations. Two states, with deperturbed B-values of $B_1^0 = 1.1$ and $B_2^0 = 0.9\ \text{cm}^{-1}$, cross at exactly $J = 15$. The filled circles correspond to a homogeneous interaction where the interaction matrix element is $H_{12} = 10.0\ \text{cm}^{-1}$ for all J-values. The open circles correspond to a heterogeneous interaction with

$$H_{12}(\mathrm{J}) = 0.6455\,[J(J+1)]^{1/2},$$

which is chosen so that, at the crossing point ($J = 15$), the perturbation matrix elements for both the homogeneous and heterogeneous interactions are identical. The level-shift plot for the homogeneous interaction is symmetric about $J = 15$ and asymptotically approaches $|\delta E| = 0$ at

Note that since $\delta B_{\Delta v}$ and $\delta G_{\Delta v}$ usually have the same sign, the slopes of these constant Δv perturbation tie lines will be smaller than both B_{main} and $B_{\mathrm{perturber}}$. Thus, once several families of perturbers are identified using clues 1–3, the remaining perturbers can be tentatively grouped by inspecting the E_{vJ} versus $J(J+1)$ plot for approximately horizontal tie lines.

Even after the perturbers are grouped into classes and the relative vibrational numbering within a class established, it may still be premature to apply the matrix element method. Knowledge of the perturber's electronic symmetry is necessary, because this determines the J-dependence of the perturbation matrix element. The matrix element at the perturbation culmination that one determines from a local graphical treatment of the J-levels near the crossing can be quite different from the one obtained from a least-squares fit of all J-values to a model Hamiltonian. It is the value of the matrix element in the deperturbation model, not the local magnitude of the matrix element, to which the matrix element method applies. In order to illustrate this point, three types of perturbations of the SiO $A^1\Pi$ state will be discussed: $A^1\Pi \sim e^3\Sigma^-$, a J-independent interaction; $A^1\Pi \sim D^1\Delta$, an explicitly J-dependent interaction; $A^1\Pi \sim d^3\Delta$, an implicitly J-dependent interaction.

On Fig. 4.7 the $A^1\Pi \sim e^3\Sigma^-$ crossings appear in groups of three. The f-parity levels of $A^1\Pi$ (Q-branch lines of the $A^1\Pi$–$X^1\Sigma^+$ system) are perturbed by a single class of $e^3\Sigma^-$ levels; these are the f-parity, $\Omega = \Sigma = 1$ F_2 levels. The local matrix element at the $A^1\Pi \sim e^3\Sigma^-(F_2)$ crossing, determined from one-half the interpolated closest approach of main and extra lines, is exactly equal to the matrix element in the least-squares model Hamiltonian,

$$A_{10^-} = \langle {}^3\Sigma_1^- f | \mathbf{H}^{\mathrm{SO}} | {}^1\Pi f \rangle.$$

The e-parity levels of $A^1\Pi$ (R and P branch lines of A–X) are perturbed by two

---

Fig. 4.8 (*Legend continued*)
high J. The level shifts for the heterogeneous interaction are not symmetric about $J = 15$ and will asymptotically approach $|\delta E| = (0.6455)^2/\Delta B = 2.08\ \mathrm{cm}^{-1}$. The reduced term values (see Fig. 2.16 and Section 3.3.1),

$$T_{iJ} - J(J+1)(B_1^0 + B_2^0)/2,$$

where T_{1J} and T_{2J} are the perturbed term values, are plotted versus $J(J+1)$ and form high- and low-energy series of levels. The high-energy series is nominally state 2 at $J < 15$ and goes smoothly over into nominally state 1 at $J > 15$ (vice versa for the low-energy series). The deperturbed term values are shown as a pair of straight lines that cross at $J = 15$. The effective B-value plot for the high-energy series of levels goes from $\sim 0.91\ \mathrm{cm}^{-1}$ at low J to $\sim 1.10\ \mathrm{cm}^{-1}$ at high J (vice versa for the low-energy series). The $\Delta_2 F(J)$ values asymptotically approach B_1^0 and B_2^0 at high J, regardless of the perturbation mechanism. The B-value plots cross at $(B_1^0 + B_2^0)/2$ (see Fig. 3.1), but the J-value of the crossing for the heterogeneous case is ~ 0.5 J-unit lower than that of the actual level crossing.

classes of $e^3\Sigma^-$ levels; these are the e-parity, about 50–50 mixed† $|\Omega| = |\Sigma| = 1$ and $\Omega = \Sigma = 0$ F_1 and F_3 levels,

$$\begin{aligned} |^3\Sigma^-, J, F_1\rangle &\simeq (2)^{-1/2}[|^3\Sigma_0^-, J\rangle + |^3\Sigma_{|\Omega|=1}^-, J\rangle] \\ |^3\Sigma^-, J, F_3\rangle &\simeq (2)^{-1/2}[|^3\Sigma_0^-, J\rangle - |^3\Sigma_{|\Omega|=1}^-, J\rangle]. \end{aligned} \tag{4.1.56}$$

Since only the $^3\Sigma_1^-$ character of the F_1 and F_3 $e^3\Sigma^-$ components can interact via $\mathbf{H}^{\mathrm{SO}}$ with $A^1\Pi$, the local matrix element at the e-parity crossings is a factor of $(2)^{-1/2}$ smaller than the matrix element in the model Hamiltonian,

$$H(^1\Pi_e \sim {}^3\Sigma_e^-, F_1 \quad \text{or} \quad F_3) = (2)^{-1/2} A_{10^-}. \tag{4.1.57}$$

A_{10^-} may be obtained from a local graphical treatment of any one of the $^1\Pi \sim {}^3\Sigma^-$ (F_1, F_2, F_3) crossings, provided it is certain that the perturber is a component of a $^3\Sigma^-$ state and proper account is taken of the $(2)^{-1/2}$ factor.

Table 4.1 shows that the vibrational numbering of the SiO $e^3\Sigma^-$ state is unambiguously determined by the matrix element method. The weighted rms deviations from the supposedly constant average electronic factor for the various trial numberings are 55% ($v_{\mathrm{pert}} + 3$), 26% ($v_{\mathrm{pert}} + 1$), 5% (v_{pert}), 37% ($v_{\mathrm{pert}} - 1$), and 31% ($v_{\mathrm{pert}} - 3$). One expects the matrix element method to work because the $A^1\Pi$ and $e^3\Sigma^-$ potential curves cross near $v_A = 2.5 \sim v_e = 9$ ($E_C \simeq 44{,}600$ cm^{-1}, $R_C = 1.50$ Å), which is within the range of vibrational levels sampled (Field *et al.*, 1976). As expected, A_{10}^- goes through a maximum at an energy, $v_A = 5 \sim v_e = 13$, slightly above the A–e curve crossing.

The $A^1\Pi \sim D^1\Delta$ interaction is explicitly J-dependent,

$$\langle ^1\Delta, J|\mathbf{H}^{\mathrm{ROT}}|^1\Pi, J\rangle = -\beta_{12}[J(J+1) - 2]^{1/2}. \tag{4.1.58}$$

There are two crossings for each $v_A \sim v_D$ pair, one for each parity. Since neither the $A^1\Pi$ nor the $D^1\Delta$ state has appreciable Λ-doubling, the e- and f-parity perturbations culminate at approximately the same J value. The local matrix elements for the e- and f-parity crossings should be identical and yield β_{12} upon

† The 50–50 mixing of $\Omega = 0$ and $|\Omega| = 1$ basis functions shown in Eq. (4.1.56) is a limit reached when $BJ \gg \lambda$. First-order perturbation theory for the correction to the case (a) wavefunctions implies that the mixing coefficient of $|^3\Sigma_{|\Omega|=1}\rangle$ in the nominal $|^3\Sigma_0\rangle$ function is

$$\begin{aligned} C_{01} &= \frac{\langle ^3\Sigma_0|\mathbf{H}'|^3\Sigma_{|\Omega|=1}\rangle}{\langle ^3\Sigma_1|\mathbf{H}^0|^3\Sigma_1\rangle - \langle ^3\Sigma_0|\mathbf{H}^0|^3\Sigma_0\rangle} \\ &= \frac{-B_\Sigma 2[J(J+1)]^{1/2}}{(2B_\Sigma - \frac{4}{3}\lambda) - (0B_\Sigma + \frac{2}{3}\lambda)} \\ &= 2[J(J+1)]^{1/2}\frac{B_\Sigma}{2\lambda - 2B_\Sigma} \end{aligned}$$

which implies that the mixing is essentially complete when $J > |\lambda/B| - 1$.

Table 4.1
Calculated Electronic Factors for Trial Vibrational Numberings of the $e^3\Sigma^-$ Perturber of SiO $A^1\Pi$ (cm^{-1})

Isotopic species	v_Σ^a	v_Π	$\langle {}^1\Pi, v_\Pi \lvert \mathbf{H}^{SO} \rvert {}^3\Sigma_1^-, v_\Sigma \rangle / \langle v_\Pi \lvert v_\Sigma \rangle$ $v_\Sigma + 3$	$v_\Sigma + 1$	v_Σ	$v_\Sigma - 1$	$v_\Sigma - 3$
$Si^{16}O$	9	2	225(2)[b]	44.6(4)[b]	25.1(2)[b]	17.1(2)[b]	20.0(2)[b]
	14	6	87(4)	68.8(28)	23.7(10)	29.4(12)	33.6(13)
	15	7	63(1)	24.6(2)	23.3(2)	40.4(4)	24.2(2)
$Si^{18}O$	8	1	297(10)	51.5(15)	26.2(8)	15.5(5)	9.8(3)
	13	5	109(5)	32.1(14)	23.7(11)	23.9(11)	67.5(30)
	14	6	81(1)	28.0(3)	23.6(2)	29.1(3)	32.5(3)

[a] Vibrational quantum number of $e^3\Sigma^-$ state as determined from isotope shifts.
[b] Numbers in parentheses are five-standard-deviation uncertainties in the last digits of the fitted perturbation matrix element.

division by $[J(J+1) - 2]^{1/2} \simeq J$. The crucial problem here is the ability to distinguish a perturbation by $^1\Delta$ from one by another double-parity state, such as $^1\Pi$, $^3\Pi_2$, $^3\Delta_1$, $^3\Delta_2$, or $^3\Delta_3$, for which the J-dependence is not $\approx J^1$. For SiO, A$\sim$D perturbations occur over a range of J-values, 11.4–112.9, that is larger than the range of derived β_{12} values, 0.039–0.084 cm^{-1}, from which the J-dependence of the local matrix element has been removed; success of the matrix element method may therefore be construed as proof of both the absolute vibrational numbering and the electronic assignment.

The $A^1\Pi \sim d^3\Delta$ interaction is explicitly J-independent,

$$\langle {}^3\Delta_1, J \lvert \mathbf{H}^{SO} \rvert {}^1\Pi \rangle = A_{12}, \tag{4.1.59}$$

but implicitly J-dependent because the strengths of the local interactions with the three spin-components of $^3\Delta$ depend on the $\lvert {}^3\Delta_1 \rangle$ basis function character, C_{1i}, present in the $\lvert {}^3\Delta, J, F_i \rangle$ eigenfunctions,

$$\lvert {}^3\Delta_1, J, F_i \rangle = C_{1i} \lvert {}^3\Delta_1 \rangle + C_{2i} \lvert {}^3\Delta_2 \rangle + C_{3i} \lvert {}^3\Delta_3 \rangle \tag{4.1.60}$$

$$\langle {}^1\Pi, J \lvert \mathbf{H}^{SO} \rvert {}^3\Delta, J, F_i \rangle = C_{1i} A_{12} \tag{4.1.61}$$

(for $^3\Delta$, $N = J - 2 + i$, nominal $\Omega = i$ for $A_\Delta > 0$ and nominal $\Omega = 4 - i$ for $A_\Delta < 0$). The $\lvert {}^3\Delta_1 \rangle$ mixing coefficients depend on J, A_Δ, and B_Δ, for example, using first-order perturbation theory for the correction to the wavefunction

$$\lvert C_{12} \rvert = \left| \frac{\langle {}^3\Delta_1 \lvert \mathbf{H} \rvert {}^3\Delta_2 \rangle}{\langle {}^3\Delta_1 \lvert \mathbf{H}^0 \rvert {}^3\Delta_1 \rangle - \langle {}^3\Delta_2 \lvert \mathbf{H}^0 \rvert {}^3\Delta_2 \rangle} \right|$$
$$= \left| \frac{\langle {}^3\Delta_1 \lvert (-B\mathbf{J}^+\mathbf{S}^-) \rvert {}^3\Delta_2 \rangle}{\{-2A_\Delta + B_\Delta[J(J+1) - 3]\} - \{0A_\Delta + B_\Delta[J(J+1) + 1]\}} \right|,$$

and neglecting the $4B_\Delta$ in the denominator,

$$|C_{12}| \approx [2J(J+1) - 4]^{1/2} \frac{B_\Delta}{2A_\Delta}. \tag{4.1.62}$$

Analytic expressions for the mixing coefficients are given Kovács (1969, pp. 68–70). In order to apply the matrix element method to a $^3\Delta$ perturber, it is necessary to observe perturbations by at least two of the three $^3\Delta$ spin components and to infer the relative vibrational numbering of these spin components. There are generally four possible relative numberings, corresponding to whether A_Δ is positive or negative, and whether $|2A_\Delta|$ is larger or smaller than $\omega_e(^3\Delta)$. If the electronic configuration of a plausible $^3\Delta$ perturber is known, the sign and magnitude of A_Δ can be estimated with sufficient accuracy to eliminate all but one of these relative numberings. Success of the matrix element method for $^1\Pi \sim {}^3\Delta$ perturbations implies determination of both the absolute vibrational numbering and the diagonal spin–orbit constant of the $^3\Delta$ state. For SiO, the matrix element method suggests a vibrational numbering in agreement with that deduced from isotope shifts even though the $d^3\Delta \sim A^1\Pi$ curve crossing occurs just at the highest vibrational levels sampled, $v_A = 8 \sim v_d = 19$, $E_C \simeq 49{,}000\ cm^{-1}$, $R_C = 1.42$ Å).

The $A^1\Pi \sim (C^1\Sigma^-, D^1\Delta, d^3\Delta$, and $e^3\Sigma^-)$ perturbations of the $^{28}Si^{32}S$ and $^{30}Si^{32}S$ molecules (Harris *et al.*, 1982) are analogous to those observed for SiO except that the vibrational quantum numbers of the SiS perturbing levels are considerably higher than those for SiO, $v = 14$–31 versus $v = 6$–19. The quality of perturbation data is generally insufficient to determine more than the minimum number of equilibrium constants, T_e, ω_e, $\omega_e x_e$, B_e, and α_e. This is true even for SiS $A^1\Pi$, where the number of perturbations (83) and range of vibrational levels sampled is far greater than for SiO $A^1\Pi$. Neglect of higher-order constants (especially γ_e) and the length of the extrapolation to $v = -\frac{1}{2}$ lead to a large uncertainty in R_e. This is particularly serious when the perturbation matrix elements sampled do not include the first maximum, slightly above the curve crossing (E_C, R_C), as is the case for SiS $A^1\Pi$ (v = 0–11). For an inner-wall curve crossing, an error in R_C can cause a large change in the apparent location of (E_C, R_C). Consequently, the matrix element method cannot distinguish between the $v-1$, v, and $v+1$ absolute vibrational numberings. In addition, *ab initio* values of the electronic perturbation parameters (Robbe *et al.*, 1981) cannot be used to resolve this numbering ambiguity because, even with the correct vibrational assignment, extrapolation uncertainty could cause the calculated vibrational overlap to be erroneous by a factor of two.

It must be emphasized that the reason the matrix element method begins to fail for SiS $A^1\Pi$ has nothing to do with the strength of the perturbation interactions or the degree to which the perturbers remain isolated from each other. In general, as the interactions between basis states become stronger, one must increase the dimension of the effective Hamiltonian used to account for the observed details. This does not necessarily imply a vast increase in the number of variable perturbation parameters, provided that the potential energy curves of all but one of the interacting states are known well enough so that a deperturbation procedure, based on matrix element factorization, can be initiated. One example will be discussed briefly, BaS $A^1\Sigma^+ \sim A'^1\Pi \sim a^3\Pi$ (Cummins *et al.*, 1981)]. A similar case, the NO $B^2\Pi \sim C^2\Pi \sim D^2\Sigma^+$ interaction, was discussed in Section 3.5.

The lowest electronic states of the BaS molecule are summarized in Fig. 4.9. Figure 4.10 shows the $A'^1\Pi$ and $a^3\Pi_i$ levels responsible for perturbations of BaS $A^1\Sigma^+(v = 0\text{–}2)$ [Cummins *et al.*, 1981]. The spin–orbit constant $A(a^3\Pi) = -186\ \text{cm}^{-1}$ is almost as large as $\omega_e(a^3\Pi) = 260\ \text{cm}^{-1}$, and the $A'^1\Pi \sim a^3\Pi_1$ spin–orbit interaction, $A_{11} = 178\ \text{cm}^{-1}$, is only slightly smaller than the deperturbed separation of these two states, $\Delta T_e = 260\ \text{cm}^{-1}$. Figure 4.10 shows that the $a^3\Pi$ $(v = 12) \sim A'^1\Pi$ $(v = 12)$ perturbation complex spans more than $500\ \text{cm}^{-1}$ and is interleaved with components of the $v = 10\text{–}14$ $A' \sim a$ complexes. Each nominal $A'^1\Pi$, $v_{A'}$ and $a^3\Pi_\Omega$, v_a level is a complicated mixture

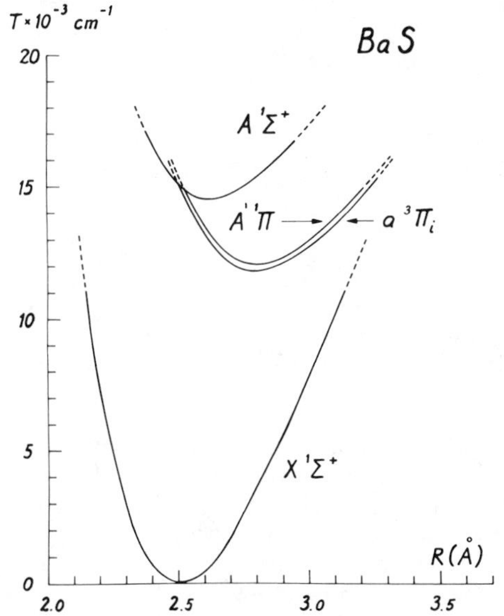

Fig. 4.9 Potential energy curves for the $X^1\Sigma^+$, $a^3\Pi_i$, $A'^1\Pi$, and $A^1\Sigma^+$ electronic states of the BaS molecule. Solid lines indicate the energy range over which the potentials are determined by Cummins *et al.* (1981).

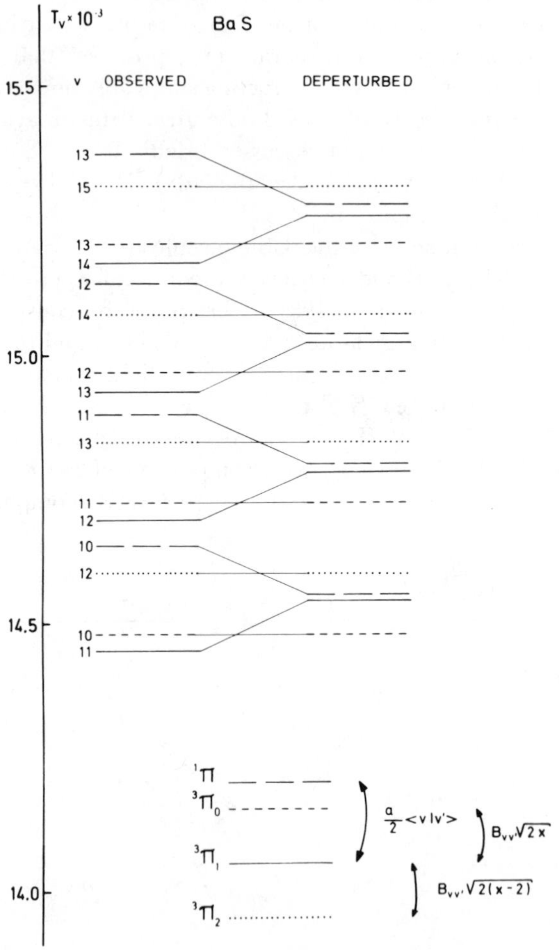

Fig. 4.10 Interleaved $A'^1\Pi$ and $a^3\Pi$ levels of BaS. Energy-level diagram showing observed and deperturbed energies of the rotationless vibrational levels of the BaS $A'^1\Pi$ and $a^3\Pi_i$ states that interact with $v = 0$–2 of $A^1\Sigma^+$. The grouping together of $^3\Pi_2$ $(v + 1)$, $^1\Pi$ $(v - 1)$, $^3\Pi_1$ $(v + 1)$, and $^3\Pi_0$ (v) levels is suggestive of Hund's case (c) coupling. The interactions between case (*a*) basis functions are shown schematically at the bottom of the figure. Levels are indicated as follows: $^1\Pi$, long dashes; $^3\Pi_0$, short dashes; $^3\Pi_1$, solid lines; $^3\Pi_2$, dotted lines.

Table 4.2
Perturbation Parameters for the BaS Molecule

(a) $A^1\Sigma^+ \sim A'^1\Pi$ orbital perturbation parameters

$A^1\Sigma^+(v)$	$A'^1\Pi(v')$	$B_{vv'}(\mathrm{cm}^{-1})$	b^a	$\bar{R}_{vv'}$(Å)
0	10	0.01131	1.08(2)	2.530
1	11	0.01963	1.11(1)	2.533
2	12	0.02470	1.07(2)	2.536
			Average: 1.09(1)[a,b]	

(b) $A^1\Sigma^+ \sim a^3\Pi$ spin–orbit perturbation parameters

$A^1\Sigma^+(v)$	$a^3\Pi_0(v')$	$\langle v\|v'\rangle$	$a_+(\mathrm{cm}^{-1})^a$	$\bar{R}_{vv'}$(Å)
0	10	0.11177	189.6(1)	2.530
0	11	−0.07977	176.4(6)	2.513
1	12	−0.15080	183(4)	2.516
2	13	−0.20760	182(4)	2.519
			Average: 183(5)	

[a] 1σ uncertainty in parentheses.

[b] This value of b does not allow for J-independent admixture of $|^3\Pi_1\rangle$ character into the nominal $^1\Pi$ eigenfunction,

$$|^1\Pi\rangle_{\text{nominal}} \simeq 0.89|^1\Pi_1\rangle + 0.46|^3\Pi_1\rangle.$$

Thus, the fully deperturbed value of b is

$$b = \frac{1.09}{0.89} = 1.21.$$

of $A'^1\Pi$, $a^3\Pi_{0,1,2}$ basis functions belonging to several vibrational quantum numbers. A partial prediagonalization procedure was employed in the fitting of the BaS $A^1\Sigma^+$ perturbations. Since a $|^1\Sigma^+\rangle$ basis function interacts only with $|^1\Pi\rangle$ and $|^3\Pi_0\rangle$ basis functions, the interaction of each $A^1\Sigma^+, v_\Sigma$ level with a nominal $^{2S+1}\Pi_{\Omega,v_\Pi}$ level was represented by a sum of two contributions, interactions with $|a^3\Pi_0, v_a = v_\Pi\rangle$ and $|A'^1\Pi, v_{A'} = v_\Pi\rangle$. Since the $A'^1\Pi$ and $a^3\Pi$ potentials are nearly identical, it is a good approximation to neglect all $v_a \neq v_{A'}$ interactions and consider only the dominant vibrational component of the nominal $^{2S+1}\Pi_{\Omega,v_\Pi}$ function. The success of this approach is indicated by the constancy of the a and b parameters derived from the observed $A \sim a$ and $A \sim A'$ perturbations (Table 4.2). It is important to note that this deperturbation was aided considerably by knowledge of the lower vibrational levels of the $A'^1\Pi$ and $a^3\Pi$ states derived from $B^1\Sigma^+$–$A'^1\Pi$ and $B^1\Sigma^+$–$a^3\Pi_1$ fluorescence spectra.

4.2 Order of Magnitude of Electrostatic Perturbation Parameters: Interactions between Valence and Rydberg States of the Same Symmetry

In light molecules, homogeneous perturbations due to electrostatic interaction (the e^2/r_{ij} operator) are often very strong. Some values of the electrostatic interaction have been listed in Table 2.5 for the case where H^e values, representing the interaction between crossing diabatic curves, are particularly large ($400\ \text{cm}^{-1} < H^e < 10{,}000\ \text{cm}^{-1}$). The large sizes of these tabulated H^e values indicate that an adiabatic approach to these interactions would be more convenient, but the diabatic picture is equally valid and often more insightful.

4.2.1 VALENCE AND RYDBERG STATES

Most frequently interactions between states of the same electronic symmetry are between a Rydberg state and a valence state and are often referred to in the literature as valence ~ Rydberg mixing. In the following, a distinction is drawn between these two classes of electronic states.

First, the valence-shell states or valence states are those associated with electronic configurations in which all occupied molecular orbitals are comparable to or smaller in size than the typical 2 Å molecular internuclear distance. These valence molecular orbitals are constructed from atomic orbitals with principal quantum number, n, less than or equal to the maximum n involved in the ground state configurations of the constituent atoms. Transitions from the molecular ground state to an excited valence state frequently involve promotion of an electron from a bonding to an antibonding valence orbital. In such a case, R_e of this excited valence state will be larger and ω_e smaller than for the ground state.

Second, molecular Rydberg states are excited states similar to atomic Rydberg states. The excited electron is in an orbital that can be constructed from atomic orbitals with principal quantum numbers larger than those of the ground state of the constituent atoms (see Section 7.2). The first Rydberg state of H_2 has $n = 2$, whereas $n = 3$ for the lowest Rydberg states of molecules composed of atoms from the first row of the periodic table. The mean radius and other properties of a Rydberg orbital may be estimated using formulas derived for hydrogenic orbitals (Condon and Shortley, 1953, p. 117) by simply replacing the principal quantum number, n, by $n^* \equiv n - a$, where a is the quantum defect. The mean radius in atomic units (1 a.u. = 0.52917 Å) is

$$\langle r \rangle_{n^*,l} = \frac{n^{*2}}{Z}\left\{1 + \frac{1}{2}\left[1 - \frac{l(l+1)}{n^{*2}}\right]\right\}, \tag{4.2.1}$$

where Z is the charge of the ion core ($Z = 1$ for neutral molecules). $\langle r \rangle$ rapidly becomes very large as n^* increases. For example,

$$\langle r \rangle_{3p} = 6.9 \text{ a.u.} = 3.7 \text{ Å} \qquad \text{with} \quad a_p = 0.7$$

$$\langle r \rangle_{4f} = 18 \text{ a.u.} = 9.5 \text{ Å} \qquad \text{with} \quad a_f = 0.0$$

Thus, the Rydberg electron neither strongly perturbs the core molecular orbitals nor contributes to bonding; consequently, Rydberg potential curves are very similar to the potential curve of the ion to which the Rydberg series converges. The energy levels follow the Rydberg formula:

$$E_{\mathrm{R}} = E_{\mathrm{ion}} - \frac{\mathscr{R}}{n^{*2}} = E_{\mathrm{ion}} - \frac{\mathscr{R}}{(n-a)^2}. \tag{4.2.2}$$

where R is the Rydberg constant ($= 109{,}737.318$ cm^{-1}).

The spin–orbit multiplet splitting of Rydberg states may originate from the outer Rydberg electron or the ion core ($\Lambda_{\text{core}} \geq 1$, $S_{\text{core}} \geq \frac{1}{2}$). The contribution from the Rydberg orbital ($Z = 1$) is proportional to $\langle r^{-3} \rangle$

$$\langle r^{-3} \rangle = \frac{Z^3}{n^{*3}(l+1)(l+\frac{1}{2})l} \propto (n^*)^{-3}. \tag{4.2.3}$$

Consequently, the spin–orbit constants for $\lambda \neq 0$ Rydberg states will be smaller than those of the valence states constructed from orbitals of smaller n^* values. In Table 4.3, spin–orbit constants of valence and Rydberg states of the same molecule are compared.

Another readily estimable quantity for Rydberg states is the energy interval between same-$nl\lambda$ states of different multiplicities. This splitting is governed by an exchange integral, K, between a compact inner orbital and a diffuse Rydberg orbital. To a good approximation, $K \propto (n^*)^{-3}$. This can be used to distinguish a Rydberg state from a valence state. For example, in N_2 the singlet–triplet interval is 10^4 cm^{-1} for the $\sigma_g\pi_g\, a^1\Pi_g$ and $B^3\Pi_g$ valence states and 870 cm^{-1} for the $\sigma_g 4p\sigma_u\, c_4'{}^1\Sigma_u^+$ and $D^3\Sigma_u^+$ Rydberg states. (However, K can also be very small for valence states of highly ionic molecules where the orbitals involved are highly localized on separate atoms.) The $(n^*)^{-3}$ dependence of A and K has been used in Section 2.4.2 to illustrate the convergence of $^1\Pi$ and $^3\Pi$ states to the $^2\Pi_{3/2}$ and $^2\Pi_{1/2}$ ion-core substates.

Note that a Rydberg state is generally well described by a single electronic configuration of the form

$$(\mathrm{AB}^+)(nl\lambda)$$

Table 4.3
Comparison of Spin–Orbit Coupling Constants of Rydberg and Valence States (cm^{-1})

Molecule	Rydberg state	Rydberg orbital	A(atom)[a]	A_R(obs)	Valence state	A_V(obs)
NO	$C^2\Pi$	$3p\pi$	$2.17(3p_O)$	3.2[b]	$B^2\Pi$	28[e]
NO	$H^2\Pi$	$3d\pi$	$0.02(3d_O)$	0.96[c]	$L^2\Pi$	−104[e]
PO	$D^2\Pi$	$4p\pi$	$6.13(4p_P)$	25.1[d]	$B'^2\Pi$ ($v = 24$)	32.3[d]

[a] The difference between the spin–orbit constants of the main constituent *atomic* orbital, A(atom, nl), in the Rydberg molecular orbital and that of the molecular Rydberg state, A_R, is due to "penetration" of the Rydberg MO into the molecular core, i.e., to the contribution of the $(n-1)$ atomic orbital responsible for the orthogonality between the Rydberg MO and the molecular core orbitals.
[b] Ackermann and Miescher (1968).
[c] Suter (1969).
[d] Couet (1968).
[e] Dressler and Miescher (1965).

where (AB^+) is the configuration of the core-ion, $nl\lambda$ is the special notation reserved for Rydberg orbitals, and n has the value that appears in Eq. (4.2.2). l is only an approximate quantum number because the molecular potential in which the Rydberg electron moves is slightly distorted from a spherical potential.

4.2.2 DIFFERENT CLASSES OF VALENCE~RYDBERG MIXING

Two types of mixing between Rydberg and valence states can be distinguished:

1. *The configurations of the Rydberg and valence states differ by two orbitals.* In the diabatic approximation, an electrostatic interaction takes place and mixes the vibrational levels of these two states. In the adiabatic representation, the coefficients of the two configurations vary with R, and, through the d/dR operator, this R-dependence of the configuration mixing coefficients is the origin of nonzero matrix elements between adiabatic states (see Section 2.3). For example, at short internuclear distance the O_2 valence $B^3\Sigma_u^-$ and Rydberg $B'^3\Sigma_u^+$ states arise from a mixture of the $\pi_u^3\pi_g^3$ and $\pi_u^4\pi_g 3p\pi_u$ configurations (Katayama *et al.*, 1977).

2. *The configurations of the Rydberg and valence states differ by only one orbital.* The adiabatic representation is appropriate. In this picture, only the orbital, not the configuration mixing coefficients, changes with R. For the lowest Rydberg

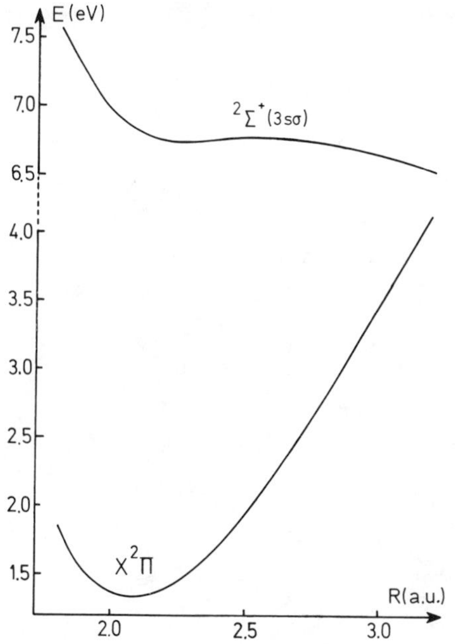

Fig. 4.11 Calculated $^2\Sigma^+$ $(3s\sigma)$ state of CH. (Lefebvre-Brion, unpublished calculation.)

state, this phenomenon has been called "Rydbergization" because the molecular orbital that, at dissociation, is an atomic orbital becomes a Rydberg molecular orbital at small internuclear distances (Mulliken, 1976). The character of the orbital changes smoothly with R. Rydbergization occurs in the first Rydberg state of many AH hydrides. The orbital, which is $3s_A$ at short R, becomes at intermediate R an antibonding orbital $(2p\sigma_A - 1s_H)$, and finally becomes localized on the $1s_H$ orbital at infinite R. The resulting state is, in general, unbound. Examples include CH $\pi^0 3s\sigma\,^2\Sigma^+$ (see Fig. 4.11), OH $\pi^2 3s\sigma\,^2\Sigma^-$ (Lefebvre-Brion, 1973) and NH $\pi^1 3s\sigma\,^1\Pi$ (Mulliken, 1972) states. This explains the fact that the first states of the Rydberg $s\sigma$ series have not been observed in CH, NH, and OH.

The existence of states complementary to those for which Rydbergization occurs as R decreases, in which the molecular orbital becomes a Rydberg orbital at large R, has been suggested by *ab initio* calculations. For example, in O_2 the configuration $\pi_u^4\pi_g 3p\sigma_u$ gives rise to a $^3\Pi_u$ state that is a Rydberg state at small R, but since the $3p\sigma_u$ orbital becomes a $\sigma_u 2p$ antibonding orbital at large R (see correlation diagram of Herzberg, 1950, p. 329), the state is repulsive and responsible, in part, for predissociation of the $B^3\Sigma_u^-$ state. The complementary $^3\Pi_u$ state is probably the $F^3\Pi_u$ state observed by Chang and Ogawa (1972).

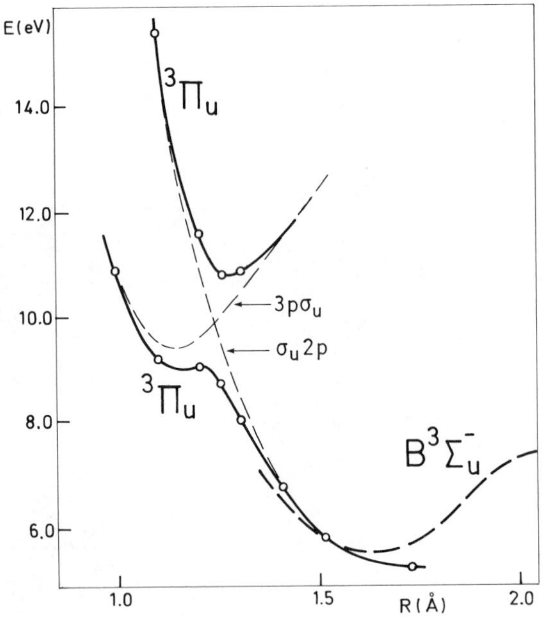

Fig. 4.12 Rydbergized and complementary $^3\Pi_u$ states in O_2 arising from configurations with a $\pi_u^4\pi_g$ core and a σ_u external orbital. The solid lines depict the adiabatic potential curves calculated by Buenker and Peyerimhoff (1975). The lower curve is the Rydbergized state, which has an external orbital that evolves from a $3p\sigma_u$ Rydberg orbital at small R to a $\sigma_u 2p$ antibonding orbital at large R. The upper curve is the complementary state, associated with an external orbital that is $\sigma_u 2p$ at small R and $3p\sigma_u$ at large R. The dashed lines represent the diabatic states that would be obtained if the external orbital were frozen as R varies.

The isoconfigurational $^1\Pi_u$ state seems also to have been observed (Katayama *et al.*, 1977). The energy separation between the Rydbergized and the complementary state is generally large: 8500 cm^{-1} for the $^3\Pi_u$ states of O_2 (see Fig. 4.12), 5000 cm^{-1} for the $^2\Sigma^+$ states of NO [the $\sigma^2 3p\sigma$ configuration of the $D^2\Sigma^+$ state becomes the $\sigma^2\sigma^*$ configuration to which the repulsive $A'^2\Sigma^+$ state has been attributed (Gallusser, 1976)].† In principle, a weak coupling can occur between the pair of complementary states, which in the adiabatic picture results from the d/dR operator acting on the *orbital* and not on the configuration-mixing coefficients. However, at present information about complementary

† The complementary state may be located above the ionization limit. It may then appear as a "shape resonance" in the photoionization cross-section (Morin *et al.*, 1982).

states is insufficient to provide examples of such perturbations (or predissociations).

This second class of valence ~ Rydberg interaction (configurations differing by one orbital) is often confused with the first class. Some theoreticians consider that the diabatic picture is valid for states belonging to configurations which differ by a single orbital, but it is a very poor approximation for this case because H^e is too large. The adiabatic picture is directly obtained in the single-configuration approximation. To compute an *ab initio* diabatic potential curve, it would be necessary to freeze the external orbital as R varies. In the following subsection, only the first class of valence ~ Rydberg mixing, for which the configurations differ by two electrons, is considered.

4.2.3 ELECTROSTATIC PERTURBATIONS BETWEEN VALENCE AND RYDBERG STATES

Some of the highest energy valence states lie at similar energies to those of Rydberg states and numerous valence ~ Rydberg perturbations result. Table 4.4 reviews several cases treated in the literature. The values of H^e are obtained either by deperturbation or by *ab initio* calculations. Prediction of the strength of H^e is, in general, difficult. A detailed numerical example was discussed in Section 2.3.2 for the case of interacting NO $^2\Delta$ states. The interaction strength between pairs of states belonging to the same pair of configurations but to different symmetries can vary by an order of magnitude. In the case of the O_2 $\pi_u^3\pi_g^3$ and $\pi_u^4\pi_g\ 3p\pi_u$ configurations, the interaction between the $^3\Sigma_u^-$ states is 10 times larger than that between the $^3\Delta_u$ states (see Table 4.4). This results from partial cancellation between two relatively large integrals for $^3\Delta_u$ states that add in the $^3\Sigma^-$ case [cf. Eq. (2.3.8)]. For molecules composed of atoms from the first row of the periodic table, the diabatic potential curves of valence and Rydberg states generally have very different spectroscopic constants. This is particularly true when the valence state is an ionic state (examples include perturbations in H_2, HF, F_2, Cl_2). For heavier molecules, the potential curves of valence and Rydberg states become very similar and the deperturbation process becomes more difficult (e.g., SiO, PO).

The interaction between a valence state and successive Rydberg states of the same series (same values of l and λ, but different n) decreases proportionally to $(n^*)^{-3/2}$. Consequently, the product $H_n^e \times (n^*)^{3/2}$ should be constant (cf. Section 7.3). Table 4.5 shows how well this law is obeyed for the NO molecule. This is one of the basic concepts of the multichannel quantum defect theory (Dill and Jungen, 1980). Thus, the adiabatic picture is the best starting point in some cases for treating the interaction between a valence state and the first Rydberg state of a given series (for example, the E, F; G, K interaction in H_2).

Table 4.4
Some Values of the Electrostatic Parameter

Molecule	Symmetry	Valence states: Name	Valence states: Configuration	Rydberg states: Name	Rydberg states: Configuration	$H^e(\text{cm}^{-1})$[a]	Ref.[b]
N_2	$^1\Pi_u$	b	$\sigma_u\pi_u^4\sigma_g^2\pi_g$	c	$\sigma_u^2\pi_u^4\sigma_g\,3p\pi_u$	1100	1
		b	$\sigma_u\pi_u^4\sigma_g^2\pi_g$	o	$\sigma_u^2\pi_u^3\sigma_g^2\,3s\sigma_g$	1000	
	$^1\Sigma_u^+$	b′	$\sigma_u^2\sigma_g^2\pi_u^3\pi_g$	c′	$\sigma_u^2\sigma_g\pi_u^4\,3p\sigma_u$	900	
CO	$^3\Sigma^+$	a′	$\sigma^2\pi^3\pi^*$	b	$\sigma\pi^4\,3s\sigma_g$	700	2
NO	$^2\Sigma^+$	I	$\sigma\pi^{*2}$	H	$\sigma^2\,3d\sigma$	700	3
		I	$\sigma\pi^{*2}$	E	$\sigma^2\,4s\sigma$	50	
		I	$\sigma\pi^{*2}$	M	$\sigma^2\,4p\sigma$	100	
O_2	$^3\Sigma_u^-$	b′	$\pi_u^3\pi_g^3$		$\pi_u^4\pi_g\,3p\pi_u$	(4000)	4
	$^3\Delta_u$		$\pi_u^3\pi_g^3$		$\pi_u^4\pi_g\,3p\pi_u$	(320)	
BeH	$^2\Sigma^+$		$\sigma\sigma^{*2}$		$\sigma^2\,3s\sigma$	(660)	5
PO	$^2\Sigma^+$	F	$\sigma\pi^{*2}$	A	$\sigma^2\,4s\sigma$	410	6
		F	$\sigma\pi^{*2}$	G	$\sigma^2\,4p\sigma$	540	

[a] Values in parentheses are calculated values.
[b] References: (1) Stahel *et al.* (1983); (2) from Stepanov (1940) and RKR curve calculations; (3) Gallusser (1976); (4) Buenker and Peyerimhoff (1975); (5) Lefebvre-Brion and Colin (1977); (6) Ngo *et al.* (1974).

This adiabatic picture becomes increasingly difficult to apply for the interaction of higher vibrational levels of the valence state with the higher-n members of the Rydberg series because the strength of the valence ~ Rydberg interactions become weaker.

Table 4.5
Dependence on n^* of the Electrostatic Parameter

Molecule	Symmetry	Configuration of: Valence state	Configuration of: Rydberg state	$H_n^e(\text{cm}^{-1})$	$n^{*3/2}H_n^e$ (a.u.)
NO	$^2\Pi$[a]	B $\pi^3\pi^{*2}$	C $\pi^4 3p\pi$	1383	0.0208
			K $\pi^4 4p\pi$	804	0.0214
			Q $\pi^4 5p\pi$	595	0.0237
		L $\pi^3\pi^{*2}$	C $\pi^4 3p\pi$	550	0.0083
			K $\pi^4 4p\pi$	250	0.0066
			Q $\pi^4 5p\pi$	200	0.0080
	$^2\Delta$[b]	B′ $\sigma\pi^{*2}$	F $\sigma^2 3d\delta$	450	0.0103
			N $\sigma^2 4d\delta$	400	0.0142

[a] Gallusser and Dressler, 1982.
[b] Lefebvre-Brion, 1969.

4.2.4 ELECTROSTATIC PERTURBATIONS BETWEEN RYDBERG STATES CONVERGING TO DIFFERENT STATES OF THE ION

Electrostatic perturbations can occur between two Rydberg states that belong to series converging to different states of the ion. One example, the interaction between $^1\Pi_u$ states of the N_2 molecule (Stahel *et al.*, 1983), has been studied. The $c^1\Pi_u$ state has the configuration $\sigma_u^2\sigma_g\pi_u^4 3p\pi_u$ and belongs to a series converging to the $X^2\Sigma_g^+$ ground state of N_2^+. The $o^1\Pi_u$ state has the configuration $\sigma_u^2\sigma_g^2\pi_u^3 3s\sigma_g$ and belongs to a series converging to the $A^2\Pi_u$ state of N_2^+. These two Rydberg states are strongly perturbed by the $b^1\Pi_u$ valence state, but they also interact with each other. The H^e effective interaction parameter between the two Rydberg states has been found to be about 100 cm^{-1}. The same type of interaction is responsible for the autoionization of the upper members of Worley's third series (of which $o^1\Pi_u$ is the first member) by the continuum of the $X^2\Sigma_g^+$ state of N_2^+ (see Section 7.7).

It is possible for Rydberg states of the same symmetry, converging to different multiplet states of the same ionic configuration, to interact strongly. In such a case, the configurations of the interacting Rydberg states appear to differ by only the Rydberg orbital, but the multiplicities of the ion cores are different. The interacting states differ by two *spin*-orbitals, the Rydberg spin-orbital and one of the core spin-orbitals. The electrostatic interaction is expressed in terms of two-electron integrals, as in Eq. (2.3.8), where the ϕ_a and ϕ_b orbitals are identical. This type of interaction can give rise to autoionization: for example, the autoionization of the Rydberg states of O_2 converging to the O_2^+ $\sigma_g\pi_g^2$ $B^2\Sigma_g^-$ state by the ionization continuum of the $\sigma_g\pi_g^2$ configuration of the $b^4\Sigma_g^-$ state of O_2^+ (Dehmer and Chupka, 1975). The interaction between two $^3\Sigma_u^-$ Rydberg states belonging to series that converge to these $^2\Sigma_g^-$ and $^4\Sigma_g^-$ ion core states is given by

$$\langle (b^4\Sigma_g^-)np\sigma_u|\mathbf{H}^{el}|(B^2\Sigma_g^-)n'p\sigma_u\rangle = \frac{2\times 2^{1/2}}{3}[\langle \pi_g np\sigma_u|\pi_g n'p\sigma_u\rangle - \langle \sigma_g np\sigma_u|\sigma_g n'p\sigma_u\rangle],$$

where the wavefunctions are

$$|(b^4\Sigma_g^-)np\sigma_u\rangle = 12^{-1/2}|np\sigma_u\sigma_g\pi_g^+\pi_g^-|[-3\beta\alpha\alpha\alpha + \alpha\alpha\beta\alpha + \alpha\alpha\alpha\beta + \alpha\beta\alpha\alpha]$$

$$|(B^2\Sigma_g^-)n'p\sigma_u\rangle = 6^{-1/2}|n'p\sigma_u\sigma_g\pi_g^+\pi_g^-|[-2\alpha\beta\alpha\alpha + \alpha\alpha\alpha\beta + \alpha\alpha\beta\alpha].$$

(See Table 2.4 for the case of four open subshells.)

It was suggested in Section 4.2.2 that a valence antibonding orbital can become, at small R, a Rydberg orbital of the same symmetry (Rydbergization).

Table 4.6
Spin–Orbit Constants of Atoms and Ions (cm^{-1})[a]

(a) $\zeta(np)$

				B^+	15.2	C^+	42.7	N^+	87.5	O^+	*169	F^+	327.1
				$2s2p$		$2p$		$2p^2$		$2p^3$		$2p^4$	
Li	0.2	Be	2.0	B	10.7	C	29.0	N	*73.3	O	151	F	269.3
$2p$		$2s2p$		$2s^22p$		$2p^2$		$2p^3$		$2p^4$		$2p^5$	
		Be^-	0.7	B^-	5.3	C^-	—	N^-	55	O^-	121	F^-	—
		$2s^22p$		$2s^22p^2$				$2p^4$		$2p^5$			
				Al^+	124.9	Si^+	191.3	P^+	313.5	S^+	*482.8	Cl^+	664
				$3s3p$		$3p$		$3p^2$		$3p^3$		$3p^4$	
Na	11.5	Mg	40.5	Al	74.7	Si	148.9	P	*275.2	S	382.4	Cl	587.3
$3p$		$3s3p$		$3s^23p$		$3p^2$		$3p^3$		$3p^4$		$3p^5$	
		Mg^-	13.0	Al^-	48	Si^-	—	P^-	188	S^-	326	Cl^-	—
		$3s^23p$		$3s^23p^2$				$3p^4$		$3p^5$			
		Ca^+	149	Ga^+	920	Ge^+	1178	As^+	1693	Se^+	[2310]	Br^+	2560
		$4p$		$4s4p$		$4s^24p$		$4p^2$		$4p^3$		$4p^4$	
K	38	Ca	87	Ga	(464)	Ge	940	As	*1500	Se	1690	Br	2457
$4p$		$4s4p$		$4s^24p$	551	$4p^2$	*880	$4p^3$	(1201)	$4p^4$	(1658)	$4p^5$	(2215)
							(800)						
				Ga^-	387	Ge^-	—	As^-	1000	Se^-	1519	Br^-	—
				$4s^24p^2$				$4p^4$		$4p^5$			
		Sr^+	534	In^+	2368	Sn^+	2834	Sb^+	[3700]				
		$5p$		$5s5p$		$5s^25p$		$5p^2$					
Rb	158	Sr	387	In	1475	Sn	*2097.3	Sb	*3400	Te	(3383)	I	(4303)
$5p$		$5s5p$		$5p$	(1183)	$5p^2$	(1855)	$5p^3$	(2593)	$5p^4$		$5p^5$	5068
								Sb^-	2000	Te^-	3338		
								$5p^4$		$5p^5$			
		Ba^+	1127					Bi^+	[11,540]				
		$6p$						$6p^2$					
Cs	369	Ba	832	Tl	5195	Pb	*7294	Bi	*10,100	Po	(8608)	At	(10,607)
$6p$		$6s6p$		$6p$	(3410)	$6p^2$		$6p^3$		$6p^4$		$6p^5$	

Thus, the valence state can be considered as the first member of a Rydberg series converging to an excited state of the ion. The π_g antibonding orbital of N_2 becomes similar to a $3d\pi_g$ Rydberg orbital near R_e (Mulliken, 1976). The $N_2\, b'^1\Sigma_u^+$ valence state, arising from the configuration $\sigma_g^2\pi_u^3\pi_g$, could thus be considered as the first member of the series "$d\pi$" converging to the $N_2^+\ A^2\Pi_u$ state. (Unfortunately, this valence state is not satisfactorily described by this single configuration.) With this point of view, the $(n^*)^{-3/2}$ matrix element scaling rule, which is valid for Rydberg ~ Rydberg interactions, could also be applied to valence states to predict interaction parameters for perturbations and/or autoionizations in the upper members of the series of which the first member is a valence state (see Section 7.7).

Table 4.6 (Continued).

(b) $\zeta(nd)$

$N = 1$		$N = 2$		$N = 3$		$N = 4$		$N = 5$		$N = 6$		$N = 7$		$N = 8$		$N = 9$	
		Ti^{2+}	119	V^{2+}	167	Cr^{2+}	233	Mn^{2+}	317	Fe^{2+}	509	Co^{2+}	539	Ni^{2+}	677	Cu^{2+}	833
Sc^+	56	Ti^+	96	V^+	152	Cr^+	224	Mn^+	392	Fe^+	416	Co^+	536	Ni^+	672	Cu^+	824
	79		117								409				670		827
Sc	(77)	Ti	(123)	V	(177)	Cr	(243)	Mn	(325)	Fe	(417)	Co	(530)	Ni	(663)	Cu	(818)
$3d$		$3d^2$		$3d^3$		$3d^4$		$3d^5$		$3d^6$		$3d^7$		$3d^8$		$3d^9$	
		Zr^+	363	Nb^+	502	Mo^+	663			Ru^+	1052	Rh^+	1278	Pd^+	1527	Ag^+	[1830]
Y	(261)	Zr	(387)	Nb	(524)	Mo	(677)	Tc	(850)	Ru	(1038)	Rh	(1253)	Pd	(1495)	Ag	(1767)
$4d$		$4d^2$		$4d^3$		$4d^4$		$4d^5$		$4d^6$		$4d^7$		$4d^8$		$4d^9$	
Lu	(1158)	Hf	1219	Ta	1699	W	2085	Re	2545	Os	3045	Ir	3617	Pt	4221	Au	(5097)
$5d$		$5d^2$	(1578)	$5d^3$	(1995)	$5d^4$	(2432)	$5d^5$	(2901)	$5d^6$	(3378)	$5d^7$	(3905)	$5d^8$	(4475)	$5d^9$	

(c) $\zeta(nf)$

$N = 1$		$N = 2$		$N = 3$		$N = 4$		$N = 5$		$N = 6$		$N = 7$		$N = 8$		$N = 9$		$N = 10$		$N = 11$		$N = 12$		$N = 13$	
		Ce^+	526	Pr^+	649	Nd^+	772	Pm^+	911	Sm^+	1039			Gd^+	1347	Tb^+	1578	Dy^+	1791	Ho^+	2044	Er^+	2216	Tm^+	2505
La	(556)	Ce	(687)	Pr	(820)	Nd	(957)	Pm	(1103)	Sm	(1263)	Eu	(1438)	Gd	(1611)	Tb	(1802)	Dy	(2010)	Ho	(2233)	Er	(2471)	Tm	(2728)
$4f$		$4f^2$		$4f^3$		$4f^4$		$4f^5$		$4f^6$		$4f^7$		$4f^8$		$4f^9$		$4f^{10}$		$4f^{11}$		$4f^{12}$		$4f^{13}$	

[a] Values in parentheses are calculated values from Froese-Fischer (1972, 1977). Other calculated Hartree–Fock values can be found in Fraga *et al.*, 1976, Table VII (2).

For $\zeta(np)$, the experimental values are deduced from the observed splitting, $\Delta E = E_{J_{\max}} - E_{J_{\min}}$, of the lowest-energy $L > 0$ multiplet listed in Moore's (1949) tables of atomic energy levels. For both $p^N\,{}^3P$ ($N = 2$ or 4) and ${}^2P$ ($N = 1$ or 5),

$$\zeta(np) = \tfrac{2}{3}|\Delta E|.$$

Values with asterisks are, for $2p$ and $3p$, interpolated values from Ishiguro and Kobori (1967) and, for np $n > 3$, are deduced from a matrix diagonalization fit to observed levels which exhibit coupling intermediate between L–S and j–j (Condon and Shortley, 1953). Values in brackets are from Sobel'man (1972).

For the $\zeta(nd)$ table, the first value listed for the $3d$ positive ion corresponds to the $3d^N4s$ configuration where N is the number of $3d$ electrons in the neutral atom (Johansson *et al.*, 1980). The second value is obtained by a fit of the observed levels from the $3d^{N+1}$, $3d^N4s$, and $3d^{N-1}4s^2$ configurations (Shadmi *et al.*, 1968) and corresponds to the $3d^N4s$ value. The values for doubly positive ions are for the $3d^N$ configuration (Shadmi *et al.*, 1969). Values in brackets are calculated for $d^9\,{}^2D$ states using

$$\zeta(nd) = \tfrac{2}{5}|E_{J_{\max}} - E_{J_{\min}}|$$

and energy levels from Moore (1949).

For $\zeta(4d)$, the three configurations $4d^{N+1}$, $4d^N5s$, and $4d^{N-1}5s^2$ of the positive ions are fitted (Shadmi, 1961). The values listed correspond to the $4d^N5s$ configuration.

For $\zeta(5d)$, values are obtained by fitting the $5d^N6s^2$, $5d^{N+1}6s$, and $5d^{N+2}$ configurations of the neutral atoms. The values for $\zeta(5d)$ depend strongly on the number of d-electrons and the values given are for the $5d^N6s^2$ configuration (Wyart, 1978).

For $\zeta(4f)$, the values tabulated for the positive ions are deduced from a fit to experimental energy levels belonging to $4f^N5d$ and $4f^N6s$ configurations and should not differ significantly from the values for the $4f^N6s^2$ configurations for the neutral atoms (Wyart and Bauche-Arnoult, 1981).

Values of $\zeta(5f)$ are not tabulated here. See Blaise *et al.* (1980).

Values for negative ions are taken from Sambe and Felton (1976) and Hotop and Lineberger (1975).

Table 4.4 collects values for the electronic part of the electrostatic interaction. It is clear that this interaction will give rise either to perturbations or to predissociations depending on whether the interaction occurs between discrete states or discrete and continuum states. In the above example concerning the N_2 $^1\Pi_u$ states of Worley's third series, it was shown that the electrostatic interaction can also give rise to autoionizations.

4.3 Order of Magnitude of Spin Parameters

In Section 2.4.2, spin–orbit matrix elements are expressed, in the single-configuration approximation, in terms of molecular spin–orbit parameters. These molecular parameters can also be related to atomic spin–orbit parameters. In Table 4.6, some values are given for atomic spin–orbit constants, $\zeta(nl)$. Sections 4.3.1–4.3.3 describe semiempirical methods for estimating molecular spin–spin parameters and both diagonal and off-diagonal spin–orbit parameters.

4.3.1 DIAGONAL SPIN–ORBIT PARAMETERS

A semiempirical calculation of molecular spin–orbit constants can be made, using the method of Ishiguro and Kobori (1967).

In the simple case where the π molecular orbitals can be expanded in terms of one atomic p orbital on each atom, we have

$$|\pi\rangle = C_{\mathrm{A}}|p\pi_{\mathrm{A}}\rangle + C_{\mathrm{B}}|p\pi_{\mathrm{B}}\rangle \qquad \text{for the } \pi \text{ bonding orbital,}$$

$$|\pi^*\rangle = C_{\mathrm{A}^*}|p\pi_{\mathrm{A}}\rangle + C_{\mathrm{B}^*}|p\pi_{\mathrm{B}}\rangle \qquad \text{for the } \pi^* \text{ antibonding orbital,}$$

and

$$\begin{aligned} a_\pi &= \langle \pi|\mathbf{H}^{\mathrm{SO}}|\pi\rangle = C_{\mathrm{A}}^2\zeta_{\mathrm{A}}(p) + C_{\mathrm{B}}^2\zeta_{\mathrm{B}}(p) \\ a_{\pi^*} &= \langle \pi^*|\mathbf{H}^{\mathrm{SO}}|\pi^*\rangle = C_{\mathrm{A}^*}^2\zeta_{\mathrm{A}}(p) + C_{\mathrm{B}^*}^2\zeta_{\mathrm{B}}(p), \end{aligned} \tag{4.3.1}$$

where $\zeta_{\mathrm{A}}(p)$ and $\zeta_{\mathrm{B}}(p)$ are the atomic spin–orbit parameters given in Table 4.6,

$$\zeta_K(p) = \left\langle p_K \left| \frac{\alpha^2}{2} \frac{Z_{\mathrm{eff},K}}{r_K^3} \right| p_K \right\rangle.$$

This semiempirical approximation neglects two-center integrals such as

$$\left\langle p_{\mathrm{A}} \left| \frac{\alpha^2}{2} \frac{Z_{\mathrm{eff},K}}{r_K^3} \right| p_{\mathrm{B}} \right\rangle, \qquad K = \mathrm{A, B},$$

which, due to the r_K^{-3} dependence of the spin–orbit operator, are very small. Their contribution is always less than 10% of the one-center contribution. Even in *ab initio* computations of spin–orbit constants using Eq. (2.4.2), the two-center integrals are frequently neglected because the contributions of the one- and the two-electron two-center integrals tend to cancel each other (see, for example, Richards *et al.*, 1981). Note that C_K^2 is slightly different from the fraction of the electron located on the K atom, Mulliken's "atomic population" (Mulliken, 1955). The Mulliken populations are given in this case by

$$C_A^2 + C_A C_B S_{AB} \qquad \text{the atom A population}$$

$$C_B^2 + C_A C_B S_{AB} \qquad \text{the atom B population}$$

where

$$S_{AB} = \langle p\pi_A | p\pi_B \rangle, \qquad \text{the atomic overlap integral.}$$

Normalization of the molecular orbital requires that

$$C_A^2 + C_B^2 + 2C_A C_B S_{AB} = 1.$$

For homonuclear molecules,

$$|\pi_u\rangle = C_u[|p\pi_A\rangle + |p\pi_B\rangle] \qquad \text{with} \quad C_u = [2(1 + S_{AB})]^{-1/2} \leq 2^{-1/2}$$

and

$$|\pi_g\rangle = C_g[|p\pi_A\rangle - |p\pi_B\rangle] \qquad \text{with} \quad C_g = [2(1 - S_{AB})]^{-1/2} \geq 2^{-1/2}.$$

If the overlap integral is neglected, $C_u = C_g = 2^{-1/2}$ and

$$a(\pi_u) = a(\pi_g) = \zeta_A(p). \tag{4.3.2}$$

This gives [see Eq. (2.4.12)]

$$A(B^3\Pi_g, \sigma_g\pi_g, N_2) = \frac{a(\pi_g)}{2} \simeq \frac{\zeta_N(p)}{2} = 36.6\ \text{cm}^{-1}\ (\text{exp: } 42.2\ \text{cm}^{-1}).^\dagger$$

† Note that neglect of S_{AB} leads to a slight underestimate for $a(\pi_g)$ and overestimate for $a(\pi_u)$. Using

$$A(W^3\Delta_u, \pi_u^3\pi_g, N_2) = [a(\pi_g) - a(\pi_u)]/4 = 5.7\ \text{cm}^{-1}$$

and the semiempirical value of $a(\pi_g) = 84.5\ \text{cm}^{-1}$ from $A(B^3\Pi_g, \sigma_g\pi_g, N_2)$, one obtains

$$a(\pi_u) = 61.7\ \text{cm}^{-1} \qquad (\text{from} \quad \zeta(2p)\text{: } 73.3\ \text{cm}^{-1}).$$

Similarly, for the O_2 molecule, the value of $a(\pi_g) = 184\ \text{cm}^{-1}$ deduced from

$$A(F^3\Pi_u, \pi_u^4\pi_g 3p\sigma_u) = 92\ \text{cm}^{-1}$$

is slightly larger than the value of $\zeta_O(2p) = 151\ \text{cm}^{-1}$. Also, the value of $a(\pi_u) = 108\ \text{cm}^{-1}$ obtained from the equality

$$A(A'^3\Delta_u, \pi_g^3\pi_u^3)\text{exp} = -73\ \text{cm}^{-1} = -\tfrac{1}{4}[a(\pi_g) + a(\pi_u)]$$

Similarly,

$$A(\mathrm{A}^2\Pi_u, \pi_u^3, \mathrm{N}_2^+) = -a(\pi_u) = -[\zeta_{\mathrm{N}}(2p) + \zeta_{\mathrm{N}^+}(p)]/2$$
$$= -80\ \mathrm{cm}^{-1} \qquad (\mathrm{exp}: -80\ \mathrm{cm}^{-1}).$$

Also,

$$A(\mathrm{b}^3\Pi_g, \sigma_g\pi_g, \mathrm{P}_2) = +138\ \mathrm{cm}^{-1}\ (\mathrm{exp}: +131\ \mathrm{cm}^{-1})$$

(Brion *et al.*, 1977) and

$$A(\mathrm{X}^2\Pi_u, \pi_u^3, \mathrm{P}_2^+) = -294\ \mathrm{cm}^{-1}\ (\mathrm{exp}: -260 \pm 20\ \mathrm{cm}^{-1})$$

(Malicet *et al.*, 1976). This last example illustrates that the spin–orbit constants increase as the weight of the molecule increases.

For heteronuclear molecules, *ab initio* calculations of the self-consistent molecular orbitals are usually necessary to obtain the values of the C_K coefficients. The values of the spin–orbit constants depend on the effective charge on each atom, a positive charge causing a contraction of the atomic orbitals and an increase in ζ(atomic); a negative charge causes an opposite effect. In the simple case of an expansion of the molecular orbitals in terms of a large number of atomic orbitals associated with a single l-value on each atom (Lefebvre-Brion and Moser, 1966),

$$|\pi\rangle = \sum_{i=1}^{r} C_{i\mathrm{A}}|il\pi_{\mathrm{A}}\rangle + \sum_{j=1}^{s} C_{j\mathrm{B}}|jl\pi_{\mathrm{B}}\rangle$$

$$a_\pi = \left[\sum_i^r\left(C_{i\mathrm{A}}^2 + \sum_{i\neq j}^{r} C_{i\mathrm{A}}C_{j\mathrm{A}}S_{i\mathrm{A},j\mathrm{A}}\right)\right]\zeta_{\mathrm{A}}(l) + \left[\sum_j^s\left(C_{j\mathrm{B}}^2 + \sum_{k\neq j}^{s} C_{j\mathrm{B}}C_{k\mathrm{B}}S_{j\mathrm{B},k\mathrm{B}}\right)\right]\zeta_{\mathrm{B}}(l) \tag{4.3.3}$$

where the S terms are the overlap between Slater-type atomic orbitals located on the same atom. More sophisticated expressions for a_π are required when a molecular orbital has mixed-l or mixed-n character.

It is interesting to relate the spin–orbit constants for corresponding states of molecules having the same number of valence electrons (isovalent series). An example is given for molecules with 10 valence electrons for the $^3\Pi$ state of the $\sigma\pi^*$ configuration for which

$$A(^3\Pi, \sigma\pi^*) = \tfrac{1}{2}a_{\pi^*} = \tfrac{1}{2}(C_{\mathrm{A}}^{*2}\zeta_{\mathrm{A}} + C_{\mathrm{B}}^{*2}\zeta_{\mathrm{B}}).$$

---

is smaller than $\zeta_{\mathrm{O}}(2p)$. However, this semiempirical method should not be expected to give more than a best guess for the value of the molecular spin–orbit constant. This method assumes that the open-shell atomic orbitals are not significantly altered (in the core region) by the molecular field and that the single-configuration approximation is valid.

Table 4.7
Spin–Orbit Constants for Isovalent Molecules (cm^{-1})[a]

	$^3\Pi$ ($\sigma\pi^*$ configuration)			$^3\Delta$ ($\pi^3\pi^*$ configuration)			
	Calculated[a]						
	SCF	CI	Exp.	Semiempirical calculated	Exp.	$\zeta_A(np)$ (less electronegative atom)	$\zeta_B(np)$ (more electronegative atom)
CO	35.0	39.5[b]	41.51[c]	−16	−16[c]	$\zeta_C(2p) = 24$	$\zeta_O(2p) = 150$
CS	64.5	92.5[d]	92.76[e]	−45	−61[e]		$\zeta_S(3p) = 382$
CSe	(200)	(330)[d]	326.0[f]	(−208)	—		$\zeta_{Se}(4p) = 1690$
CTe	(420)	(690)[d]	—	(−420)	—		$\zeta_{Te}(5p) = 3380$
SiO	58.5	61.5[g]	73.0[h]	0	8[l]	$\zeta_{Si}(3p) = 150$	
SiS	78	85[i]	<96[j]	−43	—		
SiSe	(170)	(203)[i]	—	(−285)	—		
SiTe	(270)	(340)[i]	—	(−600)	—		
GeO	(420)	(423)[i]	—	(146)	—	$\zeta_{Ge}(4p) = 940$	
GeS	(440)	(447)[i]	—	(103)	—		
PN	88[k]		—	31[k]	—	$\zeta_P(3p) = 275$	$\zeta_N(2p) = 73$
AsN	(670)[k]		—	(190)[k]	—	$\zeta_{As}(4p) = 1500$	

[a] Values in parentheses are values estimated from calculations on lighter molecules.
[b] Hall *et al.* (1973); [c] Field *et al.* (1972); [d] Robbe and Schamps (1976); [e] Cossart and Bergeman (1976); [f] Lebreton *et al.* (1973); [g] Robbe *et al.* (1979); [h] Bredohl *et al.* (1974); [i] Robbe *et al.* (1981); [j] Linton (1980); [k] Gottscho *et al.* (1978); [l] Field *et al.* (1976).

Since, by definition, the π^* antibonding orbital has a higher energy than that of the π bonding orbital, it is reasonable to expect that the π^* orbital will be concentrated on the less electronegative atom, but this localization is by no means complete. Table 4.7 shows that the dominant atomic character of the molecular π^* orbital is in qualitative agreement with this picture. In contrast to the case for the carbon and silicon compounds, the less electronegative atom in the germanium compounds has the larger atomic spin–orbit constant. This explains the large value predicted for the $A(^3\Pi)$ spin–orbit constant of GeO and GeS. The calculated values given in Table 4.7 have been obtained by more sophisticated calculations than the semiempirical estimates. The importance of configuration interaction must be stressed, especially for heavy molecules. A small mixing with a configuration in which the π orbital is present will give a large contribution to the spin–orbit constant if this π orbital is concentrated on the heavier atom. Furthermore, it must be kept in mind that second-order effects, neglected in the results of simple calculations listed in Table 4.7, can be important for heavy molecules, leading to an asymmetric splitting of the $^3\Pi$ states.

Another example is illustrated by the $^3\Delta$ states of the same type of molecules. The $^3\Delta$ state has the $\sigma^2\pi^3\pi^*$ configuration, and the spin–orbit constant is given by

$$A(^3\Delta) = \tfrac{1}{2}\langle ^3\Delta_3|\mathbf{H}^{\mathrm{SO}}|^3\Delta_3\rangle = \tfrac{1}{4}(a_{\pi^*} - a_\pi). \qquad (4.3.4)$$

Using the expressions for

$$a_\pi = C_{\mathrm{A}}^2\zeta_{\mathrm{A}} + C_{\mathrm{B}}^2\zeta_{\mathrm{B}}$$

$$a_{\pi^*} = C_{\mathrm{A}}^{*2}\zeta_{\mathrm{A}} + C_{\mathrm{B}}^{*2}\zeta_{\mathrm{B}} \simeq C_{\mathrm{B}}^2\zeta_{\mathrm{A}} + C_{\mathrm{A}}^2\zeta_{\mathrm{B}},$$

one expects, as previously, that if A is the more electropositive atom, then $C_{\mathrm{B}}^2 > C_{\mathrm{A}}^2$, since the antibonding π^* orbital tends to be more localized on this atom. Thus,

$$|A(^3\Delta)| = -\frac{(C_{\mathrm{A}}^2 - C_{\mathrm{B}}^2)}{4}|\zeta_{\mathrm{A}} - \zeta_{\mathrm{B}}|. \qquad (4.3.5)$$

If the more electropositive atom has the smaller $\zeta_{\mathrm{A}}(np)$ value, $A(^3\Delta)$ will be negative, and vice versa. For homonuclear molecules, Eq. (4.3.5) predicts a very small value for $A(^3\Delta)$; for example, $A(\mathrm{W}^3\Delta, \mathrm{N}_2) = 5.82\ \mathrm{cm}^{-1}$ (Effantin *et al.*, 1979). Note that, because

$$C_{\mathrm{A}}^2 - C_{\mathrm{B}}^2 = (C_{\mathrm{A}}^2 + C_{\mathrm{A}}C_{\mathrm{B}}S_{\mathrm{AB}}) - (C_{\mathrm{B}}^2 + C_{\mathrm{A}}C_{\mathrm{B}}S_{\mathrm{AB}}),$$

$C_{\mathrm{A}}^2 - C_{\mathrm{B}}^2$ is simply the difference between the Mulliken populations of the two atoms in either the π or π^* orbital. Using Eq. (4.3.5) and the spin–orbit parameters listed in Table 4.7, for carbon compounds the atomic population of π^* on carbon is semiempirically calculated to be 0.75 for CO and 0.84 for CS, leaving 0.25 or 0.16 for the more electronegative atom. The atomic population difference is thus approximately 0.5, and this justifies the formula for $A(^3\Delta)$ recommended by Field *et al.* (1976). For silicon and germanium compounds, the atomic population is 0.87 on the less electronegative atom and 0.13 on the other atom. Thus the orbital localization is slightly larger ($C_{\mathrm{A}}^2 - C_{\mathrm{B}}^2 = 0.74$). Using Eq. (4.3.5), semiempirical estimates of $A(^3\Delta)$ for $\pi^3\pi^*$ configurations are given in Table 4.7 and shown to be in good agreement with experiment.

4.3.2 OFF-DIAGONAL SPIN–ORBIT PARAMETERS

A similar semiempirical method can be used to estimate off-diagonal spin–orbit parameters. Consider the a_+ parameter, which occurs for the example in Eq. (2.4.18):

$$a_+ = \langle \pi^+|\hat{a}\mathbf{l}^+|\sigma\rangle.$$

Assuming simple expressions for molecular orbitals,

$$|\pi^+\rangle = C_A|p\pi_A^+\rangle + C_B|p\pi_B^+\rangle$$

and

$$|\sigma\rangle = C'_A|p\sigma_A\rangle + C'_B|p\sigma_B\rangle + d_A|s\sigma_A\rangle + d_B|s\sigma_B\rangle.$$

By definition,

$$\mathbf{l}^+|p\sigma_K\rangle = 2^{1/2}|p\pi_K\rangle \qquad \text{and} \qquad \mathbf{l}^+|s\sigma_K\rangle = 0.$$

Neglecting the two-center integrals,

$$\begin{aligned}\langle\pi^+|\hat{a}\mathbf{l}^+|\sigma\rangle &= 2^{1/2}\{C_A C'_A\langle p_A|\hat{a}|p_A\rangle + C_B C'_B\langle p_B|\hat{a}|p_B\rangle\} \\ &= 2^{1/2}\{C_A C'_A\zeta_A(p) + C_B C'_B\zeta_B(p)\}.\end{aligned} \qquad (4.3.6)$$

Note that $\langle\sigma|\hat{a}\mathbf{l}^-|\pi^+\rangle = \langle\pi^+|\hat{a}\mathbf{l}^+|\sigma\rangle = a_+$, since the spin–orbit operator is Hermitian. In the homonuclear case, if the $s\sigma$ contribution to the σ orbital and the S_{AB} overlap are neglected, then $C_A = C'_A = 2^{-1/2}$, $C_B = C'_B = \pm 2^{-1/2}$ (where the $+$sign is for u orbitals and the $-$sign for g orbitals), and

$$\langle\pi_g^+|\hat{a}\mathbf{l}^+|\sigma_g\rangle = \langle\pi_u^+|\hat{a}\mathbf{l}^+|\sigma_u\rangle = 2^{1/2}\zeta_A(p). \qquad (4.3.7)$$

This result may be extended to the case of a molecular orbital expanded in terms of many atomic orbitals similarly to that for diagonal spin–orbit constants [see Eq. (4.3.3)]. The spin interactions in the CO molecule are an example of such a calculation (Hall *et al.*, 1973). If the different excited states of CO are represented by the same set of molecular orbitals, many off-diagonal spin–orbit interactions can be expressed in terms of the same molecular spin–orbit parameter. Perturbations known for many years (Schmid and Gerö, 1937) occur between the vibrational levels of the $a^3\Pi$ and $A^1\Pi$ states, which belong to the $\pi^4\sigma\pi^*$ configuration, and those of the $a'^3\Sigma^+$, $e^3\Sigma^-$, $d^3\Delta$, $I^1\Sigma^-$, and $D^1\Delta$ states of the $\pi^3\sigma^2\pi^*$ configuration (Fig. 2.17). These interactions are expressed in terms of a_+ and b parameters in Table 4.8. Figure 4.13 illustrates the evaluation of $a_+(a^3\Pi \sim a'^3\Sigma^+)$ both semiempirically, using a minimal SCF basis set, and fully *ab initio*, using configuration–interaction wave functions and different molecular orbital basis sets for the two states. This figure shows surprising agreement between the semiempirical method and the experimental value, when both calculations are performed at the R-centroid of the intersection between the potential curves of the two interacting states. Note the large R-variation of the a_+ parameter, which implies that the value of a_+ will depend on the location of the curve crossings for different pairs of electronic states. This R-dependence of a_+ is related to the R-variation of the atomic orbital composition of the σ and π molecular orbitals (see Section 4.6).

Table 4.8
Electronic Perturbation Parameters in the Single Configuration Approximation[a,b]

$$\left\langle {}^3\Pi_{1_f^e}, v \left\| \left[\sum_{i=1}^{6} \hat{a}_i \vec{\mathbf{l}}_i \cdot \vec{\mathbf{s}}_i + B(\mathbf{L}^+\mathbf{S}^- + \mathbf{L}^-\mathbf{S}^+) \right] \right\| {}^3\Sigma^{+e}_{1f}, v' \right\rangle = \tfrac{1}{4} a_+ \langle v|v'\rangle - B_{vv'} b$$

$$\left\langle {}^3\Pi_{1_f^e}, v \left\| \left[\sum_{i=1}^{6} \hat{a}_i \vec{\mathbf{l}}_i \cdot \vec{\mathbf{s}}_i + B(\mathbf{L}^+\mathbf{S}^- + \mathbf{L}^-\mathbf{S}^+) \right] \right\| {}^3\Sigma^{-e}_{1f}, v' \right\rangle = -\tfrac{1}{4} a_+ \langle v|v'\rangle + B_{vv'} b$$

$$\left\langle {}^3\Pi_{1_f^e}, v \left\| \left[\sum_{i=1}^{6} \hat{a}_i \vec{\mathbf{l}}_i \cdot \vec{\mathbf{s}}_i + B(\mathbf{L}^+\mathbf{S}^- + \mathbf{L}^-\mathbf{S}^+) \right] \right\| {}^3\Delta_{1_f^e}, v' \right\rangle$$
$$= \tfrac{1}{4} 2^{1/2} a_+ \langle v|v'\rangle - 2^{1/2} B_{vv'} b$$

$$\left\langle {}^3\Pi_0 f, v \left| \sum_{i=1}^{6} \hat{a}_i \vec{\mathbf{l}}_i \cdot \vec{\mathbf{s}}_i \right| {}^1\Sigma^-_0 f, v' \right\rangle = \tfrac{1}{4} 2^{1/2} a_+ \langle v|v'\rangle$$

$$\left\langle {}^3\Pi_{2_f^e}, v \left| \sum_{i=1}^{6} \hat{a}_i \vec{\mathbf{l}}_i \cdot \vec{\mathbf{s}}_i \right| {}^1\Delta_{2_f^e}, v' \right\rangle = \tfrac{1}{4} 2^{1/2} a_+ \langle v|v'\rangle$$

$$\left\langle {}^1\Pi_{1_f^e}, v \left| \sum_{i=1}^{6} \hat{a}_i \vec{\mathbf{l}}_i \cdot \mathbf{s}_i \right| {}^3\Sigma^{+e}_{1f}, v' \right\rangle = \tfrac{1}{4} a_+ \langle v|v'\rangle$$

$$\left\langle {}^1\Pi_{1_f^e}, v \left| \sum_{i=1}^{6} \hat{a}_i \vec{\mathbf{l}}_i \cdot \vec{\mathbf{s}}_i \right| {}^3\Sigma^{-e}_{1f}, v' \right\rangle = -\tfrac{1}{4} a_+ \langle v|v'\rangle$$

$$\left\langle {}^1\Pi_{1_f^e}, v \left| \sum_{i=1}^{6} \hat{a}_i \vec{\mathbf{l}}_i \cdot \vec{\mathbf{s}}_i \right| {}^3\Delta_{1_f^e}, v' \right\rangle = -\tfrac{1}{4} 2^{1/2} a_+ \langle v|v'\rangle$$

$$\langle {}^1\Pi_1 f, v | -B(\mathbf{J}^+\mathbf{L}^- + \mathbf{J}^-\mathbf{L}^+) | {}^1\Sigma^-_0 f, v' \rangle = -B_{vv'} b X^{1/2}$$

$$\langle {}^1\Pi_{1_f^e}, v | -B(\mathbf{J}^+\mathbf{L}^- + \mathbf{J}^-\mathbf{L}^+) | {}^1\Delta_{2_f^e}, v' \rangle = B_{vv'} b (X-2)^{1/2}$$

[a] The $^3\Pi$ and $^1\Pi$ states belong to the $1\pi^4 5\sigma 2\pi$ configuration. The $^3\Sigma^+$, $^3\Sigma^-$, $^3\Delta$, $^1\Sigma^-$, and $^1\Delta$ states belong to $1\pi^3 5\sigma^2 2\pi$. The one-electron orbital integrals, a_+ and b, and the vibrational integral, $B_{vv'}$, are defined as follows:

$$a_+ = \langle 1\pi | \hat{a} \mathbf{l}^+ | 5\sigma \rangle \qquad b = \langle 1\pi | \mathbf{l}^+ | 5\sigma \rangle \qquad B_{vv'} = \left\langle v \left| \frac{\hbar^2}{2\mu R^2} \right| v' \right\rangle \qquad X = J(J+1)$$

[b] Not all possible matrix elements for $^3\Delta \sim {}^3\Pi$ and $^3\Sigma^\pm \sim {}^3\Pi$ perturbations are given. The others may be obtained by applying the Wigner–Eckart theorem and using the above results to define the reduced matrix elements.

4.3.3 SPIN–SPIN PARAMETERS

The direct spin–spin parameter has been expressed in Section 2.4.5 in terms of the atomic η parameter. Its value is given in Table 4.9 for some atoms. It decreases rapidly with atomic weight, and for second-row atoms and beyond, the second-order spin–orbit contribution to the effective spin–spin constant is

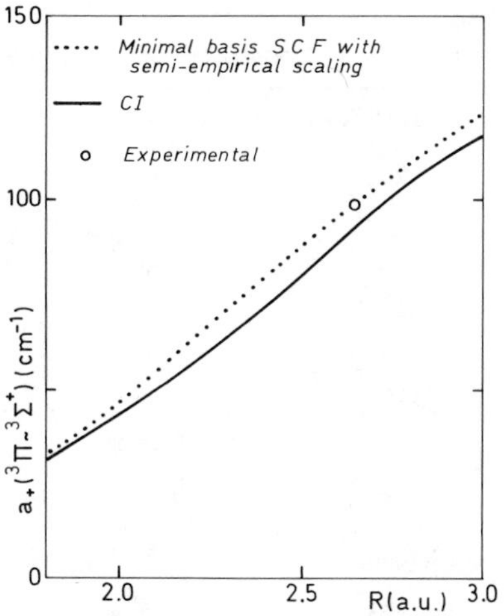

Fig. 4.13 CO $a^3\Pi_1 \sim a'^3\Sigma_1^+$ spin–orbit parameters calculated at various levels of approximation. [From Schamps (1973).]

Table 4.9
Values of the Atomic Spin–Spin Parameter η (cm^{-1})[a]

	C	N	O	F
2p	0.08	(0.14)	0.27	(0.35)
	Si	P	S	Cl
3p	(0.02)	(0.03)	(0.04)	(0.06)
	Ge	As	Se	Br
4p	(0.01)	(0.02)	(0.03)	(0.04)

[a] Values in parentheses are calculated values from Froese-Fischer (1972). Experimental values for C and O are taken from Kayama and Baird (1967).

dominant. This contribution can be estimated using the semiempirical formulas given in Section 2.4.4. For the lowest $^3\Sigma^-$ state of the π^2 configuration one obtains [Eq. (2.4.35)], for example,

$$\lambda^{\mathrm{SO}} = \frac{[\langle ^3\Sigma_0^- |\mathbf{H}^{\mathrm{SO}}|\, ^1\Sigma_0^+\rangle]^2/2}{E^0(^1\Sigma^+) - E^0(^3\Sigma_0^-)} \simeq \frac{2A(^3\Pi, \sigma\pi^3)^2}{E^0(^1\Sigma^+) - E^0(^3\Sigma_1^-)}. \tag{4.3.8}$$

However, Eq. (4.3.8) is not applicable if $A(^3\Pi, \sigma\pi^3)$ is unknown or if the $^3\Sigma_0^- \sim {}^1\Sigma^+$ spin–orbit interaction is comparable to $E^0(^1\Sigma^+) - E^0(^3\Sigma_0^-)$, the $^1\Sigma^+/^3\Sigma_0^-$ energy separation excluding the second-order spin–orbit contribution. When experimental $A(^3\Pi)$ values are not available, they may be

Table 4.10
Effective Spin–Spin Constants for $^3\Sigma^-$ States of π^2 Configurations (cm^{-1})

	$\langle ^1\Sigma^+ \vert \mathbf{H}^{SO} \vert ^3\Sigma_0^- \rangle$[a]	$[E(^1\Sigma^+) - E(^3\Sigma_0^-)]_{obs}$	$[E^0(^1\Sigma^+) - E^0(^3_0\Sigma^-)]_{depert}$	λ^{SO}(calc.)[b]	λ_e(exp)
O_2	94[c] (95)	13,120.9[c]	13113	1.35 (1.38)	1.98[c]
SO	159[c] (172)	10,469.3[c]	10448	4.84 (5.66)	5.26[c]
S_2	231[c] (240)	7961 ± 15[d]	7914	13.49 (14.6)	11.82[c]
SeO	(630)	9685.4[c]	9354	(84.1)[k]	82.95[c]
SeS	((650))	7794 ± 15[e]	7400	((112))[k]	98.4[f]
Se_2	(1060)	7957.2[g]	6933	(310)[k]	256.0[g]
TeO	(1223)	9966 ± 10[h]	8390	(343)[k]	394 ± 5[h]
TeS	((1181))	8457 ± 10[h]	6801	((388))[k]	414 ± 3[h]
TeSe	((1591))	8785 ± 5[i]	6313	((720))[k]	618 ± 2[i]
Te_2	(2122)	9600.0[j]	5650	(1296)[k]	987.4[c]

[a] Values of $\langle ^1\Sigma^+ \vert \mathbf{H}^{SO} \vert ^3\Sigma_0^- \rangle$ without parentheses are derived from the spin–orbit constant of the $\pi_u^4\pi_g\sigma_u\ ^3\Pi_u$ state. Values in single parentheses are calculated using atomic orbital mixing coefficients $C_g = 0.792$ for the π_g^* orbital of homonuclear molecules or $C_O = 0.71$ and $C_X = 0.837$ for the π' orbital of monoxide molecules, XO. Values in double parentheses are calculated using C = 0.792 for both nonoxygen atoms.

[b] Direct spin–spin contributions have been calculated to be 0.75 cm^{-1} (O_2), 0.28 cm^{-1}(SO), and 0.13 cm^{-1}(S_2) (Wayne and Colbourn, 1977).

[c] Huber and Herzberg (1979). [d] Barnes *et al.* (1979). [e] Winter *et al.* (1980). [f] MacDonald *et al.* (1982). [g] Prosser *et al.* (1982).

[h] Winter *et al.* (1982). [i] Fink, private communication. [j] Vergès *et al.* (1982).

[k] Calculated from level shifts obtained by diagonalizing the 2 × 2 $^1\Sigma \sim {}^3\Sigma_0^-$ matrix.

estimated semiempirically as described in Section 4.3.2 using the following formulas. For homonuclear molecules,

$$A(\sigma_g\pi_g\ ^3\Pi) = \tfrac{1}{2}\langle \pi_g \vert \mathbf{H}^{SO} \vert \pi_g \rangle = C_g^2 \zeta_A(p).$$

For heteronuclear AB molecules,

$$A(\sigma\pi^*\ ^3\Pi) = \tfrac{1}{2}\langle \pi^* \vert \mathbf{H}^{SO} \vert \pi^* \rangle = [C_A^2\zeta_A(p) + C_B^2\zeta_B(p)]/2.$$

When the semiempirical A value is comparable to the observed $^1\Sigma^+$, $^3\Sigma_0^-$ energy separation, an exact 2 × 2 diagonalization is required to obtain a calculated value of λ^{SO} (or to compute a semiempirical value of A from the observed λ_e value). The deperturbed energy separation is obtained from

$$[E^0(^1\Sigma^+) - E^0(^3\Sigma_0^-)]_{\text{depert}} = [E(^1\Sigma^+) - E(^3\Sigma_0^-)]_{\text{obs}} - 2\Delta$$

where

$$\Delta = [E(^3\Sigma_1^-) - E(^3\Sigma_0^-)]_{\text{obs}}.$$

This assumes that the $^3\Sigma_1^-$ level is not shifted from its deperturbed position by spin–orbit interactions with $^3\Sigma^+$ or $^3\Pi$ states from other configurations. The λ^{SO} (calculated) values in Table 4.10, obtained from

$$\lambda^{\text{SO}}(\text{calc}) = \{(\langle ^3\Sigma_0^- |\mathbf{H}^{\text{SO}}| ^1\Sigma^+\rangle)^2 + [E^0(^1\Sigma^+) - E^0(^3\Sigma_0^-)]^2_{\text{depert}}/4\}^{1/2}/2,$$

compare favorably with experimental values,

$$\lambda(\exp) = [E(^3\Sigma_1^-) - E(^3\Sigma_0^-)]_{\text{obs}}/2.$$

4.4 Order of Magnitude of Rotational Perturbation Parameters

The operator $\mathbf{L}^+ = \sum_i \mathbf{l}_i^+$, where $\mathbf{l}$ is the electronic angular momentum, $\mathbf{l}_{\text{G}}$, defined with respect to the center of mass G of the molecule,

$$\vec{\mathbf{l}}_{\text{G}} = \vec{\mathbf{r}}_{\text{G}} \times \vec{\mathbf{p}},$$

where $\mathbf{p}$ is the momentum of the electron. Because, in general, atomic orbitals are centered at the nucleus of the respective atom, one can write

$$\vec{\mathbf{l}}_{\text{G}} = \vec{\mathbf{l}}_{\text{A}} - \frac{m_{\text{B}}}{m_{\text{A}} + m_{\text{B}}}(\vec{R} \times \vec{\mathbf{p}}) = \vec{\mathbf{l}}_{\text{B}} + \frac{m_{\text{A}}}{m_{\text{A}} + m_{\text{B}}}(\vec{R} \times \vec{\mathbf{p}}) \qquad (4.4.1)$$

where m_{A} and m_{B} are the atomic masses and $\vec{R}$ is the internuclear distance vector pointing from A to B.

This expression shows clearly that $\vec{\mathbf{l}}_{\text{G}}$ depends on internuclear distance and that one-electron two-center integrals *cannot be neglected* in calculating matrix elements of $\mathbf{l}_{\text{G}}^+$ (Colbourn and Wayne, 1979). This situation is quite different from that for the off-diagonal spin–orbit parameters. The b parameters are given (Section 4.3.2) by:

$$\begin{aligned} b = \langle \pi^+ |\mathbf{l}_{\text{G}}^+| \sigma\rangle = {} & C_{\text{A}}C_{\text{A}'} \langle p\pi_{\text{A}} |\mathbf{l}_{\text{G}}^+| p\sigma_{\text{A}}\rangle + C_{\text{B}}C_{\text{B}'} \langle p\pi_{\text{B}} |\mathbf{l}_{\text{G}}^+| p\sigma_{\text{B}}\rangle \\ & + C_{\text{A}}C_{\text{B}'} \langle p\pi_{\text{A}} |\mathbf{l}_{\text{G}}^+| p\sigma_{\text{B}}\rangle + C_{\text{A}'}C_{\text{B}} \langle p\pi_{\text{B}} |\mathbf{l}_{\text{G}}^+| p\sigma_{\text{A}}\rangle \end{aligned} \qquad (4.4.2)$$

Table 4.11
Some Spin–Orbit and Orbital Perturbation Matrix Elements

	CO[a]	CS[b]	SiO[c]	SiS[d]	CS^{+}[e]
	$^1\Sigma^- \sim {}^1\Pi$	$^3\Sigma^- \sim {}^3\Pi$	$^1\Sigma^- \sim {}^1\Pi$	$^1\Sigma^- \sim {}^1\Pi$	$^2\Pi \sim {}^2\Sigma^+$
b (unitless)	0.2	0.14	0.71	0.357	−0.08
	$^3\Sigma^- \sim {}^1\Pi$	$^3\Sigma^- \sim {}^3\Pi$	$^3\Sigma^- \sim {}^1\Pi$	$^3\Sigma^- \sim {}^1\Pi$	$^2\Pi \sim {}^2\Sigma^+$
a_+ (cm^{-1})	100	312	100	362	290

[a] Field *et al.* (1972).
[b] Cossart *et al.* (1977).
[c] Field *et al.* (1976).
[d] Harris *et al.* (1982).
[e] Gauyacq and Horani (1978).

where $\mathbf{l}_G^+$ must be replaced by its form given in Eq. (4.4.1). Consequently, a semiempirical evaluation of b is never valid. As distinct from $\hat{a}\mathbf{l}^+$, $\mathbf{l}_G^+$ does not act only in the neighborhood of each nucleus. Only if the σ and π molecular orbitals are well-represented by *one p* atomic orbital centered at the center of mass, (i.e., if there is no mixing with orbitals with $l_G > 1$) is it possible to say that $|b| \leq 2^{1/2}$. Since, for AH hydride molecules, the center of mass is located very near the heavy atom nucleus, $\vec{\mathbf{l}}_A \simeq \vec{\mathbf{l}}_G$ and semiempirical relationships between b and l_A are approximately valid. This, in part, explains the success of the pure precession approximation (Section 4.5) for AH molecules.

It has been pointed out (Robbe and Schamps, 1976) that there is no general reason that the $\hat{a}\mathbf{l}^+$ and $\mathbf{l}^+$ matrix elements should be simply proportional to each other. For $\hat{a}\mathbf{l}^+$, the contribution from the atom with the larger atomic spin–orbit parameter is dominant; for $\mathbf{l}^+$, the p orbitals of both atoms play equal roles. Table 4.11 compares some values for matrix elements of these two operators. For a_+, the values of CO and SiO are similar, as are the values for CS and SiS. For b, the strong R-dependence of this parameter prohibits any simple predictions. Even the sign changes when one passes from CS to CS$^+$.

4.5 Pure Precession Approximation

The hypothesis of pure precession (Van Vleck, 1929) is often used in the estimation of Λ-doubling constants. These constants, in the o, p, q notation suggested by Mulliken and Christy (1931), are introduced into the effective Hamiltonian matrix for Π-states in order to represent the J- and Ω-dependence of the splitting between the e- and f-symmetry Λ-components caused by the interaction between the Π-state and energetically remote Σ-states (Zare *et al.*, 1973; Horani *et al.*, 1967; Brown and Merer, 1979).

This discussion is intended to distinguish the levels of approximation often hidden behind the name "pure precession." For clarity, only $^2\Sigma \sim {}^2\Pi$ interactions will be discussed. The considerably more complicated $^3\Sigma \sim {}^3\Pi$ case is treated by Brown and Merer (1979).

The o, p, and q constants may be defined in a form that is semiempirically evaluable by second-order perturbation theory (see the discussion of the Van Vleck transformation in Section 3.2). Using the notation suggested by Rostas *et al.* (1974) and Merer *et al.* (1975),

$$o_v^{\Pi}(^2\Sigma^s)^{\dagger} = \tfrac{1}{4}\sum_{^2\Sigma,v'} \frac{(-1)^s |\langle ^2\Pi, v|\sum_i \hat{a}_i \mathbf{l}_i^+ \mathbf{s}_i^- |^2\Sigma^s, v'\rangle|^2}{E_{\Pi,v} - E_{\Sigma,v'}} \tag{4.5.1a}$$

$$p_v^{\Pi}(^2\Sigma^s) = (-1)^s \langle ^2\Pi, v|\textstyle\sum_i a_i \mathbf{l}_i^+ \mathbf{s}_i^- |^2\Sigma^s, v'\rangle \times 2\sum_{^2\Sigma,v'} \frac{\left\langle ^2\Sigma^s, v'\left|\frac{\hbar^2}{2\mu R^2}\sum_i \mathbf{l}_i^-\right|^2\Pi, v\right\rangle}{E_{\Pi,v} - E_{\Sigma,v'}} \tag{4.5.2a}$$

$$q_v^{\Pi}(^2\Sigma^s) = 2\sum_{^2\Sigma,v'} \frac{(-1)^s |\langle ^2\Pi, v|(\hbar^2/2\mu R^2)\sum_i \mathbf{l}_i^+|^2\Sigma^s, v'\rangle|^2}{E_{\Pi,v} - E_{\Sigma,v'}} \tag{4.5.3a}$$

where s is zero for $^2\Sigma^+$ states and 1 for $^2\Sigma^-$ states and the operators are understood to act on unsymmetrized, signed-Ω basis functions. Since remote levels of both $^2\Sigma^+$ and $^2\Sigma^-$ states contribute to the Λ-doubling of a given $^2\Pi$ level, the o, p, and q parameters may be partitioned into parts arising from $^2\Sigma^+$ and $^2\Sigma^-$ states, as follows:

$$o_v^{\Pi} = o_v^{\Pi}(^2\Sigma^+) + o_v^{\Pi}(^2\Sigma^-) \tag{4.5.1b}$$

$$p_v^{\Pi} = p_v^{\Pi}(^2\Sigma^+) + p_v^{\Pi}(^2\Sigma^-) \tag{4.5.2b}$$

$$q_v^{\Pi} = q_v^{\Pi}(^2\Sigma^+) + q_v^{\Pi}(^2\Sigma^-). \tag{4.5.3b}$$

In the *unique perturber* approximation (Zare *et al.*, 1973), only one $^2\Sigma$ electronic state is considered to be responsible for the Λ-doubling of the $^2\Pi$ state. If the interacting $^2\Pi$ and $^2\Sigma$ states belong to the π^1 and σ^1 configurations, then the matrix elements of the exact many-electron wavefunctions may be replaced by the one-electron matrix elements (see Section 2.5.4),

$$a_+ = \langle \pi^+ |\hat{a}\mathbf{l}^+|\sigma\rangle$$

$$b = \langle \pi^+ |\mathbf{l}^+|\sigma\rangle$$

† $o_v^{\Pi}(^2\Sigma^s)$ is included even though it makes no contribution to the Λ-doubling in $^2\Pi$ states.

and

$$\left\langle {}^2\Pi, v \left| \sum_i \hat{a} \mathbf{l}_i^+ \mathbf{s}_i^- \right| {}^2\Sigma^+, v' \right\rangle = a_+(\bar{R}_{vv'})\langle v|v'\rangle \tag{4.5.4}$$

$$\left\langle {}^2\Pi, v \left| (\hbar^2/2\mu R^2) \sum_i \mathbf{l}_i^+ \right| {}^2\Sigma^+, v' \right\rangle = b(\bar{R}_{vv'})B_{vv'}, \tag{4.5.5}$$

where a_+ and b are dependent on the R-centroid which is v'-dependent, unlike the situation for interactions between all near-degenerate pairs of vibrational levels belonging to two potentials that cross at R_C. Inserting Eqs. (4.5.4) and (4.5.5) into (4.5.1a) (4.5.2a), and (4.5.3a),

$$o_v^\Pi = \frac{1}{4} \sum_{v'} \frac{[a_+(\bar{R}_{vv'})\langle v|v'\rangle]^2}{E_{\Pi,v} - E_{\Sigma,v'}} \tag{4.5.1c}$$

$$p_v^\Pi = 2 \sum_{v'} \frac{a_+(\bar{R}_{vv'}) b(\bar{R}_{vv'})\langle v|v'\rangle B_{vv'}}{E_{\Pi,v} - E_{\Sigma,v'}} \tag{4.5.2c}$$

$$q_v^\Pi = 2 \sum_{v'} \frac{[b(\bar{R}_{vv'})B_{vv'}]^2}{E_{\Pi,v} - E_{\Sigma,v'}}. \tag{4.5.3c}$$

If one makes the generally very bad approximation that a_+ and b are independent of $\bar{R}_{vv'}$ (thus defining $\bar{a}_+$, $\bar{b}$), then the perturbation summations

$$o_v^\Pi = \tfrac{1}{4} \bar{a}_+^2 \sum_{v'} \frac{[\langle v|v'\rangle]^2}{E_{\Pi,v} - E_{\Sigma,v'}} \tag{4.5.1d}$$

$$p_v^\Pi = 2\bar{a}_+ \bar{b} \sum_{v'} \frac{\langle v|v'\rangle B_{vv'}}{E_{\Pi,v} - E_{\Sigma,v'}} \tag{4.5.2d}$$

$$q_v^\Pi = 2\bar{b}^2 \sum_{v'} \frac{(B_{vv'})^2}{E_{\Pi,v} - E_{\Sigma,v'}} \tag{4.5.3d}$$

may be evaluated numerically. This approach should account for the v-dependence (over a small range of v, especially at low v, see Wicke *et al.*, 1972) but not the magnitudes of o_v^Π, p_v^Π, and q_v^Π.

In the special case when the unique interacting pair of ${}^2\Pi \sim {}^2\Sigma^+$ states have identical potential energy curves, then the v' summations reduce to a single $v = v'$ term and

$$o_v^\Pi = \tfrac{1}{4} \frac{[a_+(\bar{R}_{vv})]^2}{\Delta E_{\Pi\Sigma}} \tag{4.5.1e}$$

$$p_v^\Pi = 2 \frac{a_+(\bar{R}_{vv}) b(\bar{R}_{vv}) B_v}{\Delta E_{\Pi\Sigma}} \tag{4.5.2e}$$

$$q_v^\Pi = 2B_v^2 \frac{[b(\bar{R}_{vv})]^2}{\Delta E_{\Pi\Sigma}} \tag{4.5.3e}$$

where

$$\Delta E_{\Pi\Sigma} = E_{\Pi,v} - E_{\Sigma,v}.$$

At this level of approximation, a convenient estimate of o_v^{Π} is obtained from

$$o \simeq p^2/4q.$$

Even in this special case, the $\bar{R}$-dependence of a_+ and b cannot be ignored.

An important and frequently encountered result, often mistakenly taken as evidence for "pure precession," is that, in the *unique perturber*, *identical potential curve* limit, the effective spin–rotation constant of the $^2\Sigma^+$ state, γ_v, is equal to p_v^{Π}. This result is a direct consequence of the second-order perturbation theoretical definition of the contribution of a $^2\Pi$ state to the spin–rotation splitting in a $^2\Sigma$ state (see Section 2.5.4):

$$\begin{aligned}\gamma_v^{\text{eff}} &= -p_v^{\Sigma^+}(^2\Pi)\\ &= 2\sum_{^2\Pi,v'} \frac{\langle ^2\Sigma^+, v|\sum_i \hat{a}_i \mathbf{l}_i^- \mathbf{s}_i^+|^2\Pi, v'\rangle\langle ^2\Pi, v'|(\hbar^2/2\mu R^2)\sum_i \mathbf{l}_i^+|^2\Sigma^+, v\rangle}{E_{\Pi,v} - E_{\Sigma,v'}}\end{aligned} \quad (4.5.6)$$

which, in the unique perturber, identical potential limit reduces to

$$\gamma_v^{\text{eff}} = -p_v^{\Sigma^+}(^2\Pi) = +p_v^{\Pi}(^2\Sigma). \quad (4.5.7)$$

Finally, the *pure precession approximation* requires, in addition to the unique perturber, identical potential assumptions, that the interacting $^2\Pi$ and $^2\Sigma$ states are each well described by a single configuration, that these configurations are identical except for a single spin-orbital, and that this spin-orbital is a pure atomic $|nl\lambda\rangle$ orbital so that

$$\mathbf{l}^{\pm}|nl\lambda\rangle = [l(l+1) - \lambda(\lambda \pm 1)]^{1/2}|nl\lambda \pm 1\rangle. \quad (4.5.8)$$

Then, for $l = 1$ (a p-complex),

$$a_+ = \langle \pi^+|\hat{a}\mathbf{l}^+|\sigma\rangle = \zeta(np)2^{1/2} = A_v^{\Pi}2^{1/2} \quad (4.5.9)$$

$$b = \langle \pi^+|\mathbf{l}^+|\sigma\rangle = 2^{1/2} \quad (4.5.10)$$

and, for the case of "*simple* pure precession" where all orbitals except the unique π/σ pair are completely full,

$$o_v^{\Pi}(^2\Sigma^+) = \tfrac{1}{2}A_v^2/\Delta E_{\Pi\Sigma} \quad (4.5.1\text{f})$$

$$p_v^{\Pi}(^2\Sigma^+) = 4A_vB_v/\Delta E_{\Pi\Sigma} \quad (4.5.2\text{f})$$

$$q_v^{\Pi}(^2\Sigma^+) = 4B_v^2/\Delta E_{\Pi\Sigma} \quad (4.5.3\text{f})$$

$$p_v^{\Pi}/q_v^{\Pi} = A_v/B_v \quad (4.5.11)$$

Similar equations for o, p, and q may be derived for "generalized pure precession" situations involving two or more open shells. The validity of these pure precession equations depends on many assumptions, yet experimentalists are very quick to use them to "explain" the sign or even the magnitude of observed Λ-doubling and spin–rotation constants or to use these constants to "infer" l_{eff} values for valence-shell σ and π orbitals.

For the OH radical, the values of p and q for the $X^2\Pi(\sigma^2\pi^3)$ ground state can be attributed to a unique perturber interaction with the $A^2\Sigma^+(\sigma\pi^4)$ state. The pure precession approximation simply ignores the contribution of the $\sigma 1s_H$ atomic orbital to the $p\sigma$ molecular orbital. For all hydrides, the $\sigma 1s_H$ orbital makes a negligible contribution to $\mathbf{H}^{SO}$ and $\mathbf{BL}^+$ matrix elements. For OH, the H-atom contributions to a_+ and b are 0.04 and 1%, respectively (Hinkley *et al.*, 1972).

For the CH radical, the simple Eqs. (4.5.4) and (4.5.5) are not valid for the interaction of $X^2\Pi(\sigma^2\pi)$ with the $B^2\Sigma^-(\sigma\pi^2)$ and $C^2\Sigma^+(\sigma\pi^2)$ states. Although the $B^2\Sigma^-$ and $C^2\Sigma^+$ states are well represented by the single $\sigma\pi^2$ configuration, both states must be written as sums of Slater determinants,

$$|^2\Sigma^-_{1/2}\rangle = 6^{-1/2}[2|\pi^+\alpha\pi^-\alpha\sigma\beta| - |\pi^+\alpha\pi^-\beta\sigma\alpha| - |\pi^+\beta\pi^-\alpha\sigma\alpha|] \quad (4.5.12)$$

$$|^2\Sigma^+_{1/2}\rangle = 2^{-1/2}[|\pi^+\alpha\pi^-\beta\sigma\alpha| - |\pi^+\beta\pi^-\alpha\sigma\alpha|], \quad (4.5.13)$$

and the matrix elements become (Hinkley *et al.*, 1972)

$$\left\langle {}^2\Pi_{1/2}, v \left| \sum_i \hat{a}_i \mathbf{l}_i^+ \mathbf{s}_i^- \right| {}^2\Sigma^-, v' \right\rangle = +(6)^{-1/2} a_+ \langle v|v'\rangle \quad (4.5.14)$$

$$\left\langle {}^2\Pi_{1/2}, v \left| \sum_i \hat{a}_i \mathbf{l}_i^+ \mathbf{s}_i^- \right| {}^2\Sigma^+, v' \right\rangle = +(2)^{-1/2} a_+ \langle v|v'\rangle \quad (4.5.15)$$

$$\left\langle {}^2\Pi_{3/2}, v \left| \mathbf{B} \sum_i \mathbf{l}_i^+ \right| {}^2\Sigma^-_{1/2}, v' \right\rangle = +(\tfrac{3}{2})^{1/2} b B_{vv'} \quad (4.5.16)$$

$$\left\langle {}^2\Pi_{3/2}, v \left| \mathbf{B} \sum_i \mathbf{l}_i^+ \right| {}^2\Sigma^+_{1/2}, v' \right\rangle = -(2)^{-1/2} b B_{vv'}. \quad (4.5.17)$$

This generalization to cases where the interacting Π and Σ states are derived from more complex configurations than π^1 and σ^1 is the basis for *generalized pure precession*. The p and q parameters for CH $X^2\Pi$ are

$$p_v^\Pi(^2\Sigma^-) = -B_v b a_+/\Delta E_{\Pi\Sigma} \quad (4.5.18a)$$

$$p_v^\Pi(^2\Sigma^+) = -B_v b a_+/\Delta E_{\Pi\Sigma} \quad (4.5.18b)$$

$$q_v^\Pi(^2\Sigma^-) = -3B_v^2 b^2/\Delta E_{\Pi\Sigma} \quad (4.5.18c)$$

$$q_v^\Pi(^2\Sigma^+) = B_v^2 b^2/\Delta E_{\Pi\Sigma}. \quad (4.5.18d)$$

Predictions based on generalized pure precession depend on the same assumptions as ordinary pure precession. The idea of generalized pure precession is especially useful for treating interactions between Rydberg states built upon non-$^1\Sigma^+$ ion cores.

For SiH, the repulsive $^2\Sigma^-$ state makes the dominant contribution to the Λ-doubling of the $X^2\Pi(\sigma^2\pi)$ ground state (Cooper and Richards, 1981). In order to use the Eqs. (4.5.1a)–(4.5.3a) definitions of p^Π and q^Π and the Eqs. (4.5.14)–(4.5.17) matrix elements, it is necessary to make assumptions about the R-dependence of a_+ and b and to replace the summation over vibrational levels of $^2\Sigma^-$ by an integration over the vibrational continuum [see Eq. (6.3.8)].

In the case of highly polar, heteronuclear, nonhydride diatomic molecules, if the interacting states belong to configurations differing by a single orbital that is highly localized on the same atom, then the p^Π, q^Π, and γ^Σ constants are likely to have values close to the pure precession predictions. However, this must not be taken as evidence that the interacting states are components of an l-complex. For example, the $A^2\Pi(\pi^1)$ and $B^2\Sigma^+(\sigma^1)$ states of the alkaline earth monohalides certainly belong to the unique perturber, identical potential limit and satisfy the requirement that

$$p^\Pi = \gamma^\Sigma.$$

Furthermore, using

$$p_v^\Pi = 2A_v B_v l_{\text{eff}}(l_{\text{eff}} + 1)/\Delta E_{\Pi\Sigma} \qquad (4.5.19)$$

to define l_{eff}, one finds $l_{\text{eff}} = 1.05$, 1.12, 1.23, and 1.38 for CaF, CaCl, CaBr, and CaI, respectively (Bernath, 1980). This suggests that the $A^2\Pi$ and $B^2\Sigma^+$ states of the CaX molecules are of predominantly Ca^+ $4p$ character, but with a small admixture of Ca^+ $3d$. This use of Eq. (4.5.19) to define l_{eff} is unsound because it assumes that the amount of $p \sim d$ mixing is the same for both π and σ orbitals. In fact, the $p \sim d$ mixing is much more extensive in σ than π orbitals. This $p \sim d$ mixing is the mechanism by which the metal-centered orbital polarizes to avoid the negatively charged X^- ion and is much more important for $p\sigma$ than $p\pi$.

It is dangerous to use the satisfaction of a pure precession prediction as evidence for the validity of the numerous pure precession assumptions, especially when there is good reason to suspect that the assumptions are invalid (HCl^+; Brown and Watson, 1977). Likewise, it is risky to use the magnitudes or signs of second-order parameters, such as p, q, or γ, to infer configurational parentage, molecular orbital composition, or whether a multiplet $\Lambda > 0$ state is regular or inverted (BeF; Cooper *et al.*, 1981).

For valence states of nonhydride molecules, there is no reason to expect that the generalized pure precession approximation should be valid. In contrast, Rydberg orbitals, because of their large size and nonbonding, single-center, near-spherical, atomic-like character, are almost invariably well-described by the pure precession picture in terms of nl-complexes.

4.6 *R*-Dependence of the Spin Interaction Parameters

Spin–orbit, spin–spin, and spin–rotation constants are found experimentally to depend on v and J. This reflects an implicit R-dependence of the electronic matrix elements. Such effects may be calculated *ab initio* and then compared against the experimentally determined R-variation inferred from the v, J dependence of spin constants. The most sophisticated $f(v, J) \leftrightarrow F(R)$ inversion methods are those of Watson (1979), Coxon (1975), and Bessis *et al.* (1984).

There is no reason to expect that diagonal and off-diagonal spin constants will vary with R in the same way, except in some special cases. Consider first the R-dependence of the diagonal spin–orbit constants. This variation originates from two sources:

1. variation of a specific open-shell $\lambda \neq 0$ molecular orbital with R, and
2. variation of the dominant configuration with R.

In Fig. 4.14, the calculated R-dependence of the spin–orbit constant of the

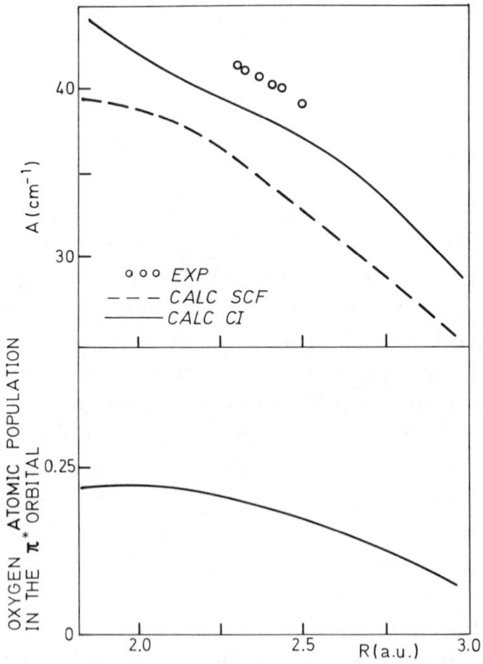

Fig. 4.14 R-variation of the spin–orbit constant for the CO $a^3\Pi$ state. Experimental values from Field *et al.* (1972), calculated values from Hall *et al.* (1973).

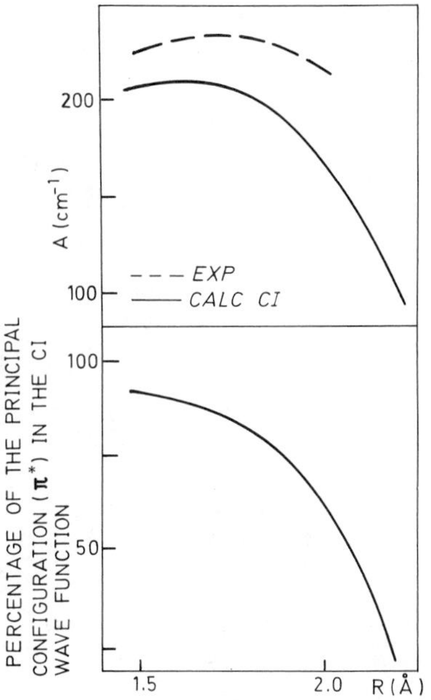

Fig. 4.15 *R*-variation of the spin–orbit constant for the PO $X^2\Pi$ state. Experimental values from Verma and Singhal (1975), calculated values from Roche and Lefebvre-Brion (1973).

$a^3\Pi$ state of CO, $A(a^3\Pi)$, is shown. This constant, in the single-configuration approximation, is related to the one-electron matrix element, $\langle\pi^*|\hat{a}\mathbf{l}_z|\pi^*\rangle = 2A(a^3\Pi)$. The main contribution to the spin–orbit constant comes from the oxygen atom, even though $C_C^2 > C_O^2$ [compare $\zeta_C(2p) = 30\ \text{cm}^{-1}$ to $\zeta_O(2p) = 150\ \text{cm}^{-1}$] and the value of A varies with R roughly as the fractional oxygen $2p$ atomic population in the π^* orbital. At large internuclear distances, the π^* orbital becomes a carbon $2p\pi$ atomic orbital. For CO $a^3\Pi$ there is no significant difference between the values of $A(a^3\Pi)$ calculated using single-configuration (SCF) or many-configuration (CI) wavefunctions.

In Fig. 4.15, the calculated variation of the spin–orbit constant of the $X^2\Pi$ state of PO is shown to be in good agreement with experiment. In this case, the dominant configuration at equilibrium internuclear distance is π^* where the π^* molecular orbital has primarily $P(3p)$ character. This character does not change appreciably with R. In contrast, the percentage of the π^* configuration in the CI wavefunction diminishes very rapidly with R to allow PO $X^2\Pi$ to dissociate into the correct separated atom states, and its R-dependence is mirrored by that of the $A(X^2\Pi)$ constant.

Finally, in a third example, both types of A versus R variation occur, the first at small internuclear separation, the second at large separation. Figure 4.16 illustrates the variation of A for the OH $X^2\Pi$ state (Coxon and Foster, 1982). Near R_e, the dominant configuration is $\sigma^2\pi^3$. The magnitude of the spin–orbit

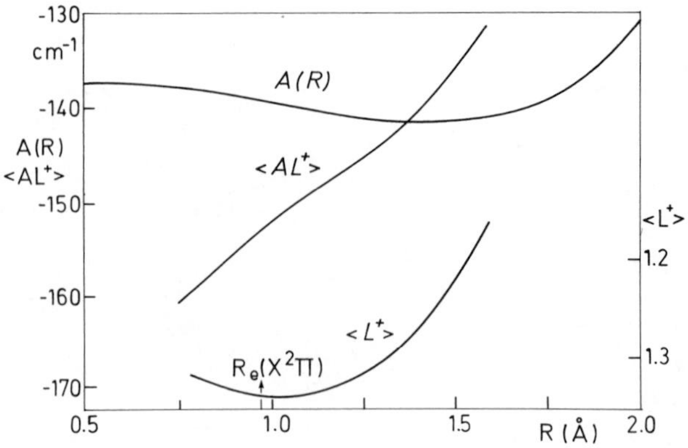

Fig. 4.16 Diagonal $[A(R)]$ and off-diagonal spin-orbit ($\langle AL^+\rangle$) and orbital ($\langle L^+\rangle$) perturbation parameters for the OH $X^2\Pi$ state (from Coxon and Foster, 1982) and the OD $X^2\Pi \sim A^2\Sigma^+$ interaction. (From Coxon and Hammersley, 1975).

constant associated with the π orbital, $A(X^2\Pi) = -a(\pi)$, increases slightly with R from $a(\pi) = \zeta_{O^-}(2p) = 120\ \text{cm}^{-1}$, the value of the constant for the oxygen negative ion, to a $(\pi) = \zeta_O(2p) = 150\ \text{cm}^{-1}$, that of the neutral oxygen atom. Near the $O(^3P) + H(^2S)$ dissociation limit, the spin–orbit splitting of OH $X^2\Pi$ should correspond to the $^3P_1$–$^3P_2$ splitting of the oxygen-atom ^3P state. Consequently, $|A(X^2\Pi)|$ decreases at large R, going to the asymptotic value $\frac{2}{3}\zeta_O = 100\ \text{cm}^{-1}$ (Langhoff *et al.*, 1982).

Except in the previous example, only the variation of A in the neighborhood of R_e has been considered. The correlation of molecular spin–orbit splittings to those of atoms at infinite internuclear distance is very difficult to establish. It is necessary to take into account not only the configuration mixing due to electronic correlation but also mixing due to off-diagonal matrix elements of the spin–orbit operator between near-degenerate states of identical Ω. This corresponds to working in a (J_1, J_2) coupling scheme (Herzberg, 1950, p. 319) where Ω is the sum $|M_{J_1} + M_{J_2}|$ for the two atoms (see Fig. 4.17).

Assuming that an electronic property varies linearly with both R (near R_e) and v (near $v = 0$), approximate formulas relating the R- and v-dependence can be deduced. For example, if one assumes

$$A(v) = A_e - \alpha_A(v + \tfrac{1}{2}) = \langle v|A(R)|v\rangle, \tag{4.6.1}$$

where

$$A(R) = A_e + \left(\frac{dA}{dR}\right)_{R=R_e}(R - R_e), \tag{4.6.2}$$

the $\Delta v = 0$ matrix element of $(R - R_e)^m$ between vibrational functions

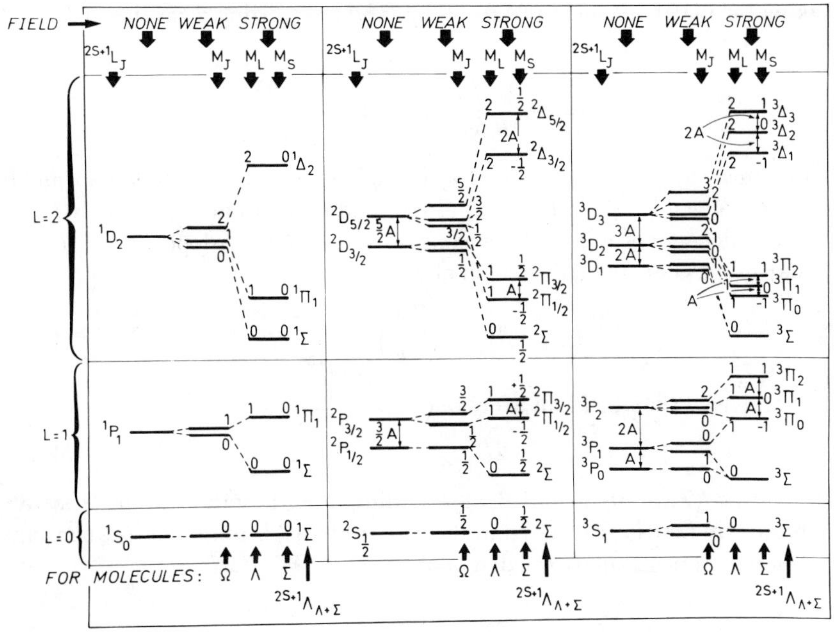

Fig. 4.17 Correlation diagram showing the effect of an electric field on atomic L–S terms (from Jevons, 1932). This electric field could be externally applied, splitting the atomic terms into J, M_J (weak field) or M_L, M_S (strong field) components, or it could be the cylindrically symmetric molecular field associated with a nearby atom, causing the atomic terms to be split into case (c) J_a, Ω (weak field) or case (a) Λ, S (strong field) terms. Two effects are illustrated: the correlation of atomic spin–orbit splittings [shown in terms of $A \equiv \zeta(L, S)$] with molecular spin–orbit intervals (ΛA), and the splitting of atomic L–S–J terms into molecular Λ–S–Ω states. For example, $\mathrm{Ar}(^1S) + \mathrm{Na}(^2P)$ correlates with $^2\Pi$ and $^2\Sigma^+$ states of NaAr. This sort of diagram provides a framework for understanding the ligand field splitting of an $\mathrm{M}^+(d^N$ or $f^N)$ L–S core into MX molecular states by $\mathrm{X}^-(^1S)$ (see Section 2.2.1). The Σ^+/Σ^- symmetry of molecular states depends on whether $L + \Sigma_{i=1}^N l_i$ is even or odd (Herzberg, 1950, p. 318). The correlations between the ^{3}D $|JM_J\rangle$ and $^3\Pi\ |\Omega\rangle$ components are as follows: $|30\rangle \leftrightarrow |0^\mp\rangle$, $|22\rangle \leftrightarrow |2\rangle$, $|21\rangle \leftrightarrow |1\rangle$, $|20\rangle \leftrightarrow |0^\pm\rangle$ (the $+/-$ symmetry of $\Omega = 0$ states is determined by whether $J + \Sigma_i l_i$ is even/odd).

belonging to the same electronic state can be expressed in terms of Dunham coefficients from the expansion of $V(R)$ in powers through the cubic term of $(R - R_e)/R_e$ (for example, see Table I of James, 1964):

$$\langle v|(R - R_e)|v\rangle = -\frac{3a_1 R_e B_e}{\omega_e}(v + \tfrac{1}{2}) \tag{4.6.3}$$

and

$$a_1 = -\frac{\alpha_e \omega_e}{6B_e^2} - 1. \tag{4.6.4}$$

Taking the matrix element of Eq. (4.6.2) and recognizing α_A as the coefficient of $-(v+\frac{1}{2})$,

$$\alpha_A = -\frac{R_e}{2B_e}\left(\frac{dA}{dR}\right)_{R=R_e}\left(\alpha_e + \frac{6B_e^2}{\omega_e}\right). \tag{4.6.5}$$

If the linear approximation for $A(v)$ is inadequate, more complete formulas have been given by De Santis *et al.* (1973) and by Zaidi and Verma (1975).

The R-dependence of A also accounts for the slight J-dependence of $A(v)$ similar to the centrifugal distortion of $B(v)$,

$$A_D \equiv 2A_J = 4\left(\frac{dA}{dR}\right)_{R=R_e}\frac{B_e^2}{\omega_e^2}R_e. \tag{4.6.6}$$

Alternatively (Veseth, 1970; Merer, 1972),

$$A_D = -2D_e\alpha_A(\alpha_e + 6B_e^2/\omega_e)^{-1}. \tag{4.6.7}$$

The recipe (Zare *et al.*, 1973) for adding the A_D term to the effective Hamiltonian matrix [expressed in the case (a) basis] is to replace the constant B_v where it appears along the diagonal ($\langle J\Omega\Lambda\Sigma|\mathbf{H}|J\Omega\Lambda\Sigma\rangle$) by

$$B_v + A_D\Lambda\Sigma$$

and by

$$B_v + A_D\Lambda(\Omega - \Lambda \pm \tfrac{1}{2})$$

where B_v appears in the $\langle J\Omega \pm 1\Lambda\Sigma \pm 1|\mathbf{H}|J\Omega\Lambda\Sigma\rangle$ location. Recall that, for $^2\Pi$ states, the A_D and γ^{Π} (spin–rotation) parameters (see Section 2.4.3) are statistically correlated and cannot be simultaneously determined. The problem of separating A_D from γ is discussed by Brown and Watson (1977).

The R-dependence of matrix elements of the $\mathbf{l}_{zi}\mathbf{s}_{zi}$ part of $\mathbf{H}^{SO}$ can be related to the R-variation of diagonal spin–orbit constants. For example, if Eq. (4.6.2) is used, and by analogy with Eq. (2.4.14),

$$\begin{aligned}&\langle{}^1\Pi_1, v=0|\mathbf{H}^{SO}|{}^3\Pi_1, v=1\rangle\\ &\approx A_e\langle v=0|v=1\rangle + \left(\frac{dA}{dR}\right)_{R=R_e}\langle v=0|(R-R_e)|v=1\rangle.\end{aligned} \tag{4.6.8}$$

The first term is zero if the two potential curves of the isoconfigurational $^1\Pi_1$ and $^3\Pi_1$ states are assumed to be similar (cf. Eq. 2.4.14). The second term can be evaluated from Eq. (4.6.5) if the α_A constant is known from the $A(v)$ variation. An example of such a nonzero interaction matrix element (Eq. 4.6.8) is given for DCl (Huber and Alberti, 1983). Also see Section 2.4.2.

The variation of orbitally or configurationally off-diagonal spin–orbit perturbation parameters cannot be related to that of a diagonal spin–orbit constant because the off-diagonal matrix element is taken between two different orbitals which do not necessarily have the same R-dependence. This is illustrated by Figure 4.16 where the variation with R of the $\langle A^2\Sigma^+|\mathbf{H}^{SO}|X^2\Pi\rangle$ matrix element for OD is given. In the single configuration approximation, the $A^2\Sigma^+$ state belongs to $\sigma\pi^4$. The electronic part of the interaction matrix element between this state and the $X^2\Pi_i(\sigma^2\pi^3)$ state is then equal to $\langle\pi^+|\hat{a}\mathbf{l}^+|\sigma\rangle$ and its absolute value decreases with R (and v) because, in order to dissociate correctly, the σ orbital becomes a $1s_H$ orbital at large R. This R-variation is in the opposite sense to that of $A(X^2\Pi)$.

In practice, the R-dependence of off-diagonal constants is less important than that of diagonal constants. In Section 2.3.1 it was shown that if the potential curves of the interacting states cross (at R_C), the value of the electronic part of the off-diagonal interaction between *near-degenerate* vibrational levels is equal to the value computed at R_C,

$$H_{1,v_i;\,2,v_j} = H^e_{12}(R_C)\langle v_i|v_j\rangle.$$

Thus, even though H^e_{12} may vary rapidly with R, perturbations will not be very sensitive to this variation.

The R- or v-dependence of second-order constants (o, p, q, γ, etc.) is more important than that of the off-diagonal perturbation parameters because, in their definitions as sums over the interactions with *energetically remote levels* [Eqs. (4.5.1a)–(4.5.3a)], the off-diagonal matrix elements are taken between nondegenerate vibrational levels. Thus the R-centroid, $\bar{R}_{vv'}$, will vary with v and v' and will not be equal to R_C. Since a_+ and b are not independent of $\bar{R}_{vv'}$ and $\bar{R}_{vv'}$ is not independent of v' at fixed v, it is not permissible to remove the electronic factor from the summation over remote perturbers. Furthermore, the denominator of each term in the perturbation summation depends on an energy difference. This can cause a strong v, J-dependence of the second-order constants, especially when the vibrational levels of the state under consideration are near the energy of the curve crossing, E_C. It is frequently necessary to introduce centrifugal distortion correction terms for p and q, p_D and q_D, similar to the A_D term (for OH, see Coxon and Foster, 1982).

4.7 Beyond the Single-Configuration Approximation

Until this point, the single-configuration approximation for each electronic state has been assumed.† The $C_2\, b^3\Sigma_g^- \sim X^1\Sigma_g^+$ interaction is one example where it is necessary to represent one of the interacting states by a mixture of

† This assumption is central to the discussion of perturbation phenomena from a diabatic point of view. In contrast, the adiabatic approach is intrinsically a mixed-configuration picture.

two configurations. Another example involving predissociation is discussed in Section 6.9.1.

For many years, the ground state of the C_2 molecule had been assumed to be a $^3\Pi_u$ state, in part because the relative energies of the manifolds of singlet and triplet states were unknown. The discovery of perturbations between the $X^1\Sigma_g^+$ and $b^3\Sigma_g^-$ states (Ballik and Ramsay, 1963) allowed the relative energies of singlet and triplet states to be determined.

The C_2 $b^3\Sigma_g^-$ state belongs predominantly to the $1\pi_u^2\ 3\sigma_g^2$ configuration, which differs by more than one orbital from the $1\pi_u^4\ 3\sigma_g^0$ dominant configuration of the $X^1\Sigma_g^+$ state. Nevertheless, the matrix element $\langle {}^1\Sigma_g^+ |\mathbf{H}^{SO}| {}^3\Sigma_g^- \rangle$ is nonzero if $X^1\Sigma_g^+$ contains a significant admixture of the $1\pi_u^2 3\sigma_g^2$ configuration. The ${}^1\Sigma_g^+(\pi^2) \sim {}^3\Sigma_g^-(\pi^2)$ interaction matrix element is given by Eq. (2.4.16). As the mixing coefficient of the $1\pi_u^2\ 3\sigma_g^2$ configuration in the $X^1\Sigma_g^+$ wavefunction is about 0.16, the electronic matrix element is

$$0.16\ \zeta_C(2p) = 4.6\ \text{cm}^{-1},$$

which compares quite favorably with the experimental value, 4.8 cm^{-1} (Langhoff *et al.*, 1977).

4.8 Identification and Location of Metastable States by Perturbation Effects

The location of metastable states is often made possible by observation of intercombination systems. For example, the intensity of the Cameron system in $CO(a^3\Pi - X^1\Sigma^+)$ is borrowed from singlet transitions through spin–orbit mixing (see Section 5.4). When intercombination transitions are not observed, it is often possible to locate the states of different multiplicity from that of the ground state by observation of spin–orbit $\Delta S \neq 0$ perturbations.

The example of the C_2 molecule given in the previous section is a very important one. A similar result has been demonstrated in the case of the MgO molecule (Ikeda *et al.*, 1977). Thanks to perturbations in the $v \geq 3$ levels of the $X^1\Sigma^+$ state by the $a^3\Pi$ state, the electronic ground state has been definitely proven to be $X^1\Sigma^+$ and the $a^3\Pi$ has been found to lie above the $X^1\Sigma^+$ state at $T_e = 2623$ cm^{-1}.

It was possible to locate the lowest singlet states of OH^+ and OD^+, $a^1\Delta$ and $b^1\Sigma^+$, from their perturbations of the $A^3\Pi_i - X^3\Sigma^-$ system (Merer *et al.*, 1975). The X, a, and b states all belong to the $\sigma^2\pi^2$ configuration, whereas $A^3\Pi_i$ belongs to $\sigma\pi^3$. $^3\Pi$ states are both sensitive detectors of perturbers as well as good discriminants of perturber symmetry. The vibrational numbering of the

$b^1\Sigma^+$ state was proven by the absence of a perturbation in $A^3\Pi$ ($v = 0$) of OH^+ and OD^+.

The NO spectrum is one of the most completely known among all diatomic molecules. However, it is only recently (Miescher, 1980) that it has been possible to locate the quartet states precisely with respect to the well-known doublet states. Small perturbations between the $B^2\Pi$ state ($v = 1$) and the $b^4\Sigma^-$ state ($v = 0$) had been predicted by Miescher on the basis of *ab initio* calculations (reported by Field *et al.*, 1975) to be characterized by interaction matrix elements on the order of $5\,cm^{-1}$. Such perturbations have been observed (Miescher, 1980) and the location of quartet states is now known with a precision of a few reciprocal centimeters.

Heavy molecules are, in principle, more favorable for detecting spin–orbit perturbations and, in this way, locating metastable states. In the NS molecule, which is isovalent with NO, a perturbation matrix element between $b^4\Sigma^-$ and $B^2\Pi$ of 8 cm^{-1} has allowed the $^4\Sigma^-$ state to be located (Jenouvrier and Pascat, 1980). In the NSe molecule, the $^4\Pi$ state has been detected by its interaction with $^2\Pi$ states (Daumont *et al.*, 1976).

The details of the location and perturbations of metastable states are important for understanding and possibly utilizing collision-induced energy transfer between metastable and short-lived, strongly radiating levels (see Section 5.5.4).

References

Ackermann, F., and Miescher, E. (1968), *Chem. Phys. Lett.* **2**, 351.

Albritton, D. L., Harrop, W. J., Schmeltekopf, A. L., and Zare, R. N. (1973), *J. Mol. Spectrosc.* **46**, 25.

Almkvist, G., and Lagerqvist, A. (1950), *Ark. Fys.* **2**, 233.

Ballik, E. A., and Ramsay, D. A. (1963), *Astrophys. J.* **137**, 61, 84.

Barnes, I., Becker, K. H., and Fink, E. H. (1979), *Chem. Phys. Lett.* **67**, 314.

Bernath, P. F. (1980), Ph.D. thesis, MIT, Cambridge, Massachusetts.

Bessis, N., Hadinger, G., and Tergiman, Y. S. (1984), *J. Mol. Spectrosc.* **107**, 343.

Blaise, J., Wyart, J. F., Conway, J. G., and Worden, E. F. (1980), *Phys. Scr.* **22**, 224.

Bredohl, H., Cornet, R., Dubois, I., and Remy, F. (1974), *J. Phys. B*, **7**, L66.

Brion, J., Malicet, J., and Merienne-Lafore, M. F. (1977), *Can. J. Phys.* **55**, 68.

Brown, J. M., and Merer, A. J. (1979), *J. Mol. Spectrosc.* **74**, 488.

Brown, J. M., and Watson, J. K. G. (1977), *J. Mol. Spectrosc.* **65**, 65.

Brown, J. D., Burns, G., and LeRoy, R. J., (1973), *Can. J. Phys.* **51**, 1664.

Buenker, R. J., and Peyerimhoff, S. D. (1975), *Chem. Phys. Lett.* **34**, 225.

Chang, H. C., and Ogawa, M. (1972), *J. Mol. Spectrosc.* **44**, 405.

Child, M. S., (1974), *in* "Molecular Spectroscopy" (R. F. Barrow, D. A. Long, and J. Sheridan, eds.), vol. 2, Chem. Soc., London; Specialist Periodical Report.

Colbourn, E. A., and Wayne, F. D. (1979), *Mol. Phys.* **37**, 1755.

Condon, E. U., and Shortley, G. H. (1953), "The Theory of Atomic Spectra," Cambridge Univ. Press, London and New York.
Connor, J. N. L. (1981), *J. Chem. Phys.* **74**, 1047.
Cooper, D. L., and Richards, W. G. (1981), *J. Chem. Phys.* **74**, 96.
Cooper, D. L., Prosser, S. J., and Richards, W. G. (1981), *J. Phys. B* **14**, 487.
Coquart, B., DaPaz, M., and Prud'homme, J. C. (1974), *Can. J. Phys.* **52**, 177.
Cossart, D., and Bergeman, T. M. (1976), *J. Chem. Phys.* **65**, 5462.
Cossart, D., Horani, M., and Rostas, J. (1977), *J. Mol. Spectrosc.* **67**, 283.
Couet, C. (1968), *J. Chim. Phys. Phys. Chim Biol.* **65**, 1241.
Coxon, J. A. (1975), *J. Mol. Spectrosc.* **58**, 1.
Coxon, J. A., and Foster, S. C. (1982), *J. Mol. Spectrosc.* **91**, 243.
Coxon, J. A., and Hammersley, R. E. (1975), *J. Mol. Spectrosc.* **58**, 29.
Cummins, P. G., Field, R. W., and Renhorn, I. (1981), *J. Mol. Spectrosc.* **90**, 327.
Daumont, D., Jenouvrier, A., and Pascat, B. (1976), *Can. J. Phys.* **54**, 1292.
Dehmer, P. M., and Chupka, W. A. (1975), *J. Chem. Phys.* **62**, 4525.
De Santis, D., Lurio, A., Miller, T. A., and Freund, R. S. (1973), *J. Chem. Phys.* **58**, 4625.
Dill, D., and Jungen, C. (1980), *J. Phys. Chem.* **84**, 2116.
Dressler, K., and Miescher, E. (1965), *Astrophys. J.* **141**, 1266.
Dunham, J. L. (1932), *Phys. Rev.* **41**, 721.
Effantin, C., d'Incan, J., Bacis, R., and Vergès, J. (1979), *J. Mol. Spectrosc.* **76**, 204.
Elander, N., Hehenberger, M., and Bunker, P. R. (1979), *Phys. Scr.* **20**, 631.
Felenbok, P. (1963), *Ann. Astrophys.* **26**, 393.
Field, R. W. (1974), *J. Chem. Phys.* **60**, 2400.
Field, R. W., Wicke, B. G., Simmons, J. D., and Tilford, S. G. (1972), *J. Mol. Spectrosc.* **44**, 383.
Field, R. W., Gottscho, R. A., and Miescher, E. (1975), *J. Mol. Spectrosc.* **58**, 394.
Field, R. W., Lagerqvist, A., and Renhorn, I. (1976), *Phys. Scr.* **14**, 298.
Fraga, S., Saxena, K. M. S., Karwowski, J. (1976), "Handbook of Atomic Data," Physical Sciences Data 5, Elsevier, Amsterdam.
Froese-Fischer, C. (1972), *At. Data* **4**, 301.
Froese-Fischer, C. (1977), "The Hartree-Fock Method for Atoms," Wiley, New York.
Gallusser, R. (1976), thesis, Physical Chemistry Laboratory, ETH Zurich, Switzerland.
Gallusser, R., and Dressler, K. (1982), *J. Chem. Phys.* **76**, 4311.
Gauyacq, D., and Horani, M. (1978), *Can. J. Phys.* **56**, 587.
Ghosh, S., Nagaraj, S., and Verma, R. D. (1975), *Can. J. Phys.* **54**, 695.
Gottscho, R. A., Field, R. W., and Lefebvre-Brion, H. (1978), *J. Mol. Spectrosc.* **70**, 420.
Gottscho, R. A., Field, R. W., Dick, K. A., and Benesch, W. (1979), *J. Mol. Spectrosc.* **74**, 435.
Hall, J. A., Schamps, J., Robbe, J. M., and Lefebvre-Brion, H. (1973), *J. Chem. Phys.* **59**, 3271.
Harris, S. M., Gottscho, R. A., Field, R. W., and Barrow, R. F. (1982), *J. Mol. Spectrosc.* **91**, 35.
Hinkley, R. K., Hall, J. A., Walker, T. E. H., and Richards, W. G. (1972), *J. Phys. B* **5**, 204.
Horani, M., Rostas, J., and Lefebvre-Brion, H. (1967), *Can. J. Phys.* **45**, 3319.
Hotop, H., and Lineberger, W. C. (1975), *J. Phys. Chem. Ref. Data* **4**, 539.
Huber, K. P., and Alberti, F. (1983). *J. Mol. Spectrosc.* **97**, 387.
Huber, K. P., and Herzberg, G. (1979), "Constants of Diatomic Molecules," Van Nostrand-Reinhold, New York.
Hultin, M., and Lagerqvist, A. (1951), *Ark. Fys.* **2**, 471.
Ikeda, T., Wong, N. B., Harris, D. O., and Field, R. W. (1977), *J. Mol. Spectrosc.* **68**, 452.
Ishiguro, E., and Kobori, M. (1967), *J. Phys. Soc. Jpn.* **22**, 263.
James, T. C. (1964), *J. Chem. Phys.* **41**, 631.
Jenouvrier, A., and Pascat, B. (1980), *Can. J. Phys.* **58**, 1275.
Jevons, W. (1932), "Band Spectra of Diatomic Molecules", Univ. Press, Cambridge, England.

Johansson, S., Litzén, U., Sinzelle, J., and Wyart, J. F. (1980), *Phys. Scr.* **21**, 40.
Kaiser, E. W. (1970), *J. Chem. Phys.* **53**, 1686.
Katayama, D. H., Ogawa, S., Ogawa, M., and Tanaka, Y. (1977). *J. Chem. Phys.* **67**, 2132.
Katriel, J. (1970), *J. Phys. B* **3**, 1315.
Kayama, K., and Baird, J. C. (1967), *J. Chem. Phys.* **46**, 2604.
Kirschner, S. M., and Watson, J. K. G. (1973), *J. Mol. Spectrosc.* **47**, 234.
Kirschner, S. M., and Watson, J. K. G. (1974), *J. Mol. Spectrosc.* **51**, 321.
Kosman, W. M., and Hinze, J. (1975), *J. Mol. Spectrosc.* **56**, 93.
Kotlar, A. J., Field, R. W., Steinfeld, J. I., and Coxon, J. A. (1980), *J. Mol. Spectrosc.* **80**, 86.
Kovács, I. (1969), "Rotational Structure in the Spectra of Diatomic Molecules," American Elsevier, Elsevier, New York.
Kratzer, A. (1920), *Z. Phys.* **3**, 289.
Lagerqvist, A., and Miescher, E. (1958), *Helv. Phys. Acta* **31**, 221.
Lagerqvist, A., Lind, E., and Barrow, R. F. (1950), *Proc. Phys. Soc., London, Sect. A.* **63**, 1132.
Langer, R. E. (1937), *Phys. Rev.* **51**, 669.
Langhoff, S. R., Sink, M. L., Pritchard, R. H., Kern, C. W., Strickler, S. J., and Boyd, M. J. (1977), *J. Chem. Phys.* **67**, 1051.
Langhoff, S. R., Sink, M. L., Pritchard, R. H., and Kern, C. W. (1982), *J. Mol. Spectrosc.* **96**, 200.
Lebreton, J., Bosser, G., and Marsigny, L. (1973), *J. Phys. B* **6**, L226.
Lefebvre-Brion, H. (1969), *Can. J. Phys.* **47**, 541.
Lefebvre-Brion, H. (1973), *J. Mol. Struct.* **19**, 103.
Lefebvre-Brion, H., and Colin, R. (1977), *J. Mol. Spectrosc.* **65**, 33.
Lefebvre-Brion, H., and Moser, C. M. (1966), *J. Chem. Phys.* **44**, 2951.
LeRoy, R. J. (1973), *in* "Molecular Spectroscopy" (R. F. Barrow, D. A. Long, and D. J. Millen, eds.) Vol. 1, Chemical Society, London, Specialist Periodical Report, Chap. 3, p. 113.
Linton, C. (1980), *J. Mol. Spectrosc.* **80**, 279.
MacDonald, C. A., Eland, J. H. D., and Barrow, R. F. (1982), *J. Phys. B* **15**, L93.
Malicet, J., Brion, J., and Guenebaut, M. (1976), *Can. J. Phys.* **54**, 907.
Mantz, A. W., Watson, J. K. G., Rao, K. N., Albritton, D. L., Schmeltekopf, A. L., and Zare, R. N. (1971), *J. Mol. Spectrosc.* **39**, 180.
Merer, A. J. (1972), *Mol Phys.* **23**, 309.
Merer, A. J., Malm, D. N., Martin, R. W., Horani, M., and Rostas, J. (1975), *Can. J. Phys.* **53**, 251.
Merzbacher, E. (1970), "Quantum Mechanics," Wiley, New York.
Miescher, E. (1980), *J. Chem. Phys.* **73**, 3088.
Miller, W. H. (1968), *J. Chem. Phys.* **48**, 464.
Miller, W. H. (1971), *J. Chem. Phys.* **54**, 4174.
Miller, S. C., and Good, R. H. (1953), *Phys. Rev. A* **91**, 174.
Moore, C. E. (1949), "Atomic Energy Levels," Nat. Bur. Stand. Circular 257, Washington, D.C.
Morin, P., Nenner, I., Adam, M. Y., Hubin-Franskin, M. J., Delwiche, J., Lefebvre-Brion, H., and Giusti-Suzor, A. (1982), *Chem. Phys. Lett.* **92**, 609.
Mulliken, R. S. (1955), *J. Chem. Phys.* **23**, 1833.
Mulliken, R. S. (1972), *Chem. Phys. Lett.* **14**, 141.
Mulliken, R. S. (1976), *Acc. Chem. Res.* **9**, 7.
Mulliken, R. S., and Christy, A. (1931), *Phys. Rev.* **38**, 87.
Ngo, T. A., DaPaz, M., Coquart, B., and Couet, C. (1974), *Can. J. Phys.* **52**, 154.
Pekeris, C. L. (1934), *Phys. Rev.* **45**, 98.
Prosser, S. J., Barrow, R. F., Effantin, C., d'Incan, J., and Vergès, J. (1982), *J. Phys. B* **15**, 4151.
Richards, W. G., Trivedi, H. P., and Cooper, D. L. (1981), "Spin-Orbit Coupling in Molecules," Oxford Univ. Press (Clarendon), London and New York.
Robbe, J. M., and Schamps, J. (1976), *J. Chem. Phys.* **65**, 5420.

Robbe, J. M., Schamps, J., Lefebvre-Brion, H., and Raseev, G. (1979), *J. Mol. Spectrosc.* **74**, 375.
Robbe, J. M., Lefebvre-Brion, H., and Gottscho, R. A. (1981), *J. Mol. Spectrosc.* **85**, 215.
Roche, A. L., and Lefebvre-Brion, H. (1973), *J. Chem. Phys.* **59**, 1914.
Rostas, J., Cossart, D., and Bastien, J. R. (1974), *Can. J. Phys.* **52**, 1274.
Sambe, H., and Felton, R. H. (1976), *Chem. Phys.* **13**, 299.
Schamps, J. (1973), thesis, Université des Sciences et Techniques de Lille, Lille, France (unpublished).
Schamps, J. (1977), *J. Quant. Spect. Radiat. Transfer* **17**, 685.
Schmid, R., and Gerö, L. (1937), *Z. Phys.* **106**, 205.
Sennesal, J. M., Robbe, J. M., and Schamps, J. (1981), *Chem. Phys.* **55**, 49.
Shadmi, Y. (1961), *Bull. Res. Counc. Isr. Sect. F* **9**, 141.
Shadmi, Y., Oreg, J., and Stein, J. (1968), *J. Opt. Soc. Am.* **58**, 909.
Shadmi, Y., Caspi, E., and Oreg, J. (1969), *J. Res. Nat. Bur. Stand, Sect. A* **73**, 173.
Shore, B. W. (1973), *J. Chem. Phys.* **59**, 6450.
Sobel'man, I. I. (1972), "Introduction to the Theory of Atomic Spectra," Pergamon, Oxford.
Stahel, D., Leoni, M., and Dressler, K. (1983), *J. Chem. Phys.* **79**, 2541.
Stepanov, B. I. (1940), *J. Phys. (USSR)* **2**, 205.
Suter, R. (1969), *Can. J. Phys.* **47**, 881.
Tellinghuisen, J. (1973), *Chem. Phys. Lett.* **18**, 544.
Tellinghuisen, J. (1984), *J. Mol. Spectrosc.* **103**, 455.
Van Vleck, J. H. (1929), *Phys. Rev.* **33**, 467.
Vergès, J., Effantin, C., Babaky, O., d'Incan, J., Prosser, S. J., and Barrow, R. F. (1982), *Phys. Scr.* **25**, 338.
Verma, R. D. (1971), *Can. J. Phys.* **49**, 279.
Verma, R. D., and Dixit, M. N. (1968), *Can. J. Phys.* **46**, 2079.
Verma, R. D., and Singhal, R. S. (1975), *Can. J. Phys.* **53**, 411.
Veseth, L. (1970), *J. Phys. B* **3**, 1677.
Vidal, C. R., and Scheingraber, H. (1977), *J. Mol. Spectrosc.* **65**, 46.
Watson, J. K. G. (1979), *J. Mol. Spectrosc.* **74**, 319.
Wayne, F. D., and Colbourn, E. A. (1977), *Mol. Phys.* **34**, 1141.
Wicke, B. G., Field, R. W., and Klemperer, W. (1972), *J. Chem. Phys.* **56**, 5758.
Winter, R., Barnes, I., Fink, E. H., Wildt, J., and Zabel, F. (1980), *Chem. Phys. Lett.* **73**, 297.
Winter, R., Barnes, I., Fink, E. H., Wildt, J., and Zabel, F. (1982), *J. Mol. Struct.* **80**, 75.
Wyart, J. F. (1978), *Phys. Scr.* **18**, 87.
Wyart, J. F., and Bauche-Arnoult, C. (1981), *Phys. Scr.* **22**, 583.
Zaidi, H. R., and Verma, R. D. (1975), *Can. J. Phys.* **53**, 420.
Zare, R. N. (1964), *J. Chem. Phys.* **40**, 1934.
Zare, R. N., Schmeltekopf, A. L., Harrop, W. J., and Albritton, D. L. (1973), *J. Mol. Spectrosc.* **46**, 37.

Chapter 5

Effects of Perturbations on Transition Intensities

5.1 Intensity Factors

There are a variety of properties related to transition intensities that can be affected by perturbations. The most important of these are the following isolated-molecule quantities.

1. Transition moment for electric dipole transitions,

$$\mu_{i,v_i;j,v_j} = \langle i, v_i | \boldsymbol{\mu} | j, v_j \rangle \tag{5.1.1}$$

$$\vec{\boldsymbol{\mu}} \equiv e \sum_{i=1}^{n} \vec{\mathbf{r}}_i, \tag{5.1.2}$$

where n is the number of electrons. The notation for the *electronic transition moment function*, recommended by Whiting *et al.* (1980), is

$$R_e(\bar{R}_{v_i,v_j}) = \langle i, v_i | \boldsymbol{\mu} | j, v_j \rangle / \langle v_i | v_j \rangle \tag{5.1.3}$$

if the R-centroid approximation is valid. This approximation is not always valid for transition intensities (Noda and Zare, 1982; Tellinghuisen, 1984; see also Section 5.4). The symbol μ, with or without subscripts, will imply a matrix element of $\boldsymbol{\mu}$ between electronic–vibration functions, whereas R_e will imply an electronic matrix element. Note that μ is a joint property of two levels.

2. Radiative lifetime, τ_i. τ_i is a property of a single level, whereas μ_{ij} describes a transition between two levels. $1/\tau_i$ is the rate at which the population of level i decays,

$$N_i(t) = N_i^0 \exp(-t/\tau_i), \tag{5.1.4}$$

where N_i^0 is the population of level i at $t = 0$.

3. Transition probabilities, such as the *Einstein spontaneous emission coefficient*, A_{ij}, defined so that, in the absence of collisions and nonradiative decay processes (see Chapters 6 and 7),

$$\tau_i^{-1} = \sum_j A_{ij}. \tag{5.1.5}$$

Note that, in the absence of collisions, the population of a single level decays as a single exponential, despite the appearance of a summation over individual Einstein A-coefficients in Eq. (5.1.5). Even if one monitors fluorescence at the frequency of the $i \to n$ transition or the appearance of level n, the decay constant will be τ_i^{-1} and not A_{in}.
A_{ij} is related to μ_{ij} by

$$A_{ij} = |\mu_{ij}|^2 \nu_{ij}^3 [8\pi^2/(3\hbar\varepsilon_0)] \tag{5.1.6a}$$

$$= 3.137 \times 10^{-7} |\mu_{ij}|^2 \nu_{ij}^3 \quad \text{s}^{-1} \tag{5.1.6b}$$

where μ is in Debyes ($1\ \text{D} = 0.3935$ a.u. $= 1 \times 10^{-18}$ esu · cm $= 3.336 \times 10^{-30}$ Coulomb · m), ε_0 is the permittivity of vacuum ($8.8542 \times 10^{-12}\ \text{s}^4 \cdot \text{A}^2 \cdot \text{kg}^{-1} \cdot \text{m}^{-3}$), and the transition frequency, ν_{ij}, is in reciprocal centimeters.

Other intensity factors include:

4. Einstein stimulated emission/absorption, B_{ij},

$$B_{ij} = |\mu_{ij}|^2/(6\hbar^2\varepsilon_0 c) \tag{5.1.7a}$$

where $B_{ij}\rho(\nu_{ij})$ is the $i \to j$ transition rate for a single molecule stimulated by a radiation field with energy density ρ (energy per unit volume per unit frequency) and frequency centered at the molecular resonance frequency, ν_{ij}.† For μ in Debye units and ρ in $\text{J}/(\text{m}^3 \cdot \text{cm}^{-1})$,

$$B_{ij} = 4.056 \times 10^{17}\,|\mu_{ij}|^2 \quad \text{J}^{-1} \cdot \text{m}^3 \cdot \text{cm}^{-1} \cdot \text{s}^{-1}. \tag{5.1.7b}$$

5. Oscillator strength, f_{ij},

$$f_{ij} = |\mu_{ij}|^2 \nu_{ij}[4\pi m_e c/(3\hbar e^2)] \tag{5.1.8a}$$

where m_e and e are the mass and charge of the electron and, for μ in Debye and $\nu_{ij}(\text{cm}^{-1})$,

$$f_{ij} = 4.703 \times 10^{-7}\, \nu_{ij}|\mu_{ij}|^2. \tag{5.1.8b}$$

6. Cross-section, $\sigma_{ij}(\nu)$ and σ_{ij}^0,

$$\sigma_{ij}(\nu) \equiv \sigma_{ij}^0 g(\nu - \nu_{ij}) \tag{5.1.9}$$

where one must be cautious to discover whether the *distributed* cross-section (g is the lineshape function, ν_{ij} is line-center) is normalized such that σ_{ij}^0 is the peak,

$$\sigma_{ij}^0 \equiv \sigma_{ij}(\nu_{ij}), \qquad g(0) \equiv 1$$

[units of both σ_{ij}^0 and $\sigma_{ij}(\nu)$ are area per unit frequency], or integrated,

$$\sigma_{ij}^0 \equiv \int_{-\infty}^{\infty} \sigma_{ij}(\nu)\, d\nu, \qquad \int_{-\infty}^{\infty} g(\nu - \nu_{ij})\, d\nu \equiv 1$$

[units of σ_{ij} and $\sigma_{ij}^0(\nu)$ are, respectively, area and area per unit frequency] cross-section. If σ_{ij}^0 is an integrated cross-section, then

$$\sigma_{ij}^0 = 2.688 \times 10^{-16}\, \nu_{ij}|\mu_{ij}|^2 \quad \text{cm}^2 \tag{5.1.10a}$$

for μ in Debyes and $\nu_{ij}(\text{cm}^{-1})$. If σ_{ij}^0 is the peak cross-section of a Lorentzian line

† It is implicit in the definition of the Einstein B-coefficient that $\rho(\nu)$ is constant over the linewidth of the molecular transition.

of FWHM $= \Delta\nu$ cm^{-1}, then

$$\sigma^0_{ij} = 1.711 \times 10^{-16}(\nu_{ij}/\Delta\nu)\,|\mu_{ij}|^2 \quad \text{cm}^2/\text{cm}^{-1}. \tag{5.1.10b}$$

As a point of reference, the following quantities correspond to a fully allowed transition, $f = 1$, at $\nu_{ij} = 20{,}000$ cm^{-1}:

$$\mu_{ij} = 10.3 \quad \text{D}$$

$$A_{ij} = 2.67 \times 10^8 \text{s}^{-1}$$

$$\tau_i = 3.75 \times 10^{-9} \quad \text{s} \qquad \left(\text{if } \sum_k A_{ik} \approx A_{ij}, \text{ i.e. all } A_{ik} \ll A_{ij} \text{ for } k \neq j\right)$$

$$B_{ij} = 4.31 \times 10^{19} \quad \text{J}^{-1}\cdot\text{m}^3\cdot\text{cm}^{-1}\cdot\text{s}^{-1}$$

$$\sigma^0_{ij} = 5.72 \times 10^{-10} \quad \text{cm}^2$$

$$\Delta\nu = 1.42 \times 10^{-3} \quad \text{cm}^{-1}$$

$$\sigma_{ij}(0) = 2.57 \times 10^{-7} \quad \text{cm}^2/\text{cm}^{-1}.$$

A final quantity, the line strength, is defined assuming isotropic excitation and/or unpolarized radiation,

$$S_{i,v',J';j,v'',J''} \equiv \sum_{M',M''} |\langle i, v', J', M'|\boldsymbol{\mu}|j, v'', J'', M''\rangle|^2 \tag{5.1.11}$$

$$S_{i,v',J';j,v'',J''} = q_{v'v''}|R_e^{i,j}|^2\,\underline{S}_{J'J''} \tag{5.1.12}$$

where q is the Franck–Condon factor, $|\langle v'|v''\rangle|^2$ and $\underline{S}_{J'J''}$ is the Hönl–London rotational linestrength factor (Whiting, *et al.*, 1980). Definitions and interrelationships between intensity factors are given by Tatum (1967), Whiting and Nicholls (1974), and Schadee (1978). The definitions of transition moments and intensity factors suggested by Whiting *et al.* (1980) are adopted in this chapter, except in Sections 5.3 and 5.4 where the factorization of Eq. (5.1.12) is not possible.

There are also several perturbation-sensitive nonisolated molecule properties, such as collision-induced population anomalies (Radford and Broida, 1963) (Section 5.5.4) and differential pressure and power (Gottscho and Field, 1978) broadening effects (Section 5.5.1). Since all of the isolated molecule properties are explicitly related to μ, the following discussion focuses on μ. Note, however, that the nature of a perturbation related intensity anomaly is profoundly dependent on whether $|\mu|^2$, τ, or (A, B, f, σ) is being measured and on the state selectivity and spectral resolution of the specific experiment.

In this chapter, discussion will be restricted to intensity effects arising from bound $\sim$ bound interactions. Bound $\sim$ free interactions are discussed in Chapters 6 and 7 on predissociation and autoionization.

5.2 Intensity Borrowing

5.2.1 PERTURBATIONS BY STATES WITH "INFINITE" RADIATIVE LIFETIME; SIMPLE INTENSITY BORROWING

Let the vibronic basis functions 1 and 2 correspond to a mutually interacting pair of electronic levels,

$$|1\rangle \quad \text{with} \quad E_1^0 = \langle 1|\mathbf{H}^0|1\rangle$$

$$|2\rangle \quad \text{with} \quad E_2^0 = \langle 2|\mathbf{H}^0|2\rangle$$

$$H_{12} = \langle 1|\mathbf{H}|2\rangle.$$

The simplest and most common situation is that basis functions 1 and 2 have, respectively, appreciable and negligible transition probability into a lower energy basis function 0,

$$I_{10}^0 \gg I_{20}^0.$$

I_{20}^0 will be negligible whenever the lifetime of basis function 2 is infinite,

$$\tau_2^0 = \left(\sum_i A_{2i}^0\right)^{-1} \to \infty$$

or, less stringently, whenever

$$(A_{20}^0)^{-1} \to \infty$$

(i.e., τ_2^0 is not infinite). The former case corresponds to Eq. (5.2.11). The latter case, often misleadingly referred to as "perturbation by a level without oscillator strength" (meaning $f_{20}^0 = 0$ but not that $f_{2i}^0 = 0$ for all i), corresponds to Eq. (5.2.11′).

The result of the H_{12} interaction is two mixed eigenstates, the higher-energy eigenstate, $|+\rangle$, corresponding to the eigenenergy

$$E_+ = \overline{E^0} + [(\Delta E^0/2)^2 + H_{12}^2]^{1/2}, \tag{5.2.1a}$$

and the lower energy eigenstate, $|-\rangle$, corresponding to

$$E_- = \overline{E^0} - [(\Delta E^0/2)^2 + H_{12}^2]^{1/2} \tag{5.2.1b}$$

$$\overline{E^0} = (E_1^0 + E_2^0)/2 \tag{5.2.2}$$

$$\Delta E^0 = E_1^0 - E_2^0 \tag{5.2.3}$$

and, if $E_1^0 > E_2^0$, the energy shift is

$$\delta E = E_+ - E_1^0 = -(E_- - E_2^0) \tag{5.2.4}$$

$$\delta E = [(\Delta E^0/2)^2 + H_{12}^2]^{1/2} - \Delta E^0/2. \tag{5.2.5}$$

The eigenfunctions are

$$|+\rangle = C_{1+}|1\rangle + C_{2+}|2\rangle \tag{5.2.6a}$$

where, by the normalization requirement,

$$C_{2+} = \pm[1 - C_{1+}^2]^{1/2}, \qquad (5.2.6b)^\dagger$$

and, by the orthogonality requirement,

$$|-\rangle = \mp[1 - C_{1+}^2]^{1/2}|1\rangle + C_{1+}|2\rangle \qquad (5.2.6c)^\dagger$$

$$C_{1+} = 2^{-1/2}[1 + (\Delta E^0/2)[(\Delta E^0/2)^2 + H_{12}^2]^{-1/2}]^{1/2} \geq 0. \tag{5.2.7}$$

Note that, whenever $\Delta E^0 > 0$, regardless of the magnitude or sign of H_{12}, $C_{1+} \geq 2^{-1/2}$ and thus $|+\rangle$ and $|-\rangle$ are respectively the "nominal" perturbed $|'1'\rangle$ and $|'2'\rangle$ functions.

Both $|+\rangle$ and $|-\rangle$ have nonzero transition probability into level 0,

$$I_{+0} \propto |\langle + |\boldsymbol{\mu}| 0\rangle|^2$$

$$I_{+0} = C_{1+}^2 I_{10}^0 \tag{5.2.8a}$$

$$I_{-0} = (1 - C_{1+}^2) I_{10}^0. \tag{5.2.8b}$$

In the absence of the $1 \sim 2$ interaction ($H_{12} = 0$), only *one* line, $1 \leftrightarrow 0$, of intensity I_{10}^0 is observable. When the interaction is turned on ($H_{12} \neq 0$), *two* transitions become observable. These are called main and extra lines. For the specific case of a two-level interaction for which $I_{10}^0 \gg I_{20}^0$ (and $E_1^0 > E_2^0$), the level (E_+) giving rise to the main line is, by definition, the level with larger basis function 1 character. Thus the main line is necessarily more intense than the extra line. In addition, the main and extra lines lie above and below the calculated position of the (nonexistent) deperturbed $E_1^0 \leftrightarrow E_0$ transition, with the main line lying closer than the extra line to $E_1^0 - E_0$.

The occurrence of two perturbed lines where there would be only one unperturbed line is where the idea of extra lines originates. The metaphor, intensity borrowing, arises from the conservation of total intensity,

$$I_{+0} + I_{-0} = I_{10}^0, \tag{5.2.9}$$

which is a necessary consequence of a simple two-level interaction (provided

† The correct sign choice for the $[1 - C_{1+}^2]^{1/2}$ term (C_{2+}) in Eqs. (5.2.6b and c) is the top sign for $H_{12} > 0$. The opposite choice would cause $|+\rangle$ to belong to the eigenenergy E_-.

$I_{10}^0 \gg I_{20}^0$). Basis function 1 character is conserved, as are the properties of basis function 1.

This is where important differences in the effect of perturbations on measured intensities and lifetimes become evident. Let $\tau_2^0 = \infty$ (the related case of $A_{20}^0 = 0$ and τ_2^0 finite will be discussed below).

In a high resolution experiment (transitions into levels E_+ and E_- are resolved), one finds

$$I_{+0} = C_{1+}^2 I_{10}^0 \tag{5.2.8a}$$

$$\tau_+ = \tau_1^0 C_{1+}^{-2} \tag{5.2.10a}$$

$$\tau_- = \tau_1^0 (1 - C_{1+}^2)^{-1} \tag{5.2.10b}$$

$$1/\tau_+ + 1/\tau_- = 1/\tau_1^0. \tag{5.2.11}$$

Note that Eq. (5.2.9) requires that intensity be conserved whereas Eq. (5.2.11) requires that $1/\tau$, not τ, be conserved. In a low-resolution experiment (levels E_+ and E_- not resolved), measurement of integrated intensity would show no perturbation effect, whereas a lifetime measurement (assuming broad bandwidth, nonsaturating excitation) would show biexponential decay,

$$I(t) = I_{+0}(t) + I_{-0}(t) \tag{5.2.12}$$

$$I_{+0}(t) = I_{10}^0 C_{1+}^2 \exp[-t(C_{1+}^2/\tau_1^0)] \tag{5.2.13a}$$

$$I_{-0}(t) = I_{10}^0 (1 - C_{1+}^2)\exp[-t(1 - C_{1+}^2)/\tau_1^0]. \tag{5.2.13b}$$

Note that the intensity decay from either mixed level separately is single-exponential [Eqs. (5.2.13a) and (5.2.13b)] but the unresolved fluorescence decay is biexponential (Eq. 5.2.12). Failure to recognize this biexponential character can yield an effective τ-value that depends artificially on the relative weighting of early versus late fluorescence both in the actual experiment and the fitting algorithm. When fluorescence from the extensively perturbed CO $A^1\Pi$ $v = 0$, 1, and 6 levels is excited by electron bombardment (Imhof and Read, 1971) or by 15-Å bandwidth synchrotron radiation (Field *et al.* 1983), highly nonexponential decay is observed. Also, if collisions transfer population from a single, selectively populated level into nearby levels (a process facilitated by perturbation, see Section 5.5.4), multiexponential decay is observed (Banic *et al.*, 1981).

The main line can have intensity and lifetime in the ranges

$$I_{10}^0 \geq I_{+0} \geq 0.5 I_{10}^0 \tag{5.2.14}$$

$$\tau_1^0 \leq \tau_+ \leq 2\tau_1^0. \tag{5.2.15}$$

However, the effect of the perturbation on the extra line appears much more dramatic than a mere factor of 2 when one recalls that $I_{20}^0 = 0$ and $\tau_2^0 = \infty$.

The intensity and radiative decay rate of an extra line can be increased by many orders of magnitude, whereas, for the main line, the perturbation decreases the intensity and decay rate by at most a factor of 2.

When $I_{20}^0 = 0$ but $\tau_2^0 \simeq \tau_1^0$ (e.g., suppose levels 2 and 1 are, respectively, upper levels of allowed triplet–triplet and singlet–singlet systems), all of the perturbation effects on intensities [Eqs. (5.2.8), (5.2.9), and (5.2.14)] are unchanged from the $\tau_2^0 = \infty$ case, but the perturbation effects on lifetimes are slightly different:

$$\tau_+ = [(1 - C_{1+}^2)/\tau_2^0 + C_{1+}^2/\tau_1^0]^{-1} \tag{5.2.10a'}$$

$$\tau_- = [(1 - C_{1+}^2)/\tau_1^0 + C_{1+}^2/\tau_2^0]^{-1} \tag{5.2.10b'}$$

$$1/\tau_+ + 1/\tau_- = 1/\tau_1^0 + 1/\tau_2^0 \tag{5.2.11'}$$

$$I_{+0}(t) = I_{10}^0 C_{1+}^2 \exp(-t/\tau_+) \tag{5.2.13a'}$$

$$I_{-0}(t) = I_{10}^0 (1 - C_{1+}^2) \exp(-t/\tau_-). \tag{5.2.13b'}$$

In contrast to the simple two-level problem, when level 1 is perturbed by many levels that do not have oscillator strength to level 0, the intensity and lifetime of nominal level 1 can be altered by far more than a factor of 2. Such a multistate perturbation is unusual for a diatomic molecule, but for polyatomic molecules with their huge density of vibronic perturbing levels, lifetime lengthening by factors much larger than 2 can occur. Douglas (1966) explained the paradoxical observation of lifetimes considerably longer than those predicted on the basis of integrated absorption strength, which is insensitive to perturbation effects. Recall that, in a low-resolution absorption experiment (i.e., absorption integrated over an entire band system), perturbation effects on intensities vanish whereas radiative decay becomes slower than τ_1^0 and nonexponential.

There are several reasons why a perturbing level might have negligible oscillator strength ($A_{20}^0 \ll A_{10}^0$) to the level 0 of the monitoring transition. For example, the $A^1\Pi$ state of CO is perturbed by $a'^3\Sigma^+$, $e^3\Sigma^-$, $d^3\Delta$, $I^1\Sigma^-$, and $D^1\Delta$ states (see Fig. 2.17). Transitions from the a′, e, and d states into $X^1\Sigma^+$ are spin-forbidden ($\Delta S \neq 0$), whereas the D–X and I–X transitions, although spin-allowed, are forbidden by $\Delta\Lambda = 0, \pm 1$ and $\Sigma^+ \nleftrightarrow \Sigma^-$ electric dipole selection rules, respectively. Although the a′, e, and d states (unlike the D and I states) all have fully allowed transitions to the $a^3\Pi$ state, because of the ν^3 factor in the Einstein A-coefficient their radiative lifetimes are at least 10^2 times longer than $\tau(A^1\Pi) \simeq 10$ ns, so the perturbations of the CO $A^1\Pi$ state may be treated as if $\tau_2^0 = \infty$.

In the spectrum of the CN molecule, two electric dipole allowed band systems, $B^2\Sigma^+ - X^2\Sigma^+$ and $A^2\Pi - X^2\Sigma^+$, are observed in two quite separate wavelength regions. B–X bands do not appear in the red region where A–X is found because Franck–Condon factors restrict strong B–X bands to $\Delta v = 0$, ± 1 transitions. This leads to an unusual situation for the CN $A^2\Pi$ ($v_A = 10$) $\sim$ $B^2\Sigma^+$ ($v_B = 0$) perturbation, where, when monitoring the A–X 10–4 band, the perturbing level has a shorter intrinsic lifetime, $\tau_B^0 \ll \tau_A^0$, than the perturbed level. Even so, because the B–X 0–4 band has such a small Franck– Condon factor, $q_{0,4}^{BX} = 2.1 \times 10^{-5}$, the $B^2\Sigma^+$ $v_B = 0$ "extra" level must borrow transition strength from $A^2\Pi$ ($v_A = 10$), $q_{10,4}^{AX} = 3.5 \times 10^{-2}$.

The N_2^+ ion, which is isoelectronic with CN, has similar $B^2\Sigma^+ \sim A^2\Pi$ perturbations. Dufayard *et al.* (1974) have measured radiative lifetimes for individual rotational lines of the $B^2\Sigma_u^+$ ($v_B = 1$) $\sim A^2\Pi_u$ ($v_A = 11$) perturbation complex. The lifetimes of the perturbation-free $B^2\Sigma_u^+$ levels ($\tau_B^0 \simeq 60$ ns) are much shorter than those of $A^2\Pi_u$ ($\tau_A^0 \simeq 7$ μs), and perturbed main lines of the B–X system are observed to have lifetimes as long as 95 ns. There is good qualitative agreement between measured lifetimes and those calculated using Eq. (5.2.10a) and B$\sim$A mixing coefficients derived from the molecular constants of Gottscho *et al.* ((1979). Provorov *et al.* (1977) and Girard *et al.* (1982) have measured single-rotational-level lifetimes for the CO $e^3\Sigma^-$ ($v_e = 1$) $\sim A^1\Pi$ ($v_A = 0$) $\sim d^3\Delta_1$ ($v_d = 4$) perturbation complex. Again, the agreement shown in Fig. 5.1 between measured (Maeda and Stoicheff, 1984; Girard *et al.*, 1982) and calculated (Field, 1971; Field *et al.*, 1983) perturbed lifetimes is satisfactory.

It is almost invariably true that mixing coefficients, derived from measured frequencies of correctly assigned main and extra lines, provide at least as reliable a measure of the lifetime and relative oscillator strength of all perturbed main and extra lines as direct radiative decay measurements. This statement is supported indirectly by Fig. 5.2, which illustrates several perturbations of the ^{30}SiS $A^1\Pi$ ($v = 5$) level (Harris *et al.*, 1982). Even though the percent $^1\Pi$ character drops to a minimum of less than 30%, no systematic deviations between the wave numbers of the observed and calculated levels is evident. From Eqs. (5.2.4), (5.2.5), and (5.2.7), one obtains

$$C_{1+}^2 = 1 - \frac{\delta E}{E_+ - E_-},$$

which implies that a 0.05-cm^{-1} error in a measured or computed transition frequency leads to an error of

$$\delta C_{1+}^2 \frac{0.05}{E_{\text{main}} - E_{\text{extra}}} < \frac{0.05}{\Delta E^0}$$

in the fractional character. If $E_+ - E_-$ were as small as 1 cm^{-1}, the errors in C_{1+}^2 and τ_+ would be only 5% and $0.05\,\tau_1^0$, respectively.

The names "main" and "extra" should be applied only to transitions, not to levels. For example, the nominal $|'A^2\Pi', v_A = 10\rangle$ levels associated with main

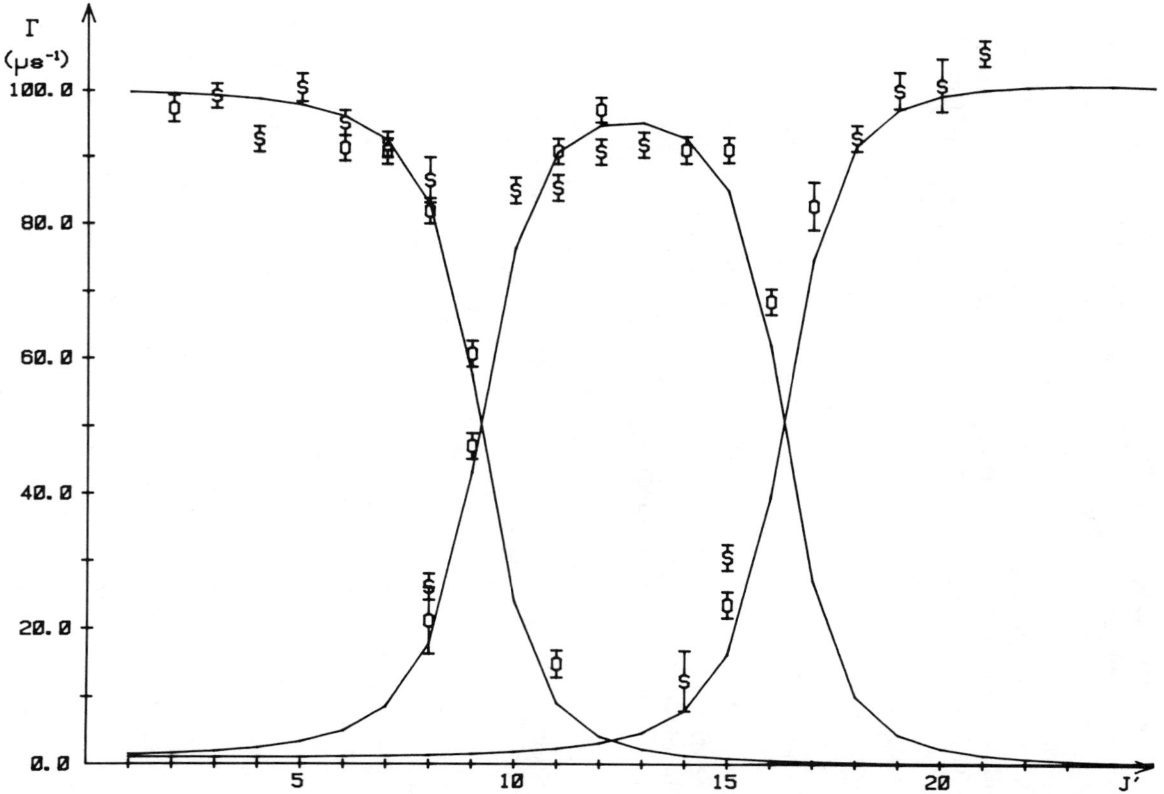

Fig. 5.1 Radiative decay rates (Γ/τ) for single e-parity rotational levels of CO $A^1\Pi$ ($v = 0$). The effects of perturbations by $e^3\Sigma^-$ ($v = 1$) are evident near $J = 9$ (F_1) and $J = 16$ (F_3). The points (O and S refer to branches in the two-photon spectrum) are measured and the solid curves depict values calculated for the nominal $A^1\Pi$, $c^3\Sigma^-$ (F_1), and $e^3\Sigma^-$ (F_3) levels from the deperturbed $\tau^0_{v=0} = 9.9$ ns of Field *et al.* (1983) and mixing coefficients from the deperturbation analysis (Field, 1971). [From Girard *et al.* (1982).]

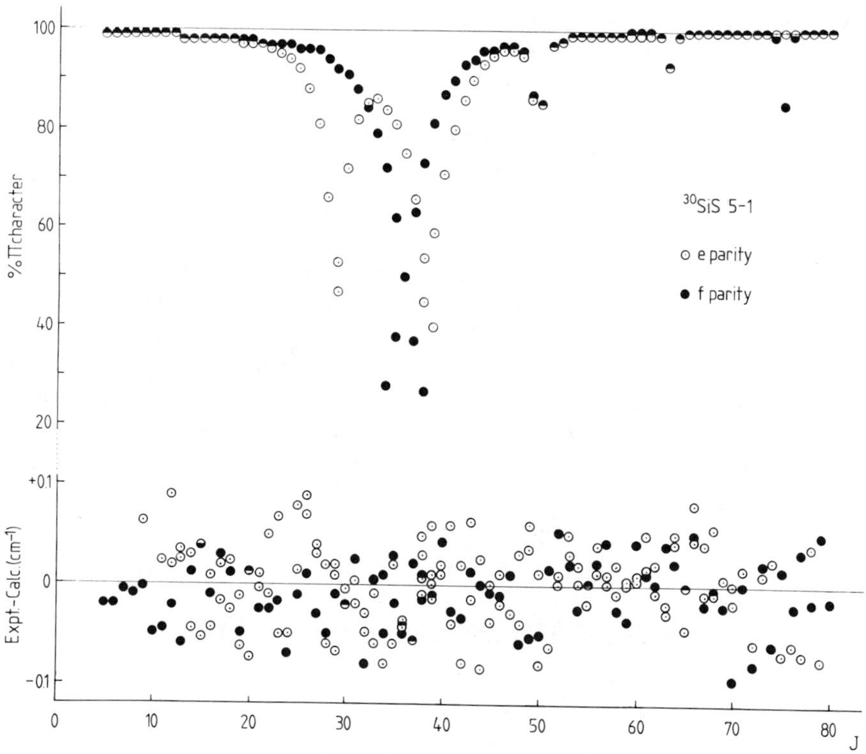

Fig. 5.2 Mixing fractions in $^{30}Si^{32}S$ $A^1\Pi$ ($v = 5$). The deperturbation model fits all observed lines, main and extra, perturbed and unperturbed, in the spectrum without systematic residuals. Mixing coefficients computed from such a deperturbation model should be more accurate than from measurements of radiative lifetimes. [From Harris *et al.* (1982).]

lines (i.e., stronger oscillator strength than extra lines) in the CN ($A, v_A = 10 \sim B, v_B = 0) \rightarrow X, v_X = 4$ band give rise to extra lines in the ($B, v_B = 0 \sim A, v_A = 10) \rightarrow X, v_X = 0$ band. Similarly, for a CO $A^1\Pi$–$d^3\Delta_1$ perturbation, the nominal $^1\Pi$ levels correspond to main lines in the $A^1\Pi$–$X^1\Sigma^+$ system and extra lines in the $d^3\Delta$–$a^3\Pi$ system.

It is not only the names "main" and "extra" or "perturbed" and "perturbing" that depend on which of two possible electronic transitions has oscillator strength in the spectral region being monitored; *the vibrational (Franck–Condon) and rotational (Hönl–London) linestrength factors that are applicable for both main and extra lines are those associated with the electronic–vibrational band with nonzero oscillator strength.* Complications occur when both bands of a perturbation complex have oscillator strength (as could happen for the CN $B \sim A \rightarrow X$ $0 \sim 10 \rightarrow 3$ band, but not for the CO singlet$\sim$triplet perturbations in the

A–X or d–a bands). These effects are discussed in Section 5.3. For the general $1 \sim 2 \rightarrow 0$ problem where the electronic $1 \rightarrow 0$ system is allowed and $2 \rightarrow 0$ is forbidden, the linestrength factors defined by Eq. (5.1.8) are

$$S^{\text{main}}_{1,v_1,J';0,v_0,J''} = [C_{1,\text{main}}(J)]^2 q_{v_1 v_0} |R_e^{1,0}|^2 \underline{S}_{J'J''} \tag{5.2.16a}$$

$$S^{\text{extra}}_{1,v_1,J';0,v_0,J''} = [1 - C_{1,\text{main}}(J)^2] q_{v_1 v_0} |R_e^{1,0}|^2 \underline{S}_{J'J''} \tag{5.2.16b}$$

$$S^{\text{extra}}_{1J';0J''}/S^{\text{main}}_{1J';0J''} = [C_{1,\text{main}}(J)]^{-2} - 1, \tag{5.2.17}$$

where $q_{v_1 v_0}$, $R_e^{1,0}$, and $\underline{S}_{J'J''}$ are the Franck–Condon, electronic transition moment, and Hönl–London factors for the unperturbed $1 \leftrightarrow 0$ electronic system.

Even if the mixed level has only 1% basis function 1 character, if $I^0_{10} \gg 100 I^0_{20}$, then the relevant Hönl–London, Franck–Condon, and transition moment factors for the nominal $2 \leftrightarrow 0$ transition are those of level 1, not those of level 2. The electronic and vibrational wavefunctions of basis function 2 are *completely irrelevant* to linestrengths in the $2 \rightarrow 0$ band system except insofar as they determine the magnitudes of H_{12} matrix elements.

5.2.2 MULTISTATE DEPERTURBATION; THE NO $^2\Pi$ STATES

The examples already cited from the spectra of CO, CN, and N_2^+ involve rather weak perturbations with highly J-dependent level shifts and mixing coefficients. The $^{14}N^{16}O$ $C^2\Pi$, $v_C = 3 \sim B^2\Pi$, $v_B = 15$ Rydberg ~ non-Rydberg (R ~ NR) interaction is so strong (interaction strength $H_{C3;B15} = 256\ \text{cm}^{-1}$; deperturbed energy, spin–orbit, and rotational constants $E^0_{B15} = 59{,}478\ \text{cm}^{-1}$, $A^0_{B15} = 53\ \text{cm}^{-1}$, $B^0_{B15} = 0.912\ \text{cm}^{-1}$, $E^0_{C3} = 59{,}373\ \text{cm}^{-1}$, $A^0_{C3} = 8\ \text{cm}^{-1}$, $B^0_{C3} = 1.903\ \text{cm}^{-1}$) that the result is two vibrational levels with rotational structure that, at first sight, appears to be perturbation-free (observed constants: $E^{\text{obs}}_{B15} = 59{,}663\ \text{cm}^{-1}$, $A^{\text{obs}}_{B15} = 38\ \text{cm}^{-1}$, $B^{\text{obs}}_{B15} = 1.280\ \text{cm}^{-1}$, $E^{\text{obs}}_{C3} = 59{,}224\ \text{cm}^{-1}$, $A^{\text{obs}}_{C3} = 33\ \text{cm}^{-1}$, $B^{\text{obs}}_{C3} = 1.360\ \text{cm}^{-1}$) (Gallusser and Dressler, 1982; Lagerqvist and Miescher, 1958). Every rotational level of this B ~ C complex is significantly mixed. Treated (inappropriately) as a two-level $B\,(v = 15) \sim C\,(v = 3)$ interaction, the fractional admixed perturber character increases smoothly from ~40% at $J = \frac{1}{2}$ of $^2\Pi_{1/2}$ to 50% near $J = 10$ where the deperturbed B and C levels cross and then decreases slowly to ~20% by $J = 20.5$. However, the NO $B^2\Pi \sim C^2\Pi$ interaction is so strong that it is insufficient to think of pairwise $v_B \sim v_C$ vibrational level interactions. Gallusser and Dressler (1982) have simultaneously deperturbed the NO $B^2\Pi$, $C^2\Pi$, $K^2\Pi$, $L^2\Pi$, and $Q^2\Pi$ states (see Fig. 5.3) by diagonalizing two diabatic perturbation matrices (one for $^2\Pi_{1/2}$, the other for $^2\Pi_{3/2}$) of

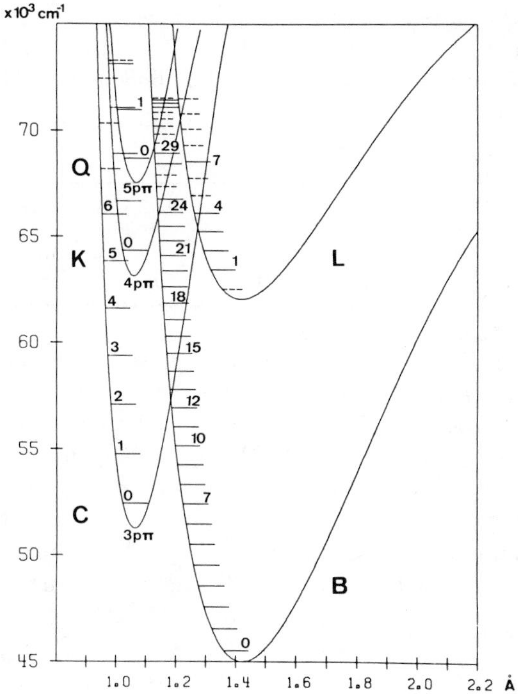

Fig. 5.3 Deperturbed potential curves for the mutually interacting $^2\Pi$ valence and Rydberg states of NO. [From Gallusser and Dressler (1982).]

dimension 69 that include the $v_B = 0$–37, $v_C = 0$–9, $v_K = 0$–4, $v_L = 0$–11, and $v_Q = 0$–3 levels. One product of this deperturbation is a set of calculated nonrotating-molecule absorption oscillator strengths (from $X^2\Pi, v_X = 0$) for the 69 × 2 vibronic eigenstates of the perturbation matrices corresponding to each of the three isotopic molecules $^{14}N^{16}O$, $^{15}N^{16}O$, and $^{14}N^{18}O$ (Gallusser and Dressler, 1982). Although computationally complex, it is conceptually simple, explaining all observables in terms of five diabatic Born–Oppenheimer potential energy curves, deperturbed electronic and vibrational properties of these five electronic states, and six R-independent $R \sim NR$ electronic interaction parameters.

Gallusser and Dressler (1982) calculate that the deperturbed, nonrotating molecule oscillator strengths for several $B^2\Pi \sim C^2\Pi$ mixed pairs of vibrational levels differ by more than a factor of 40. Thus the $I_{10}^0 \gg I_{20}^0$ approximation is valid and the NO $C^2\Pi \sim B^2\Pi \rightarrow X^2\Pi$ $v_C = 2 \sim v_B = 12 \rightarrow v_X = 0$ and $v_C = 3 \sim v_B = 15 \rightarrow v_X = 0$ perturbation complexes may be understood in terms of simple intensity borrowing.

The $^{14}N^{16}O$ C∼B←X (3∼15, 0) and B∼C←X (15∼3, 0) bands appear with nearly equal intensity in the absorption spectrum of Lagerqvist and Miescher (1958). Gallusser and Dressler (1982) find, for the $^2\Pi_{1/2}$ levels of the nonrotating molecule:

Band	Deperturbed	Perturbed (observed)
C–X (3, 0)	$f = 2.62 \times 10^{-3}$	$f = 0.98 \times 10^{-3}$
B–X (15, 0)	$f = 0.06 \times 10^{-3}$	$f = 0.71 \times 10^{-3}$.

The nearly identical perturbed f-values are consistent with the observation of similar absorption intensities. However, the fact that the sum of the deperturbed f-values, 2.68×10^{-3}, is different from the sum of the perturbed f-values, 1.69×10^{-3}, indicates that $C^2\Pi, v_C = 3$ character is distributed over several nominal $B^2\Pi$ vibrational levels, not simply the closest lying level, $v_B = 15$. Assuming that, at $J = \frac{1}{2}$, the nominal $v_C = 3$ and $v_B = 15$ levels obtain all of their oscillator strength from the $C^2\Pi_{1/2}$ $v_C = 3$ basis function, the percent $v_C = 3$ character may be deduced to be 37 and 27% in the C∼B and B∼C levels, respectively. This means that the nominal $C^2\Pi_{1/2}, v_C = 3, J = \frac{1}{2}$ level is composed 63% of basis functions other than $v_C = 3$ (27% is $v_B = 15$, leaving 36% to be distributed among other basis functions, of which $v_B = 14$ and 16 are probably most important).

The $^{14}N^{16}O$ C∼B←X (2∼12, 0) and B∼C←X (12∼2, 0) bands illustrate one of the difficulties in comparing measured (Bethke, 1959) and calculated (Gallusser and Dressler, 1982) perturbed oscillator strengths.

Band	Deperturbed	Calculated $J = \frac{1}{2}$	Observed $\Omega = \frac{1}{2}$ and $\frac{3}{2}$, room temperature
$B^2\Pi_{1/2}$–$X^2\Pi_{1/2}(12, 0)$	$f = 0.05 \times 10^{-3}$	2.45×10^{-3}	2.31×10^{-3}
$C^2\Pi_{1/2}$–$X^2\Pi_{1/2}(2, 0)$	$f = 5.29 \times 10^{-3}$	2.59×10^{-3}	2.74×10^{-3}
sum	$f = 5.34 \times 10^{-3}$	5.04×10^{-3}	5.05×10^{-3}

Because of the large matrix element ($H_{B12,C2} = 161\ \text{cm}^{-1}$) and close energy proximity of the interacting basis functions ($\Delta E^0 = 117\ \text{cm}^{-1}$), the B∼C mixing should be 74%/26% using Eq. (5.2.7) and assuming a perhaps too simple two-level interaction. The near equality of the calculated oscillator strengths is an excellent reminder that the $v_B = 12 \sim v_C = 2$ interaction is not a simple two-level problem, even though the deperturbed and calculated oscillator strength sums are nearly equal. Evidently, the nominal $v_B = 12$ level borrows some of its oscillator strength from the remote $v_C = 1$ and 3 basis functions. The excellent agreement between the calculated and observed oscillator strength sums is a simple consequence of $|R_e^{BX}|^2 \ll |R_e^{CX}|^2$ and the use by Gallusser and Dressler (1982) of Bethke's (1959) measured *absolute* oscillator strengths to normalize the calculated *relative* oscillator strengths. The important point is that $f_{calc} - f_{obs}$ is $+0.14 \times 10^{-3}$ and -0.15×10^{-3} for the nominal B–X and C–X bands, respectively. This reflects that Bethke's measurements correspond to an average over a room temperature distribution of $\Omega = \frac{1}{2}, \frac{3}{2}$ and J-levels and that the calculated oscillator strengths are for $\Omega = \frac{1}{2}$ and $J = \frac{1}{2}$. However, the simple two-level $v_B = 12 \sim v_C = 2$ mixing should decrease rapidly as J increases (the mixing with remote vibrational levels will be approximately J-independent), going from 74%/26% at $J = \frac{1}{2}$ to 84%/16% at $J = 10.5$. This means that, as observed, the measured oscillator strength for a thermal J-distribution should be slightly smaller than the $J = \frac{1}{2}$ calculated value for the nominal $v_B = 12$ level and larger by the same amount for the nominal $v_C = 2$ level.

This NO B$\sim$C example demonstrates that oscillator strengths for a perturbed band can be temperature-dependent, even though the f-value of a single resolved line and the oscillator strength sum over a group of interacting levels (bands) must be temperature independent. Other quantities, such as radiative lifetimes, will also exhibit temperature dependences when the measurement involves an average over a thermal distribution of levels. Although it is common practice to speak of the oscillator strength and radiative lifetime of an entire vibrational level, perturbations can introduce unexpected obstacles to the measurement of such intensity factors.

5.3 Interference Effects

The discussion in Section 5.2.1 was based on the assumption of either $\tau_2^0 = \infty$ or $I_{20}^0 = 0$. When two interacting basis functions have comparable transition probabilities to a common level, $I_{10}^0 \approx I_{20}^0$, then the perturbed intensities, I_{+0} and I_{-0}, exhibit a form of anomalous behavior that is explicable as a quantum-mechanical interference effect.

The transition probability is proportional to $|\mu_{ij}|^2$. By analogy with the discussion of the $1 \sim 2 \rightarrow 0$ problem discussed in Section 5.2.1, but with $\mu_{20}^0 \neq 0$,

$$\mu_{+0} = C_{1+}^* \mu_{10}^0 + C_{2+}^* \mu_{20}^0$$

$$\mu_{+0} = C_{1+}^* \mu_{10}^0 \pm (1 - |C_{1+}|^2)^{1/2} \mu_{20}^0. \qquad (5.3.1)^\dagger$$

† The $\pm$ signs in Eqs. (5.3.1) and (5.3.2) depend on whether H_{12} is positive (top sign) or negative (bottom sign). $C_{1+} \geq 0$ is assumed.

Equation (5.3.1) is written as if μ_{10}^0, μ_{20}^0, and C_{1+} were complex quantities. It is always possible to choose a self-consistent set of phases such that all off-diagonal matrix elements of **H**, transition moments (μ_{ij}), and mixing coefficients are real numbers (Whiting and Nicholls, 1974; Hougen, 1970). Therefore complex conjugation (*) and absolute value signs will be suppressed in the following equations. The μ^2 values for $+\leftrightarrow 0$ and $-\leftrightarrow 0$ transitions are

$$\mu_{+0}^2 = C_{1+}^2(\mu_{10}^0)^2 + (1 - C_{1+}^2)(\mu_{20}^0)^2 \pm 2C_{1+}(1 - C_{1+}^2)^{1/2}\mu_{10}^0\mu_{20}^0 \quad (5.3.2a)^\dagger$$

$$\mu_{-0}^2 = (1 - C_{1+}^2)(\mu_{10}^0)^2 + C_{1+}^2(\mu_{20}^0)^2 \mp 2C_{1+}(1 - C_{1+}^2)\mu_{10}^0\mu_{20}^0. \quad (5.3.2b)^\dagger$$

The first two terms in Eq. (5.3.2) are both positive and are simply the deperturbed transition moments weighted by the fractional basis function character. These are the simple intensity borrowing terms discussed in Section 5.2.1. The third term, the *interference term, may be positive or negative.* Note that

$$\mu_{+0}^2 + \mu_{-0}^2 = (\mu_{10}^0)^2 + (\mu_{20}^0)^2,$$

which implies that transition probability is conserved and that the interference term,

$$\Theta_{+0} \equiv 2C_{1+}(1 - C_{1+}^2)^{1/2}\mu_{10}^0\mu_{20}^0, \quad (5.3.3)$$

for the $+\leftrightarrow 0$ transition is equal in magnitude and opposite in sign to that for $-\leftrightarrow 0$. The effect of the interference term is to transfer probability from one transition to another in a manner that depends on the relative *signs* of off-diagonal matrix elements. Over and above the usual intensity borrowing terms, probability is transferred from the main to the extra line, or vice versa, such that

$$\Theta_{+0} = -\Theta_{-0}. \quad (5.3.4)$$

The magnitude of the intensity transfer is dependent on $|H_{12}|$, and the maximum value is reached when

$$C_{1+} = 2^{-1/2}$$

and is

$$|\Theta_{+0}| = |\Theta_{-0}| \le |\mu_{10}^0\mu_{20}^0|. \quad (5.3.5)$$

For the special case,

$$\mu_{10}^0 = +\mu_{20}^0, \qquad C_{1+} = +2^{-1/2}, \qquad \text{and} \quad E_1^0 > E_2^0,$$

$$\mu_{+0}^2 = 2(\mu_{10}^0)^2 \qquad \text{and} \qquad (\mu_{-0})^2 = 0.$$

† The $\pm$ signs in Eqs. (5.3.1) and (5.3.2) depend on whether H_{12} is positive (top sign) or negative (bottom sign). $C_{1+} \ge 0$ is assumed.

The direction of the intensity transfer depends on the sign of $H_{12}\mu_{10}^0\mu_{20}^0$. Since, for an explicitly defined phase choice, it is often possible to deduce the sign of the electronic factor of H_{12} from knowledge of the electronic configurations of the interacting states, the sign of an intensity interference effect can be useful for determining the relative signs of two transition moments.

5.3.1 PERTURBATIONS BETWEEN STATES OF THE SAME SYMMETRY; VIBRATIONAL-BAND INTENSITY ANOMALIES

When two states of the same symmetry interact, interference affects the $\Delta J = 0, \pm 1$ and $\Delta\Omega = 0, \pm 1$ rotational transitions out of a common mixed eigenstate in exactly the same way. This is in contrast to the behavior described in Sections 5.3.2 and 5.3.3 in which intensity is transferred between rotational branches or Ω-subbands. For transitions between unperturbed levels, the linestrength factor [Eq. (5.1.12)] is expressed as a product of three terms, each of which is the absolute magnitude squared of an off-diagonal matrix element. This factorization of $S_{i,v_i,J';0,v_0,J''}$ is no longer possible when the ith level is a $1 \sim 2$ mixed eigenstate. However, when the interacting 1 and 2 states have identical symmetry ($\Delta\Omega = \Delta\Sigma = \Delta\Lambda = \Delta S = 0, \Sigma^{\pm} \leftrightarrow \Sigma^{\pm}$), the Hönl–London rotational factor can be factored out of $S_{\pm,J';0,v_0,J''}$,

$$S_{+,J';0,v_0,J''} = \underline{S}_{J'J''}\{C_{1+}^2(\mu_{10}^0)^2 + (1 - C_{1+}^2)(\mu_{20}^0)^2 \pm 2C_{1+}(1 - C_{1+}^2)^{1/2}\mu_{10}^0\mu_{20}^0\}. \tag{5.3.6}$$

(The implicit phase choice is always $C_{1+} \geq 0$ and the sign choice for the interference term is top sign for $H_{12} \geq 0$.) The expression for $S_{-,J';0,v_0,J''}$ is identical to Eq. (5.3.6) except the locations of C_{1+}^2 and $(1 - C_{1+}^2)$ are interchanged and the interference term appears with $\mp$ signs.

The sign of the interference term may be determined experimentally from the ratio of the difference and sum of the intensities of main and extra lines,

$$\frac{S_{+,J';0,v_0,J''} - S_{-,J';0,v_0,J''}}{S_{+,J';0,v_0,J''} + S_{-,J';0,v_0,J''}} \equiv \frac{\Delta S}{\sum S}$$
$$= \frac{(2C_{1+}^2 - 1)[(\mu_{10}^0)^2 - (\mu_{20}^0)^2] \pm 4C_{1+}(1 - C_{1+}^2)^{1/2}\mu_{10}^0\mu_{20}^0}{(\mu_{10}^0)^2 + (\mu_{20}^0)^2}. \tag{5.3.7}$$

If the measured $\Delta S/\sum S$ intensity ratio is plotted versus the quantity $(2C_{1+}^2 - 1)$ obtained from the deperturbation calculation, the $\Delta S/\sum S$ value may be interpolated to the position of the level crossing ($E_1^0 = E_2^0$), at which point

$$2C_{1+}^2 - 1 = 0$$

$$\left(\frac{\Delta S}{\sum S}\right)_{F_1^0 = F_2^0} = \{\text{sign}(H_{12})\} \frac{2\mu_{10}^0 \mu_{20}^0}{(\mu_{10}^0)^2 + (\mu_{20}^0)^2} \quad (5.3.8)$$

where

$$\mu_{10}^0 = \langle v_1 | v_0 \rangle R_e^{10}$$

$$\mu_{20}^0 = \langle v_2 | v_0 \rangle R_e^{20}.$$

None of the five signed quantities (H_{12}, $\langle v_1|v_0\rangle$, $\langle v_2|v_0\rangle$, R_e^{10}, and R_e^{20}) that determine the sign of the intensity interference term are directly derivable from experiment. The two vibrational overlap factors may be calculated provided that the potential energy curves for the three electronic states have been adequately determined from independent experimental measurements. The H_{12} perturbation matrix element is actually the product of a calculable $\langle v_1|v_2\rangle$ vibrational factor and an electronic term. This means that each vibrational wavefunction appears exactly twice in the intensity interference term. Although the absolute phases of the three vibrational and three electronic wavefunctions are indeterminate, the sign of the interference term is a measurable quantity and must therefore be calculable *ab initio*. It is a straightforward matter to calculate the sign and magnitude of $H_{ij}^e(R)$ and $R_e^{i,j}(R)$ electronic quantities, but it is essential that the same phase choice be made for each computation. It is conceivable that knowledge of correct *ab initio* signs of $R_e^{i,j}$ and H_{jk}^e electronic quantities combined with observed intensity interference effects could distinguish between a correct and an incorrect deperturbation model.

One of the most dramatic manifestations of an interference effect is the vanishing of a line or of an entire band that, on the basis of known Franck–Condon factors and inappropriately simple intensity borrowing ideas, should be quite intense (see Fig. 5.4). This effect can easily be mistaken as an accidental predissociation (Sections 6.5 and 6.11). Yoshino *et al.* (1979) have studied the valence ~ Rydberg N_2 $b'^1\Sigma_u^+ \sim c_4'^1\Sigma_u^+$ perturbations. Abrupt decreases in emission intensity for c_4'–$X^1\Sigma_g^+$ ($v' = 1$ and 4) and $b' - X$ ($v' = 4$) bands had been attributed to weak predissociation rather than perturbation effects (Gaydon, 1944; Lofthus, 1957; Tilford and Wilkinson, 1964; Wilkinson and Houk, 1956). The b' ($v = 4$) ~ c_4' ($v = 1$) and b' ($v = 13$) ~ c_4' ($v = 4$) deperturbation models of Yoshino *et al.* (1979) provide a predissociation-free unified account of both level shift and intensity effects. Weak predissociation effects cannot be ruled out, but are not needed to account for the present experimental observations.

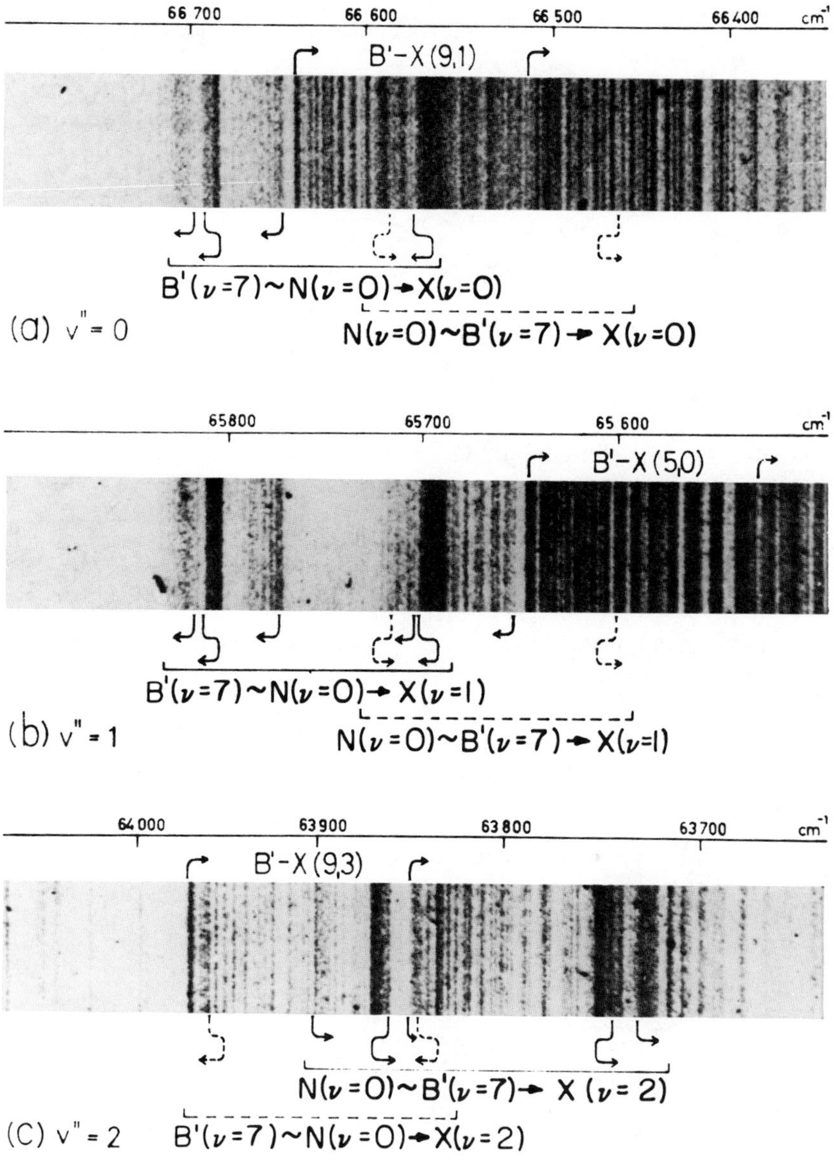

Fig. 5.4 The vanishing of an entire band as a result of a homogeneous perturbation. The emission spectra from the valence ~ Rydberg $^{14}N^{16}O$ $B'^2\Delta$ ($v' = 7$) ~ $N^2\Delta$ ($v' = 0$) perturbation complex into $X^2\Pi$ $v'' = 0$, 1, and 2 (segments a, b, and c, respectively) illustrate the effect of a sign reversal on a near-perfect cancellation of B′–X and N–X transition amplitudes. The N ($v = 0$) level crosses B′ ($v = 7$) from below at $J \simeq 8.4$. The interaction matrix element, $H_{B'7;N0} = 45\ \mathrm{cm}^{-1}$, splits the B′ ~ N mixed levels into two series of rotational levels in which same-J levels are separated by at least 90 cm^{-1}. The higher-energy series of levels, nominally $B'^2\Delta$ for $J' < 8.5$ and nominally $N^2\Delta$ for $J' > 8.5$, appears with enhanced intensity (constructive interference, highlighted by solid arrows below the spectra marking head-like features) in emission into $v'' = 0$ and $v'' = 1$ [labeled B′ ($v = 7$) ~ N ($v = 0$) → X ($v = v''$) on the figure], but with diminished intensity (destructive interference—the weakened features are marked by dotted arrows below the spectra) in emission into $v'' = 2$. The intensity interference affects the lower energy series of levels [labeled N ($v = 0$) ~ B′ ($v = 7$) → X ($v = v''$) on the figure] in the opposite sense. [From Jungen (1966).]

It will be useful to derive the condition for the vanishing of transition probability for the general case

$$\mu_{10}^{0} \neq \mu_{20}^{0}$$

where

$$\mu_{i0}^{0} \equiv R_{e}^{i,0}\langle v_i | v_0 \rangle.$$

From Eq. (5.3.1), the μ_{+0} transition moment will vanish if

$$\frac{C_{1+}}{C_{2+}} = -\frac{\mu_{20}^{0}}{\mu_{10}^{0}}, \tag{5.3.9}$$

where $C_{1+} \geq 0$ by definition and C_{2+} is understood to have the same sign as H_{12}. Thus Eq. (5.3.9) can be satisfied only if $H_{12}\mu_{10}^{0}\mu_{20}^{0} < 0$.

Now, to obtain a relationship between C_{1+} and the observable separation between main and extra lines, $E_{+} - E_{-}$, recall that if

$$|+\rangle = C_{1+}|1\rangle + C_{2+}|2\rangle \tag{5.3.10}$$

is the eigenfunction belonging to the E_{+} eigenenergy (which is true as long as $C_{1+}C_{2+}H_{12} > 0$), then

$$|-\rangle = -C_{2+}|1\rangle + C_{1+}|2\rangle \tag{5.3.11}$$

is the eigenfunction belonging to E_{-}. Left-multiplying Eq. (5.3.10) by $\langle 1|\mathbf{H}$, one obtains

$$\begin{aligned} E_{+}C_{1+} &= C_{1+}E_{1}^{0} + C_{2+}H_{12} \\ E_{+} &= E_{1}^{0} + \frac{C_{2+}}{C_{1+}}H_{12}, \end{aligned} \tag{5.3.12}$$

and similarly left-multiplying Eq. (5.3.11) by $\langle 1|\mathbf{H}$, one obtains

$$E_{-} = E_{1}^{0} - \frac{C_{1+}}{C_{2+}}H_{12}, \tag{5.3.13}$$

and thus

$$E_{+} - E_{-} = \left[\frac{C_{2+}}{C_{1+}} + \frac{C_{1+}}{C_{2+}}\right]H_{12}. \tag{5.3.14a}$$

Alternatively, a similar expression,

$$E_{1}^{0} - E_{2}^{0} = \left[\frac{C_{1+}}{C_{2+}} - \frac{C_{2+}}{C_{1+}}\right]H_{12}, \tag{5.3.14b}$$

is obtained from Eq. (5.3.12) and by left-multiplying Eq. (5.3.10) by $\langle 2|\mathbf{H}$.

Inserting Eq. (5.3.9) into Eq. (5.3.14a) gives the general condition for the vanishing of the transition between level 0 and the *higher-energy* member of the pair of 1 ~ 2 interacting levels,

$$[E_+ - E_-]_{\mu_{+0}=0} = -\left(\frac{\mu_{10}^0}{\mu_{20}^0} + \frac{\mu_{20}^0}{\mu_{10}^0}\right) H_{12}. \tag{5.3.15}$$

Inserting Eq. (5.3.9) into Eq. (5.3.14b) gives

$$[E_1^0 - E_2^0]_{\mu_{+0}=0} = \left[\frac{\mu_{10}^0}{\mu_{20}^0} - \frac{\mu_{20}^0}{\mu_{10}^0}\right] H_{12} \tag{5.3.16}$$

which determines whether the zero in μ_{+0} occurs before $(E_1^0 > E_2^0)$ or after $(E_1^0 < E_2^0)$ the level crossing $(E_1^0 = E_2^0)$. Equation (5.3.15) can be satisfied whenever

$$\mu_{10}^0 \mu_{20}^0 H_{12} < 0$$

and Eq. (5.3.16) can be satisfied for $E_1^0 > E_2^0$ whenever

$$|\mu_{20}^0| > |\mu_{10}^0|.$$

The four possible locations of the intensity null occur as follows:

$$\begin{array}{llll} E_+ & E_1^0 > E_2^0 & \mu_{10}^0\mu_{20}^0 H_{12} < 0 & |\mu_{20}^0| > |\mu_{10}^0| \\ E_+ & E_1^0 < E_2^0 & \mu_{10}^0\mu_{20}^0 H_{12} < 0 & |\mu_{20}^0| < |\mu_{10}^0| \\ E_- & E_1^0 > E_2^0 & \mu_{10}^0\mu_{20}^0 H_{12} > 0 & |\mu_{20}^0| < |\mu_{10}^0| \\ E_- & E_1^0 < E_2^0 & \mu_{10}^0\mu_{20}^0 H_{12} > 0 & |\mu_{20}^0| > |\mu_{10}^0| \end{array} \tag{5.3.17}$$

Equation (5.3.16) was first derived by Dressler (1970). Whenever it is possible to derive intensity parameters from measurements of frequency intervals, far greater accuracy is obtainable than from direct intensity measurements. Equation (5.3.15) allows extremely precise measurement of μ_{10}^0/μ_{20}^0 from the observed $E_+ - E_-$ difference at the J-value of the exact intensity null.

Figure 5.5 illustrates the vanishing of intensity at a homogeneous perturbation. Three cases are shown:

1. Simple intensity borrowing: $\mu_{10}^0 = 0$. Intensity vanishes when $E_+ - E_- \to +\infty$; in the higher-energy eigenvalue when $E_1^0 \gg E_2^0$ and in the lower-energy eigenvalue when $E_1^0 \ll E_2^0$.
2. Interference at the crossing point $(E_1^0 = E_2^0)$:

$$|\mu_{10}^0| = |\mu_{20}^0|.$$

If $\mu_{10}^0\mu_{20}^0 H_{12} > 0$, intensity vanishes for the lower energy eigenvalue at $E_+ - E_- = 2|H_{12}|$.

3. General case: $|\mu_{10}^0| \neq |\mu_{20}^0|$. Example shows $\mu_{10}^0\mu_{20}^0 H_{12} > 0$ and $|\mu_{20}^0| > |\mu_{10}^0|$.

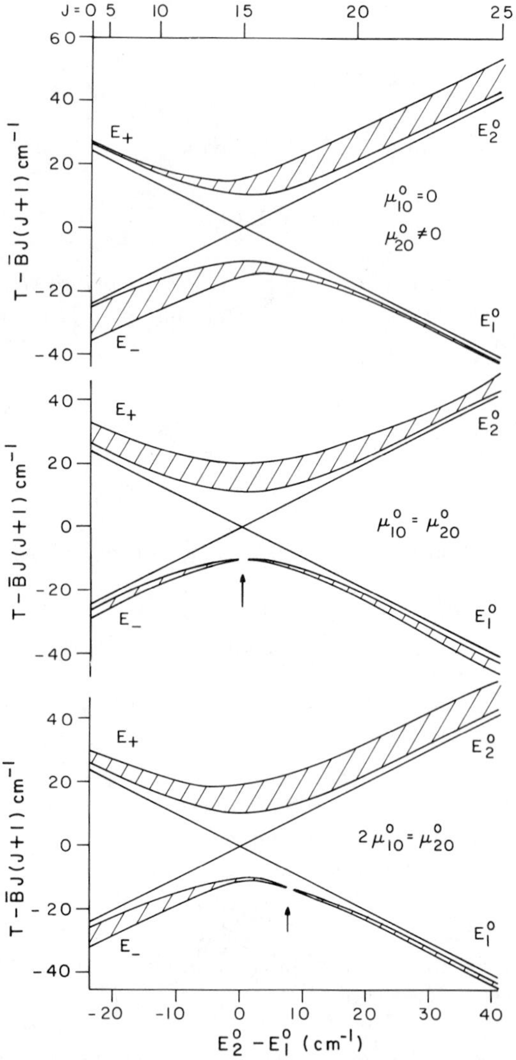

Fig. 5.5 Schematic illustration of intensity interference effects. The straight lines are the basis function (i.e., deperturbed) reduced term values, $E_i^0 - \bar{B}J(J+1)$, where $\bar{B} = (B_1^0 + B_2^0)/2$. The level crossing occurs at $J = 15$ ($B_1^0 = 0.9\ \text{cm}^{-1}$, $B_2^0 = 1.1\ \text{cm}^{-1}$, $H_{12} = 10\ \text{cm}^{-1}$). The inner edges and widths of the solid curves depict respectively the energies of the perturbed levels, E_+ and E_- [reduced by $\bar{B}J(J+1)$], and the intensities of the transitions into E_+ and E_-. The top frame illustrates simple intensity borrowing, $\mu_{10}^0 = 0$. The middle frame shows the vanishing of intensity at the crossing point ($J = 15$), $E_1^0 = E_2^0$, when $\mu_{10}^0 = \mu_{20}^0$. The bottom frame exemplifies the general case, $\mu_{10}^0 \neq \mu_{20}^0 \neq 0$ (here $2\mu_{10}^0 = \mu_{20}^0$), where the location of the single-intensity vanishing point ($J \simeq 17.3$) is given by Eq. (5.3.15) or (5.3.16). See also Fig. 4.8 for other properties of the same homogeneous perturbation illustrated here. [After Dressler (1970).]

In Figure 5.5, the basis function energies E_1^0 and E_2^0 are shown as straight lines, plotted as

$$E_i^0 - \frac{E_1^0 + E_2^0}{2} \qquad \text{versus} \quad E_2^0 - E_1^0.$$

The eigenenergies and transition strengths are shown as double solid lines, plotted as $E_\pm - \bar{E}^0$ versus ΔE^0, where the width of the shaded region depicts the intensity of the $E_\pm \leftrightarrow E_0$ transition.

Interference effects are not limited to simple two-level interactions. Stahel *et al.* (1983) have performed a multilevel deperturbation ($^1\Sigma$-matrix: 41 levels; $^1\Pi$-matrix: 30 levels) of the N_2 $^1\Sigma_u^+$ and $^1\Pi_u$ states. Prior to the work of Dressler (1969), the numerous known N_2 vibronic levels of $^1\Sigma_u^+$ and $^1\Pi_u$ symmetry between 100,000 and 120,000 cm^{-1} had not been convincingly organized into separate electronic states. Dressler (1969) showed that the chaotic level and intensity patterns resulted from homogeneous perturbations of the b$^1\Pi_u$ and b$'^1\Sigma_u^+$ valence states by the c$^1\Pi_u$ (N_2^+ X$^2\Sigma_g^+$, $3p\pi$), o$^1\Pi_u$(N_2^+ A$^2\Pi_u$, $3s\sigma$), e$^1\Pi_u$(X, $4p\pi$), c$'^1\Sigma_u^+$(X, $3p\sigma$), and e$'^1\Sigma_u^+$(X, $4p\sigma$) Rydberg states.

Figure 5.6 illustrates a pronounced intensity interference effect for the two $^1\Pi_u$ levels at 104,700 and 105,346 cm^{-1}. These levels are nominally b$^1\Pi_u$ ($v=5$ and 6); however both levels have significant c$^1\Pi_u$ ($v=0$) and o$^1\Pi_u$ ($v=0$) character. The dotted tie-lines on the figure depict the deperturbed intensity distributions for transitions from X$^1\Sigma_g^+$ ($v = 0$) (Franck–Condon factor times R-independent electronic transition moment); the heavy vertical lines are the observed (perturbed) relative intensities. The small intensities of the nominal b–X (5, 0) and (6, 0) bands, despite the large μ_{bX}^0 and μ_{cX}^0 transition moments for the b$^1\Pi_u$ and c$^1\Pi_u$ basis components of the two observed levels, are a result of destructive interference between b–X and c–X transition amplitudes. The nominal c$^1\Pi_u$ ($v = 0$) level lies below the b($v = 5$ and 6) levels, and the interference effect is such that the intensity null occurs in the upper (E_+) component. This implies that

$$\mu_{\mathrm{bX}}^0 \mu_{\mathrm{cX}}^0 H_{\mathrm{bc}} < 0$$

for the $v_{\mathrm{b}} = 5, 6$, $v_{\mathrm{c}} = 0$, and $v_{\mathrm{X}} = 0$ basis levels. Intensity is transferred from the nominal b–X (5, 0) and (6, 0) bands into the nominal c–X (0, 0) band. A similar explanation accounts for the excess intensity in the b–X (3, 0) and (4, 0) bands. The c–X (0, 0) transition amplitude interferes with the b–X amplitude, again in the sense

$$\mu_{\mathrm{bX}}^0 \mu_{\mathrm{cX}}^0 H_{\mathrm{bc}} < 0,$$

with the result that intensity is transferred into the lower (E_-) component. For a more complete discussion of intensity interference effects of the type shown in Fig. 5.6 as well as of other types, see Stahel *et al.* (1983).

The intensity interference pattern illustrated by Fig. 5.6 for bound–bound

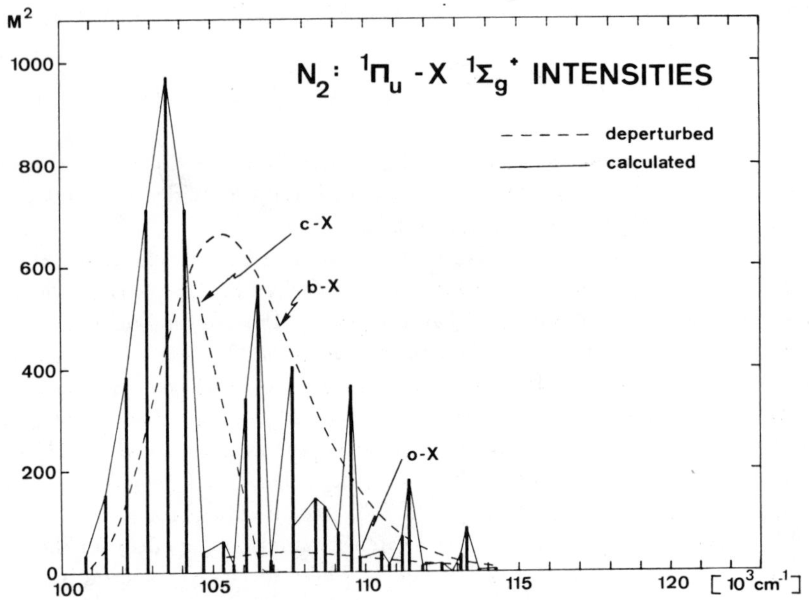

Fig. 5.6 Calculated perturbed and deperturbed intensities for $N_2\ {}^1\Pi_u \leftarrow X^1\Sigma_g^+$ $(v = 0)$ transitions. The valence $b^1\Pi_u$ and Rydberg $c^1\Pi_u(3p\pi)$ and $o^1\Pi_u$ (N_2^+ $A^2\Pi_u$, $3s\sigma$) states interact strongly, giving rise to a chaotic appearing spectrum. Stahel *et al.* (1983) have deperturbed the $N_2\ {}^1\Sigma_u^+$ and ${}^1\Pi_u$ states. The dotted lines show the Franck–Condon envelopes [from $X^1\Sigma_g^+$ $(v'' = 0)$] for the deperturbed states ($X^2\Sigma_g^+$–core Rydberg sharply peaked at $v' = 0$, valence and $A^2\Pi$–core Rydberg broadly peaked at high v'). The solid lines are the calculated intensities for transitions into strongly mixed eigenstates. Note the near-perfect intensity cancellation for the three transitions into b∼c mixed levels near 105,000 cm^{-1}. [From Stahel *et al.* (1983)].

transitions is very similar to the Beutler–Fano lineshape for bound–free transitions discussed in Sections 6.7 and 7.8. The 101,000–106,000 cm^{-1} region of Fig. 5.6 is a band-by-band rather than a continuous representation of a Fano profile with $q < 0$ [see Fig. 6.15 and compare Eq. (6.7.6) to (5.3.17)].

5.3.2 $\Delta\Lambda = \pm 1$ PERTURBATIONS; ROTATIONAL–BRANCH INTENSITY ANOMALIES

There are two classes of transitions between pure case (a) basis functions. *Parallel* and *perpendicular* transitions are readily distinguished by their characteristic rotational branch intensity patterns. $\Delta\Lambda = 0$ transitions are called parallel bands and have R and P branches of comparable intensity and a weak or absent Q branch. $\Delta\Lambda = \pm 1$ transitions are perpendicular bands and have a strong Q branch and R and P branches of approximately half the intensity of the Q branch.

When a $\Delta\Lambda = \pm 1$ perturbation occurs, the PQR intensity pattern for transitions terminating on the mixed levels is not necessarily intermediate between the parallel and perpendicular patterns. For example, consider a perturbation in the upper electronic state which affects the J' level. The intensity ratio for the $P(J'+1)$ and $R(J'-1)$ lines can depart from the approximate 1:1 ratio typical of both parallel and perpendicular unperturbed transitions. In fact, one or the other line can vanish. Furthermore, in a fluorescence progression from a mixed upper level, $v'J'$, the $P(J'+1)$-line can be anomalously weak relative to $R(J'-1)$ in some (v', v'') bands and anomalously strong in others. The sign and magnitude of the effect can vary with v'' even though the perturbation is in the v' level!

The interference effects discussed in Section 5.3.1 could be explained without detailed examination of rotational line strength factors. Equation (5.3.6) shows that, for perturbations between states of the same symmetry, transition intensities can be expressed as a product of the usual Hönl–London rotational factor times a sum of terms involving products of electronic and vibrational matrix elements. The interference effects were of electronic–vibrational origin, and it was unnecessary to consider the signs of matrix elements of the rotational part of the wavefunction, $|\Omega JM\rangle$. The interference effects resulting from $\Delta\Lambda = \pm 1$ perturbations always affect the intensities of $\Delta J = +1$ and -1 transitions out of each mixed level in equal and opposite amounts, because the transition amplitude phases for P and R lines of parallel transitions are identical while those for P and R lines of perpendicular transitions are opposite. If $\| \sim \perp$ interference is constructive for a P line, it will inevitably be destructive for the R line (Klynning, 1974).

The operators responsible for electric dipole transitions are molecule-fixed components of $\boldsymbol{\mu}$ [Eq. (5.1.2)]. $\boldsymbol{\mu}$ operates exclusively on the spatial coordinates of the electrons, and hence has selection rules $\Delta S = \Delta\Sigma = 0$ and, because it is a vector operator with respect to the spatial coordinates (see Section 2.4.5), $\Delta\Lambda = \Delta\Omega = 0, \pm 1$. $\Delta\Lambda = 0$ transitions arise from the $\boldsymbol{\mu}_z$ operator component. The relationship between $\Lambda \rightarrow \Lambda$ and $-\Lambda \rightarrow -\Lambda$ transition amplitudes may be derived from

$$\boldsymbol{\sigma}_v(xz)\,|n\Lambda^s\rangle = (-1)^{\Lambda+s}|n, -\Lambda^s\rangle \qquad \boldsymbol{\sigma}_v(xz)\boldsymbol{\mu}_z = \boldsymbol{\mu}_z$$

[see Eqs. (2.2.22), (2.2.25), and (2.2.32), and also Hougen (1970)] by applying $\boldsymbol{\sigma}_v$ to the two wavefunctions and to the operator in $\langle n'\Lambda^{s'}|\boldsymbol{\mu}_z|n''\Lambda^{s''}\rangle$ to give

$$\langle n'\Lambda^{s'}|\boldsymbol{\mu}_z|n''\Lambda^{s''}\rangle = (-1)^{s'+s''}\langle n', -\Lambda^{s'}|\boldsymbol{\mu}_z|n'', -\Lambda^{s''}\rangle \equiv \mu_{\|}\,\delta_{s's''}. \quad (5.3.18)^\dagger$$

The $\Sigma^+ \nleftrightarrow \Sigma^-$ transition selection rule is proved by Eq. (5.3.18), since

$$\langle \Sigma^+|\boldsymbol{\mu}_z|\Sigma^-\rangle = -\langle \Sigma^+|\boldsymbol{\mu}_z|\Sigma^-\rangle = 0.$$

† n is a shorthand for the electronic name and the vibrational quantum number of the state; s is the Σ^+/Σ^- symmetry, and $s = 0$ for all states except Σ^- states, for which $s = 1$.

$\Delta\Lambda = \pm 1$ transitions arise from

$$\boldsymbol{\mu}^{\pm} \equiv \boldsymbol{\mu}_x \pm i\boldsymbol{\mu}_y$$

and

$$\langle n'(\Lambda \pm 1)^{s'}|\boldsymbol{\mu}^{\pm}|n''\Lambda^{s}\rangle = -(-1)^{s+s'}\langle n'(-\Lambda \mp 1)^{s'}|\boldsymbol{\mu}^{\mp}|n'', -\Lambda^{s}\rangle \equiv 2^{1/2}\mu_{\perp} \tag{5.3.19}$$

where the relationship between the $+\Lambda$ and $-\Lambda$ matrix elements is given by Eq. (2.2.32) and

$$\boldsymbol{\sigma}_v\boldsymbol{\mu}^{\pm} = \boldsymbol{\mu}^{\mp}\boldsymbol{\sigma}_v.$$

Hougen (1970) has shown that it is always possible to define phases so that $\mu_{\parallel}$, $\mu_{\perp}$, and all perturbation mixing coefficients are real quantities. This fact has not always been recognized in previous work (Dieke, 1941) and has given rise to some errors in the literature.

Since transitions are observed in a space fixed rather than molecule-fixed reference system, it is necessary to compute matrix elements of space-fixed components of $\boldsymbol{\mu}$, for example (for $\Delta M_J = 0$ transitions),

$$\boldsymbol{\mu}_Z = \tfrac{1}{2}\boldsymbol{\alpha}_Z^{+}\boldsymbol{\mu}^{-} + \tfrac{1}{2}\alpha_Z^{-}\boldsymbol{\mu}^{+} + \boldsymbol{\alpha}_Z^{z}\boldsymbol{\mu}_z, \tag{5.3.20}$$

where the $\boldsymbol{\alpha}_I^{j}$ are direction cosine operator components (see Section 1.3.1) and

$$\boldsymbol{\alpha}_Z^{\pm} \equiv \boldsymbol{\alpha}_Z^{x} \pm i\boldsymbol{\alpha}_Z^{y}.$$

Matrix elements of $\boldsymbol{\mu}_Z$ may be factored into empirically definable, *electronic transition-specific* molecule-fixed, nonrotating molecule ($|n\Lambda^{s}S\Sigma\rangle$) matrix elements ($\mu_{\perp}$ or $\mu_{\parallel}$ parameters) and *universal*, tabulated (Hougen, 1970, p. 31 and Table 1.1), rotating molecule ($|\Omega JM\rangle$) matrix elements of $\boldsymbol{\alpha}$. Although all matrix elements of $\boldsymbol{\alpha}$ are tabulated, it is convenient to derive the relationship between (Ω', Ω'') and $(-\Omega', -\Omega'')$ matrix elements. Following Hougen (1970),

$$\boldsymbol{\sigma}_v\boldsymbol{\alpha}_Z^{z} = -\boldsymbol{\alpha}_Z^{z}\boldsymbol{\sigma}_v \qquad \boldsymbol{\sigma}_v\boldsymbol{\alpha}_Z^{\pm} = -\boldsymbol{\alpha}_Z^{\mp}\boldsymbol{\sigma}_v$$

$$\boldsymbol{\sigma}_v|\Omega JM\rangle = (-1)^{J-\Omega}|-\Omega JM\rangle \qquad \text{(see Eq. 2.2.30)}$$

therefore, for $\Delta\Omega = 0$ transitions (parallel since $\Delta\Omega = \Delta\Lambda$)

$$\langle \Omega J'M|\boldsymbol{\alpha}_Z^{z}|\Omega J''M\rangle = (-1)^{J'+J''-2\Omega+1}\langle -\Omega J'M|\boldsymbol{\alpha}_Z^{z}|-\Omega J''M\rangle. \tag{5.3.21a}$$

$(-1)^{J'+J''-2\Omega+1}$ is respectively $+1$, -1, $+1$ for P, Q, R transitions. For $\Delta\Omega = \pm 1$ transitions (perpendicular),

$$\langle \Omega \mp 1, J'M|\boldsymbol{\alpha}_Z^{\pm}|\Omega J''M\rangle = (-1)^{J'+J''-2\Omega\pm 1+1}\langle -\Omega \pm 1, J'M|\boldsymbol{\alpha}_Z^{\mp}|\Omega J''M\rangle. \tag{5.3.21b}$$

$(-1)^{J'+J''-2\Omega\pm 1+1}$ is respectively -1, $+1$, -1 for P, Q, R transitions.

Combining Eqs. (5.3.19)–(5.3.21), one obtains for $\Delta\Omega = 0$ transitions,

$$\begin{aligned}
&\langle n'\Lambda S\Sigma|\langle\Omega J'M|\boldsymbol{\mu}_Z|\Omega J''M\rangle|n''\Lambda S\Sigma\rangle \\
&\quad = \mu_{\|}(n'\Lambda, n''\Lambda)\langle\Omega J'M|\boldsymbol{\alpha}_Z^z|\Omega J''M\rangle \\
&\quad = (-1)^{J'+J''-2\Omega+1}\langle n', -\Lambda S, -\Sigma| \\
&\qquad \times \langle -\Omega J'M|\boldsymbol{\mu}_Z|-\Omega J''M\rangle|n'', -\Lambda, S, -\Sigma\rangle
\end{aligned} \tag{5.3.22a}$$

and for $\Delta\Omega = \pm 1$ transitions,

$$\begin{aligned}
&\langle n'(\Lambda \pm 1)^{s'}S\Sigma|\langle\Omega \pm 1, J'M|\boldsymbol{\mu}_Z|\Omega J''M\rangle|n''\Lambda^{s''}S\Sigma\rangle \\
&\quad = 2^{-1/2}\mu_{\perp}[n'(\Lambda \pm 1)^{s'}, n''\Lambda^{s''}]\langle\Omega \pm 1, J'M|\boldsymbol{\alpha}_Z^{\mp}|\Omega J''M\rangle \\
&\quad = (-1)^{J'+J''-2\Omega\pm 1+2+s'+s''}\langle n', -\Lambda \mp 1, S, -\Sigma| \\
&\qquad \times \langle -\Omega \mp 1, J'M|\boldsymbol{\mu}_Z|-\Omega J''M\rangle|n'', -\Lambda, S, -\Sigma\rangle.
\end{aligned} \tag{5.3.22b}$$

When one forms e/f-symmetrized basis functions [Eqs. (2.2.34), (2.2.37), (2.2.38)] by taking linear combinations of the form

$$2^{-1/2}[|\Omega JM\rangle|n\Lambda S\Sigma\rangle \pm |-\Omega JM\rangle|n, -\Lambda, S, -\Sigma\rangle],$$

one finds that, for both parallel and perpendicular transitions, R and P transitions are $e \leftrightarrow e$ or $f \leftrightarrow f$ and Q transitions are $e \leftrightarrow f$ or $f \leftrightarrow e$.

For parallel transitions, matrix elements of $\boldsymbol{\mu}_Z\boldsymbol{\alpha}_Z^z$ evaluated in the symmetrized $(e/f, |\Omega|, |\Lambda|)$ and unsymmetrized (signed Λ, Ω) basis sets are identical and equal to $\mu_{\|}\alpha_Z^z(\Omega J'M; \Omega JM)$. For perpendicular transitions between two non-Σ states, matrix elements of $\frac{1}{2}(\boldsymbol{\mu}^+\boldsymbol{\alpha}_Z^- + \boldsymbol{\mu}^-\boldsymbol{\alpha}_Z^+)$ evaluated in the symmetrized and unsymmetrized basis sets are identical and equal to $2^{-1/2}\mu_{\perp}$ α_Z^- $(\Omega + 1, J'M, \Omega JM)$ where, from the Eq. (5.3.19) definition,

$$\mu_{\perp} \equiv +2^{-1/2}\langle n', \Lambda + 1|\boldsymbol{\mu}^+|n\Lambda\rangle \qquad \text{for} \quad \Lambda > 0.$$

For perpendicular transitions involving one $\Sigma_0^{\pm}$ state, matrix elements of $\boldsymbol{\mu}_Z$ evaluated in the unsymmetrized basis are smaller, by a factor of $2^{1/2}$, than the corresponding symmetrized matrix element. In the unsymmetrized basis, matrix elements of $\boldsymbol{\mu}_Z$ for $\Pi_{\Omega'} - \Sigma_{\Omega}^{\pm}$ transitions are equal to $2^{-1/2}\mu_{\perp}$ α_Z^- $(\Omega + 1, J', M; \Omega JM)$ times a phase factor as in the following table.

Transition	$\Omega' = 1 + \|\Sigma\|$	$\Omega' = 1 - \|\Sigma\|$
Even multiplicity		
R_{ff}^{ee} and P_{ff}^{ee}	$+1$	$\pm(-1)^{-S+s+1/2}$
Q_{fe}^{ef}	$+1$	$\mp(-1)^{-S+s+1/2}$
Odd multiplicity		
R_{ff}^{ee} and P_{ff}^{ee}	$+1$	$\pm(-1)^{-S+s}$
Q_{fe}^{ef}	$+1$	$\mp(-1)^{-S+s}$

The difference in phase factor for $\Omega' = 1 - |\Sigma|\,\Pi_{\Omega'}$–$\Sigma^+_\Omega$ versus $\Pi_{\Omega'}$–Σ^-_Ω transitions is the source of the $\mu_\perp \sim \mu_\perp$ interference effect discussed in Section 5.3.3.

Equation (5.3.22) can be simplified, for the case of isotropically oriented molecules, unpolarized radiation, and zero external magnetic or electric fields, by "summing" over M (see Hougen, 1970, p. 39).† The resultant M-independent $\langle \Omega' J' |\boldsymbol{\alpha}| \Omega J\rangle$ direction cosine matrix elements are listed in Table 5.1. Note that the $\boldsymbol{\alpha}^\pm$ $\Delta\Omega = \mp 1$ matrix elements have opposite signs for P versus R transitions, whereas the $\boldsymbol{\alpha}^z$ $\Delta\Omega = 0$ matrix elements have the same signs for P and R transitions.

It is possible to express the intensities of all rotational branches of any unperturbed electric dipole allowed ${}^{2S+1}\Lambda'$–${}^{2S+1}\Lambda''$ transition in terms of a single $\mu_\perp$ or $\mu_\parallel$ parameter. This is true even for transitions in which one or both of the electronic states is in an intermediate case (a)–(b) Hund's coupling and the transition intensities cannot be reduced to simple closed form expressions. For example, the individual line intensities in all 54 rotational branches of a ${}^3\Delta$–${}^3\Pi$ transition are proportional to a single $\mu_\perp$ parameter. The only nonzero case (a) *basis state transition moments* are for $\Delta\Omega = \Delta\Lambda = +1$, and their values are $2^{-1/2}\mu_\perp$ times the factor listed in the table below, which we will refer to as Eq. (5.3.23).

Transition	$R_{ee}(J)$ or $R_{ff}(J)$	$Q_{ef}(J)$ or $Q_{fe}(J)$	$P_{ee}(J)$ or $P_{ff}(J)$
${}^3\Delta_3$–${}^3\Pi_2$	$-\left[\frac{(J+3)(J+4)}{3(J+1)}\right]^{1/2}$	$+\left[\frac{(2J+1)(J-2)(J+3)}{3J(J+1)}\right]^{1/2}$	$+\left[\frac{(J-2)(J-3)}{3J}\right]^{1/2}$
${}^3\Delta_2$–${}^3\Pi_1$	$-\left[\frac{(J+2)(J+3)}{3(J+1)}\right]^{1/2}$	$+\left[\frac{(2J+1)(J-1)(J+2)}{3J(J+1)}\right]^{1/2}$	$+\left[\frac{(J-1)(J-2)}{3J}\right]^{1/2}$
${}^3\Delta_1$–${}^3\Pi_0$	$-\left[\frac{(J+1)(J+2)}{3(J+1)}\right]^{1/2}$	$+\left[\frac{(2J+1)J(J+1)}{3J(J+1)}\right]^{1/2}$	$+\left[\frac{J(J-1)}{3J}\right]^{1/2}$

The *eigenstate transition moments* may be obtained from Eq. (5.3.23) by finding the eigenfunctions of the ${}^3\Delta$ and ${}^3\Pi$ effective Hamiltonian matrices for each value of the good quantum numbers J and e/f. For example, the nominal ${}^3\Delta_{3e}$ and

† In the absence of an external field, M is a good quantum number. Since $\mathbf{H}$ is a scalar operator, the $\mathbf{H}$-matrix factors into $2J + 1$ identical M-blocks for each J. In addition, the M-dependence of $\boldsymbol{\mu}_Z$ matrix elements is contained in a (J', J'', M)-dependent factor that is identical for all initial and final electronic basis functions. The average over the M-dependence of *transition probabilities* is accomplished by squaring the M-dependent *transition amplitude* factor, summing over M, and taking the square root.

Table 5.1
M-Independent $\langle \Omega' J' | \boldsymbol{\alpha} | \Omega J \rangle$ Direction Cosine Matrix Elements

	R $J' = J + 1$	Q $J' = J$	P $J' = J - 1$
$\alpha^z(\Omega' = \Omega)$	$\left[\frac{(J+\Omega+1)(J-\Omega+1)}{3(J+1)}\right]^{1/2}$	$\Omega\left[\frac{2J+1}{3J(J+1)}\right]^{1/2}$	$\left[\frac{(J+\Omega)(J-\Omega)}{3J}\right]^{1/2}$
$\alpha^{\pm}(\Omega' = \Omega \mp 1)$	$\pm\left[\frac{(J\mp\Omega+1)(J\mp\Omega+2)}{3(J+1)}\right]^{1/2}$	$\left[\frac{(2J+1)(J\pm\Omega)(J\mp\Omega+1)}{3J(J+1)}\right]^{1/2}$	$\mp\left[\frac{(J\pm\Omega)(J\pm\Omega-1)}{3J}\right]^{1/2}$

$^3\Pi_{0e}$ eigenfunctions are

$$|'^3\Delta_{3e'}, J'\rangle = C^{\Delta}_{3e,3e}(J')|^3\Delta_{3e}, J'\rangle + C^{\Delta}_{3e,2e}(J')|^3\Delta_{2e}, J'\rangle + C^{\Delta}_{3e,1e}(J')|^3\Delta_{1e}, J'\rangle \tag{5.3.24a}$$

$$|'^3\Pi_{0e'}, J''\rangle = C^{\Pi}_{0e,2e}(J'')|^3\Pi_{2e}, J''\rangle + C^{\Pi}_{0e,1e}(J'')|^3\Pi_{1e}, J''\rangle + C^{\Pi}_{0e,0e}(J'')|^3\Pi_{0e}, J''\rangle \tag{5.3.24b}$$

where the mixing coefficients $C^{\Lambda}_{\Omega,\Omega'}(J)$ give the $|\Lambda_\Omega, J\rangle$ basis function character present in the nominal $|'\Lambda_\Omega{}'J\rangle$ eigenstate. Thus, the nominally forbidden $^3\Delta_3$–$^3\Pi_0\, P_{ee}(J)$ transition probability is proportional to

$$|\langle'^3\Delta'_{3e}, J-1|\boldsymbol{\mu}|'^3\Pi'_{0e}, J\rangle|^2$$

$$= \mu_\perp^2/2\left\{+C^{\Delta}_{3e,3e}(J-1)C^{\Pi}_{0e,2e}(J)\left[\frac{(J-2)(J-3)}{3J}\right]^{1/2} + C^{\Delta}_{3e,2e}(J-1)C^{\Pi}_{0e,1e}(J)\left[\frac{(J-1)(J-2)}{3J}\right]^{1/2} + C^{\Delta}_{3e,1e}(J-1)C^{\Pi}_{0e,0e}(J)\left[\frac{J(J-1)}{3J}\right]^{1/2}\right\}^2. \tag{5.3.25}$$

The signs and magnitudes of the $C^{\Delta}_{\Omega,\Omega'}$ and $C^{\Pi}_{\Omega,\Omega'}$ factors depend in a complicated way on the molecular constants that define the effective Hamiltonians, but most importantly on the sign and magnitude of A_Λ/B_Λ.

The transition amplitudes in Eq. (5.3.25) interfere with each other systematically, in a way that implicitly reflects the fine structures of the two multiplet states. At very low J, when both states are at the case (a) limit, there will be 18 strong branches (exclusively $\Delta\Omega = \Delta\Lambda = +1$); at very high J, when both states are at the case (b) limit, there will again be 18 strong branches (exclusively $\Delta N = \Delta J = 0, \pm 1$), few of which are continuations of the strong low-J branches. The branch intensity patterns display the same pattern-forming rotational quantum number [J, N, J, N^+ in cases (a), (b), (c), (d)], as does the energy-level structure (see Fig. 2.1). This sort of systematic interference effect is important for transitions into Rydberg complexes. The NO $nf \leftarrow X^2\Pi$ system (Jungen and Miescher, 1969) is a beautiful example of this effect.

Before discussing the intensity interference effects resulting from the multiple $\Delta\Lambda = \pm 1$ perturbations in an nf-complex ($f\phi \sim f\delta \sim f\pi \sim f\sigma$), it is useful to analyze several examples of isolated $\Delta\Lambda = \pm 1$ perturbations:

1. a $^1\Pi \sim \Sigma^+$ perturbation;
2. a case (c) $\Omega = \frac{1}{2}$ state;
3. a forbidden $^3\Sigma^-$–$^1\Sigma^+$ transition; and
4. an np Rydberg complex.

Consider transitions between a pair of mutually interacting $^1\Sigma^+$ and $^1\Pi$ upper states and an unperturbed $^1\Sigma^+$ lower state. The $^1\Sigma^+–^1\Sigma^+$ system has only one P and one R branch, and the transition amplitudes are

$$R(J) \quad +\mu_{\parallel}\left(\frac{J+1}{3}\right)^{1/2} \qquad P(J) \quad +\mu_{\parallel}\left(\frac{J}{3}\right)^{1/2}. \tag{5.3.26}$$

The $^1\Pi–^1\Sigma^+$ transition has only three branches, a Q branch (from $^1\Pi_f$ levels) and R and P branches (from $^1\Pi_e$ levels). The $^1\Pi$ e/f basis functions are [Eq. (2.2.38)]

$$|^1\Pi(^e_f), J'\rangle = 2^{-1/2}[|J', \Lambda = \Omega = 1|\rangle \pm |J', \Lambda = \Omega = -1\rangle],$$

and the $^1\Pi–^1\Sigma^+$ transition amplitudes are

$$\begin{aligned} &R(J) \qquad -\mu_{\perp}\left(\frac{J+2}{3}\right)^{1/2} \\ &Q(J) \qquad +\mu_{\perp}\left(\frac{2J+1}{3}\right)^{1/2} \\ &P(J) \qquad +\mu_{\perp}\left(\frac{J-1}{3}\right)^{1/2}. \end{aligned} \tag{5.3.27}$$

Only the $^1\Pi_e$ levels can mix with $^1\Sigma^+$. The eigenfunctions are

$$\begin{aligned} |'^1\Pi'_e, J\rangle &= C_{\Pi\Pi}(J)\,|^1\Pi_e, J\rangle + C_{\Pi\Sigma}(J)\,|^1\Sigma^+, J\rangle \\ |'^1\Sigma^{+\prime}, J\rangle &= C_{\Sigma\Pi}(J)\,|^1\Pi_e, J\rangle + C_{\Sigma\Sigma}(J)\,|^1\Sigma^+, J\rangle \\ |'^1\Pi'_e, J\rangle &= |^1\Pi_f, J\rangle, \end{aligned} \tag{5.3.28}$$

where

$$\begin{aligned} C_{\Sigma\Sigma}(J) &= \quad C_{\Pi\Pi}(J) \\ C_{\Sigma\Pi}(J) &= -C_{\Pi\Sigma}(J) = \mp[1 - C_{\Pi\Pi}(J)^2]^{1/2}. \end{aligned}$$

The perturbed transition amplitudes are given below, referred to as Eq. (5.3.29).

	$'^1\Pi' \rightarrow {}^1\Sigma^+$	
$R(J)$	$-C_{\Pi\Pi}(J+1)\mu_{\perp}\left(\frac{J+2}{3}\right)^{1/2} + C_{\Pi\Sigma}(J+1)\mu_{\parallel}\left(\frac{J+1}{3}\right)^{1/2}$	
$Q(J)$	$\mu_{\perp}\left(\frac{2J+1}{3}\right)^{1/2}$	(5.3.29a)
$P(J)$	$+C_{\Pi\Pi}(J-1)\mu_{\perp}\left(\frac{J-1}{3}\right)^{1/2} + C_{\Pi\Sigma}(J-1)\mu_{\parallel}\left(\frac{J}{3}\right)^{1/2}$	

	$'^1\Sigma^{+\prime} \to {}^1\Sigma^+$
$R(J)$	$+C_{\Pi\Sigma}(J+1)\mu_\perp\left(\frac{J+2}{3}\right)^{1/2} + C_{\Pi\Pi}(J+1)\mu_\parallel\left(\frac{J+1}{3}\right)^{1/2}$
$Q(J)$	0
$P(J)$	$-C_{\Pi\Sigma}(J-1)\mu_\perp\left(\frac{J-1}{3}\right)^{1/2} + C_{\Pi\Pi}(J-1)\mu_\parallel\left(\frac{J}{3}\right)^{1/2}.$

(5.3.29b)

If $\mu_\perp = 0$, then the spectrum contains four lines for each J'-value, a pair of main $'^1\Sigma^{+\prime}$–$^1\Sigma^+$ R and P lines and a pair of extra R and P lines. The ratio of transition probabilities for the two main lines terminating in a common J' level is

$$\frac{I[R(J-1)]}{I[P(J+1)]} = \frac{J}{J+1}, \tag{5.3.30}$$

which is identical to the R/P ratio for the extra lines and to the ratio expected for an unperturbed $^1\Sigma^+$–$^1\Sigma^+$ transition. If $\mu_\parallel = 0$, then the spectrum contains five lines for each J', an unperturbed Q line, a pair of main $'^1\Pi'$–$^1\Sigma^+$ R and P lines, and a pair of extra R and P lines. The ratio of transition probabilities for the main lines is

$$\frac{I[R(J-1)]}{I[P(J+1)]} = \frac{J+1}{J}, \tag{5.3.31}$$

which is identical to the R/P ratio for the extra lines and to the ratio expected for an unperturbed $^1\Pi$–$^1\Sigma^+$ transition. In addition, there is a sum rule (for $\mu_\parallel = 0$),

$$\begin{aligned} I[Q(J)] = I[R_{\text{main}}(J-1)] + I[R_{\text{extra}}(J-1)] + I[P_{\text{main}}(J+1)] \\ + I[P_{\text{extra}}(J+1)]. \end{aligned} \tag{5.3.32}$$

If

$$C_{\Pi\Pi}\mu_\perp \simeq C_{\Pi\Sigma}\mu_\parallel,$$

then the $'^1\Pi'$–$^1\Sigma^+$ R/P ratio is approximately zero. Similarly, if

$$C_{\Pi\Pi}\mu_\parallel \simeq C_{\Pi\Sigma}\mu_\perp,$$

then the $'\Sigma^{+\prime}$–$^1\Sigma^+$ R/P ratio is very large. If $\mu_\parallel = \mu_\perp$, then the higher-energy P line and the lower-energy R line (or vice versa depending on the sign of the interaction matrix element between $^1\Pi$ and $^1\Sigma^+$, $H_{\Pi\Sigma}$) vanish at the $^1\Pi$–$^1\Sigma^+$ level crossing, J_0. Regardless of the signs and relative magnitudes of $\mu_\parallel$ and $\mu_\perp$, the sense of the R/P intensity anomaly in the nominal $'^1\Pi'$–$^1\Sigma^+$ transition will always be opposite (but not necessarily equal in magnitude) to that for the same J'-value in $'^1\Sigma'-^1\Sigma^+$. The sense of this anomaly will reverse, within a given *nominal* band, for J-values below and above J_0.

Gottscho *et al.* (1978) observed R/P intensity anomalies in the BaO $C^1\Sigma^+ \rightarrow X^1\Sigma^+$ fluorescence spectrum. Along a (v', v'') progression, the magnitude and sense of the anomalies exhibited a strong dependence on the $X^1\Sigma^+$ state v'' level, even though the $^1\Pi \sim {}^1\Sigma^+$ mixing was in the $C^1\Sigma^+$ state. For several v'' members of the C–X (v', v'') progression, either the $P(J' + 1)$ or the $R(J' - 1)$ line was undetectable, even though the remaining line was quite strong. Since the fractional R/P intensity anomaly,

$$\frac{I[R(J-1)] - I[P(J+1)]}{I[R(J-1)] + I[P(J+1)]} \simeq \pm \frac{2C_{\Sigma\Sigma}(J)[1 - C_{\Sigma\Sigma}^2(J)]^{1/2}\mu_{\parallel}\mu_{\perp}}{C_{\Sigma\Sigma}^2(J)\,\mu_{\parallel}^2 + [1 - C_{\Sigma\Sigma}^2(J)]\,\mu_{\perp}^2}, \qquad (5.3.33)$$

depends on v'' (the only v''-dependent terms are $\mu_{\parallel}$ and $\mu_{\perp}$), the intensity anomalies could be used to characterize the otherwise unobserved $^1\Pi$-perturber. A similar R/P intensity anomaly has been observed in fluorescence from Se_2 (Gouedard and Lehmann, 1976).

Case (c) $\frac{1}{2}$–$\frac{1}{2}$ transitions have been discussed in detail by Kopp and Hougen (1967) [where the definition of $\mu_{\perp}$ differs by a factor of $2^{1/2}$ from that used here and by Hougen (1970)]. Case (c) corresponds to the strong spin–orbit limit where $\Delta\Omega = 0$ spin–orbit perturbations have destroyed both S and Λ as useful quantum numbers. For simplicity, the $\frac{1}{2}$–$\frac{1}{2}$ example to be discussed here will be a J-independent spin–orbit mixed $^2\Sigma^+ \sim {}^2\Pi_{1/2}\,\Omega' = \frac{1}{2}$ upper level and a pure $^2\Sigma^+\ \Omega'' = \frac{1}{2}$ lower level. $\frac{1}{2}$–$\frac{1}{2}$ transitions consist, in general, of six rotational branches: R_{ee}, R_{ff}, Q_{ef}, Q_{fe}, P_{ee}, and P_{ff}. Although a general case (c) $\Omega = \frac{1}{2}$ state can be composed of an unspecified mixture of $^2\Sigma^+$, $^2\Sigma^-$, $^2\Pi_{1/2}$, $^4\Sigma^+$, $^4\Sigma^-$, $^4\Pi$, $^4\Delta$, etc. case (a) basis states, the simplified example treated here contains all of the features of rotational linestrength anomalies in a $\frac{1}{2}$–$\frac{1}{2}$ case (c) transition.

Just as for the $^1\Pi \sim {}^1\Sigma^+ \rightarrow {}^1\Sigma^+$ example, two transition moments are needed to account for the rotational linestrengths in the $^2\Pi_{1/2} \sim {}^2\Sigma^+ \rightarrow {}^2\Sigma^+$ transition. Since case (c) implies that

$$\left|\frac{\langle {}^2\Pi_{1/2}|\mathbf{H}^{SO}|{}^2\Sigma^+\rangle}{E^0_{{}^2\Pi_{1/2}} - E^0_{{}^2\Sigma^+}}\right| \gg 1, \qquad (5.3.34)$$

the $^2\Pi \sim {}^2\Sigma$ mixing can be considered to be J-independent and the mixing coefficients, $C_{\Pi\Pi}(J)$ and $C_{\Pi\Sigma}(J)$, can be absorbed into the transition moment factors, $\mu_{\parallel}$ and $\mu_{\perp}$. However, unlike the $^1\Pi \sim {}^1\Sigma^+$ example, the mixed state is composed entirely of basis functions belong to the same Ω-value; and, since the $\boldsymbol{\alpha}$ rotational matrix elements depend exclusively on J and Ω, the rotational linestrength factors can be reduced to simple closed form expressions (Table 5.2).

Table 5.2
Rotational Linestrengths for Case (c) $\frac{1}{2}$–$\frac{1}{2}$ Transitions

Transition	$\lvert\langle n'\frac{1}{2}J'\lvert\boldsymbol{\mu}\rvert n''\frac{1}{2}J\rangle\rvert^2$
$P_{ee}(J)$	$\frac{(J+\frac{1}{2})(J-\frac{1}{2})}{3J}[\mu_\parallel - 2^{-1/2}\mu_\perp]^2$
$P_{ff}(J)$	$\frac{(J+\frac{1}{2})(J-\frac{1}{2})}{3J}[\mu_\parallel + 2^{-1/2}\mu_\perp]^2$
$Q_{ef}(J)$	$\frac{(J+\frac{1}{2})}{6J(J+1)}[\mu_\parallel + (2J+1)2^{-1/2}\mu_\perp]^2$
$Q_{fe}(J)$	$\frac{(J+\frac{1}{2})}{6J(J+1)}[\mu_\parallel - (2J+1)2^{-1/2}\mu_\perp]^2$
$R_{ee}(J)$	$\frac{(J+\frac{1}{2})(J+\frac{3}{2})}{3(J+1)}[\mu_\parallel + 2^{-1/2}\mu_\perp]^2$
$R_{ff}(J)$	$\frac{(J+\frac{1}{2})(J+\frac{3}{2})}{3(J+1)}[\mu_\parallel - 2^{-1/2}\mu_\perp]^2$

If $\mu_\perp = 0$, the usual ${}^2\Sigma^+$–${}^2\Sigma^+$ intensity pattern is obtained: two P and two R branches of almost equal intensities at high J and two very weak Q branches. If $\mu_\parallel = 0$, the usual ${}^2\Pi$–${}^2\Sigma^+$ intensity pattern emerges: two P and two R branches of comparable intensities and two Q branches that are stronger than each of the other four branches by a factor of two. If $2^{-1/2}\mu_\perp = \mu_\parallel$, then the P_{ee} and R_{ff} branches vanish and Q_{fe} is weaker than Q_{ef} by a factor $[J/(J+1)]^2$, which is significantly different from unity only at low J. If $2^{-1/2}\mu_\perp = -\mu_\parallel$, the pattern reverses ($e \leftrightarrow f$). Departures from the intensity pattern predicted by Table 5.2 could result from J-dependent $-B(R)\mathbf{J}\cdot(\mathbf{L}+\mathbf{S})$ mixing with $\Omega = \frac{3}{2}$ states or J-dependent $[E^0_{{}^2\Pi_{1/2}}(J) - E^0_{{}^2\Sigma^+}(J)]$ energy denominators [Eq. (5.3.34)].

${}^3\Sigma^-$–${}^1\Sigma^+$ transitions are "doubly forbidden" in the sense that they violate two selection rules, $\Delta S = 0$ and $\Sigma^+ \nleftrightarrow \Sigma^-$. However, this term is misleading because only one perturbing state is required to lend intensity to this nominally forbidden transition (Watson, 1968; Hougen, 1970). There are four relevant perturbation mechanisms, each involving a $\Delta\Omega = 0$ spin–orbit interaction:

$$\left.\begin{array}{l} {}^3\Sigma^- \sim {}^1\Pi \\ {}^3\Sigma^- \sim {}^1\Sigma^+ \end{array}\right\} \text{upper-state perturbations,}$$

$$\left.\begin{array}{l} {}^1\Sigma^+ \sim {}^3\Pi \\ {}^1\Sigma^+ \sim {}^3\Sigma^- \end{array}\right\} \text{lower-state perturbations.}$$

The case (b) limit ($J \gg |\lambda/B|$) eigenstates are

$$|'^3\Sigma_e^-{}'F_1\rangle = \left(\frac{J+1}{2J+1}\right)^{1/2}[|^3\Sigma_{1e}^-\rangle + \alpha|^1\Pi_e\rangle] + \left(\frac{J}{2J+1}\right)^{1/2}[|^3\Sigma_0^-\rangle + \beta|^1\Sigma^+\rangle] \tag{5.3.35a}$$

$$|'^3\Sigma_f^-{}'F_2\rangle = [|^3\Sigma_{1f}^-\rangle + \alpha|^1\Pi_f\rangle] \tag{5.3.35b}$$

$$|'^3\Sigma_e^-{}'F_3\rangle = -\left(\frac{J}{2J+1}\right)^{1/2}[|^3\Sigma_{1e}^-\rangle + \alpha|^1\Pi_e\rangle] + \left(\frac{J+1}{2J+1}\right)^{1/2}[|^3\Sigma_0^-\rangle + \beta|^1\Sigma^+\rangle] \tag{5.3.35c}$$

$$|'^1\Sigma_e^{+}{}'\rangle = |^1\Sigma^+\rangle + \gamma|^3\Sigma_0^-\rangle + \delta|^3\Pi_{0e}\rangle. \tag{5.3.36}$$

(Watson (1968), using a spherical tensor phase convention [see Eqs. (2.4.40b) and (2.4.40c)] for $\mu_\perp$ of opposite sign to that used here and by Hougen (1970), has treated the general intermediate case (a)–(b) $^3\Sigma^- - {}^1\Sigma^+$ problem.)

There are four transition moments,

$$\mu_{3\perp} \equiv \langle ^3\Sigma^- |\boldsymbol{\mu}| ^3\Pi\rangle \tag{5.3.37a}$$

$$\mu_{3\parallel} \equiv \langle ^3\Sigma^- |\boldsymbol{\mu}| ^3\Sigma^-\rangle \tag{5.3.37b†}$$

$$\mu_{1\perp} \equiv \langle ^1\Pi |\boldsymbol{\mu}| ^1\Sigma^+\rangle \tag{5.3.37c}$$

$$\mu_{1\parallel} \equiv \langle ^1\Sigma^+ |\boldsymbol{\mu}| ^1\Sigma^+\rangle, \tag{5.3.37d†}$$

and five possible rotational branches with linestrength factors given in Table 5.3 where

$$\mu_\parallel \equiv \beta\mu_{1\parallel} + \gamma\mu_{3\parallel} \tag{5.3.38a}$$

$$\mu_\perp \equiv \alpha\mu_{1\perp} + \delta\mu_{3\perp}. \tag{5.3.38b}$$

The pattern of interference effects is identical to the $^1\Pi \sim {}^1\Sigma^+$ example discussed earlier.

It has been pointed out by Johns (1974) and Steimle *et al.* (1982) that the energy-level structure of a $^1\Pi \sim {}^1\Sigma^+$ perturbation complex is identical to that of a $^3\Sigma^-$ state provided that

$$B(^1\Pi) = B(^1\Sigma^+) = B(^3\Sigma^-) \equiv B \tag{5.3.39a}$$

$$E^0(^1\Pi) - E^0(^1\Sigma^+) = -B + 2\lambda(^3\Sigma^-) - 3\gamma(^3\Sigma^-) \tag{5.3.39b}$$

† In the case where the $^3\Sigma^-$ upper state is the same state mixed into $'^1\Sigma^{+\prime}$, $\mu_{3\parallel}$ is the electric dipole moment of the $^3\Sigma^-$ state; similarly, $\mu_{1\parallel}$ can be the dipole moment of $^1\Sigma^+$ (see Section 5.4).

$$E \equiv E^0(^1\Pi) = E^0(^3\Sigma^-) + B + \tfrac{2}{3}\lambda - \tfrac{7}{3}\gamma \tag{5.3.39c}$$

$$\langle ^1\Pi_e | BL^+ | ^1\Sigma^+ \rangle = B \tag{5.3.39d}$$

$$q(^1\Pi) = \gamma(^3\Sigma^-)/2. \tag{5.3.39e}$$

These requirements are automatically satisfied for a Rydberg p-complex because

$$\langle \pi | l^+ | \sigma \rangle = 2^{1/2},$$

and, since Rydberg electrons do not contribute significantly to bonding, the potential curves for the $^1\Sigma$ and $^1\Pi$ components of a p-complex should be identical so that

$$B(^1\Pi) = B(^1\Sigma) = \langle v_\Pi | B(R) | v_\Sigma \rangle \, \delta_{v_\Pi v_\Sigma}.$$

The $^3\Sigma^-$ and p-complex structures resemble each other because both consist of one unit of spin or electronic angular momentum (**S** or **L**) coupled to the nuclear rotation (**R**). However, since $\boldsymbol{\mu}$ operates exclusively on electron spatial coordinates, any resemblance between the rotational-branch intensity patterns for $^3\Sigma^- - {}^1\Sigma^+$ and p-complex$-{}^1\Sigma^+$ transitions would seem to be coincidental. A $^3\Sigma^- - {}^1\Sigma^+$ transition will look exactly like a p-complex$-{}^1\Sigma^+$ transition if, in addition to satisfying Eqs. (5.3.39), the σ-orbital of the $^1\Sigma^+$ state is predominantly of $s\sigma$ united atom character. Then the transition moment ratio

Table 5.3
Rotational Linestrengths for $^3\Sigma^- - {}^1\Sigma^+$ Transitions

$^OP(J)(F_1)$	$\left\{\mu_\parallel\left[\frac{J(J-1)}{3(2J-1)}\right]^{1/2} + \mu_\perp\left[\frac{J(J-1)}{3(2J-1)}\right]^{1/2}\right\}^2$
$^QP(J)(F_3)$	$\left\{\mu_\parallel\left[\frac{J^2}{3(2J-1)}\right]^{1/2} - \mu_\perp\left[\frac{(J-1)^2}{3(2J-1)}\right]^{1/2}\right\}^2$
$^QQ(J)(F_2)$	$\left[\mu_\perp\left(\frac{2J+1}{3}\right)^{1/2}\right]^2$
$^QR(J)(F_1)$	$\left\{\mu_\parallel\left[\frac{(J+1)^2}{3(2J+3)}\right]^{1/2} - \mu_\perp\left[\frac{(J+2)^2}{3(2J+3)}\right]^{1/2}\right\}^2$
$^SR(J)(F_3)$	$\left\{\mu_\parallel\left[\frac{(J+1)(J+2)}{3(2J+3)}\right]^{1/2} + \mu_\perp\left[\frac{(J+1)(J+2)}{3(2J+3)}\right]^{1/2}\right\}^2$

will be

$$\frac{\mu_{\|}}{\mu_{\perp}} \simeq \frac{\langle p\sigma|\boldsymbol{\mu}_z|s\sigma\rangle}{2^{-1/2}\langle p\pi|\boldsymbol{\mu}^+|s\sigma\rangle} = \frac{\begin{pmatrix}1 & 0 & 1\\ 0 & 0 & 0\end{pmatrix}}{\begin{pmatrix}1 & 0 & 1\\ -1 & 0 & 1\end{pmatrix}} = -1, \tag{5.3.40a}$$

which is the atomic transition moment ratio (see Section 2.4.5). Similarly, if the σ orbital is approximately a $d\sigma$ orbital, then

$$\frac{\mu_{\|}}{\mu_{\perp}} \simeq \frac{\langle p\sigma|\boldsymbol{\mu}_z|d\sigma\rangle}{2^{-1/2}\langle p\pi|\boldsymbol{\mu}^+|d\sigma\rangle} = +2, \tag{5.3.40b}$$

and the transition will look like an $np \leftarrow nd\sigma$ transition.

Johns (1974) and Johns and Lepard (1975) have examined the intensity pattern of several bands in the Worley–Jenkins $np \leftarrow \mathrm{X}^1\Sigma_g^+$ Rydberg series of N_2. The $n = 6$ and 8 members of this series are clear examples of well-behaved p-complexes with $E(^1\Sigma)$–$E(^1\Pi)$ separations of 92 cm^{-1} (48B) and 40 cm^{-1} (21B), respectively. L-uncoupling effects are evident in both the level structure and intensity pattern, each suggestive of an approach to pure atomic electronic behavior. The Rydberg electron is beginning to forget that it is attached to a molecule. The energy levels follow the case (d) $BN^+(N^+ + 1)$ pattern rather than the case (a) $BJ(J + 1)$ pattern. As J increases, the five rotational branches look increasingly like O, Q, Q, Q, and S form $(N^{+\prime} - J'')$ rather than $P_{\Pi\Sigma}$, $Q_{\Pi\Sigma}$, $R_{\Pi\Sigma}$, $P_{\Sigma\Sigma}$, $R_{\Sigma\Sigma}$ $(J' - J'')$ branches. As N^+ (the rotational quantum number of the ion core) becomes the pattern-forming rotational quantum number, the three branches in the center of the complex become very compact, as if the molecular rotation were disappearing from the spectrum. This effect is heightened by the low intensity of the O and S form branches, which results from destructive interference between $\mu_{\|}$ and $\mu_{\perp}$ because Eq. (5.3.40a) is satisfied. The uncoupling of the Rydberg electron from the core causes the molecular spectrum to look like an atomic spectrum. All of the photon angular momentum goes into the Rydberg electron, hence $\Delta N^+ = 0$.

The failure of the N_2 $np \leftarrow \mathrm{X}^1\Sigma_g^+$ spectrum to go entirely over, at high J-values, into an atomic spectrum is primarily a consequence of the nonsphericity of the $\sigma_g 2p$ valence orbital (composed primarily from $2p$ orbitals on each N atom). $\sigma_g 2p$ is a bonding orbital, thus $B'' > B'$ and the Q-form branches do not collapse into a perfectly line-like feature. More importantly, the $\sigma_g 2p(3\sigma_g)$ orbital has a mixture of united atom characters, of which only $s\sigma_g$ and $d\sigma_g$ have the correct symmetry to combine with $np\lambda_u$ orbitals. Johns and Lepard (1975) conclude that the N_2 $np \leftarrow \mathrm{X}$ intensity pattern is consistent with significant united atom $s\sigma$ character in the $\sigma_g 2p$ orbital. This experimentally identified $s\sigma$ character is confirmed by calculations (Raoult *et al.*, 1983), which show 54% s character and 42% d character for this $3\sigma_g$ orbital.

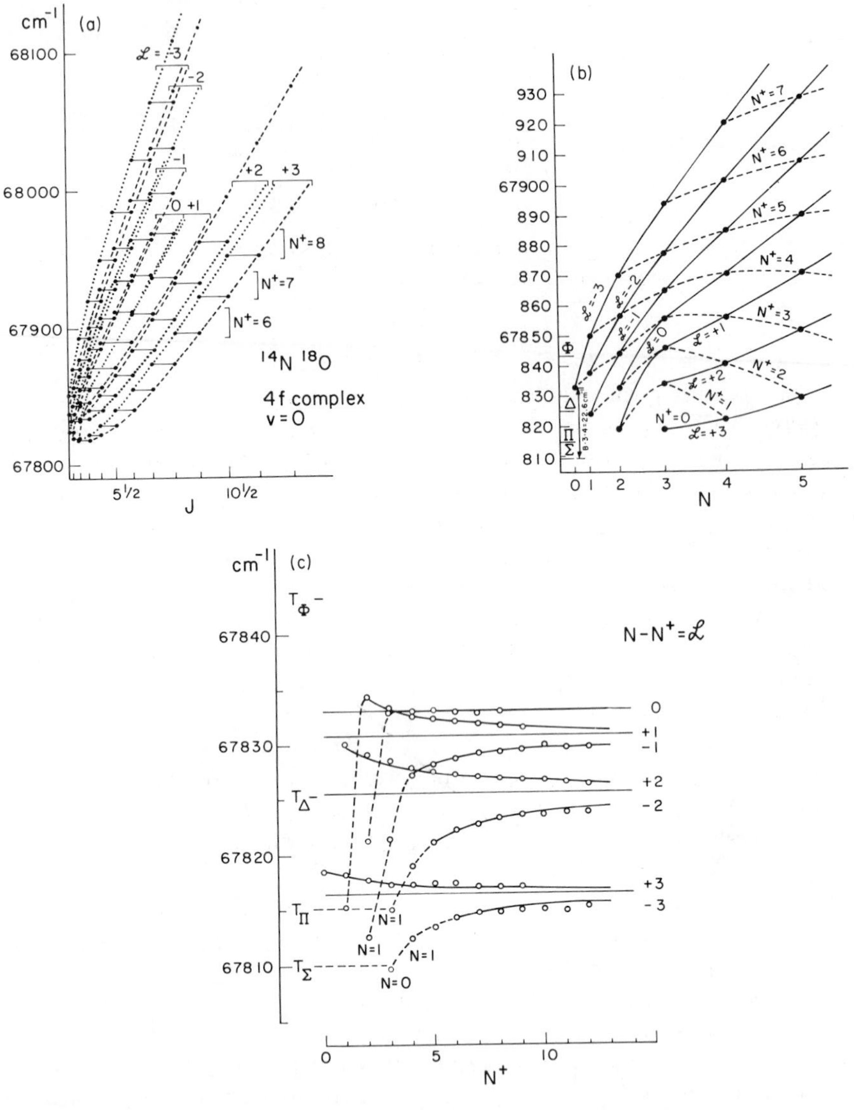

Fig. 5.7 Natural rotational quantum numbers for the NO $4f$ Rydberg complex. (a) Energy of all 14 components of the $^{14}N^{18}O$ $4f$ complex plotted versus $J(J+1)$. The components are labeled by $\mathscr{L} \equiv N - N^+$ and $\mathscr{S} \equiv J - N$. The e- and f-levels of a given $\mathscr{L}$, $\mathscr{S}$ series are connected by dashed and dotted lines, respectively. The horizontal tie lines connect the near-degenerate same-N doublet components belonging to $\mathscr{S} = \pm\frac{1}{2}$. The levels belonging to the same N^+ valueare grouped by a vertical bracket. (b) Energies are plotted versus N(N + 1). Solid curves connect the N-levels belonging to the same $\mathscr{L}$-value. The $\mathscr{S}$-doublet structure has vanished owing to the presentation versus $N(N+1)$. Levels belonging to the same value of N^+ are connected by dashed lines. (c) A reduced term value plot, $T - BN^+(N^+ + 1)$ versus N^+, displays the approach to the case (d) pattern of levels varying as $N^+(N^+ + 1)$ and split into $\mathscr{L}$ components by the weak electric field of the nonspherical core. (From Jungen and Miescher, 1969, Figs. 3-5.)

Rydberg–Rydberg transitions often appear with no more structure than a few compact Q-form branches. Dressler *et al.* (1981) have observed the NO $5g \rightarrow 4f$ bands and suggested not only that is rotation–vibration structure essentially absent, but also that similar-appearing $5g \rightarrow 4f$ bands of many molecules will occur at the same wavelength. Herzberg and Jungen (1982) have rotationally analyzed the corresponding system in H_2 and D_2. For both NO and H_2/D_2, the rotational structure collapses at high rotational quantum numbers into atom-like features that correspond to transitions in which the core angular momentum quantum number ($J^+ = R^+ = N^+$ for NO^+ $X^1\Sigma^+$, N^+ for H_2^+ $X^2\Sigma_g^+$) does not change ($\Delta N^+ = 0$).

Jungen and Miescher (1969) were the first to analyze the rotational structure of an f-complex. The case (d) picture of an nf Rydberg electron interacting with an NO^+ $X^1\Sigma^+$ ion core is far more insightful than a computationally equivalent $^2\Phi \sim {}^2\Delta \sim {}^2\Pi \sim {}^2\Sigma^+$ multistate perturbation approach. This is illustrated by Fig. 5.7, which shows how the energy-level pattern is simplified when it is plotted versus $N^+(N^+ + 1)$ rather than $J(J + 1)$. At $N^+ \geq 10$, all 14 components of the $4f$ complex have rotational energies that go as $BN^+(N^+ + 1)$. The limiting structure is well developed, showing seven pairs ($J - N = \mathscr{S} = \pm\frac{1}{2}$) of doublet components. These, in turn, are arranged in four near-degenerate $\mathscr{L} = N - N^+ = \pm 3, \pm 2, \pm 1, 0$ groups corresponding to the $nf\lambda$ components split by the molecular quadrupole moment of the NO^+ ion core. The intensity pattern in this $nf \leftarrow X^2\Pi$ system is consistent with atomic $nf\lambda_u \leftarrow 3d\pi_g$ transition moments (no adjustable intensity parameters) and suggests that the singly occupied antibonding $\pi_g 2p$ orbital in the NO $X^2\Pi$ has the predominant $3d\pi_g$ united atom character predicted by Mulliken (1964).

5.3.3 $\Sigma^+ \sim \Sigma^-$ PERTURBATIONS; SUBBAND INTENSITY ANOMALIES IN $\Sigma \leftrightarrow \Pi$ TRANSITIONS

A special class of intensity anomaly arises from $\Sigma^+ \sim \Sigma^-$ perturbations. Renhorn (1980) has shown that interference between two $\mu_\perp$ transition moments could account for the anomalous $^2\Sigma$–$^2\Pi_{3/2}/{}^2\Sigma$–$^2\Pi_{1/2}$ subband intensity ratios observed by Appleblad *et al.* (1981) in a CuO $^2\Sigma$–$X^2\Pi$ transition at 767 nm. An unperturbed $^2\Sigma \rightarrow {}^2\Pi$ transition should have approximately equal intensity Σ–$\Pi_{3/2}$ and Σ–$\Pi_{1/2}$ subbands. The $\mu_\perp \sim \mu_\parallel$ interference effects discussed in Section 5.3.2 only affect R/P intensity ratios within a subband without changing the total subband intensity.

The effect identified by Renhorn can occur for Π–Σ and Σ–Π transitions with $S \geq \frac{1}{2}$ and $\Sigma \neq 0$ and has the appearance of an intensity transfer from all rotational branches of the Σ–$\Pi_{\Omega = 1 + |\Sigma|}$ subband into the corresponding branches of the Σ–$\Pi_{\Omega=1-|\Sigma|}$ subband. In fact, the intensity transfer is between corresponding subbands of Σ^+–Π and Σ^-–Π transitions.

The subband intensity anomaly arises from $\Delta S = 0, \Delta\Omega = 0, \Omega \neq 0, \Sigma^+ \sim \Sigma^-$ spin–orbit perturbations combined with the opposite behavior of the phase factors for Σ^+–$\Pi_{\Omega=1-|\Sigma|}$ versus Σ^-–$\Pi_{\Omega=1-|\Sigma|}$ transitions (Section 5.3.2). The nominal Σ^+ and Σ^- eigenstates are

$$
\begin{aligned}
|'\Sigma^{+\prime}, F_i, J\rangle &= \sum_{\Omega=-S}^{S} C^+_{i\Omega}[(1-\beta^2)^{1/2}|\Sigma^+_\Omega, J\rangle + \beta|\Sigma^-_\Omega, J\rangle] \\
|'\Sigma^{-\prime}, F_i, J\rangle &= \sum_{\Omega=-S}^{S} C^-_{i\Omega}[-\beta|\Sigma^+_\Omega, J\rangle + (1-\beta^2)^{1/2}|\Sigma^-_\Omega, J\rangle],
\end{aligned}
\tag{5.3.41}
$$

where the $C^\pm_{i\Omega}$ are mixing coefficients for $\Sigma^\pm$ states expressed in the case (a) basis and β is the Ω-independent $\Sigma^+ \sim \Sigma^-$ mixing coefficient,

$$
\beta \equiv \frac{\langle \Sigma^+_\Omega | \mathbf{H}^{\mathrm{SO}} | \Sigma^-_\Omega \rangle}{E^0(\Sigma^+) - E^0(\Sigma^-)}[1 - \delta_{\Omega,0}]. \tag{5.3.42}
$$

The $1 - \delta_{\Omega,0}$ factor reflects the parity forbiddenness of $\Sigma^+_0 \sim \Sigma^-_0$ mixing. For a transition into a case (a) Π_Ω state, only the $|\Sigma| = \Omega + 1$ and $|\Sigma| = \Omega - 1$ character of the Σ-state contributes to the transition probability, thus:

$$
\begin{aligned}
&\mu('\Sigma^+(F_i)' - \Pi_{\Omega=1+|\Sigma|}) \\
&\quad = 2^{-1/2}\langle \Omega = 1 + |\Sigma|, J_\Pi | \boldsymbol{\alpha}^- | |\Sigma|, J_\Sigma \rangle C^+_{i|\Sigma|}(J_\Sigma)[(1-\beta^2)^{1/2}\mu_{\perp+} + \beta\mu_{\perp-}]
\end{aligned}
\tag{5.3.43a}
$$

$$
\begin{aligned}
&\mu('\Sigma^-(F_i)' - \Pi_{1+|\Sigma|}) \\
&\quad = 2^{-1/2}\langle 1 + |\Sigma|, J_\Pi | \boldsymbol{\alpha}^- | |\Sigma|, J_\Sigma \rangle C^-_{i|\Sigma|}(J_\Sigma)[-\beta\mu_{\perp+} + (1-\beta^2)^{1/2}\mu_{\perp-}]
\end{aligned}
\tag{5.3.43b}
$$

$$
\begin{aligned}
&\mu('\Sigma^+(F_i)' - \Pi_{1-|\Sigma|}) \\
&\quad = 2^{-1/2}\langle 1 - |\Sigma|, J_\Pi | \boldsymbol{\alpha}^- | -|\Sigma|, J_\Sigma \rangle C^+_{i,-|\Sigma|}(J_\Sigma)[(1-\beta^2)^{1/2}\mu_{\perp+} - \beta\mu_{\perp-}]
\end{aligned}
\tag{5.3.43c}
$$

$$
\begin{aligned}
&\mu('\Sigma^-(F_i)' - \Pi_{1-|\Sigma|}) \\
&\quad = 2^{-1/2}\langle 1 - |\Sigma|, J_\Pi | \boldsymbol{\alpha}^- | -|\Sigma|, J_\Sigma \rangle C^-_{i,-|\Sigma|}(J_\Sigma)[-\beta\mu_{\perp+} \\
&\qquad - (1-\beta^2)^{1/2}\mu_{\perp-}]
\end{aligned}
\tag{5.3.43d}
$$

where

$$
\mu_{\perp\pm} = 2^{-1/2}\langle \Pi | \boldsymbol{\mu}^+ | \Sigma^\pm \rangle.
$$

Equation (5.3.42) implies that transitions terminating in $\Pi_{\Omega=1}$ levels will not exhibit the $\mu_{\perp+} \sim \mu_{\perp-}$ interference effect. The interference term for the $'\Sigma^+(F_i)'-\Pi_{1+|\Sigma|}$ subbands,

$$[\langle 1+|\Sigma|, J_\Pi|\boldsymbol{\alpha}^-||\Sigma|, J_\Sigma\rangle]^2[C_{i|\Sigma|}^-(J_\Sigma)]^2(1-\beta^2)^{1/2}\beta\mu_{\perp+}\mu_{\perp-},$$

is of comparable magnitude but opposite sign to that for the $'\Sigma^-(F_i)'-\Pi_{1+|\Sigma|}$ and $'\Sigma^+(F_i)'-\Pi_{1-|\Sigma|}$ subbands. If $|\beta| \ll 2^{-1/2}$, the interference effect will be most noticeable in the $\Sigma^\pm-\Pi$ system associated with the smaller transition moment.

5.4 Forbidden Transitions; Intensity Borrowing by Mixing with a Remote Perturber

It is always possible to express the eigenfunction corresponding to a nominal i, v_i, J' electronic–vibration–rotation level as a sum of rovibronic basis functions,

$$|'i, v_i, J''\rangle = \sum_{k,v_k,\Omega_k} C_{i;k,v_k,\Omega_k}|v_k\rangle|k\Lambda_k S_k \Sigma_k\rangle|\Omega_k J' M\rangle, \qquad (5.4.1)$$

and to express the transition moment matrix element between $|'i, v_i, J''\rangle$ and an unperturbed basis function $|0, v_0, J\rangle$ as [see Eq. (5.1.12)]

$$\langle 'i, v_i, J''|\boldsymbol{\mu}|0, v_0, J\rangle = \sum_{k,v_k,\Omega_k} C_{i;k,v_k,\Omega_k}\langle v_k|v_0\rangle R_e^{k0}(R_{v_k v_0})\langle\Omega_k J'|\boldsymbol{\alpha}|\Omega_0 J\rangle. \qquad (5.4.2)$$

This equation can be generalized to include a mixed $|'0, v_0, J'\rangle$ eigenfunction. Whenever more than one term in the Eq. (5.4.2) summation is significant, the simple factorization of the transition probability into electronic (R_e^2), Franck–Condon ($q_{v'v''} = \langle v'|v''\rangle^2$), and Hönl–London rotational factors ($S_{J'\Omega',J''\Omega''}$) breaks down. In Section 5.2.1 the case was discussed in which only the $k = i$ term in the summation was nonzero. Interference effects arise when two or more terms in Eq. (5.4.2) are significant and have comparable magnitudes. This section deals with the situation where the $k = i$ term is zero and hence the transition is nominally forbidden. One example of this type, the forbidden $^3\Sigma^- - ^1\Sigma^+$ transition, was discussed in Section 5.3.2.

There has been a great deal of confusion about what Franck–Condon factors are appropriate for modelling the vibrational intensities in a fluorescence v''-progression originating from a perturbed, nominally nonfluorescing, upper level. Three cases must be considered: (1) the $|'i, v'\rangle$ vibronic level borrows all of its $i \rightarrow 0$ oscillator strength from a single, remote, electronic-vibrational level,

$|j, v_j\rangle$; (2) intensity is borrowed from all vibrational levels of a single remote electronic state; (3) more than one remote electronic state contributes to the $i \to 0$ transition probability. The Franck–Condon factors appropriate to case (1) are those for the remote perturbing $|j, v_j\rangle$ basis state ($q^{j,0}_{v_j v''}$) whereas for case (2) the $q^{i,0}_{v_i v''}$ are appropriate even though $R^{i,0}_e = 0$. In case (3), interference effects will make calculation of vibrational and rotational linestrengths difficult, but not impossible.

The phrase "remote perturber" is intended to imply that

$$|E^0_i - E^0_j| \gg |H_{ij}| \tag{5.4.3a}$$

$$|E^0_i - E^0_j| \gg |B^0_i - B^0_j| J(J+1) \tag{5.4.3b}$$

so that the C_{ij} mixing coefficients are not J-dependent. The Eq. (5.4.3) limit is frequently applicable for homogeneous interactions ($\Delta\Omega = 0$) because H_{ij} is J-independent. However, this limit can never be achieved for heterogeneous ($\Delta\Omega = \pm 1$) perturbations because H_{ij} is proportional to J. However, $\Delta\Omega = \pm 1$ mixing is often a less important intensity borrowing mechanism than $\Delta\Omega = 0$ mixing (up to some threshold J-value) and can never contribute directly to the oscillator strength of a nominally spin-forbidden transition.

Case (1). Perturbation by a single remote $|j, v_j\rangle$ vibronic state with nonzero oscillator strength to $|0, v_0\rangle$. The only nonzero term in Eq. (5.4.2) is the j, v_j term. There may be many terms with $C_{i;\,j,v_j,\Omega_j} \neq 0$ but only one term where both $R^{j,0}_e \langle v_j | v_0 \rangle$ and C are nonzero. Equation (5.4.2) becomes

$$\langle 'i, v_i, J'' | \boldsymbol{\mu} | 0, v_0, J\rangle = C_{i;\,j,v_j,\Omega_j} \langle v_j | v_0 \rangle R^{j,0}_e (R_{v_j v_0}) \langle \Omega_j J' | \boldsymbol{\alpha} | \Omega_0 J\rangle, \tag{5.4.4}$$

and the transition probabilities can be expressed in terms of

$$[\mu_{'i,v_i,J'';\,0,v_0,J}]^2 = \mu^2 q^{j,0}_{v_j v_0} S_{\Omega_j J';\,\Omega_0 J}, \tag{5.4.5}$$

where

$$\mu^2 \equiv [C_{i;\,j,v_j,\Omega_j}]^2 [R^{j,0}_e (R_{v_j v_0})]^2. \tag{5.4.6}$$

Equation (5.4.5) looks like the usual product of electronic, vibrational, and rotational factors except that μ^2 is an effective parameter, which may be strongly dependent on v_i, Ω_i, e/f, and J (rather than a function of the $R_{v_j v_0}$ R-centroid), and q and $\underline{S}$ are the Franck–Condon and Hönl–London factors appropriate for the $j, v_j, \Omega_j \to 0, v_0, \Omega_0$ transition rather than the $i, v_i, \Omega_i \to 0, v_0, \Omega_0$ transition suggested by the label of the $'i'$ state. This result is counterintuitive and has been forgotten many times in published papers.

Case (2). Perturbation by a single remote electronic state with nonzero oscillator strength to $|0, v_0\rangle$. The only nonzero terms in Eq. (5.4.2) are those for all v_j of the jth electronic state. If Eq. (5.4.3) is satisfied, the sum over remote states can be simplified by expressing the mixing coefficients as

$$C_{i;\,j,v_j,\Omega_j} = \frac{\langle i,\Omega_i|\mathbf{H}|j,\Omega_j\rangle\langle v_i|v_j\rangle}{E^0_{i,v_i} - E^0_{j,v_j}}, \tag{5.4.7}$$

assuming that $\langle i\Omega_i|\mathbf{H}|j,\Omega_j\rangle/(E^0_{i,v_i} - E^0_{j,v_j})$ is independent of v_j and equal to $H_{ij}/(E^0_{i,0} - E^0_{j,0})$, assuming that $R_e^{j,0}$ is independent of R-centroid, and recognizing that

$$\sum_{v_j}\langle v_i|v_j\rangle\langle v_j|v_0\rangle = \langle v_i|v_0\rangle \tag{5.4.8}$$

by the completeness of the $|v_j\rangle$ vibrational basis set. The transition probabilities can then be expressed as

$$|\langle 'i, v_i, J''|\boldsymbol{\mu}|0, v_0, J\rangle|^2 = \mu^2 q^{i,0}_{v_i v_0} S_{\Omega_j J';\,\Omega_0 J} \tag{5.4.9}$$

where

$$\mu^2 \equiv (H_{ij}/\Delta E^0_{ij})^2 (R_e^{j,0})^2. \tag{5.4.10}$$

The assumptions required to obtain Eqs. (5.4.9) and (5.4.10) are more drastic than those for Eqs. (5.4.5) and (5.4.6), but their failure can often be absorbed into a strongly R-centroid dependent effective μ^2 parameter. Note that Eq. (5.4.9) contains the "intuitive" Franck–Condon factors but the Hönl–London factors are those appropriate to the $\mathbf{\Omega}$-value of the remote perturber.

Case (3). Several remote electronic states contribute to the $'i' \to 0$ oscillator strength. A frequently encountered example is the nominally forbidden $^3\Pi$–$X^1\Sigma^+$ transition. The $^3\Pi$ state can become mixed with singlet states,

$$|'^3\Pi_1'\rangle = |^3\Pi_1\rangle + \frac{\langle ^3\Pi_1|\mathbf{H}^{SO}|^1\Pi_1\rangle}{E^0(^3\Pi_1) - E^0(^1\Pi_1)}|^1\Pi_1\rangle \tag{5.4.11}$$

$$|'^3\Pi_{0e}'\rangle = |^3\Pi_{0e}\rangle + \frac{\langle ^3\Pi_{0e}|\mathbf{H}^{SO}|^1\Sigma_0^+\rangle}{E^0(^3\Pi_{0e}) - E^0(^1\Sigma_0^+)}|^1\Sigma_0^+\rangle, \tag{5.4.12}$$

and the $X^1\Sigma^+$ state can be mixed with triplet states (the possibility of mixing $|^3\Sigma_0^-\rangle$ and states other than $|^3\Pi_{0e}\rangle$ into $X^1\Sigma^+$ will be ignored here),

$$|'X^1\Sigma_0^{+\prime}\rangle = |X^1\Sigma_0^+\rangle - \frac{\langle X^1\Sigma^+|\mathbf{H}^{SO}|^3\Pi_{0e}\rangle}{E^0(^3\Pi_0) - E^0(X^1\Sigma^+)}|^3\Pi_{0e}\rangle. \tag{5.4.13}$$

The $^1\Pi$ state most likely to mix with $^3\Pi_1$ is the isoconfigurational state (see Eq. 2.4.14) and

$$\langle ^3\Pi_1|\mathbf{H}^{SO}|^1\Pi\rangle = \pm A(^3\Pi).$$

Although $^3\Pi_{0e}$ can mix with many excited $^1\Sigma^+$ states, mixing with the $X^1\Sigma^+$

state introduces a novel feature, namely, the appearance of permanent electric dipole moments as well as transition moments in the intensity borrowing model. For $^3\Pi$ in the case (a) limit ($A \gg 2^{1/2}BJ$),

$$\mu_\perp \equiv 2^{-1/2}\langle '^3\Pi_1' |\boldsymbol{\mu}^+| 'X^1\Sigma^{+\prime}\rangle$$
$$= 2^{-1/2}\frac{\langle ^3\Pi_1 |\mathbf{H}^{SO}| ^1\Pi_1\rangle}{E^0(^3\Pi_1) - E^0(^1\Pi)}\langle ^1\Pi|\boldsymbol{\mu}^+|X^1\Sigma^+\rangle, \qquad (5.4.14)$$

$$\mu_\parallel \equiv \langle '^3\Pi_{0e'} |\boldsymbol{\mu}_z| 'X^1\Sigma^{+\prime}\rangle$$
$$= \frac{\langle X^1\Sigma_0^+ |\mathbf{H}^{SO}| ^3\Pi_{0e}\rangle}{E^0(^3\Pi_{0e}) - E^0(X^1\Sigma^+)}[\langle X^1\Sigma^+ |\boldsymbol{\mu}_z| X^1\Sigma^+\rangle - \langle ^3\Pi_{0e} |\boldsymbol{\mu}_z| ^3\Pi_{0e}\rangle]$$
$$(5.4.15)$$

The term in square brackets is the difference in permanent electric dipole moments of the $X^1\Sigma^+$ and $^3\Pi$ states. In the case of CO, $\mu(X^1\Sigma^+) \simeq 0$, $\mu(a^3\Pi) \simeq 1.4$ D, and $\langle X^1\Sigma^+ |\mathbf{H}^{SO}| a^3\Pi_{0e}\rangle \simeq 60$ cm^{-1} (Schamps, 1973). The $\mu_\perp$ term expresses the transfer of intensity from $^1\Pi$–$X^1\Sigma^+$ to $^3\Pi_1$–$X^1\Sigma^+$ and makes the $^3\Pi_1$–X subband look exactly like a $^1\Pi$–$^1\Sigma^+$ transition. The $\mu_\parallel$ term expresses the intensity transferred from $^3\Pi_{0e}$–$^3\Pi_{0e}$ and $^1\Sigma^+$–$^1\Sigma^+$ transitions (in this case, pure rotational and rotation–vibration transitions within the $X^1\Sigma^+$ and $a^3\Pi$ states) into the $^3\Pi_{0e}$–$X^1\Sigma^+$ subband. The $^3\Pi_{0f}$–$X^1\Sigma^+$ and $^3\Pi_2$–$X^1\Sigma^+$ transitions remain forbidden in the case (a) limit. The case (a) $^3\Pi_{0e}$–$X^1\Sigma^+$ transition looks like an $\Omega = 0 \rightarrow \Omega = 0$ transition in which the Q-branch is rigorously forbidden. For the analogous $a^3\Pi_{0e}$–$X^1\Sigma^+$ transition in CS, the Q branch appears only for $J' > 10$ and is a consequence of spin-uncoupling in the $a^3\Pi$ state (Cossart *et al.*, 1977).

When $J \neq 0$, the **S**-uncoupling operator ($-\mathbf{B}\mathbf{J}^\pm\mathbf{S}^\mp$) causes the $\boldsymbol{\Omega}$-components of the $^3\Pi$ state to mix [see Eqs. (5.3.23) and (5.3.24) for the effect of spin-uncoupling on an unperturbed $^3\Delta$–$^3\Pi$ transition],

$$|'^3\Pi'_{\Omega e/f}\ J\rangle = \sum_{\Omega'=0}^{2} C_{\Omega\Omega'}(J, e/f)|^3\Pi_{\Omega' e/f}\ J\rangle. \qquad (5.4.16)$$

The general spin-uncoupling expression for $\mu('^3\Pi'$–$^1\Sigma^+)$ is then

$$\langle '^3\Pi_{\Omega e/f'}\ J |\boldsymbol{\mu}| X^1\Sigma^+\rangle = C_{\Omega 0}(J, e)\mu_\parallel + C_{\Omega 1}(J, e/f)2^{1/2}\mu_\perp. \qquad (5.4.17)$$

When $\mu_\perp \gg \mu_\parallel$, one need only keep track of the $\boldsymbol{\Omega} = 1$ character in each of the $'\Omega' e/f$ substates. When $2^{1/2}BJ < A$, approximate expressions for C_{21} and C_{01} can be obtained from first-order perturbation theory:

$$C_{21}(J) \simeq -2^{1/2}BJ/A \simeq -C_{01}(J). \qquad (5.4.18)$$

The intensity of $^3\Pi_{0e/f}$–$X^1\Sigma^+$ and $^3\Pi_{2e/f}$–$X^1\Sigma^+$ subbands will increase as J^2, and the R : Q : P branch intensities will be 1 : 2 : 1. When $\mu_{\parallel} \gg \mu_{\perp}$, the $\Omega = 0$ character determines the subband intensity. The ${}^3\Pi_{1e}$–$X^1\Sigma^+$ subband will consist exclusively of equal-intensity R and P branches with intensity proportional to J^2. The ${}^3\Pi_{2e}$–$X^1\Sigma^+$ subband will only appear at very high J (intensity proportional to J^4). The general case of Ω, J, e/f-dependent ${}^3\Pi$–${}^1\Sigma^+$ transition probabilities and radiative lifetimes has been treated theoretically by James (1971a) and experimentally for CO by James (1971b) and Slanger and Black (1971). The radiative lifetime of each Ω, J, e/f level is different,

$$\tau_{v\Omega e/f}(J) \propto \left[\left(\sum_{v_X} q_{v_a v_X} \nu^3_{v_a v_X}\right)\left(C_{\Omega 0}(J,e)\mu_{\parallel} + C_{\Omega 1}(J,e/f)2^{1/2}\mu_{\perp}\right)^2\right]^{-1} \tag{5.4.19}$$

and it is misleading to speak of "the radiative lifetime of the vth vibrational level." The predicted J, Ω-dependence of the radiative lifetime in the CO $a^3\Pi$ state (setting $\mu_{\parallel} = 0$) is illustrated by Fig. 5.8. Measurements of τ that are partially averaged over J, Ω, e/f will not agree with each other and will exhibit unexpected temperature and pressure dependences. However, since the ${}^3\Pi \sim {}^1\Pi$ and ${}^3\Pi \sim {}^1\Sigma$ perturbation matrix elements are not J-dependent, the fractional ${}^1\Pi$ and ${}^1\Sigma^+$ characters, summed over the six Ω, e/f ${}^3\Pi$ components of a given J-value, are J-independent. This provides an unambiguous

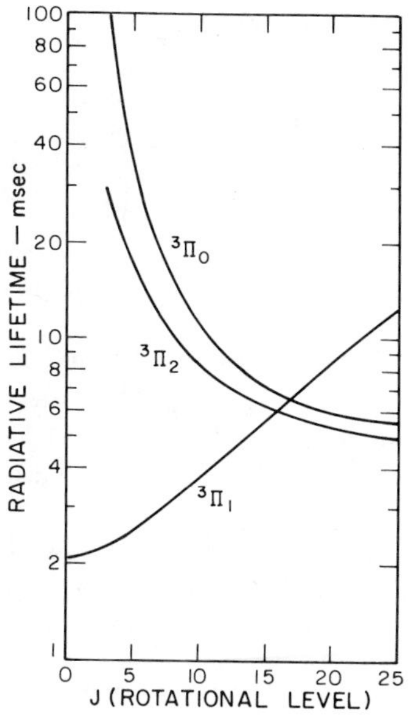

Fig. 5.8. Variation of the CO $a^3\Pi$ radiative lifetime versus J and Ω. The CO $a^3\Pi$ state borrows its oscillator strength from $A^1\Pi$ and, to a much lesser extent, from $X^1\Sigma^+$. The effective lifetime of an entire vibrational level will depend in a complex way ,on pressure, temperature, and the mode of excitation. From data of Fairbairn (1970) corrected by Slanger and Black (1971).

definition of an average, J-independent, vibrational-level radiative lifetime:

$$6/\bar{\tau}_{v'} \equiv \sum_{\Omega,e/f} 1/\tau_{v'\Omega e/f}(J). \tag{5.4.20}$$

Fournier *et al.* (1979) have observed and calculated the $\tau_\Omega(J)$ lifetimes for the CS $a\,^3\Pi$ states.

$\mu_\parallel \sim \mu_\perp$ interference effects in a $^3\Pi \sim {}^1\Sigma$ transition appear as anomalies in the ratios of R and P branch intensities, primarily in the $^3\Pi_1 - {}^1\Sigma^+$ and $^3\Pi_0 - {}^1\Sigma^+$ subbands. The Q lines can never be affected because the $\mu_\parallel$ mechanism cannot contribute to their intensity. If $|\mu_\parallel| > |\mu_\perp|$, the R/P anomalies will be most pronounced in the weaker $(^3\Pi_1 - {}^1\Sigma^+)$ subband. If $|\mu_\parallel| < |\mu_\perp|$, the anomalies will appear in the weak $^3\Pi_0 - {}^1\Sigma^+$ subband. At some J' value, either the $P(J+1)$ or the $R(J-1)$ line in the weaker subband will vanish, having transferred all of its transition probability to its same-subband partner.

The forbidden $^3\Sigma^- - {}^1\Sigma^+$ transition has been discussed earlier [Section 5.3.2, Eqs. (5.3.35)–(5.3.38), Table 5.3]. It is useful to return to a specific example here. $^3\Sigma^- - {}^1\Sigma^+$ transitions have been observed between two states belonging to the same π^2 configuration (molecules with 6 valence electrons, NH, PH, etc. and heteronuclear molecules with 12 valence electrons, SO, NF, NCl, PF). Wayne and Colbourn (1977) have discussed the spin–orbit interaction between this pair of isoconfigurational states. Inserting the first-order perturbation theory definitions of the β and δ mixing coefficients of Eq. (5.3.38a), an expression analogous to Eq. (5.4.15) is obtained,

$$\mu_\parallel = \frac{\langle {}^3\Sigma_0^- |\mathbf{H}^{\mathrm{SO}}| {}^1\Sigma^+\rangle}{E^0(^3\Sigma^-) - E^0(^1\Sigma^+)}[\langle {}^1\Sigma^+ |\boldsymbol{\mu}_z| {}^1\Sigma^+\rangle - \langle {}^3\Sigma_0^- |\boldsymbol{\mu}_z| {}^3\Sigma_0^-\rangle]. \tag{5.4.21}$$

Similarly, Eq. (5.3.38b) can be rewritten

$$\mu_\perp = \left[\frac{\langle {}^3\Sigma_1^- |\mathbf{H}^{\mathrm{SO}}| {}^1\Pi_1\rangle}{E^0(^3\Sigma^-) - E^0(^1\Pi)}\mu_\perp(^1\Pi - {}^1\Sigma) + \frac{\langle {}^1\Sigma_0^+ |\mathbf{H}^{\mathrm{SO}}| {}^3\Pi_{0e}\rangle}{E^0(^1\Sigma^+) - E^0(^3\Pi)}\mu_\perp(^3\Sigma^- - {}^3\Pi)\right]. \tag{5.4.22}$$

Using the branch intensity formulas in Table 5.3 and observed intensity ratios, a value of $\mu_\perp/\mu_\parallel$ can be obtained and compared with *ab initio* predictions. For this specific π^2 example, $\mu_\parallel$ is the difference of permanent electric dipole moments for two states belonging to the same configuration. In the same-orbital single-configuration approximation, $\mu(\pi^2\,{}^1\Sigma^+) = \mu(\pi^2\,{}^3\Sigma^-)$, and thus $\mu_\parallel$ should be very small. Wayne and Colbourn (1977) have shown that $\mu_\perp$ [Eq. (5.4.22)] involves a sum of two terms with the same sign if $^1\Sigma^+$ and $^3\Sigma^-$ belong to π^2 and $^1\Pi$ and $^3\Pi$ both belong to another configuration ($\pi^3\sigma$ or $\pi\sigma$).†

† This is true provided that both Π states lie above or below both Σ states.

$\mu_\perp \neq 0$ and $\mu_\parallel \simeq 0$ regardless of phase choice (however, see Havriliak and Yarkony (1985)]. Mixing with other remote states cannot be ignored, especially when the radiative lifetime is very long ($\tau \gtrsim 10^{-4}$ s).

Forbidden transitions borrow their oscillator strength, in principle, from an infinite number of allowed transitions. Usually, when one or two terms dominate Eq. (5.4.2), isoconfigurational estimates of perturbation parameters and dipole moments will be capable *a priori* of identifying those terms. Semiempirical parameter estimates can also provide a warning that no specific terms will dominate the phenomenological $\mu_\perp$ and $\mu_\parallel$ transition moments. Consider a typical interelectronic energy separation of 10,000 cm^{-1}, a spin–orbit matrix element of 100 cm^{-1}, and an allowed radiative lifetime of 10^{-8} s. The borrowed radiative rate would then be approximately 10^4 s^{-1}, corresponding to a lifetime of 10^{-4} s. Any electronic state with a lifetime longer than 10^{-4} s is likely to have borrowed its ability to radiate from many remote states.

5.5 Special Effects

A variety of unusual experimental schemes have been employed to detect perturbations and to characterize the perturbing state. The methods described in this section involve subjecting the molecule to external perturbations such as an intense monochromatic radiation field (Section 5.5.1), a static magnetic or electric field (Sections 5.5.2 and 5.5.3), multiple static, oscillatory, or pulsed electromagnetic fields (Sections 5.5.2 and 5.5.3), weak bimolecular collisions (Section 5.5.4) or confining the molecule in a high-pressure collisional (Section 5.5.5) or matrix (Section 5.5.6) cage. External perturbations can make observable extremely weak or exotic internal perturbations or can create intramolecular interactions that do not exist in the isolated molecule.

5.5.1 DIFFERENTIAL POWER BROADENING

Perturbations often provide information about previously unknown perturbing states. Extra lines, arising from levels with predominant perturber character, are perhaps the most information-rich features of a perturbed band spectrum. When the perturbation interaction is very weak, one is lucky to locate an extra line and even luckier to be able to prove that this line is *extra* rather than *extraneous* (rare isotope or impurity).

An optical–optical double resonance (OODR) scheme exists, utilizing two continuous-wave (cw), monochromatic, tunable lasers, whereby the rotational quantum numbers of all observed lines may be established without ambiguity, prior knowledge of B-values, trial-and-error searches for consistent combination differences, or redundant confirmation; lines that are weak because of level population may be distinguished from those that are weak because of intrinsic linestrength; and forbidden transitions may be made to appear with comparable peak intensities but considerably narrower widths than allowed transitions.

Figure 5.9 illustrates the level scheme for OODR. The PUMP laser prepares a single v', J' level of the intermediate electronic state. Since the PUMP laser spectral width (~ 1 MHz) is smaller than the Doppler width (~ 1 GHz) of the A, v', $J' \leftarrow$ X, v'', J'' transition, the prepared v', J' molecules are partially velocity-selected. Only the component of velocity along the PUMP laser propagation direction is selected. The v', J' molecules are prepared with a longitudinal velocity distribution much narrower than the full thermal distribution. Thus the linewidth of the C, v^*, $J^* \leftarrow$ A, v', J' transition, recorded

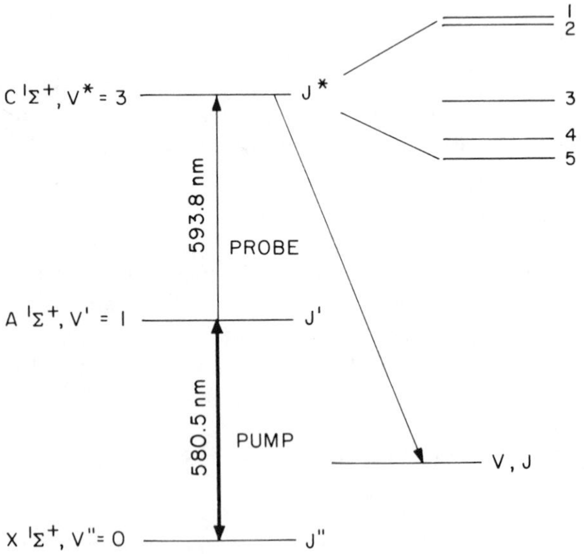

Fig. 5.9 Level scheme for optical–optical double resonance on BaO $C^1\Sigma^+ \leftarrow A^1\Sigma^+ \leftarrow X^1\Sigma^+$. The level structure for $C^1\Sigma^+$ ($v^* = 3$, $J^* = 50$) is expanded and drawn to scale illustrating the five observed $J^* = 50$ levels. Undispersed v^*, $J^* \rightarrow v$, J ultraviolet fluorescence is detected as the probe dye laser is scanned. [From Gottscho and Field (1978).]

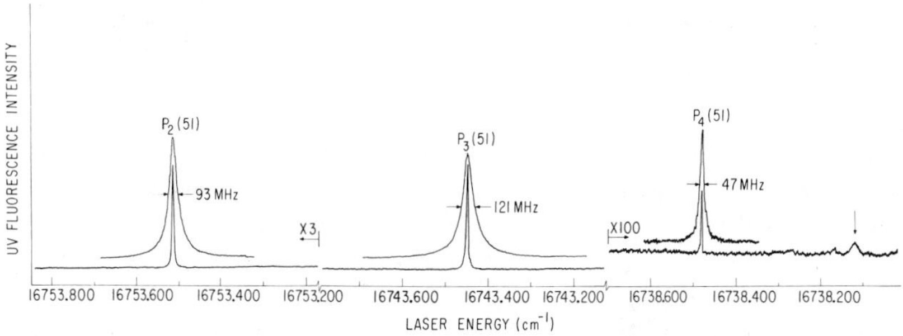

Fig. 5.10 OODR differential power broadening in BaO. The PUMP laser is tuned to the $A^1\Sigma^+ \leftarrow X^1\Sigma^+$ (1, 0) $R(50)$ line; the PROBE is scanned in the region near $P(51)$ of $C^1\Sigma^+ \leftarrow A^1\Sigma^+$ (3, 1). The main line (P_3) and two of four extra lines (P_2 and P_4) are shown. The unassigned collisional satellite line marked with an arrow has area equal to that of the P_4 extra line. [From Gottscho and Field (1978).]

by scanning the co- or counterpropagating PROBE laser and recording undispersed fluorescence at shorter wavelength than λ_{PUMP} and λ_{PROBE}, is significantly sub-Doppler. This means that differences in OODR lineshapes reflect differences in the radiative and collisional properties of the various v_i^*, J_i^* levels and, most importantly, the relative ease of power-broadening the $v_i^*, J^* \leftarrow v', J'$ transitions. Extra lines (partially forbidden transitions) power-broaden more slowly than main lines (allowed transitions) as the intensity of the PROBE laser is increased.

Figure 5.10 shows the main line and two of four extra lines observed upon excitation from BaO $A^1\Sigma^+$ ($v' = 1$, $J' = 51$) by Gottscho and Field (1978) as the PROBE laser was scanned in the region of the $C^1\Sigma^+$–$A^1\Sigma^+$ (3, 1) $P(51)$ line. A similar spectrum was obtained from $A^1\Sigma^+$ ($v = 1$, $J = 49$) near the C–A $R(49)$ line to prove that all observed lines belong to $J^* = 50$ levels. The main line, $P_3(51)$, has an FWHM of 121 MHz, considerably larger than the FWHM of 47 MHz of the weakest extra line, $P_4(51)$.

At high PROBE power where the limit of $n_{v^*,J^*} \simeq n_{v',J'}$ population saturation is approached, the peak intensities of main and extra lines would be comparable but their integrated line intensities should continue to reflect the large difference in intrinsic linestrengths. In an experiment where the resolution limit is set by the molecular linewidth rather than the measuring device, it is easier to detect a tall, narrow line than a broad line of equal integrated intensity. In fact, it is possible to imagine a situation where the PROBE laser is so intense that main lines become so broad as to become unrecognizable, leaving only sharp features associated with extra and forbidden transitions.

5.5.2 EFFECTS OF MAGNETIC AND ELECTRIC FIELDS ON PERTURBATIONS

Molecular rotational levels can be split into *M*-components and shifted in static electric and magnetic fields. Stark and Zeeman effects have been very useful in revealing otherwise unobservable perturbations and in diagnosing the electronic symmetry of previously unknown perturbers. Long before the phenomena of perturbations and the Zeeman effect were understood, there were several studies of the effect of a magnetic field on perturbed lines (Fortrat, 1913; Bachem, 1920).

Between 1929 and 1935, several perturbations in the CO $A^1\Pi$ and N_2^+ $B^2\Sigma^+$ states were studied in magnetic fields up to 36 kG (Crawford, 1929, 1934; Watson, 1932; Parker, 1933; Schmid and Gerö, 1935). Lines were observed to split, to broaden symmetrically and asymmetrically, to gain or lose intensity; each line was a special case. Spectral resolution and sensitivity were seldom adequate to resolve individual *M*-components. Although many qualitative features were satisfactorily explained, several perturbing states were conclusively but incorrectly assigned, and no quantitative theory of the effect of an external field on perturbed line positions, shapes, and intensities emerged.

Wood and Hackett (1909) observed anomalies in the magnetic rotation spectrum of the Na_2 $A^1\Sigma_u^+-X^1\Sigma_g^+$ system. [A magnetic rotation (Faraday effect) spectrum differs from an ordinary absorption spectrum in that the sample is placed between crossed polarizers and in a magnetic field directed parallel to the incident radiation.] Fredrickson and Stannard (1933) suggested that the isolated rotational transitions that displayed significant magnetic field induced polarization rotations were perturbed $A^1\Sigma_u^+ \sim b^3\Pi_{0u}-X^1\Sigma_g^+$ lines. Carroll (1937) developed the general theory of perturbed and unperturbed magnetic rotation spectra of diatomic molecules.

Radford (1961, 1962) and Radford and Broida (1962) presented a complete theory of the Zeeman effect for diatomic molecules that included perturbation effects. This led to a series of detailed investigations of the CN $B^2\Sigma^+$ $(v = 0) \sim A^2\Pi$ $(v = 10)$ perturbation in which many of the techniques of modern high-resolution molecular spectroscopy and analysis were first demonstrated: anticrossing spectroscopy (Radford and Broida, 1962, 1963), microwave optical double resonance (Evenson *et al.*, 1964), excited-state hyperfine structure with perturbations (Radford, 1964), effect of perturbations on radiative lifetimes and on inter-electronic-state collisional energy transfer (Radford and Broida, 1963). Moehlmann *et al.* (1972) observed both magnetic and electric field induced perturbations in HCP and discussed the theory of such perturbations.

A crucial feature of the CN system is that, when CN molecules are formed in a flame of < 1 torr active nitrogen plus CH_2Cl_2, the $A^2\Pi$ state is populated preferentially to $B^2\Sigma^+$. The result is a considerably larger population flow rate into the red-emitting, longer-lived $A^2\Pi$ ($v = 10$) level than the violet-emitting, shorter-lived $B^2\Sigma$ ($v = 0$) level. Figure 5.11 illustrates the complex magnetic field and pressure dependence of a portion of the $B^2\Sigma^+ - X^2\Sigma^+$ (0, 0) band. The collisional aspects of this figure are discussed in Section 5.5.4. The perturbed lines, marked on the figure by M (main) and E (extra), vary rapidly in relative intensity, lineshape, and frequency as the magnetic field strength changes. Most of this variation is caused by Zeeman tuning of the M_J components (each at a different rate) through various $\Delta M_J = 0$ $E_{A,J_A,M} = E_{B,J_B,M}$ degeneracies. As this occurs, a nominal A level (extra line) acquires appreciable B state character and becomes able to radiate in the violet. Its violet/red fluorescence branching ratio increases drastically. At the same time, a nominal B level acquires more A character, thus enabling the reaction to populate it more rapidly. However, because the Einstein A-coefficient for B–X (0, 0) violet fluorescence is much larger than for A–X (10, v″) red fluorescence, the fractional change in violet photon yield from a nominal B-level is smaller than the increase in its chemical formation rate and the net effect is an increase in both main and extra line violet fluorescence intensity.

Figure 5.11 conceals a tremendous amount of information about the many individual M_J crossings. Figure 5.12, which shows the energy versus magnetic field behavior for each M_J level at the four $B \sim A$ crossings ($J = 3.5$, 7.5, 10.5, and 15.5), reveals the true complexity typical of all perturbation plus static field problems. Figure 5.12 also suggests a variety of ways of recording spectra of field induced level crossings: scan wavelength at fixed field (high-resolution absorption or resolved fluorescence spectroscopy), scan field at fixed excitation and/or detection wavelength (anticrossing spectroscopy, laser magnetic resonance, laser Stark spectroscopy), fixed field and fixed wavelength with scanning radiofrequency (rf) or microwave radiation (rf or microwave–optical double resonance), short-pulse preparation of $\Delta M = 0$ or $\Delta M = 2$ coherent superposition state with time-resolved fluorescence decay (quantum-beat spectroscopy). The observability of double-resonance transitions or quantum-beat signals between nominal A and B levels can depend on setting the magnetic field strength to cause appreciable $A \sim B$ mixing in at least one of the two levels involved. Levy (1972) and Cook and Levy (1973a) combined a magnetic field with a perpendicular electric field in order to observe CN $B \sim A$ $\Delta M = \pm 1$ and $+ \leftrightarrow -$ level crossings which are unobservable in a magnetic field alone.

The remainder of this section is devoted to a simplified two-level treatment of the Zeeman and Stark effects in the presence of zero-field and field-dependent interactions between basis functions $|1M\rangle$ and $|2M\rangle$. In the presence of a static field directed along the space Z-axis, M_J remains a good quantum number. The Zeeman and Stark Hamiltonians involve the interaction

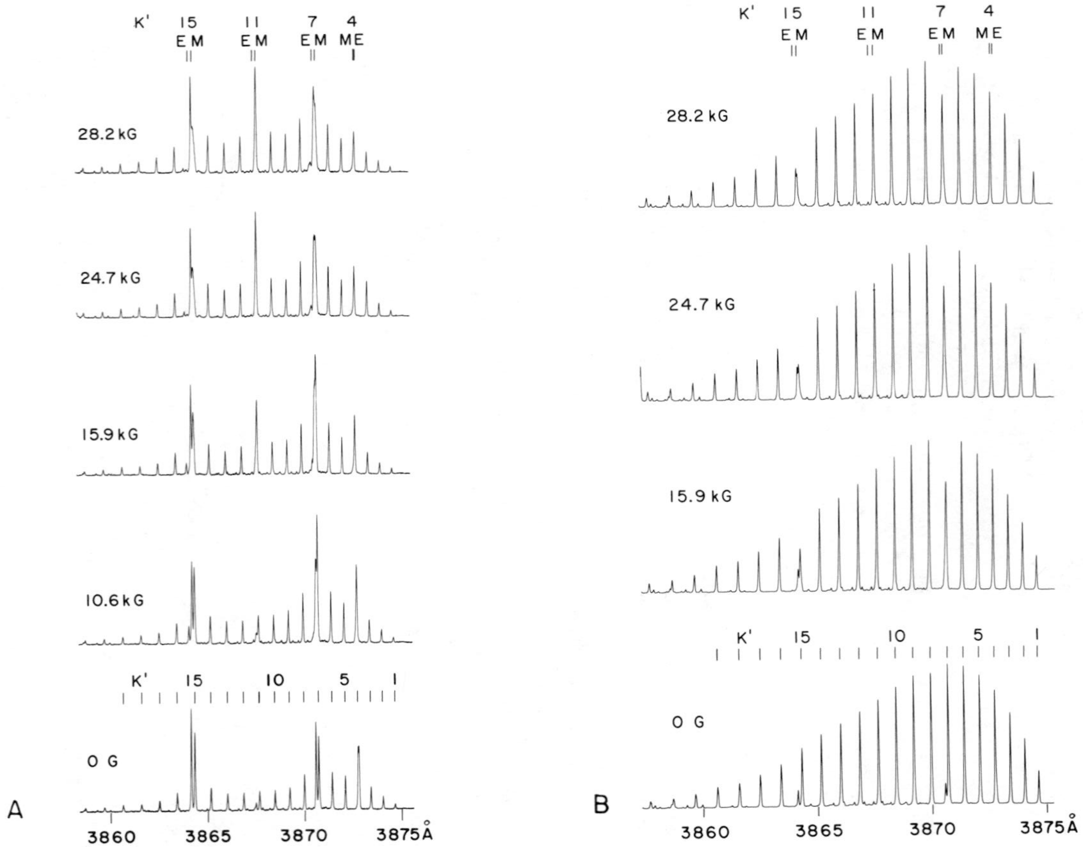

Fig. 5.11 Effect of magnetic field and pressure on the CN $B^2\Sigma^+ - X^2\Sigma^+$ (0, 0) *R* branch lines. (a) Spectra recorded at 0.6 torr in an active nitrogen plus CH_2Cl_2 flame. (b) Spectra recorded at 30 torr. Main and extra lines are marked by M and E. [From Radford and Broida (1963), in which Figs. 1 and 2 were inadvertently interchanged.]

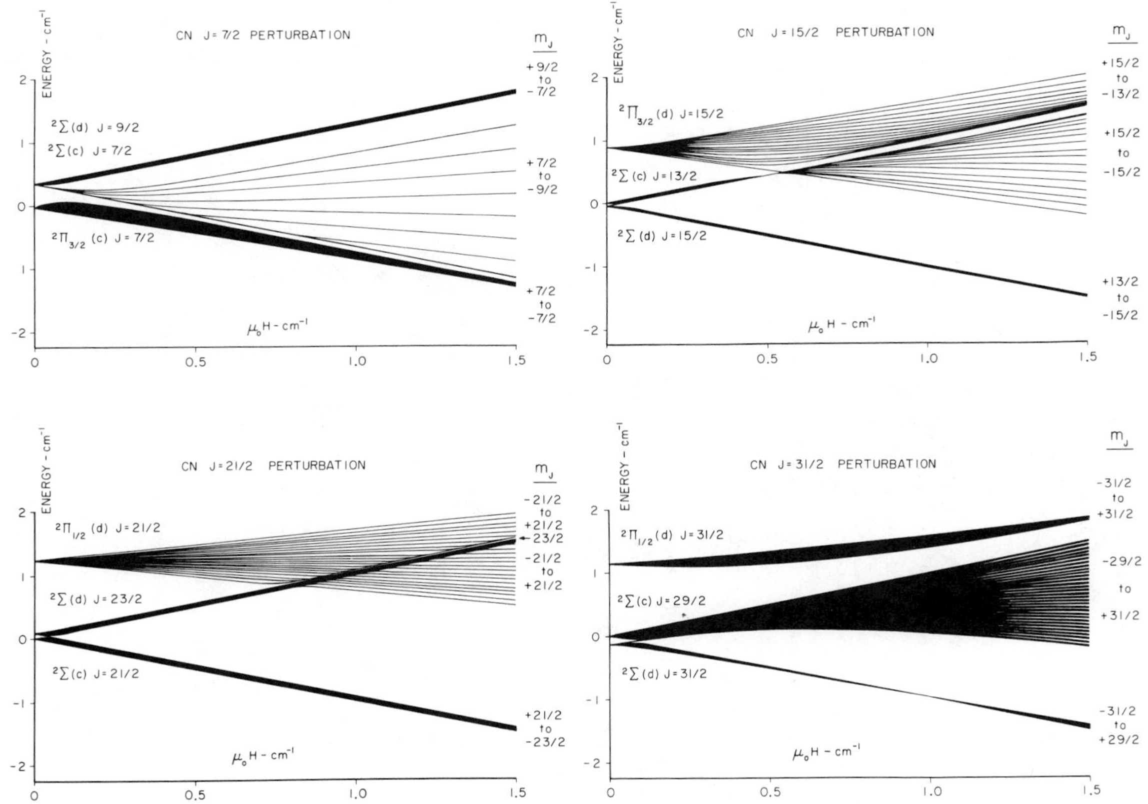

Fig. 5.12 Calculated Zeeman patterns for the perturbed rotational levels of CN $A^2\Pi$ $(v = 10) \sim B^2\Sigma^+$ $(v = 0)$. Only the $^2\Pi$ Λ-doublet and Ω-components involved in the perturbation are shown. The horizontal axis is labeled in units of μ_0 $(4.6688 \times 10^{-5}\ \mathrm{cm}^{-1}/\mathrm{G})$ multiplied by the magnetic field (H) to give reciprocal centimeters rather than gauss. [From Radford and Broida (1962).]

between a magnetic or electric dipole, $\boldsymbol{\mu}$, in the molecule-fixed axis system and the space-fixed magnetic or electric field, $\vec{\mathrm{F}}$, parallel to the laboratory direction $\hat{K}$. The interaction can be expressed in terms of direction cosines

$$\begin{aligned}\mathbf{H}^{\text{field}} &= -\vec{\boldsymbol{\mu}} \cdot \vec{\mathrm{F}} = -(F\hat{K}) \cdot (\boldsymbol{\mu}_x \hat{i} + \boldsymbol{\mu}_y \hat{j} + \boldsymbol{\mu}_z \hat{k}) \\ &= -F(\boldsymbol{\alpha}_Z^x \boldsymbol{\mu}_x + \boldsymbol{\alpha}_Z^y \boldsymbol{\mu}_y + \boldsymbol{\alpha}_Z^z \boldsymbol{\mu}_z). \end{aligned} \tag{5.5.1}$$

The $\Delta J = 0$ matrix elements of $\boldsymbol{\alpha}$ are listed in Table 1.1. All $\Delta J = 0$, $\Delta M = 0$ $\boldsymbol{\alpha}$ matrix elements are proportional to M. The following discussion deals exclusively with $\Delta J = 0$ matrix elements. Note that there are nonzero $\Delta J = \pm 1$ matrix elements of $\boldsymbol{\alpha}_Z^i$ and these all include the factor

$$\langle J + 1\, M | \boldsymbol{\alpha}_Z^i | JM \rangle \propto [(J + 1)^2 - M^2]^{1/2}. \tag{5.5.2}$$

Thus $\Delta J = \pm 1$ Stark and Zeeman matrix elements have a qualitatively different M-dependence from the $\Delta J = 0$ examples discussed below. This difference is useful for distinguishing $\Delta J = 0$ from $\Delta J = \pm 1$ field-induced interactions.

Three cases must be considered when a magnetic or electric field of magnitude F is applied and the energies of basis functions $|1M\rangle$ and $|2M\rangle$ are tuned through exact degeneracy,

$$E_{1M}^0(F) = E_{2M}^0(F),$$

where

$$\begin{aligned} E_{iM}^0(F) &= \langle iM | \mathbf{H}^{\text{zf}} + \mathbf{H}^{\text{field}} | iM \rangle \\ &= \langle iM | \mathbf{H}^{\text{zf}} | iM \rangle + \langle iM | \mathbf{H}^{\text{field}} | iM \rangle \\ &= E_i^0 + MF\mu_i. \end{aligned} \tag{5.5.3}^\dagger$$

$\mathbf{H}^{\text{zf}}$ and $\mathbf{H}^{\text{field}}$ are respectively the zero-field and field-dependent parts of $\mathbf{H}$ and $M\mu_i$ is the tuning rate for the energy of basis function $|iM\rangle$ in the field F. The Hamiltonian matrix for this $\Delta J = 0$ two-level problem is

$$\mathbf{H} = \begin{pmatrix} E_1^0 + MF\mu_1 & H_{12} + MF\mu_{12} \\ H_{12} + MF\mu_{12} & E_2^0 + MF\mu_2 \end{pmatrix}, \tag{5.5.4}$$

where

$$\langle 1M | \mathbf{H}^{\text{zf}} | 2M \rangle \equiv H_{12}, \tag{5.5.5}$$

and, for $\Delta J = 0$ matrix elements only,

$$\langle 1M | \mathbf{H}^{\text{field}} | 2M \rangle \equiv MF\mu_{12}. \tag{5.5.6}^\dagger$$

† All J-dependence is implicitly included in the μ_i and μ_{12} tuning-rate factors.

The three cases are

(1) $H_{12} \neq 0, \mu_{12} = 0$;
(2) $H_{12} = 0, \mu_{12} \neq 0$; and
(3) $H_{12} \neq 0, \mu_{12} \neq 0$.

Cases (1) and (2) correspond to simple anticrossings, whereas interference effects can occur for case (3).

The above two-level treatment conceals a simplifying approximation for the electric field case. All diagonal matrix elements of $\mathbf{H}^{\mathrm{Stark}}$ of parity basis functions are rigorously zero because $\mathbf{H}^{\mathrm{Stark}}$ has odd parity (see Section 2.2.2). However, if two zero-field basis functions of opposite parity are degenerate (or if $[E^0(+) - E^0(-)] \ll \underline{E}\langle + |\boldsymbol{\mu}| - \rangle = \underline{E}\mu_{+-}$), a nonzero electric field ($\underline{E}$) will destroy parity and cause the energies of the resultant parity-mixed functions to tune as $\pm|M|\underline{E}\mu_{+-}$. The degeneracy between the energies of the $|+, J, M\rangle$ and $|+, J, -M\rangle$ basis functions is not lifted but, for every degenerate $+M$, $-M$ pair of levels tuning to higher energy as $\underline{E}$ increases, there will be another degenerate pair with identical $|\mu_{+-}|$ tuning to lower energy. Aside from the caveat about $|M|$ versus M and some peculiarities associated with incomplete parity mixing, the two-level treatment provides a framework for understanding Stark as well as Zeeman anticrossings.

Case (1). $H_{12} \neq 0, \mu_{12} = 0$. As F is increased from zero there will be a series of $\Delta J = 0, \Delta M = 0$ anticrossings, one for each value of $|M| \neq 0$, which occur at field strengths of

$$F_{|M|} = \frac{E_1^0 - E_2^0}{M(\mu_2 - \mu_1)}. \tag{5.5.7}$$

The lowest field anticrossing will involve $|M| = J$, and the $|M|$ and $|M - 1|$ anticrossings will occur at field strengths differing by

$$F_{|M-1|} - F_{|M|} = \frac{E_1^0 - E_2^0}{\mu_2 - \mu_1} \frac{1}{M(M-1)}. \tag{5.5.8}$$

The lineshape of an isolated anticrossing is related to the variation with F of $|\langle 2M|1M, F\rangle|^2$, the fractional $|2M\rangle$ basis function character admixed into the nominal $|1M, F\rangle$ eigenfunction. The shape of an anticrossing signal depends on the specific methods of excitation and detection (Miller, 1973), but $|\langle 2M|1M, F\rangle|^2$ is a crucial factor. For case (1) crossings, each isolated anticrossing signal will have a symmetric lineshape,

$$\begin{aligned} \sigma_M(F) &\simeq 2|\langle 2M|1M, F\rangle|^2 \\ &= 1 - \left\{1 + \left[\frac{2H_{12}}{M(\mu_1 - \mu_2)(F - F_M)}\right]^2\right\}^{-1/2}, \end{aligned} \tag{5.5.9}$$

and the linewidth (FWHM), Γ_M, in units of field strength rather than energy, is

$$\Gamma_M = \left|\frac{(16/3)^{1/2}H_{12}}{M(\mu_1 - \mu_2)}\right|. \tag{5.5.10}$$

Note that the width is proportional to H_{12} and inversely proportional to $|M|$ and to the differential tuning rate, $|\mu_1 - \mu_2|$. The sharpest anticrossings are those with small H_{12} and $|M| = J$. See Fig. 5.15 for an example of ultranarrow, forbidden (small H_{12}), singlet $\sim$ triplet anticrossings in H_2. These anticrossings are observed by monitoring the intensity of the specified spectrally-selected emission line as the magnetic field strength is varied. See Section 5.5.3 for further discussion of anticrossing spectroscopy.

Case (2). $H_{12} = 0$, $\mu_{12} \neq 0$. The only difference from case (1) is that each isolated anticrossing signal does not have a symmetric lineshape. Since the coupling matrix element increases in proportion to F, these anticrossings exhibit level-shift and mixing-coefficient (lineshape) behavior versus F similar to that versus J for heterogeneous perturbations where the coupling matrix element is proportional to J. However, for $\mu_{12} \neq 0$ anticrossings the differential tuning rate is proportional to $(\mu_1 - \mu_2)\ MF$ versus F rather than to ΔBJ^2 versus J as for heterogeneous perturbations. The lineshape is

$$\sigma_M(F) \simeq 1 - \left\{1 + \left[\frac{2\mu_{12}}{(\mu_1 - \mu_2)(1 - F_M/F)}\right]^2\right\}^{-1/2}. \tag{5.5.11}$$

and the low-field halfwidth will be narrower than the high-field halfwidth. Note that the $|1M\rangle \sim |2M\rangle$ mixing does not vanish as $F \to \infty$,

$$\sigma_M(\infty) = 1 - \left[1 + \left(\frac{2\mu_{12}}{\mu_1 - \mu_2}\right)^2\right]^{-1/2}. \tag{5.5.12}$$

The linewidth

$$\Gamma_M = (16/3)^{1/2}\mu_{12}(\mu_1 - \mu_2)F_M[(\mu_1 - \mu_2)^2 - 4\mu_{12}^2/3]^{-1} \tag{5.5.13}$$

is inversely proportional to $|M|$, as for case (1), because the only M-dependent factor in Γ_M, F_M, is proportional to $|M|^{-1}$[Eq. (5.5.7)].

Case (3). $\mu_{12} \neq 0$, $H_{12} \neq 0$. The interference effect is most evident for $E_1^0 = E_2^0 = E^0$. For that situation, all M-levels are anticrossed at $F = 0$. The energies and eigenfunctions are

$E_{+,M}$, which belongs to $|+, M, F = 0\rangle = 2^{-1/2}[|1M\rangle + |2M|]$,

$E_{-,M}$, which belongs to $|-, M, F = 0\rangle = 2^{-1/2}[|1M\rangle - |2M\rangle]$,

and

$$E_{+,M} - E_{-,M} = 2H_{12}.$$

As soon as $F \neq 0$, an asymmetry develops in the energies

$$|E_{\pm,M} - E^0| \neq |E_{\pm,-M} - E^0|$$

and in the mixing fractions

$$|\langle 1M|\pm, M, F\rangle|^2 \neq |\langle 1, -M|\pm, -M, F\rangle|^2$$

because the total off diagonal matrix element has a different absolute magnitude for $M > 0$ and $M < 0$. If

$$F = -\frac{H_{12}}{M\mu_{12}},$$

then the off-diagonal matrix element vanishes for M but has the magnitude $2H_{12}$ for $-M$. This has the effect of transferring all of the interaction strength, which would normally be shared equally, from M to $-M$.

Another consequence of the interference between $\mathbf{H}^{\text{zf}}$ and $\mathbf{H}^{\text{field}}$ is manifest as an asymmetry in the behavior of rotational levels above and below a zero-field level crossing. Let $E_1^0 - E_2^0 = \Delta > 0$; then the crossings occur at

$$F_M = \frac{\Delta}{M(\mu_2 - \mu_1)}, \tag{5.5.14}$$

and, if F and $(\mu_2 - \mu_1)$ are positive, the crossings will involve the $M > 0$ levels. The width of each crossing will be approximately proportional to the total coupling matrix element at F_M,

$$\Gamma_M \approx \frac{(16/3)^{1/2}[H_{12} + \Delta\mu_{12}/(\mu_2 - \mu_1)]}{M(\mu_1 - \mu_2)}. \tag{5.5.15}$$

Now let $E_1^0 - E_2^0 = -\Delta < 0$. The crossings at positive F will involve the $M < 0$ levels, and their widths will be

$$\Gamma_{-M} \approx \frac{(16/3)^{1/2}[H_{12} - \Delta\mu_{12}/(\mu_2 - \mu_1)]}{|M|(\mu_1 - \mu_2)}. \tag{5.5.16}$$

The anticrossings will be much narrower in rotational levels either above $(E_1^0 > E_2^0)$ or below $(E_1^0 < E_2^0)$ the zero-field level crossing, depending on the relative signs of H_{12} and μ_{12}. The sign of $H_{12}\mu_{12}$ is thus an experimental observable and, provided that consistent wavefunction phases are used in computing H_{12} and μ_{12}, should be *a priori* predictable.

Rather than measuring the widths and locations of anticrossings, another kind of useful information is obtained by direct measurement of the tuning rate of each eigenstate in the applied field,

$$\frac{\partial}{\partial F}\langle 1MF|\mathbf{H}|1MF\rangle \equiv \partial E_{1MF}/\partial F, \tag{5.5.17}$$

where

$$|1MF\rangle = (1 - C_{21}^2)^{1/2}|1M\rangle + C_{21}|2M\rangle. \tag{5.5.18}$$

It is convenient to make use of the Hellmann–Feynman theorem, which states that

$$\frac{\partial E_i}{\partial \lambda} = \left\langle i \left| \frac{\partial \mathbf{H}}{\partial \lambda} \right| i \right\rangle \tag{5.5.19}$$

if $|i\rangle$ and E_i are, respectively, an eigenstate and eigenvalue of $\mathbf{H}$. For

$$\mathbf{H} = \mathbf{H}^{\text{zf}} + \mathbf{H}^{\text{field}},$$

$$\frac{\partial \mathbf{H}}{\partial F} = \mathbf{H}^{\text{field}}/F \equiv \mathbf{H}^{\text{f}}/F. \tag{5.5.20}^\dagger$$

Thus

$$\begin{aligned}\frac{\partial E_{1MF}}{\partial F} &= \langle 1MF|\mathbf{H}^{\text{f}}/F|1MF\rangle \\ &= (1 - C_{21}^2)\langle 1M|\mathbf{H}^{\text{f}}/F|1M\rangle + C_{21}^2\langle 2M|\mathbf{H}^{\text{f}}/F|2M\rangle \\ &\quad + (1 - C_{21}^2)^{1/2}C_{21}\langle 1M|\mathbf{H}^{\text{f}}/F|2M\rangle \\ &= M[\mu_1 + C_{21}^2(\mu_2 - \mu_1) + C_{21}(1 - C_{21}^2)^{1/2}\mu_{12}] \\ &\equiv Mg_1.\end{aligned} \tag{5.5.21}$$

The important point to notice is that, since the sign of the mixing coefficient C_{21} is determined by the signs of $E_1^0 - E_2^0$ and H_{12}, the field tuning rate, g_J, for an MJ level above the zero-field crossing point will be different from that for an MJ' level an equal distance below the crossing.

This asymmetric variation of g_J near the J-value of the zero-field crossing was observed by Gouedard and Lehmann (1979, 1981) in the B 0_u^+ state of $^{80}Se_2$ and is illustrated by Fig. 5.13. They measured Zeeman-tuned quantum beats between the J, M and J, $M + 2$ nominal B 0_u^+ levels. If the μ_{12} term had been zero, their g_J-values would have been symmetrically affected on either side of each J-crossing point. The sharp onset and slow disappearance of the deviation in $v' = 1$ at $J' = 69$ and in $v' = 3$ at $J' = 107$ (see Figs. 1–7 of Gouedard and Lehmann, 1979) and the sign change in $v' = 4$ at $J' = 55$ prove that the direct magnetic coupling term is nonzero. Measurement of g-values can be a very sensitive probe for perturbations, especially if the perturbation-free g_J-value is very small, as it is for the Se_2 B 0_u^+ state.

Another rather suprising effect arises from an interference between $\mathbf{H}^{\text{Zeeman}}$ and $\mathbf{H}^{\text{ROT}}$. Although this example involves predissociation, which is not

† This requires that all matrix elements of $\mathbf{H}^{\text{field}}$ be proportional to F.

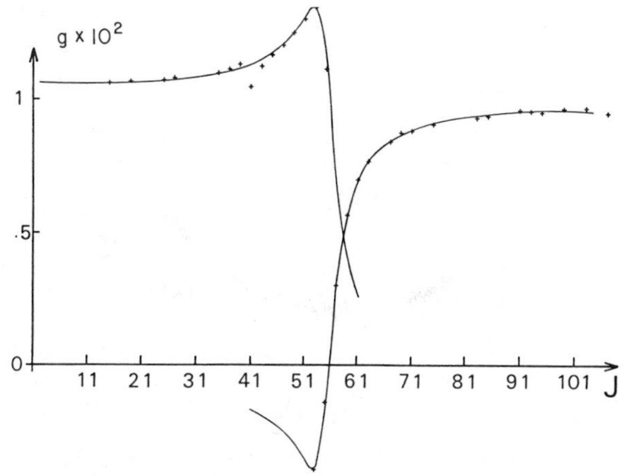

Fig. 5.13 g_J-Values for $^{80}Se_2B0_u^+$ ($v = 4$). Points are observed by $\Delta M = 2$ Zeeman quantum-beat spectroscopy. The solid curves display the g_J-values calculated from the results of a fit to the observed g values (including both main and extra lines at $J = 53$ and 55), from which g_J values for the $B0_u^+$ state and the perturbing 1_u state as well as the off-diagonal g-value (g_{12}) and $B\mathbf{J} \cdot \mathbf{J}_a$ matrix element were determined. For each J there are two calculated g_J values, the larger one corresponding to the main level. The failure of the high-J g_J values to return to the low-J value is proof that the perturbation is heterogeneous. The asymmetry (including a sign change) in the main-line g_J curve about $J = 55$ is proof that $g_{12} \neq 0$. [From Gouedard and Lehmann, (1981).]

discussed until Chapter 6, a brief description here is appropriate. Vigue *et al.* (1974) observed that when the I_2 B $0_u^+ - X^1\Sigma_g^+$ (40, 0) $R(76)$ line was excited using linearly polarized radiation with the $\underline{E}$-vector parallel to an applied magnetic field, H, the resultant fluorescence propagating along the field direction was found to be circularly polarized whenever $H \neq 0$. The degree of circular polarization depended on the magnitude of H, and the dominant polarization, σ^+ versus σ^-, depended on the sign of H. This could be explained if the $+M$ and $-M$ levels had different lifetimes. Figure 5.14 illustrates that for $H > 0$, the σ^+ polarized fluorescence ($M' - M'' = +1$, thus sampling more $M' > 0$ than $M' < 0$) has a shorter lifetime than σ^-.

This $+M/-M$ lifetime asymmetry is explained by noting that the B 0_u^+ state could interact with a 1_u state via both $\mathbf{H}^{\mathrm{ROT}}$ (the $-B\mathbf{J} \cdot \mathbf{L}$ term) and $\mathbf{H}^{\mathrm{Zeeman}}$ The coupling matrix element is a sum of two terms, one of which depends on the sign of M and the sign of the magnetic field. For $+M$, the two terms add, whereas for $-M$, they partially cancel. The fact that the perturber is a repulsive state simply means that the unbound character mixed into the nominal B 0_u^+ ($v = 40$, $J = 77$) level is larger for $+M$ than $-M$. The lifetime difference reflects the more rapid nonradiative decay (predissociation) of $+M$ than $-M$.

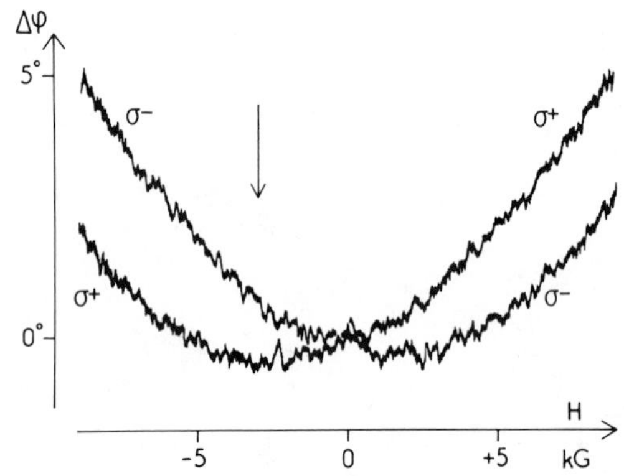

Fig. 5.14 Difference in lifetime for σ^+ versus σ^- polarized fluorescence from I_2 $B0_u^+$ ($v = 40$, $J = 77$). Lifetimes are measured by a phase-shift technique (the excitation radiation is amplitude modulated at 100 kHz) and the arrow points in the direction of increasing lifetime. Note that the σ^+/σ^- difference vanishes at $H = 0$ and reverses between $H > 0$ and $H < 0$. [From Vigue *et al.* (1975).]

5.5.3 ANTICROSSING, QUANTUM-BEAT, AND DOUBLE-RESONANCE EXPERIMENTS

When the perturbation matrix element, H_{12}, is small compared to $|E_1^0 - E_2^0|$, perturbation effects may be undetectably weak or too small to provide useful information about the perturbing state. By applying a static magnetic or electric field, it may be possible to Zeeman- or Stark-tune the $|1M\rangle$ and $|2M\rangle$ *basis* functions into the crossing region where

$$|E_{1M}^0 - E_{2M}^0| < |H_{12} + MF\mu_{12}|.$$

If this condition is met, *regardless of the size of* $H_{12} + MF\mu_{12}$, the radiative properties of both $|+, M\rangle$ and $|-, M\rangle$ *eigenfunctions* become profoundly different from those of the *basis* functions $|1M\rangle$ and $|2M\rangle$. [The Stark and Zeeman tuning coefficients are also affected; see Eq. (5.5.21).] For example, if an M_J component of a CN $A^2\Pi$ ($v = 10$) rotational level is Zeeman-tuned into resonance with a suitable ($\Delta M = 0$, $\Delta J = \pm 1, 0, + \leftrightarrow\!\!\!/ \; -$) component of $B^2\Sigma^+(v = 0)$, the radiative lifetime will decrease from $\tau_A^0 \approx 600$ ns to approximately $2\tau_B^0 \approx 130$ ns and, more importantly, the fluorescence will switch abruptly from the red to the violet spectral region.

Anticrossings enable detection of extremely weak perturbations. The energies of the triplet states of H_2 were first determined accurately relative to those of the singlets by $j^3\Delta_g \sim J^1\Delta_g$ and $r^3\Pi_g \sim R^1\Pi_g$ magnetic anticrossing spectroscopy (Miller and Freund, 1974; Jost and Lombardi, 1974). These anticrossings were observable despite the extremely small spin–orbit interaction associated with $3d\delta$ or $4d\pi$ Rydberg orbitals of the H_2 molecule. In fact, it is advantageous for H_{12} to be small because this results in extremely narrow and fully resolved anticrossings [see Eqs. (5.5.10), (5.5.13), and (5.5.16)]. Figure 5.15 shows spectra of the $\Delta M_N = 0$ anticrossings between the $i^3\Pi_g(v = 1, N = 6)$ and $W^1\Sigma_g^+(v = 1, N = 4)$ levels. This is a forbidden anticrossing because all $\Delta N = 2$ matrix elements vanish between $^3\Pi$ and $^1\Sigma$ case (b) *basis* functions. The selection rule for $\Delta S = 1$ spin–orbit interactions between case (b) basis functions is $\Delta N = 0, \pm 1$. (The hyperfine Hamiltonian gives no $\Delta N = 2$ matrix elements for para-H_2 because the two $I_H = \frac{1}{2}$ nuclear spins are paired to form a nuclear spin singlet, $I = 0$, which has zero magnetic dipole–dipole interactions with all other angular momenta.) The anticrossing becomes observable because the magnetic field mixes some $N = 5$ character into either the nominal $^3\Pi$ $N = 6$ or $^1\Sigma$ $N = 4$ level, thereby turning on a spin–orbit interaction between these levels. The coupling matrix element between the two anticrossing levels is ~ 1 MHz.

Figure 5.15 shows two complementary anticrossing spectra. Intensity is transferred from the $W^1\Sigma_g^+ – B^1\Sigma_u^+$ (1, 1) $R(3)$ line to the $i^3\Pi_g – c^3\Pi_u$ (1, 1) R(5) line. The H_2 molecules are excited by electron bombardment, resulting in slightly greater population of $W^1\Sigma_g^+$ than $i^3\Pi_g$. Each intensity decrease in the W–B line corresponds to an increase in the i–c line.

Note that the horizontal axis of Fig. 5.15 is labelled in gauss, which is *not* a unit of energy. It is a nontrivial matter to go from the anticrossing signal to a zero-field energy-level diagram, a set of Zeeman tuning coefficients, and H_{12}. The analysis of Zeeman (Stark) anticrossing spectra is always more complicated than analysis of a zero-field perturbation, in part because each J level splits into $2J + 1$ ($J + 1$ or $J + \frac{1}{2}$ for Stark because of the $+M, -M$ degeneracy) M_J-components, but primarily because at least two additional parameters must be determined, namely, the rates at which both interacting levels tune in the applied field. For most molecules, the magnetic and electric dipole moments determine the Zeeman and Stark tuning rates.

A fundamental difference exists with respect to *a priori* knowledge of electric versus magnetic dipole moments. Electric moments depend on the detailed shape of electronic wavefunctions and must always be measured experimentally or computed from accurate *ab initio* wavefunctions. In contrast, magnetic moments are directly proportional to angular momenta and may be computed from the eigenvectors of an effective Hamiltonian without knowledge of the electronic wavefunctions. In the case of electronic angular momenta, the proportionality constants g_L and g_S in

$$\boldsymbol{\mu}_L \equiv \mu_B g_L \mathbf{L} \tag{5.5.22a}$$

$$\boldsymbol{\mu}_S \equiv \mu_B g_S \mathbf{S} \tag{5.5.22b}$$

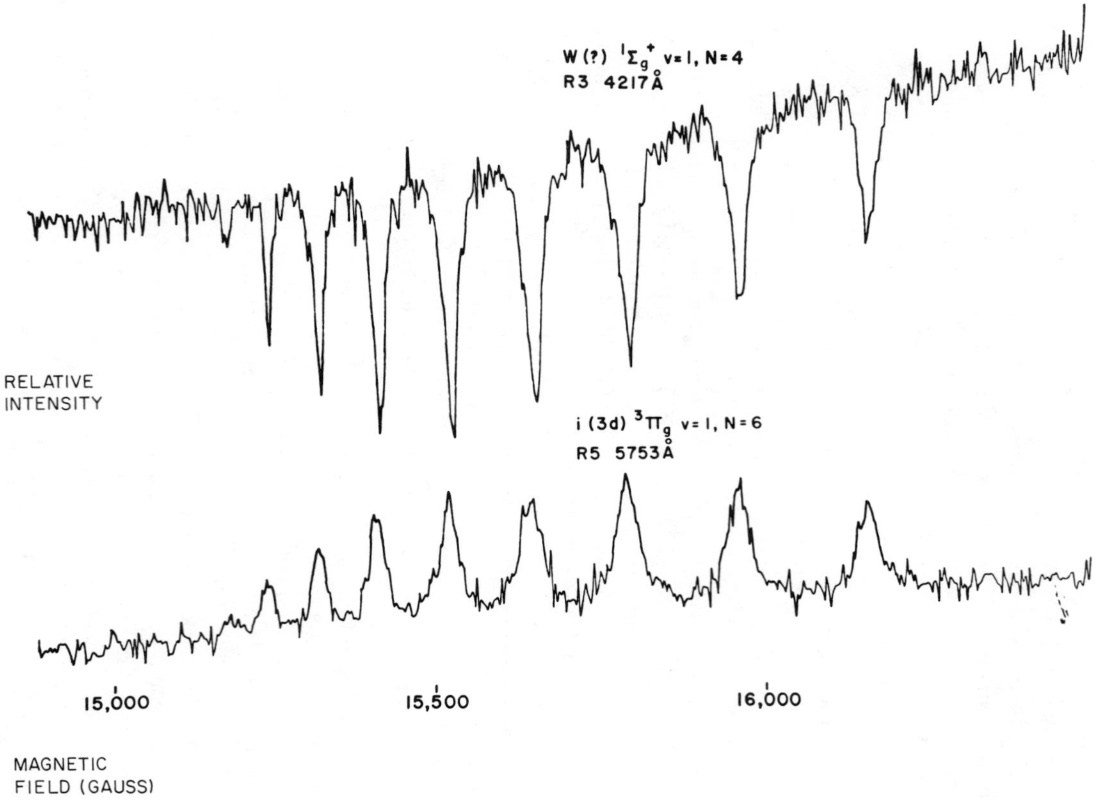

Fig. 5.15 H_2 $W^1\Sigma_g^+ \sim i^3\Pi_g$ anticrossing spectra. At each anticrossing, a decrease in the $W^1\Sigma_g^+ - B^1\Sigma_u^+$ (1, 1) $R(3)$ line corresponds to an increase in the intensity of the $i^3\Pi_g$–$c^3\Pi_u$ (1, 1) R(5) line. [From Miller and Freund (1975).]

are known exactly,

$$g_L = 1 \tag{5.5.23}$$

$$g_S = 2.00232, \tag{5.5.24}$$

and

$$\mu_B \equiv 1.39967 \text{ MHz/G}$$

is the Bohr magneton. For nuclear angular momenta, **I** and **R**, the g-values are typically much smaller (by the ratio of the proton to electron masses, $1836.11 = \mu_B/\mu_n$, where μ_n is the nuclear magneton) and are not fundamental physical quantities. However, the g_I values for all stable $I \neq 0$ nuclei have been accurately measured,

$$\boldsymbol{\mu}_I \equiv \mu_n g_I \mathbf{I} \tag{5.5.22c}$$

where the quantity $(g_I = \mu_I/\mu_n I)$ is tabulated (Townes and Schawlow, 1955, p. 644).

The only magnetic dipolar quantities that cannot be derived from *a priori* known g-values are g_R,

$$\boldsymbol{\mu}_R \equiv \mu_n g_R \mathbf{R}, \tag{5.5.22d}$$

and matrix elements of $\boldsymbol{\mu}$ between two electronic states.† See Townes and Schawlow (1955, p. 292) for an approximate calculation of g_R. There is a direct proportionality between $\mathbf{BL}^+$ and $\mathbf{BS}^+$ perturbation parameters and vibronically off-diagonal matrix elements of $\boldsymbol{\mu}$, for example,

$$\begin{gathered}\frac{\langle v_{\Lambda+1}\Lambda+1\,|\mathbf{BL}^+|\,v_\Lambda\Lambda\rangle}{\langle v_{\Lambda+1}\,|\hbar/2\mu R^2|\,v_\Lambda\rangle} = \frac{\langle v_{\Lambda+1}\Lambda+1\,|\boldsymbol{\mu}|\,v_\Lambda\Lambda\rangle}{\mu_B g_L\langle v_{\Lambda+1}|v_\Lambda\rangle}, \\ \langle\Lambda+1\,|\mathbf{L}^+|\,\Lambda\rangle = \mu_{\Lambda+1,\Lambda}/(\mu_B g_L)\end{gathered} \tag{5.5.25}$$

(Radford, 1962; Radford and Broida, 1962; Cook and Levy, 1973b.) This means that, for a rotation–electronic perturbation, knowledge of the zero-field perturbation parameter, H_{12}, implies knowledge of the off-diagonal Zeeman perturbation parameter, μ_{12}, and of the sign of μ_{12} relative to H_{12}. Any discrepancy would imply inadequacy of the deperturbation model resulting from the involvement of an additional perturbing state or the nonnegligibility of an unsuspected perturbation mechanism ($\mathbf{H}^{SO}$, $\mathbf{H}^{SS}$, $\mathbf{H}^{\text{hyperfine}}$).

Anticrossing spectroscopy is complementary to quantum-beat (QB) and radiofrequency–optical and microwave–optical double resonance (MODR) techniques. When an anticrossing is broad (because of large H_{12} or small $|\mu_1 - \mu_2|$), the useful information is concealed under a complicated line-

† $\boldsymbol{\mu}$ is total of $\boldsymbol{\mu}_L$, $\boldsymbol{\mu}_S$, $\boldsymbol{\mu}_R$, $\boldsymbol{\mu}_I$.

shape, which is a composite of anticrossings between many pairs of M-levels. A broad anticrossing does not imply broad M-levels! The splittings between M levels can be measured directly in megahertz (not in gauss or volts/centimeter, which is quite an advantage) by QB or MODR spectroscopy.

Quantum-beat spectroscopy requires preparation of a coherent superposition state, $\Psi_{+,M;-,M'}$, composed of two eigenstates, $|+,M\rangle$ and $|-,M'\rangle$, by pulsed excitation from a single initial eigenstate $|iM''\rangle$,

$$\Psi_{+,M;-,M'}(t) = \mu_{+i}|+,M\rangle \exp[t(iE_{+,M}/\hbar - 1/2\tau_+)] + \mu_{-i}|-,M'\rangle \exp[t(iE_{-,M'}/\hbar - 1/2\tau_-)], \quad (5.5.26)^\dagger$$

where τ_+ and τ_- are the spontaneous fluorescence lifetimes of $|+,M\rangle$ and $|-,M'\rangle$, and μ_{+i} and μ_{-i} are transition moments [Eq. (5.3.1)] between initial ($|iM''\rangle$) and final eigenstates. This superposition state decays by spontaneous fluorescence into $|0M'''\rangle$ at a rate proportional to

$$\begin{aligned} P(t) &= |\langle\Psi_{+,M;-,M'}(t)\,|\boldsymbol{\mu}|\,0\rangle|^2/|\langle\Psi_{+,M;-,M'}(t=0)|\Psi_{+,M;-,M'}(t=0)\rangle|^2 \\ &= f_+|\mu_{+0}|^2\exp(-t/\tau_+) + f_-|\mu_{-0}|^2\exp(-t/\tau_-) \\ &\quad + 2(f_+f_-)^{1/2}\mu_{+0}\mu_{-0}\cos[(E_+ - E_-)t/\hbar]\exp[-t(1/2\tau_+ + 1/2\tau_-)], \end{aligned} \quad (5.5.27)$$

where the fractional $|+,M\rangle$ and $|-,M'\rangle$ eigenstate characters, f_+ and f_-, are

$$f_\pm = \frac{|\mu_{\pm i}|^2}{|\mu_{+i}|^2 + |\mu_{-i}|^2}. \quad (5.5.28)$$

The fluorescence rate is modulated at the frequency splitting of eigenstates $|+,M\rangle$ and $|-,M'\rangle$. This modulation is the quantum beat. The quantum beat decays at a rate

$$1/\tau_{QB} = \tfrac{1}{2}(1/\tau_+ + 1/\tau_-), \quad (5.5.29)$$

which is the average of the two eigenstate decay rates. The depth of modulation is

$$\Phi_{QB} = \frac{2(f_+f_-)^{1/2}\mu_{+0}\mu_{-0}}{f_+|\mu_{+0}|^2 + f_-|\mu_{-0}|^2}. \quad (5.5.30)$$

The modulation depth is the ratio of the oscillatory component of $P(t)$ to the sum of the two exponentially decaying eigenstate components: 100% modulation corresponds to the case where (if $\tau_+ = \tau_-$) $P(t) = 0$ once every beat period,

† $\Psi_{+,M;-,M'}(t)$ in Eq. (5.5.26) is not normalized. The $t = 0$ normalization factor, which appears in Eq. (5.5.27), is

$$|\langle\Psi_{+,M;-,M'}(t=0)|\Psi_{+,M;-,M'}(t=0)\rangle|^{-1/2} = [|\mu_{+i}|^2 + |\mu_{-i}|^2]^{-1/2}.$$

$\Delta t = 2\pi\hbar/(E_+ - E_-)$. In the optimal case where $f_+\mu_{+0}^2 = f_-\mu_{-0}^2$, the modulation depth will be 100%.

In order to prepare $\Psi_{+,M;-,M'}$, the exciting radiation must be capable of simultaneously populating both $|+, M\rangle$ and $|-, M'\rangle$ from the common initial level $|i, M''\rangle$. Furthermore, in order to *detect* the quantum beat, the experimental apparatus must be arranged so that fluorescence from both $|+, M\rangle$ and $|-, M'\rangle$ into a common final level $|0, M'''\rangle$ is detectable. These requirements lead to quantum number, temporal, and geometric (i.e., polarization) constraints.

Typical N_2 and Nd : YAG pumped dye lasers have a pulse duration of ~6 ns FWHM. This means that quantum beats between levels split by more than $(6 \times 10^{-9}\ \text{s})^{-1} \simeq 170$ MHz will be unobservable with a preparation pulse duration longer than 6 ns. Recall that, if the two beating levels correspond to main and extra lines, the $\lesssim 170$ MHz limit restricts the perturbation interaction to $H_{12} \lesssim 85$ MHz.

The quantum number constraint leads to two usual polarization classes of QB spectra: $\Delta M = 0$ beats (polarization of excitation and detection parallel to static field, $M = M' = M'' = M'''$) and $\Delta M = 2$ beats (perpendicular polarization, $M = M' + 2 = M'' + 1 = M''' + 1$). The former type is particularly useful for examining an anticrossing between two perturbing states (provided H_{12} is small enough). The latter displays both intrastate splittings (useful for measurement of *g*-values and dipole moments) and interstate splittings.

In order to observe quantum beats, all four transition moments $\mu_{+i}, \mu_{+0}, \mu_{-i}, \mu_{-0}$ must be nonzero. If $|+, M\rangle$ and $|-, M'\rangle$ are simply field-split M-components of the same electronic state, there is no problem. However, if either $|+, M\rangle$ or $|-, M'\rangle$ is a mixed eigenstate composed of basis functions $|1M\rangle$ and $|2M\rangle$ belonging to two electronic states, for example,

$$|+, M\rangle = (1 - C_{2+}^2)^{1/2}|1M\rangle + C_{2+}|2M\rangle$$

(see Eq. 5.2.6) and

$$|-, M'\rangle = |2M'\rangle$$

where the $|1M\rangle \leftarrow |iM''\rangle$ and $|1M\rangle \rightarrow |0M'''\rangle$ transitions are forbidden, then

$$\Phi_{QB} = \frac{2C_{2+}^2\mu_{2M,i}\mu_{2M,0}\mu_{2M',i}\mu_{2M',0}}{C_{2+}^2\mu_{2M,i}^2\mu_{2M,0}^2 + \mu_{2M',i}^2\mu_{2M',0}^2} \tag{5.5.31}$$

and the QB modulation depth is approximately $2C_{2+}^2/(1 + C_{2+}^2)$. The beats vanish if $C_{2+} = 0$. Near-degeneracy between the $|1M\rangle$ and $|2M'\rangle$ basis functions is not sufficient for observation of a quantum beat. There must be a $1M \sim 2M$ perturbation to allow detection of forbidden inter-electronic-state quantum beats. A great deal of information about both electronic states and their interaction (zero-field and field-induced) is contained in the variation of QB frequency, modulation depth, and decay rate versus the applied magnetic or electric field.

Zeeman quantum-beat spectroscopy was used by Gouedard and Lehmann (1979, 1981) to measure the effect of various 1_u perturbing states on the g_J-values [Eq. (5.5.21)] of more than 150 rotational levels of the Se_2 B 0_u^+ state (see Section 5.5.2 and Fig. 5.13). In that experiment, the excitation polarization was perpendicular to the applied magnetic field so that quantum beats were observed between nominal B-state components differing in M by 2. The frequencies of these beats increase linearly from 0 MHz at 0 G until the $\Delta M = 2$ splitting falls outside the coherence width of the excitation source and the beats disappear. For each J' level of the B state, all pairs of M, $M + 2$ levels are split by the same amount† (except for J' levels near the perturbations where the Zeeman pattern will become slightly asymmetric owing to M-dependent $\mathbf{H}^{zf} \times \mathbf{H}^{Zeeman}$ interference and energy denominator effects). This near-perfect superposition of all $\Delta M = 2$ beats from a given J' level vastly enhances the sensitivity of QB spectroscopy for high J-levels. Because the beats are between levels differing in M by 2, the beats are at twice the Larmor frequency (the $\Delta M = 1$ level spacing). At high magnetic field, a mixed $|0_u M\rangle \sim |1_u M\rangle$ level could be tuned through a nominal $|\mathrm{B}\, 0_u\, M \pm 2\rangle$ level and a $\Delta M = 2$ interstate QB would result. This effect would be difficult to detect because only one of the $2J + 1$ M-levels would be affected and a lengthy search for the proper magnetic field would be required.

Stark quantum-beat (SQB) spectroscopy was used by Brieger *et al.* (1980) for LiH $A^1\Sigma^+$ and by Schweda *et al.* (1980, 1985) for BaO $A^1\Sigma^+$ to measure electric dipole moments of several rotation–vibration levels. In a $\Lambda = 0$ state, the electric-field-induced shift of a J, M-level is proportional to $\underline{E}^2/B$ because $\mathbf{H}^{Stark}$ has only $\Delta J = \pm 1$, $\Delta M = 0$ matrix elements:

$$E_{J=1,M=0} = \sum_{J'} \frac{|\langle J0|\boldsymbol{\mu}\underline{E}|J'0\rangle|^2}{E_J^0 - E_{J'}^0}$$

$$= \underline{E}^2 \mu_\Sigma^2 \left[\frac{|\langle 10|\boldsymbol{\alpha}_Z^z|20\rangle|^2}{2B_v - 6B_v} + \frac{|\langle 10|\boldsymbol{\alpha}_Z^z|00\rangle|^2}{2B_v - 0B_v} \right]$$

$$= \frac{\underline{E}^2 \mu_\Sigma^2}{B_v}\left[-\tfrac{1}{15} + \tfrac{1}{6}\right] = +\frac{\underline{E}^2 \mu_\Sigma^2}{10 B_v} \tag{5.5.32}$$

$$E_{J=1,M=1} = -\frac{\underline{E}^2 \mu_\Sigma^2}{20 B_v} \tag{5.5.33}$$

so, for an unperturbed $^1\Sigma$ state,

$$(E_{J=1,M=0} - E_{J=1,M=1})/\hbar = \frac{3}{20}\frac{\underline{E}^2 \mu_\Sigma^2}{\hbar B_v} = 7.9692 \frac{\underline{E}^2 \mu_\Sigma^2}{B_v} \ \mathrm{Hz} \tag{5.5.34}$$

† Neglecting $\Delta J = \pm 1$ matrix elements of $\mathbf{H}^{Zeeman}$.

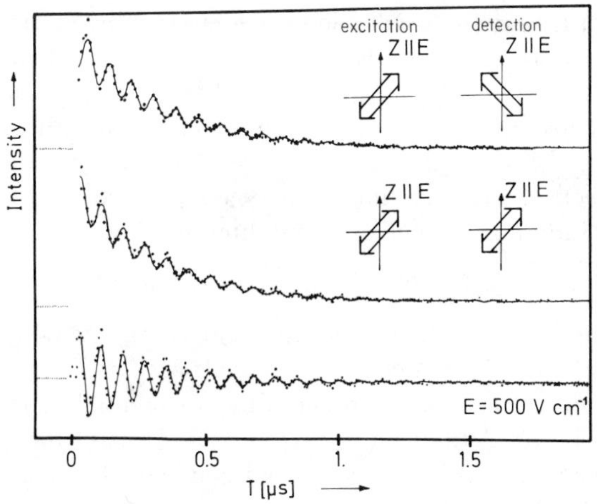

Fig. 5.16 Stark quantum beats in BaO $A^1\Sigma^+$ ($v = 2$, $J = 1$). The $J = 1$ level is excited via the $R(0)$ line by radiation from an N_2-laser-pumped dye laser. The pump radiation is linearly polarized at 45° to the $\underline{E}$-field direction in order to produce a coherent superposition of $M = 0$ with $|M| = 1$ components. The top trace shows the signal resulting when the polarization of the detected fluorescence is selected to be at 45° to $\underline{E}$ and 90° to the excitation polarization. The middle trace is for parallel excitation and detection polarizations. The bottom trace is the difference between the two detection geometries. [From Schweda *et al.* (1985).]

for $\underline{E}$ (V/cm), μ_Σ (D), and B_v (cm^{-1}). Figure 5.16 shows the SQB superimposed on the exponential decay of the fluorescence from BaO $A^1\Sigma^+$ ($v = 2$, $J = 1$) at 500 V/cm.

The BaO $A^1\Sigma^+$ state is perturbed by the $b^3\Pi$ and $A'^1\Pi$ states. The effect of perturbations on the Stark tuning coefficients of a $^1\Sigma^+$ state is more complicated than the situation described by Eq. (5.5.21) (because parity remains an almost good quantum number in $\Lambda = 0$ states at $\underline{E}$-fields that are sufficiently large to cause complete parity mixing in $|\Lambda| > 0$ states). There will be Stark tuning terms proportional to (suppressing all J, M-dependent factors)

$$(1 - C_{\Sigma\Pi}^2)^2 \underline{E}^2 \mu_\Sigma^2 / B_\Sigma \qquad \text{direct } ^1\Sigma \text{ term} \tag{5.5.35a}$$

$$C_{\Sigma\Pi}(1 - C_{\Sigma\Pi}^2)^{3/2} \underline{E}^2 \mu_\Sigma \mu_{\Sigma\Pi} / B_\Sigma \qquad \text{cross term} \tag{5.5.35b}$$

$$C_{\Sigma\Pi}^2(1 - C_{\Sigma\Pi}^2) \underline{E}^2 \mu_{\Sigma\Pi}^2 / \Delta E_{\Sigma\Pi} \qquad \text{transition moment term} \tag{5.5.35c}$$

$$C_{\Sigma\Pi}^2(1 - C_{\Sigma\Pi}^2) \underline{E}^2 \mu_\Sigma \mu_\Pi / B_\Sigma \qquad \text{borrowed } \Pi \text{ term,} \tag{5.5.35d}$$

where $C_{\Sigma\Pi}$ is the zero-field mixing coefficient, $\mu_{\Sigma\Pi}$ is the electric dipole transition moment between Σ and Π levels, and

$$\Delta E_{\Sigma\Pi} = E_{\Sigma J}^0 - E_{\Pi J'}^0.$$

Since $^1\Sigma^+ - ^3\Pi$ transitions are dipole-forbidden, the three terms involving $\mu_{\Sigma\Pi}$ are zero for the $b^3\Pi$ perturbers. However, the $\mu_{\Sigma\Pi}$ terms will not be zero for the $A^1\Sigma^+ - A'^1\Pi$ transition. The $C_{\Sigma\Pi}(1 - C_{\Sigma\Pi}^2)^{3/2}\underline{E}^2\mu_\Sigma\mu_{\Sigma\Pi}/B_\Sigma$ term is the most important perturbation-related term because it is linear rather than quadratic in $C_{\Sigma\Pi}$.

MODR spectroscopy is not limited to levels separated by $\lesssim 170$ MHz. What is needed is a radiation field of sufficient intensity that

$$\mu\underline{E}/\hbar > 1/\tau \qquad \text{or} \qquad \mu_0 g_J H/\hbar > 1/\tau,$$

so that the microwave-induced transition rate is competitive with radiative decay. MODR can be detected in a variety of ways.

Evenson *et al.* (1964) took advantage of the preferential chemical population of the CN $A^2\Pi$ state. Although the $A^2\Pi \rightarrow B^2\Sigma^+$ (10, 0) band, which occurs in the microwave region, is both electric and magnetic dipole-allowed, its small Franck–Condon factor of 0.01 makes microwave $A \rightarrow B$ transitions between unperturbed levels much weaker than pure rotational transitions within either state ($\mu_A \approx 0.6$ Debye, $\mu_B \approx 1.2$ Debye) and also weaker than interelectronic transitions involving a mixed level. When the microwave frequency is tuned to resonance with a nominal $A \rightarrow B$ transition (or a magnetic field is used to tune an M-component into resonance with a fixed microwave field), fluorescence intensity is transferred from the red (A–X) to the violet (B–X) region, exactly as in an anticrossing experiment. Since the microwave transition affects a small fraction of the chemically populated levels of the $A^2\Pi$ state, a satisfactory signal/noise is obtained by monitoring a specific B–X (0, 0) rotational line, selected using a monochromator with $\sim$0.4 Å bandpass.

Miller *et al.* (1974) used MOMRIE (microwave optical magnetic resonance induced by electrons) complementarily with anticrossing spectroscopy in experiments on He, H_2, and other molecules. Small differences in electron bombardment excitation cross-sections combined with different radiative lifetimes lead to steady-state population differences between initial and final levels of the microwave transition. By monitoring fluorescence from one of the two involved electronic states, a change in fluorescence intensity signals a microwave resonance. Electron bombardment excitation tends to produce smaller population differences than the chemical pumping scheme exploited in the CN experiments, but the MOMRIE technique is more generally applicable.

Atomic and molecular resonance lamps were used by Silvers *et al.* (1970) and Field and Bergeman (1971) to optically pump the CS $A^1\Pi(v = 0)$ level, which is perturbed by $a'^3\Sigma^+(v = 10)$. Electric-dipole-allowed Λ-doubling transitions ($\Delta J = 0, \Delta M = 0, |M| = J, e \leftrightarrow f$) in $A^1\Pi$ were observed in electric fields up to 8 kV/cm. The MODR signal was detected as a change in the polarization of undispersed side fluorescence. Stark tuning coefficients precise to ~1 part in 10^3 were obtained for $J = 3$–9. The J-dependence of the Stark coefficients, combined with $A^1\Pi \sim a'^3\Sigma^+$ mixing coefficients determined from a deperturbation analysis of the MODR Λ-doublings and the optical data of Barrow *et al.* (1960); permitted determination of the dipole moment of the perturbing $a'^3\Sigma^+$ state as well as that of the $A^1\Pi$ state.

German and Zare (1968) used an atomic lamp to populate OH $A^2\Sigma^+$ ($v = 0$, $N = 2$, $J = \frac{3}{2}$). This level is doubled by the $I = \frac{1}{2}$ proton nuclear spin. Radiofrequency magnetic dipole transitions within the Zeeman-split M_F components of the $F = 1$ and 2 levels were observed as changes in undispersed fluorescence polarization. German and Zare measured magnetic g-values that were found to be in satisfactory agreement with g_F values predicted for isolated $N = 2$, $J = \frac{3}{2}$, $F = 1$ and 2 levels of a case (b) $^2\Sigma^+$ state.

Field *et al.* (1972, 1973) used Ar^+ and cw dye lasers to observe $\Delta J = \pm 1$ MODR pure rotational transitions in the BaO $A^1\Sigma^+$ state. The MODR effect is observable either as a change in undispersed fluorescence polarization or intensity. The MODR B_v-values for $A^1\Sigma^+$($v = 0$–5, 7) show large deviations from a smooth B_v versus v variation owing to perturbations by $b^3\Pi$ and $A'^1\Pi$ levels (Field, *et al.*, 1973). No interelectronic MODR transitions were found in BaO, but several "mystery" MODR lines were reported in electronically excited NO_2 by Tanaka *et al.* (1974) and by Solarz and Levy (1973). These incompletely assigned lines are certainly perturbation-facilitated transitions between two vibronic levels. The radiative lifetimes and pressure-dependent photon yields for the initial and final levels of the MODR transition are quite dissimilar, giving rise to a complex pressure-dependence of the sign and magnitude of the MODR signal.

Recently, an extremely sensitive MODR scheme, microwave optical polarization spectroscopy (MOPS), was introduced by Ernst and Törring (1982). The most important features of MOPS are that it requires respectively 100 and 10 times lower laser and microwave intensities than MODR and results in 10 times narrower lines. This means that it will be possible to take full advantage of differential power-broadening effects (Section 5.5.1) and to utilize low-power, frequency-doubled dye lasers and low-power, broadly tunable microwave sources (backward wave oscillators) in order to gain access to and systematically study perturbations.

A perturbation acts as a window through which normally unobservable states may be viewed. Anticrossing, quantum-beat, and MODR spectroscopies are capable of examining the states in the immediate vicinity of the perturbed level. Optical–optical double resonance (OODR) allows systematic exploration of states far above and below the perturbed level. Gottscho (1979) utilized an OODR scheme to observe the BaO $a^3\Sigma^+$ state. He used two single-mode cw

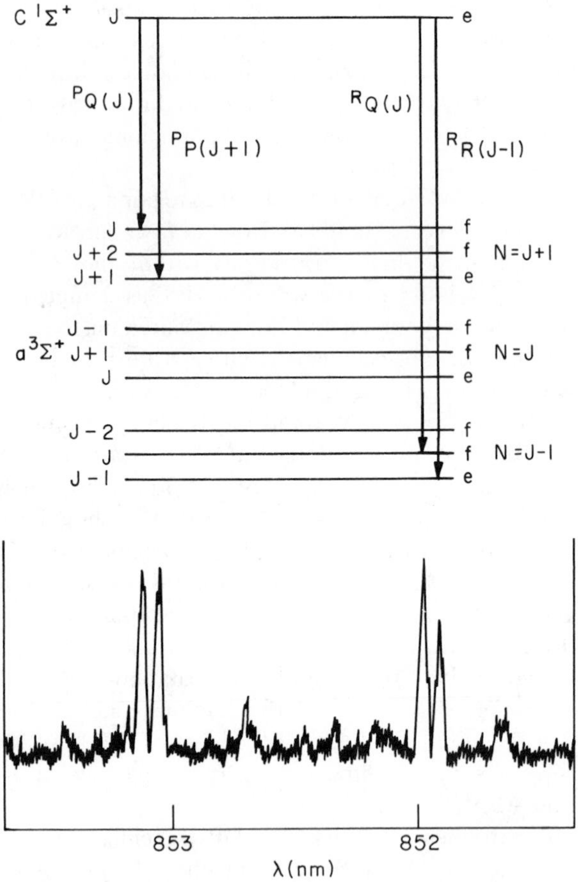

Fig. 5.17 BaO $C^1\Sigma^+ - a^3\Sigma^+$ (2, 12) fluorescence excited by OODR. The triplet contaminated $J = 16$, $v = 2$ level of $C^1\Sigma^+$ is excited using two single-mode cw dye lasers. [From Gottscho (1979).]

dye lasers to excite a singlet~triplet mixed level of the $C^1\Sigma^+$ state *via* the $A^1\Sigma^+$ state. Figure 5.17 shows the resolved fluorescence spectrum originating from the $C^1\Sigma^+$ ($v = 2$, $J = 16$) level. The fluorescence pattern immediately identifies the lower state as a $^3\Sigma^+$ state and provides approximate values for its rotational and spin–spin constants. This OODR-resolved fluorescence scheme is a rather general method for observing low-lying excited states, because, at high energies where the vibronic density of states is large, perturbations are much more common than in the lowest excited states. Thus one is likely to find the necessary type of perturbation among the high-energy states in order to gain access via fluorescence to virtually any suspected low-lying state.

In a complementary experiment. Li Li and Field (1983) have utilized Na_2 $A^1\Sigma_u^+ \sim b^3\Pi_u$ perturbations to survey, via OODR excitation spectroscopy, excited triplet *gerade* levels with T_v-values in the 33,000-cm^{-1} region. By tuning the pump laser to a transition into a singlet $\sim$ triplet mixed level, the probe laser is equally capable of exciting transitions into pure triplets or pure singlets. The two classes of excited states are experimentally separable by restricting the wavelength region of detected fluorescence and by comparing the OODR intensities obtained with the pump laser tuned to main (singlet strong, triplet weak) versus extra (singlet weak, triplet strong) lines.

5.5.4 NONTHERMAL POPULATION DISTRIBUTIONS; CHEMICAL AND COLLISIONAL EFFECTS

Perturbations affect the rates of absorption and emission of radiation in a fully understood and exactly calculable manner. They also affect the rates of chemical and collisional population/depopulation processes, but in a less completely predictable way. Perturbation effects on steady-state populations can be very large and level-specific, but both definitive state-to-state experimental measurements and a generally applicable, quantitative theory are lacking. However, although collision-induced transitions and chemical reactions are not governed by rigorous selection rules as are electric dipole transitions and perturbation interactions, some useful propensity rules have been suggested theoretically and confirmed experimentally.

Gelbart and Freed (1973) have suggested that the cross-sections for collision-induced transitions between two different electronic states, E and E', are

$$\sigma_{EJ,E'J'} \simeq \sigma_{EJ,EJ'} C_{EE'}(J')^2 + \sigma_{E'J,E'J'} C_{EE'}(J)^2 \qquad (5.5.36)$$

where $\sigma_{EJ,EJ'}$ and $\sigma_{E'J,E'J'}$ are the $J \to J'$ purely rotation-changing cross-sections within the E and E' electronic states, respectively, and $C_{EE'}(J)$ is the isolated-molecule $EJ \sim E'J$ mixing coefficient. Collision-induced inter-electronic-state transfer is viewed as a rotational relaxation process (typical values for $\sigma_{EJ,EJ\pm 1}$ and $\sum_{J'} \sigma_{EJ,EJ'}$ are 10 and 100 Å^2) reduced by the fractional mixing in either the initial or final J-level. When only one rotation–vibration level is appreciably mixed, this level should act as a gateway through which all population flows on its way from one electronic state to the other.

The Gelbart–Freed model is based on the assumption that the direct, electronically inelastic process is significantly less probable than the perturbation-facilitated one. This is probably reasonable whenever the transition is between two levels with $\Delta S \neq 0$ or a small vibrational overlap ($\langle v|v'\rangle \lesssim 10^{-2}$). In other words, the Gelbart–Freed model should be applicable when the E and E' vibronic levels are connected by a forbidden or weak electric dipole transition and the $C_{EE'}$ isolated-molecule coefficients are nonzero. The validity of the model rests on two additional requirements: the collision must be "sudden" and "weak." A collision is sudden (Mukamel, 1979) when

$$\tau_{\text{coll}} H_{EvJ,E'v'J'} < \hbar \tag{5.5.36a}$$

where τ_{coll} is the collision duration [typically 10 Å/(5 × 10^4 cm/s) = 2 ps, which puts an upper limit $H_{EvJ,E'v'J'} \ll 3\ \text{cm}^{-1}$ on the isolated-molecule perturbation matrix element]. If Eq. (5.5.36a) is not satisfied, additional terms describing the evolution of the system during the lifetime of the collision complex must be taken into account. A collision is weak (Freed and Tric, 1978) when the relative shifts of the basis function energies, $E^0_{EJ} - E^0_{E'J}$, during the collision may be neglected. Such level shifts (possibly resulting from a difference between the electric dipole moments of the $|EJ\rangle$ and $|E'J'\rangle$ basis states) would cause the mixing coefficients to evolve over a collision trajectory because of their dependence on the instantaneous distance between collision partners. After averaging over all collision trajectories, the effective mixing coefficients $\langle C_{EE'}(J)\rangle$ could become quite different from the isolated-molecule coefficients $C_{EE'}(J)$.

It is difficult to verify the predictions of the Gelbart–Freed (1973) model at the state-to-state level because it is difficult to measure the rate of the $EJ \rightarrow E'J'$ process in the presence of the considerably more rapid $EJ \rightarrow EJ'$ and $E'J \rightarrow E'J'$ processes, especially when the radiative lifetime of one of the two zero-order states is typically a factor of more than 10^2 longer than the other. Collisions equilibrate the rotational populations in the longer-lived state so rapidly that the specific $\sigma_{EJ\rightarrow E'J'}$ cross-section is not measurable by the usual selective excitation-resolved fluorescence detection technique. However, the predictions of the Gelbart–Freed (1973) model, particularly the role of gateway states, are confirmed in experimental studies of CN $A^2\Pi$ ($v = 10$) ↔ $B^2\Sigma^+$ ($v = 0$) (Pratt and Broida, 1969), BaO $b^3\Pi$ ($v = 10$) → $A^1\Sigma^+$ ($v = 1$) (Field *et al.*, 1974), and glyoxal $\tilde{A}^1A_u \rightarrow \tilde{a}^3A_u$ (Lombardi and Michel, 1985). In all three cases, the Eq. (5.5.36a) inequality is satisfied.

Grimbert et al. (1978) found a monotonic but nonlinear relationship between the observed $\sigma_{A^1\Pi J,T}$($T = a'^3\Sigma^+$, or $e^3\Sigma^-$, or $d^3\Delta$) cross-sections and the rotationally averaged mixing fractions for the $v = 0$–7 vibrational levels of CO $A^1\Pi$, $\overline{C^2_{AvJ,TJ}}$ [calculated from the mixing fractions derived from the deperturbation eigenvectors of Field, (1971)]. In this case the isolated molecule inter-state coupling is so strong that Eq. (5.5.36a) is not satisfied. A model proposed by Grimbert *et al.* (1978) explained the apparent saturation of $\sigma_{A^1\Pi J,T}$ by taking into account the finite duration of the collision.

It is traditional to distinguish between spin-conserving and nonconserving electronically inelastic processes. This is due to the much stronger effect of a light collision partner on the spatial (orbital) rather than the spin part of the electronic wavefunction. Spin-changing ($\Delta S \neq 0$) collisions are forbidden in the absence of isolated-molecule $\Delta S \neq 0$ mixing via $\mathbf{H}^{\mathrm{SO}}$, except in the case of a heavy collision partner ("external heavy atom effect," where $\mathbf{H}^{\mathrm{SO}}$ causes spin-mixing in the collision complex). Unlike the $\Delta S \neq 0$ case, it is necessary to consider both extrinsic (collision complex) and intrinsic (isolated molecule) state-mixing effects for $\Delta S = 0$ processes. For example, N_2 $W^3\Delta_u \leftrightarrow B^3\Pi_g$ (Heidner *et al.*, 1976) and N_2^+ $A^2\Pi_u \leftrightarrow X^2\Sigma_g^+$ (Katayama, 1984, 1985) collision-induced transitions are efficient processes despite the absence of intrinsic coupling owing to the rigorous $g \nleftrightarrow u$ selection rule for isolated-molecule perturbation matrix elements.

There is a striking contrast between the perturbation-facilitated CN $A^2\Pi \leftrightarrow B^2\Sigma^+$ processes observed by Pratt and Broida (1969) and the perturbation-irrelevant $A^2\Pi \leftrightarrow X^2\Sigma^+$ processes observed in CN (Katayama *et al.*, 1979), CO^+ (Katayama *et al.*, 1980), and N_2^+ (Katayama, 1984, 1985). Pratt and Broida (1969) showed that collision-induced CN $A^2\Pi$ ($v = 10$) $\leftrightarrow B^2\Sigma^+$ ($v = 0$) transfer was slower between unperturbed rotational levels than when the initial or final level (or both) was $A \sim B$ mixed. Since only a few rotational levels are appreciably mixed, the mixed levels act as gateways for $A \leftrightarrow B$ transfer. A quite different kind of rotational propensity rule was observed by Katayama (1984, 1985) for collision-induced transitions from N_2^+ $A^2\Pi_u$ ($v = 4$) into $X^2\Sigma_g^+$ ($v = 7$ and 8). The collisional transitions followed the Hönl–London factors for an electric-dipole-allowed optical transition. In fact, the rotationally selective A ($v = 4$) $\rightarrow$ X ($v = 7$) collisional process, with an energy gap of 1760 cm^{-1}, was competitive with rotation-changing collisions within $A^2\Pi$.

Bondybey and Miller (1978) and Katayama *et al.* (1979) proposed that the rates of electronically inelastic processes in the gas phase should follow a Franck–Condon rate law,

$$\sigma_{1v_1,2v_2} \approx \sigma_{12}^{\mathrm{el}} \langle v_1 | v_2 \rangle^2 e^{-\Delta E_{1v_1,2v_2}/kT} \tag{5.5.37}$$

similar to that proposed by Bondybey (1977) (see Section 5.5.6). The $\sigma_{12}^{\mathrm{el}}$ electronic factor is precisely the term assumed to be negligible by Gelbart and Freed (1973). Katayama's (1984, 1985) observations for N_2^+ $A \leftrightarrow X$ transfer are consistent with neither a perturbation gateway model [Eq. (5.5.36)] nor an energy gap law (Eq. 5.5.37).

Alexander (1982a) has proposed a dipolar model for electronically inelastic processes that exhibits limiting behavior consistent with both Eqs. (5.5.36) and (5.5.37). This model is based on a time-dependent Born approximation

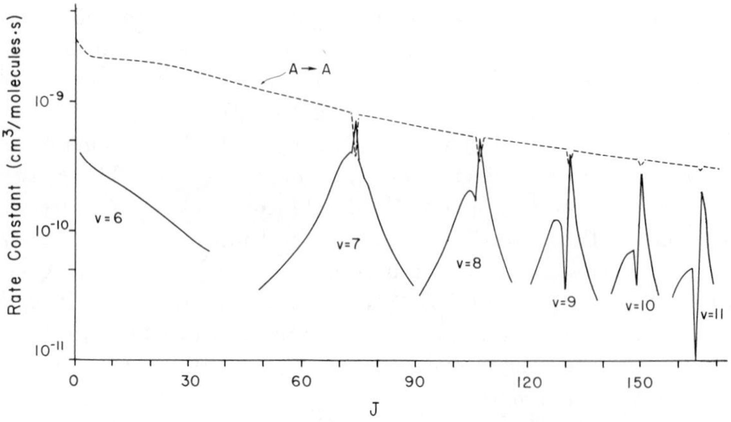

Fig. 5.18 Calculated $J \rightarrow J+1$ rate constants for CaO + N_2O collisions. The dotted curve describes transitions within the CaO nominal $A^1\Sigma^+$ $(v=0)$ level. The rates for transitions between nominal $a^3\Pi_{0e}$ (v = 6–11) and $A^1\Sigma$ $(v=0)$ levels are presented as solid curves. The $A^1\Sigma^+$ state has a larger B-value than $a^3\Pi$, hence the $A^1\Sigma^+$ $(v=0)$ level overtakes $a^3\Pi_{0e}$ $(v=7)$ at $J \simeq 73$, $v=8$ at $J \simeq 107$, etc. The downward (upward) spikes in the dotted (solid) curves occur at the level crossings. [From Alexander (1982a).]

treatment of the leading term in the multipolar expansion of the collisional interaction potential. In the case of CaO $(A^1\Sigma^+, A'^1\Pi, a^3\Pi)$ + N_2O discussed by Alexander (1982a), this leading term is the interaction between the *permanent* electric dipole moment of N_2O and the electric dipole *transition* moment between isolated CaO molecular *eigenstates*. There are several contributions to the CaO transition moments, the permanent electric dipole moments of the $A^1\Sigma^+$, $A'^1\Pi$, and $a^3\Pi$ electronic states (μ_A, $\mu_{A'}$, and μ_a) and the $A \leftrightarrow A'$ dipole allowed transition moment, $\mu_{AA'}$. A unique feature of Alexander's (1982a) model is that interference between transition amplitudes leads to an even stronger J-dependence of the transition probabilities than predicted by the Gelbart–Freed (1973) model [Eq. (5.5.36)]. The solid lines in Fig. 5.18 depict the calculated rate constants for the process

$$\text{CaO}\ ('a^3\Pi_{0e}, v', J) + N_2O \rightarrow \text{CaO}\ 'A^1\Sigma^+, v = 0', J+1) + N_2O.$$

The sharp upward spikes occur at the J-values where $A^1\Sigma^+$ $(v=0)$ crosses successive $a^3\Pi_{0e}$ vibrational levels. The sharp dips near each spike arise from destructive interference between the μ_A and μ_a $J \rightarrow J+1$ transition amplitudes,

$$\sigma_{av_aJ, Av_A=0, J+1} \propto |\langle 'a^3\Pi'_{0e} v\ J|\boldsymbol{\mu}|'A^1\Sigma^{+\prime} 0\ J+1\rangle|^2$$

$$\begin{aligned}&\propto |[1 - C_{aA}(J+1)^2]^{1/2} C_{aA}(J)\,\mu_A \\ &-[1 - C_{aA}(J)^2]^{1/2} C_{aA}(J+1)\,\mu_a \\ &+\{[1 - C_{aA}(J+1)^2]^{1/2}[1 - C_{aA}(J)^2]^{1/2} \\ &- C_{aA}(J) C_{aA}(J+1)\}\,\mu_{aA}|^2. \qquad (5.5.38a)\end{aligned}$$

where $C_{aA}(J)$ is the mixing coefficient for the $|A^1\Sigma^+, v = 0, J\rangle$ basis function in the nominal $|'a^3\Pi_{0e}, v', J\rangle$ eigenfunction. Setting the dipole transition moment $\mu_{aA} = 0$ for a $\Delta S \neq 0$ transition,

$$\begin{aligned}\sigma_{avJ,A0,J+1} \propto &[1 - C_{aA}(J)^2] C_{aA}(J+1)^2 \mu_a^2 + [1 - C_{aA}(J+1)^2] C_{aA}(J)^2 \mu_A^2 \\ &- 2[1 - C_{aA}(J)^2]^{1/2} [1 - C_{aA}(J+1)^2]^{1/2} C_{aA}(J) C_{aA}(J+1)\,\mu_a \mu_A. \qquad (5.5.38b)\end{aligned}$$

The mixing coefficient has opposite signs above and below the level crossing. This means that the $\mu_a\mu_A$ term will interfere constructively or destructively for all J-values except when J is below and $J + 1$ is above the crossing. At this point, where the coefficients of the μ_a^2, μ_A^2, and $\mu_a\mu_A$ terms approach their maximum values of $\frac{1}{4}$, $\frac{1}{4}$, and $\frac{1}{2}$, respectively, the sign of the interference effect will be opposite to that for all other values of J. Even more complicated interference patterns could occur at $A^1\Sigma^+ \sim A'^1\Pi$ perturbations because of the nonzero (but, in the case of CaO, negligibly small) $\mu_{AA'}$ transition moment term. Note that, in the case when there is no state mixing ($C_{12} = 0$) and $\mu_{12} \neq 0$, as for example N_2^+ $A^2\Pi_u \rightarrow X^2\Sigma_g^+$, an equation of the form of Eq. (5.5.38a) reduces to $\sigma_{1J,2J+1} \propto \mu_{12}^2$.

Alexander's (1982a) transition dipole model would reduce to the Gelbart–Freed (1973) model if the $\mu_a\mu_A$ term in Eq. (5.5.38b) could be neglected ($\mu_a^2 \gg \mu_A^2$ or $\mu_a^2 \ll \mu_A^2$) and if the transition moment term ($\mu_{AA'}$ or μ_{Aa}) were to vanish. Alternatively, one would obtain a model similar to Eq. (5.5.37) in the limit that both permanent moments vanish. The CN, CO^+, and N_2^+ $A^2\Pi \rightarrow X^2\Sigma^+$ transfer rates seem to follow Eq. (5.5.37) because $\mu_{AX} > \mu_A$ and μ_X. The CO $A^1\Pi \rightarrow T$ ($T = a'^3\Sigma^+$, $e^3\Sigma^-$, $d^3\Delta$) rates follow a gateway model because $\mu_{AT} = 0$ and $\mu_T > \mu_A$.

The rates of perturbation-facilitated collision-induced transitions between electronic states should exhibit strong dependences on initial and final states whenever $\mu_{12}^2 \ll \mu_1^2$ or μ_2^2 and the perturbation matrix element is small so that only a few $C_{12}(J)$ are nonnegligible. This can give rise to highly non-Boltzmann steady-state population distributions. Schemes for utilizing or manipulating these unusual distributions include electronic transition chemical lasers, magnetic or electric field switching of perturbations via anticrossings, electronic-state-specific photochemistry, and isotope separation. A chemical laser, for example, could be designed around storage of the energy released by a chemical reaction in the initially formed, energetic, nonradiating levels (typically high vibrational levels of the electronic ground state) followed by collisional transfer through a "perturbation funnel" into a perturbed level from which emission is electric-dipole-allowed. Although no such laser has been developed, several pure rotational transitions in the CH_2 $\tilde{a}\,^1A_1$ state as well as a few nominal $\tilde{a}\,^1A_1$–$\tilde{X}\,^3B_1$ transitions have been observed as stimulated emission rather than absorption lines in the laser magnetic resonance spectrum (McKellar *et al.*, 1983). The population inversions result from perturbation-facilitated rapid depopulation of the lower energy level.

The propensity rules for collision-induced transitions between electronic states and among the fine-structure components of non-$^1\Sigma^+$ states depend on the identity of the leading term in the multipole expansion of the molecule/collision-partner interaction potential. Alexander (1982a) has considered the dipole–dipole term, which includes both permanent and transition dipole contributions. The collisional propensity rules for the permanent dipole term follow from the selection rules for both perturbations and pure rotational ($\mu_{\|}$) transitions:

$$s \not\leftrightarrow a$$

$$\Delta J = \pm 1 \qquad (\Delta J = 0 \quad \text{occurs only at low} \quad J \quad \text{in} \quad \Omega \neq 0)$$

$$e \not\leftrightarrow f \qquad (\text{except for } \Delta J = 0 \quad \text{at low } J)$$

$$g \not\leftrightarrow u$$

$$\Delta S = 0, \pm 1$$

$$\Delta\Omega = \Delta\Sigma = 0 \qquad [\text{homogeneous perturbation, case (a) limit}]$$

$$\Delta\Omega = \Delta\Sigma = \pm 1 \qquad [\text{heterogeneous perturbation, case (a) limit}].$$

The collisional propensity rules for the transition dipole term follow from the selection rules for $\mu_{\|}$ or $\mu_{\perp}$ electronic transitions:

$$s \not\leftrightarrow a$$

$$\Delta J = 0, \pm 1$$

(for $\mu_{\perp}$ $\Delta J = 0$ is stronger than $\Delta J = \pm 1$, and for $\mu_{\|}$ $\Delta J \neq 0$ except at low J in $\Omega \neq 0$)

$$+ \leftrightarrow -$$

$$\Sigma^+ \not\leftrightarrow \Sigma^-$$

$$\Delta S = 0$$

$$\Delta\Omega = \Delta\Lambda = \pm 1, 0 \qquad \text{[case (a) limit]}$$

$$\Delta J = \Delta N \qquad \text{[case (b) limit]}$$

$$g \leftrightarrow u.$$

The most important difference between the permanent and transition moment terms are $e \nleftrightarrow f$ versus $+ \leftrightarrow -$ and $g \nleftrightarrow u$ versus $g \leftrightarrow u$.

Propensity rules for collision-induced transitions within non-$^1\Sigma^+$ states are discussed by Gottscho (1981), Nedelec and Dufayard (1984), Alexander (1982b) for $^2\Sigma$ and Alexander (1982c) for $^2\Pi$ states, Alexander and Dagdigian (1983) for $^3\Sigma$ and (1984) for Λ-doublet transitions in Π states, and Alexander and Pouilly (1983) for $^3\Pi$ states. The theoretical derivation of the intrastate propensity rules by Alexander and co-workers (1982b,c,1983,1984) is based on a more exact treatment of the dynamics than the interstate dipolar model discussed in the preceding paragraphs. It will be interesting to discover how the dipolar propensity rules just introduced will be modified if a similar rigorous treatment were applied to collision-induced transitions between electronic states.

5.5.5 "DEPERTURBATION" AT HIGH PRESSURE AND IN MATRICES

The purpose of a deperturbation calculation is to define a computational model that faithfully reproduces every detail of observable spectra in terms of deperturbed potential energy curves and the interaction terms that are responsible for the perturbation. Deperturbation calculations can be extremely complicated. When faced with such a calculation, one is tempted to doubt whether a specific set of deperturbed potentials has any physical significance.

Perhaps one of the most convincing demonstrations that molecules know about deperturbed potential curves and that these curves are not arbitrary constructs from an abstract and prejudiced computation is the phenomenon of pressure-induced configuration demixing (Miladi *et al.*, 1975, 1978). The electronic spectrum of the NO molecule is especially well suited to illustrate this demixing phenomenon because of the enormous structural difference between valence states (typically $\omega_e = 1000\ \text{cm}^{-1}$, $B_e = 1.1\ \text{cm}^{-1}$) and Rydberg states (typically $\omega_e = 2400\ \text{cm}^{-1}$, $B_e = 2.0\ \text{cm}^{-1}$) and the numerous valence $\sim$ Rydberg perturbations. These perturbations among the NO $^2\Pi$ states (B and L valence states, C, K, and Q Rydberg states, see Section 5.2.2) have been treated recently by Gallusser and Dressler (1982).

Because of the considerably larger size [Eq. (4.2.1)] of Rydberg ($\langle r\rangle_{3p} =$ 3.7 Å) than valence ($\langle r\rangle_{2p} = 2.6$ Å) orbitals, Rydberg states have much larger pressure broadening cross-sections than valence states, as shown in the following data at 5 bar neon (Miladi *et al.*, 1978):

Rydberg	$D^2\Sigma^+–X^2\Pi$	(0, 0)	FWHM = 1.95 cm^{-1}
Valence	$B^2\Pi–X^2\Pi$	(9, 0)	FWHM = 0.80 cm^{-1}
50/50 Mixed	$B^2\Pi$ (v = 15) ~ $C^2\Pi$ (v = 3)–$X^2\Pi$ (v = 0)		FWHM = 1.30 cm^{-1}
	$C^2\Pi$ (v = 3) ~ $B^2\Pi$ (v = 15)–$X^2\Pi$ (v = 0)		FWHM = 1.30 cm^{-1}

Miladi *et al.* (1975, 1978) have shown that, as a sample of NO is pressurized by an admixture of an inert gas, the Rydberg transitions are progressively broadened and shifted until, by ~1000 bar, the Rydberg states have all but vanished from the spectrum. This pressure-induced deperturbation process can be carried essentially to completion in inert-gas matrices (Boursey and Roncin, 1975; Boursey, 1976), where the frequencies and intensities of the observed bands correspond almost perfectly (except for a constant energy shift) with the deperturbed (Gallusser and Dressler, 1982; Jungen, 1966) quantities for valence ← $X^2\Pi$ absorption bands. Table 5.4 and Fig. 5.19 show that the

Table 5.4
The NO $B^2\Pi_{1/2}–X^2\Pi_{1/2}$ Transition in the Gas Phase and in Inert Gas Matrices

	Observed[a] gas phase		Deperturbed[b]		Observed[c] Ne matrix, 5 K			Observed[c] Ar matrix, 5 K		
v_B	E_{gas}	$f \times 10^4$[d]	E_{dep}	$f \times 10^4$	$E_{obs} - E_{dep}$	I[e]	FWHM	$E_{obs} - E_{dep}$	I	FWHM
5	50,452[f]	0.2	50,479[f]	0.1	−123[f]	0	—[f]	—[f]	—	—[f]
6	51,405	0.4	51,437	0.1	−122	0	—	−183	0	—
7	52,347	3.6	52,381	0.2	−124	1	2.0	−187	1	19
8	53,274	1.2	53,312	0.2	−128	3	3.0	−191	3	18
9	54,181	3.4	54,231	0.3	−133	6	3.4	−195	3	18
10	55,091	0.4	55,136	0.3	−132	7	3.5	−191	5	19
11	55,961	3.5	56,028	0.5	−128	7	3.8	−187	6	21
12	56,737	24.5	56,905	0.5	−126	7	3.8	−184	7	21
13	57,746	0.1	57,769	0.5	−132	8	3.8	−188	8	23
14	58,539	1.5	58,618	0.6	−135	8	4.0	−190	10	30
15	59,644	7.1	59,451	0.6	−138	9	5.0	−197	9	40
16	60,313	0.6	60,268	0.6	−141	9	5.4	−194	9	36
17	61,075	0.0	61,068	0.6	−149	10	6.5	−216	10	80
18	61,862	0.8	61,850	0.6	−160	5	38.4	(−260)	—	200
19	62,592	0.0	62,613	0.5	−170	1	—	—	—	—

[a] Tabulated as E_0 and F in Table II of Gallusser and Dressler (1982).
[b] Tabulated as H and f in Table II of Gallusser and Dressler (1982).
[c] From Table I of Boursey and Roncin (1975) but recalculated using the correct $B^2\Pi_{1/2}$ deperturbed energy.
[d] f is the oscillator strength.
[e] I is a relative intensity normalized to I = 10 for the most intense band.
[f] Energy units are reciprocal centimeters.
[g] Uncertain value.

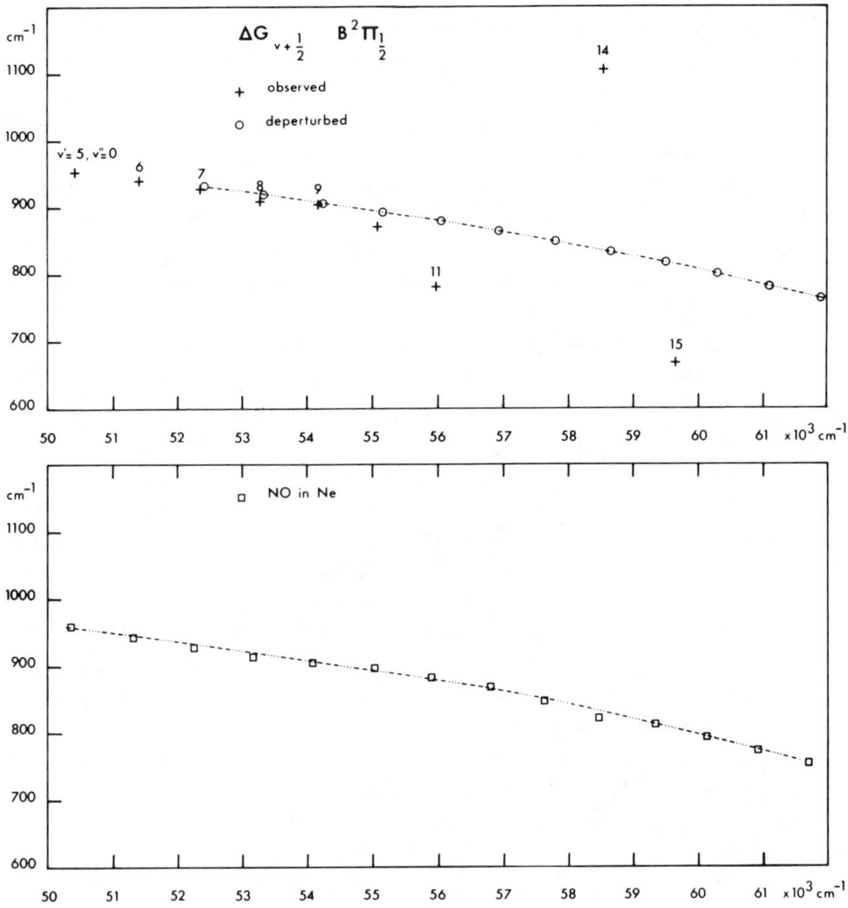

Fig. 5.19 Comparison of observed (+) and deperturbed (○) ΔG values for the NO $B^2\Pi_{1/2}$ state with ΔG values observed in a 5 K Ne matrix (□). [From Boursey and Roncin (1975).]

$B^2\Pi$–$X^2\Pi$ bands from 5 K Ne and Ar matrices display regular vibrational intervals and smoothly varying intensities, in marked contrast to the behavior of the observed gas-phase B–X bands. Buxton and Duley (1970) have observed, in the reflectance spectrum from solid N_2, a change from the gas-phase pattern of perturbations of the N_2 $b^1\Pi_u$ valence state by the $c^1\Pi_u$ and $o^1\Pi_u$ Rydberg states (Stahel *et al.*, 1983). Despite the large deperturbed oscillator strength of the $c^1\Pi_u$–$X^1\Sigma_g^+$ 0–0 band (Stahel *et al.*, 1983), there appears to be no trace of this band nor any other transitions into Rydberg levels in the solid N_2 spectrum. However, unlike the situation for the NO $B^2\Pi$–$X^2\Pi$ system, the deperturbation is not complete.

The pressure-induced and inert-gas matrix demixing phenomenon is not general, but it illustrates that the computational deperturbation process recovers physically significant potential curves. Demixing depends on the existence of two classes of interacting basis states characterized by orbitals of drastically different spatial extent. Two valence states will not be systematically demixed in a matrix, but differential matrix shifts can alter perturbation patterns (level shifts and intensity borrowing). For valence ~ Rydberg perturbations, the ease of demixing depends on the strength of the H_{12} perturbation interaction in the free molecule. The Rydberg state must either be broadened (Γ_{matrix}) or differentially shifted (δE_{matrix}) so that $\Gamma_{\text{matrix}} \gg H_{12}$ or $\delta E_{\text{matrix}} \gg H_{12}$. In no sense do the Rydberg states cease to exist; either they become so broad as to disappear from the high-resolution spectrum, or they are shifted to a higher-energy region. In the case of the NO $B^2\Pi$ state, the abrupt broadening in Ne and Ar matrices above $v = 17$ could arise from interaction with the $C^2\Pi$ Rydberg state shifted up from $T_0 = 52{,}418\ \text{cm}^{-1}$ to $T_0 \approx 61{,}000\ \text{cm}^{-1}$ (Boursey and Roncin, 1975). Chergui *et al.* (1985) have used synchrotron radiation to observe the NO $C^2\Pi$ state in an argon matrix and found the $C^2\Pi$ ($v = 1$) level near $61{,}500\ \text{cm}^{-1}$ ($v = 0$ does not perturb $B^2\Pi$ because $\langle v_C = 0 | v_B \approx 14 \rangle$ is small). The huge ($\sim 6700\ \text{cm}^{-1}$) matrix shift of the NO Rydberg $C^2\Pi$ state corresponds to a $C^2\Pi$–$X^2\Pi$ vertical excitation with respect to the matrix, in the sense that the matrix site atoms remain at the equilibrium locations that accommodate the much more compact NO $X^2\Pi$ state. Goodman and Brus (1978b) have shown that when the matrix is deformed to create a bubble cavity, the $A^2\Sigma^+$ Rydberg state is located near its gas-phase energy, $T_0 = 44{,}200\ \text{cm}^{-1}$. However, the NO $A^2\Sigma^+$ state in its bubble cavity cannot be reached directly by $\sim 44{,}200\ \text{cm}^{-1}$ vertical excitation from $X^2\Pi$.

5.5.6 MATRIX EFFECTS

When a diatomic molecule (guest) is present as a dilute impurity in an inert matrix (host), the selection rules for perturbations and electric dipole allowed transitions can be altered by guest–host interactions. For example, the inevitable absence of cylindrical symmetry ($C_{\infty v}$ or $D_{\infty h}$) at a matrix site destroys the distinction between π and δ orbitals; thus $\Delta\Lambda = 2$ transitions [S_2 and SO $A'^3\Delta$–$X^3\Sigma^-$ and $c^1\Sigma^-$–$a^1\Delta$ (Lee and Pimentel, 1978, 1979); NO $^2\Phi$–$X^2\Pi$ (Chergui *et al.*, 1984); N_2 w $^1\Delta_u$–$X^1\Sigma_g^+$ (Kunsch and Boursey, 1979)] and perturbations (Goodman and Brus, 1977) are quite common.

The precise nature of a matrix site may be inferred from the occurrence of processes that are forbidden in the gas phase, the removal of a gas-phase degeneracy (e.g., lifting of e/f-orbital degeneracy in $\Lambda > 0$ states), or observation of polarization behavior (in amorphous matrices, the degree of fluorescence polarization resulting from linearly polarized excitation or, in single-crystal hosts, the dependence of the absorption cross-section for linearly polarized radiation on the orientation of the crystal axes).

The relaxation of Σ^+/Σ^- symmetry restrictions implies a site in which there are no reflection planes containing the molecular axis. Relaxation of $\Delta\Lambda$ selection rules need not imply destruction of Σ^+/Σ^- restrictions. For example, although the O_2 $C^3\Delta_u$–$X^3\Sigma_g^-$ and $c^1\Sigma_u^-$–$a^1\Delta_g$ transition strengths are considerably enhanced in an inert matrix [for $c^1\Sigma_u^-$, Rossetti and Brus (1979) report $\tau_{\text{radiative}} \approx 2 \times 10^{-4}$ s versus 25 s gas phase], the Σ^+/Σ^- forbidden $A^3\Sigma_u^+$–$X^3\Sigma_g^-$ transition does not appear as a sharp zero phonon line except when the $A^3\Sigma_u^+$ $(v = 4)$ level is perturbed by $C^3\Delta_u(v = 6)$ [in an N_2 matrix (Goodman and Brus, 1977)].

Removal of the distinction between $\Lambda = 0$ and $\Lambda > 0$ states (and the lifting of e/f degeneracy in $\Lambda > 0$ states) implies a site with less than threefold rotation symmetry about the molecular axis (except D_{2d}). Violations of g, u symmetry restrictions are unusual in inert gas atomic matrices, but should be commonplace in polar molecular lattices. Spin selection rules are broken by the "external heavy atom effect," whereby xenon is much more effective than argon in making spin-forbidden transitions observable.

A rich variety of perturbation-mediated energy transfer phenomena are observed in inert matrices. Bondybey (1977) has shown that $B^2\Sigma^+ \leftrightarrow A^2\Pi \leftrightarrow X^2\Sigma^+$ internal conversion processes in CN are governed by an exponential energy gap law and can be much more rapid than vibrational relaxation within a single electronic state (Fig. 5.20). $B \sim A$ and $A \sim X$ perturbation strengths and vibrational overlap factors do not appear to be as important as the amount of energy that must be converted into lattice phonons. Conversely, Goodman and Brus (1978a) and Rossetti and Brus (1979) have shown that the energy gap law is violated for the anomalously slow spin-forbidden processes, of Fig. 5.21,

$$^{15}N^{16}O \qquad B^2\Pi_{1/2}\ (v = 0) \rightarrow a^4\Pi_{5/2}\ (v = 8) \qquad + \sim 20\ \text{cm}^{-1}$$

$$^{18}O_2 \qquad C^3\Delta_u\ \ (v = 0) \rightarrow c^1\Sigma^-\ \ (v = 2) \qquad + \sim 60\ \text{cm}^{-1}$$

An unusually good energy match is required before these intersystem relaxation paths compete with more rapid processes such as spin-relaxation within a multiplet state, vibrational relaxation, or spontaneous fluorescence (NO $B \rightarrow X$). It is interesting that the $B^2\Pi_{1/2} \rightarrow a^4\Pi_{5/2}$ $\Delta\Omega = 2$ process seems to require a better energy match than $\Delta\Omega = \pm 1$ or 0 processes even though the quantum numbers Ω, Λ, and Σ are destroyed by the noncylindrical matrix site.

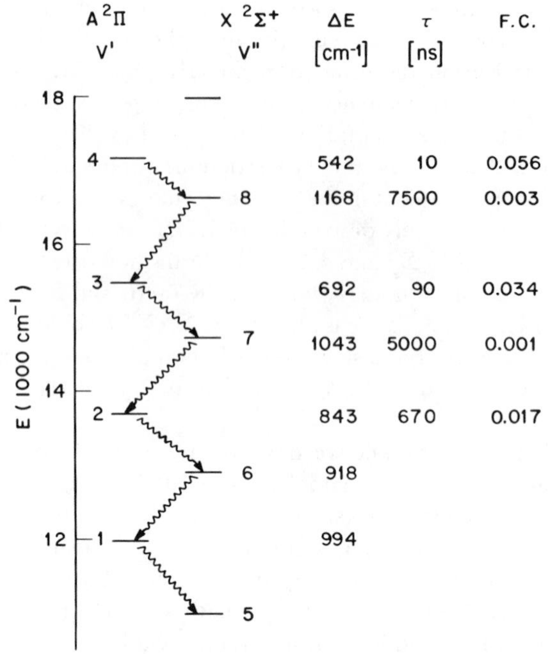

Fig. 5.20 Schematic of CN $A^2\Pi \leftrightarrow X^2\Sigma^+$ relaxation in a neon matrix. Radiationless relaxations follow the $A \leftrightarrow X$ path shown. The energy gap (ΔE), observed lifetime (τ), and Franck–Condon factors (F.C.) are also displayed. [From Bondybey (1977).]

5.5.7 MULTIPHOTON IONIZATION SPECTROSCOPY

Multiphoton ionization (MPI) spectroscopy, a recently developed technique, can provide new information about highly excited levels where many perturbations occur owing to the high density of vibronic states. In MPI spectroscopy, the frequency of a pulsed dye laser is scanned while the ions generated by multiphoton absorption are collected and detected. Large increases in the MPI signal occur when the laser frequency is resonant with an n-photon transition to an intermediate rovibronic level that is subsequently excited into the ionization continuum by absorption of m additional photons from the dye laser (or from a fixed-frequency ionizing laser). This corresponds to an $n + m$ MPI spectrum.

When a $2 + 1$-photon MPI spectrum is obtained, the intensity of an MPI line is proportional to the product of the two-photon cross-section for excitation of the intermediate level from the ground state and the one-photon ionization

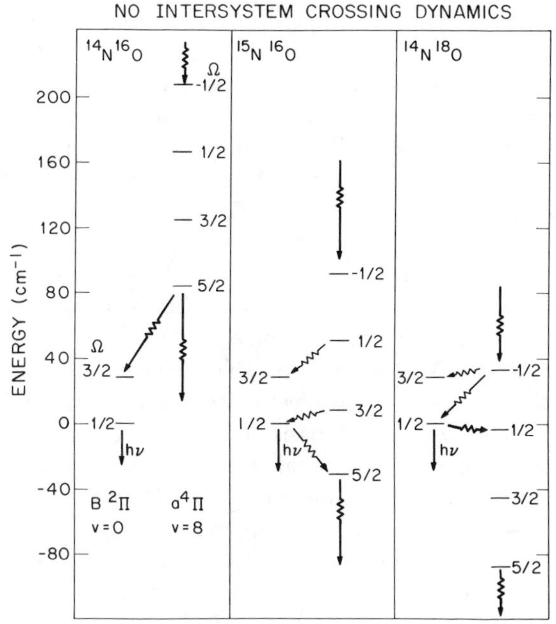

Fig. 5.21 NO $a^4\Pi \rightarrow B^2\Pi$ intersystem crossing mechanism in Ar at 5 K. Thick (thin) arrows represent high (low) quantum yield processes. Fluorescence from $^{15}N^{16}O$ $B^2\Pi_{1/2}$ is anomalously strong because of the inefficiency of $B^2\Pi_{1/2} \rightarrow a^4\Pi_{5/2}$ transfer. Fluorescence from $^{14}N^{18}O$ $B^2\Pi$ is anomalously weak because it receives population via the slow $a^4\Pi_{-1/2} \rightarrow B^2\Pi_{3/2}$ process and is efficiently depopulated by transfer from $B^2\Pi_{1/2}$ into the slightly lower-lying $a^4\Pi_{1/2}$ level. The $a^4\Pi$ levels shown are calculated from the constants of Vichon *et al.* (1978) ($T_e = 38{,}698\ \mathrm{cm}^{-1}$, $\omega_e = 1000\ \mathrm{cm}^{-1}$, $\omega_e x_e = 10.5\ \mathrm{cm}^{-1}$), $A = 41.5\ \mathrm{cm}^{-1}$, and a matrix shift of $+135\ \mathrm{cm}^{-1}$ for $a^4\Pi$ relative to $B^2\Pi$. [Revised from Goodman and Brus (1978a).]

cross-section of the intermediate level. When the intermediate level is perturbed by another state, the intensity ratio for the main and extra lines in the MPI spectrum can be very different from the corresponding ratio of two-photon cross-sections if the two interacting states have different one-photon ionization cross-sections. For example, all $^2\Pi$ Rydberg states of NO are perturbed by electrostatic interaction with the $B^2\Pi$ valence state. Ionization to the π^4 $X^1\Sigma^+$ ground state of NO^+ is a one-electron process from all Rydberg states of NO ($\pi^4\ nl\lambda$ configuration) that converge to this state. However, ionization from the $B^2\Pi$ state ($\pi^3\pi^{*2}$ configuration) is a two-electron process. Consequently, only the admixed Rydberg character of a perturbed level contributes to the ionization cross-section. The situation is similar to the "simple intensity borrowing" case discussed in Section 5.2.1.

If energy analysis of the ejected photoelectron is combined with MPI spectroscopy, additional detailed information about the structure of the intermediate state is obtained. The photoelectron spectrum, recorded at a fixed laser wavelength, provides information about the distribution of rotation–vibration levels in which the ion is formed upon photoionization of the single, resonant, intermediate level. From a series of photoelectron spectra recorded at various laser wavelengths, the MPI spectrum associated with a single ionization channel, the v^+ vibrational level, may be synthesized. The selection rule for ionization from the vth vibrational level of a Rydberg state is $v - v^+ = 0$, since the Rydberg and ion states have nearly identical potential curves. The observed v^+-ionization spectrum will sample exclusively the fractional character of the $v = v^+$ Rydberg level in the intermediate state. See Kimman *et al.* (1985) for an attempt to interpret the intensities in the NO MPI spectrum.

The importance of MPI stems from the extreme sensitivity of ionization detection. It becomes possible to detect ionization that arises from a miniscule admixture of an easily ionized basis state into a nominally nonionizable eigenstate. For example, a strong peak is obtained for formation of $v^+ = 1$ of NO^+ $X^1\Sigma^+$ upon photoionization of NO $C^2\Pi$ $(v = 1)$. However, peaks corresponding to $v^+ = 3$, 4, or 5 also appear in the photoelectron spectrum obtained from various rotational levels of $C^2\Pi$ $(v = 1)$ (White *et al.*, 1982). A qualitative interpretation can be given (ignoring autoionization effects, discussed in Chapter 7) in terms of indirect mixing of $v = 3, 4$, or 5 character of other Rydberg states into the nominal $C^2\Pi$ $(v = 1)$ level via $B^2\Pi$ $(v = 10)$ [see the analogous example illustrated by Eq. (3.4.41)]:

$$|'C^2\Pi\, v_C = 1'\rangle = |C^2\Pi\, v_C = 1\rangle + C(J)|'B^2\Pi\, v_B = 10'\rangle \qquad (5.5.39)$$

and

$$|'B^2\Pi\, v_B = 10'\rangle = |B^2\Pi\, v_B = 10\rangle + \sum_{R,v_R} C_{R,v_R}|R\,^2\Pi, v_R\rangle \qquad (5.5.40)$$

where

$$C_{R,v_R} = \frac{\langle B^2\Pi\, v_B = 10|\mathbf{H}^{el}|R\,^2\Pi\, v_R\rangle}{E^0(B^2\Pi\, v_B = 10) - E^0(R\,^2\Pi\, v_R)}. \qquad (5.5.41)$$

The ionization intensity for the $v^+ = 3$ channel should be

$$I_{v^+=3} \propto [\langle 'C^2\Pi\, v = 1'|\boldsymbol{\mu}|X^1\Sigma^+\, v^+ = 3\rangle]^2 \simeq \left[\sum_R C(J)C_{R,3}\mu^{el}_{RX}\right]^2. \qquad (5.5.42)$$

Extremely weak perturbations can be detected and characterized by the combination of optical-optical double resonance (OODR) with MPI. A tunable "pump" laser populates a rotational level of an excited state via a two- or three-photon transition. Then a second tunable laser, the "probe," ionizes the molecule via a 1 + 1 two-photon resonant process.

An OODR–MPI experiment has detected a weak, high-J perturbation between the NO $K\,^2\Pi$ $(v_K = 2)$ and $F\,^2\Delta$ $(v_F = 3)$ level (Cheung *et al.*, 1985).

This interaction between two Rydberg states of different symmetry can be explained by an indirect mechanism involving the valence $B^2\Pi$ and $B'^2\Delta$ states. The nominal $4p\pi$ $K^2\Pi$ ($v_K = 2$) level contains a small admixture of the $B^2\Pi$ ($v_B = 29$) level (see Section 5.2.2 and Fig. 5.3). A homogeneous interaction occurs between the $3d\delta$ $F^2\Delta$ ($v_F = 3$) level and $B'^2\Delta$ ($v_{B'} = 9$) (see Section 2.3.2 and Fig. 2.7). Since the $B^2\Pi_{3/2}$ and $B'^2\Delta_{3/2}$ valence substates interact via $\mathbf{H}^{SO}$ (Field *et al.*, 1975), this interaction between the $F^2\Delta(3d\delta)$ and $K^2\Pi(4p\pi)$ nominal Rydberg levels is a third-order effect where the effective interaction matrix element is given by

$$H_{\mathrm{F3,K2}} = \frac{H^{\mathrm{el}}_{\mathrm{F3,B'9}} H^{\mathrm{SO}}_{\mathrm{B'9,B29}} H^{\mathrm{el}}_{\mathrm{B29,K2}}}{(E^0_{\mathrm{F3}} - E^0_{\mathrm{B'9}})(E^0_{\mathrm{K2}} - \mathrm{E}^0_{\mathrm{B29}})}. \tag{5.5.43}$$

References

Alexander, M. H. (1982a), *J. Chem. Phys.* **76**, 429.
Alexander, M. H. (1982b), *J. Chem. Phys.* **76**, 3637.
Alexander, M. H. (1982c), *J. Chem. Phys.* **76**, 5974.
Alexander, M. H., and Dagdigian, P. J. (1983), *J. Chem. Phys.* **79**, 302.
Alexander, M. H., and Dagdigian, P. J. (1984), *J. Chem. Phys.* **80**, 4325.
Alexander, M. H., and Pouilly, B. (1983), *J. Chem. Phys.* **79**, 1545.
Appleblad, O., Lagerqvist, A., Renhorn, I., and Field, R. W. (1981), *Phys. Scr.* **22**, 603.
Bachem, A. (1920), *Z. Phys.* **3**, 372.
Banic, J. R., Lipson, R. H., Efthimiopoulos, T., and Stoicheff, B. P. (1981), *Opt. Lett.* **6**, 461.
Barrow, R. F., Dixon, R., Lagerqvist, A., and Wright, C. (1960), *Ark. Fys.* **18**, 543.
Bethke, G. W. (1959), *J. Chem. Phys.* **31**, 662.
Bondybey, V. E. (1977), *J. Chem. Phys.* **66**, 995.
Bondybey, V. E., and Miller, T. A. (1978), *J. Chem. Phys.* **69**, 3597.
Boursey, E. (1976), *J. Mol. Spectrosc.* **61**, 11.
Boursey, E., and Roncin, J.-Y. (1975), *J. Mol. Spectrosc.* **55**, 31.
Brieger, M., Hese, A., Renn, A., and Sodeik, A. (1980), *Chem. Phys. Lett.* **76**, 465.
Buxton, R. A. H., and Duley, W. W. (1970), *Phys. Rev. Lett.* **25**, 801.
Carroll, T. (1937), *Phys. Rev.* **52**, 822.
Chergui, M., Chandrasekharan, V., Böhmer, W., Haensel, R., Wilcke, H., and Schwentner, N. (1984), *Chem. Phys. Lett.* **105**, 386.
Chergui, M., Schwentner, N., Böhmer, W., and Haensel, R. (1985), *Phys. Rev. A* **31**, 527.
Cheung, W. Y., Chupka, W. A., Colson, S. D., Gauyacq, D., Avouris, P., and Wynne, J. J. (1986), *J. Phys. Chem.*
Cook, T. J., and Levy, D. H. (1973a), *J. Chem. Phys.* **59**, 2387.
Cook, T. J., and Levy, D. H. (1973b), *J. Chem. Phys.* **58**, 3547.
Cossart, D., Horani, M., and Rostas, J. (1977), *J. Mol. Spectrosc.* **67**, 283.
Crawford, F. H. (1929), *Phys. Rev.* **33**, 341.
Crawford, F. H. (1934), *Rev. Mod. Phys.* **6**, 90.
Dieke, G. H. (1941), Phys. Rev. **60**, 523.
Douglas, A. E. (1966), *J. Chem. Phys.* **45**, 1007.

Dressler, K. (1969), *Can. J. Phys.* **47**, 547.
Dressler, K. (1970), Journées d'études sur les intensités dans les spectres electroniques, Paris (unpublished).
Dressler, K., Jungen, C., and Miescher, E. (1981), *J. Phys. B* **14**, L701.
Dufayard, J., Negre, J. M., and Nedelec, O. (1974), *J. Chem. Phys.* **61**, 3614.
Ernst, W. E., and Törring, T. (1982), *Phys. Rev. A* **25**, 1236.
Evenson, K. M., and Broida, H. P. (1966), *J. Chem. Phys.* **44**, 1637.
Evenson, K. M., Dunn, J. L., and Broida, H. P. (1964), *Phys. Rev. A* **136**, 1566.
Fairbairn, A. F. (1970), *J. Quant. Spectrosc. Radiat. Transfer* **10**, 1321.
Field, R. W. (1971), Ph.D. thesis, Harvard University, Cambridge, Massachusetts.
Field, R. W., and Bergeman, T. H. (1971), *J. Chem. Phys.* **54**, 2936.
Field, R. W., Bradford, R. S., Broida, H. P., and Harris, D. O. (1972), *J. Chem. Phys.* **57**, 2209.
Field, R. W., English, A. D., Tanaka, T., Harris, D. O., and Jennings, D. A. (1973), *J. Chem. Phys.* **59**, 2191.
Field, R. W., Jones, C. R., and Broida, H. P., (1974), *J. Chem. Phys.* **60**, 4377.
Field, R. W., Benoist d'Azy, O., Lavollée, M., Lopez-Delgado, R., Tramer, A. (1983), *J. Chem. Phys.* **78**, 2838.
Fortrat, R. (1913), Compt. *Rend.* **156**, 1452.
Fournier, J., Deson, J., Vermeil, C., Robbe, J. M., and Schamps, J. (1979), *J. Chem. Phys.* **70**, 5703.
Fredrickson, W. R., and Stannard, C. R. (1933), *Phys. Rev.* **44**, 632.
Freed, K. R., and Tric, C. (1978), *Chem. Phys.* **33**, 249.
Gallusser, R., and Dressler, K. (1982), *J. Chem. Phys.* **76**, 4311.
Gaydon, A. G. (1944), *Proc. R. Soc. London* **182**, 286.
Gelbart, W. M., and Freed, K. F. (1973), *Chem. Phys. Lett.* **18**, 470.
German K., and Zare, R. (1969), *Phys. Rev. Lett.* **23**, 1207.
Girard, B., Billy, N., Vigué, J., and Lehmann, J. C., (1982), *Chem. Phys. Lett.* **92**, 615.
Goodman, J., and Brus, L. E. (1977), *J. Chem. Phys.* **67**, 1482.
Goodman, J., and Brus, L. (1978a), *J. Chem. Phys.* **69**, 1853.
Goodman, J., and Brus, L. (1978b), *J. Chem. Phys.* **69**, 4083.
Gottscho, R. A. (1979), *J. Chem. Phys.* **70**, 3554.
Gottscho, R. A. (1981), *Chem. Phys. Lett.* **81**, 66.
Gottscho, R. A., and Field, R. W. (1978), *Chem. Phys. Lett.* **60**, 65.
Gottscho, R. A., Koffend, J. B., Field, R. W., and Lombardi, J. R. (1978), *J. Chem. Phys.* **68**, 4110.
Gottscho, R. A., Field, R. W., Dick, K. A., and Benesch, W. (1979), *J. Mol. Spectrosc.* **74**, 435.
Gouedard, G., and Lehmann, J. C. (1976), *J. Phys. B* **9**, 2113.
Gouedard, G., and Lehmann, J. C. (1979), *J. Phys. (Orsay, Fr.) Lett.* **40**, L119.
Gouedard, G., and Lehmann, J. C. (1981), *Faraday Discuss. R. Soc. Chem.* **71**, 11.
Grimbert, D., Lavollée M., Nitzan, A., and Tramer, A. (1978), *Chem. Phys. Lett.* **57**, 45.
Harris, S. M., Gottscho, R. A., Field, R. W., and Barrow, R. F., (1982), *J. Mol. Spectrosc.* **91**, 35.
Havriliak, S. J., and Yarkony, D. R. (1985), *J. Chem. Phys.* **83**, 1168.
Heidner, R. F., Sutton, D. G., and Suchard, S. N. (1976), *Chem. Phys. Lett.* **37**, 243.
Herzberg, G., and Jungen, C. (1982), *J. Chem. Phys.* **77**, 5876.
Hougen, J. T. (1970). "The Calculation of Rotational Energy Levels and Rotational Line Intensities in Diatomic Molecules", Nat. Bur. Stand. (U.S.), monograph 115.
Hougen, J. T. (1972), Journées d'études sur la spectroscopie moléculaire, Tours (unpublished).
Imhof, R. E., and Read, F. H. (1971), *Chem. Phys. Lett.* **11**, 326.
James, T. C. (1971a), *J. Chem. Phys.* **55**, 4118.
James, T. C. (1971b), *J. Mol. Spectrosc.* **40**, 545.
Johns, J. W. C. (1974), *in* "Molecular Spectroscopy, Specialist Periodical Report", (R. F. Barrow, D. A. Long and D. J. Millen, eds.) Vol. 2, p. 513, The Chemical Society, London.

Johns, J. W. C., and Lepard, D. W. (1975), *J. Mol. Spectrosc.* **55**, 374.
Jost, R., and Lombardi, M. (1974), *Phys. Rev. Lett.* **33**, 53.
Jungen, C. (1966), *Can. J. Phys.* **44**, 3197.
Jungen, C. and Miescher, E. (1969), *Can. J. Phys.* **47**, 1769.
Katayama, D. (1984), *J. Chem. Phys.* **81**, 3495.
Katayama, D. (1985), *Phys. Rev. Lett.* **54**, 657.
Katayama, D., Miller, T. A., and Bondybey, V. E. (1979), *J. Chem. Phys.* **71**, 1662.
Katayama, D., Miller, T. A., and Bondybey, V. E. (1980), *J. Chem. Phys.* **72**, 5469.
Kimman, J., Lavollée, M., and Van der Wiel, M. J. (1985), *Chem. Phys.* **97**, 137.
Klynning, L. (1974), *in* "Atoms, Molecules and Lasers", p. 464, International Atomic Energy Agency, Vienna.
Kopp, I., and Hougen, J. T. (1967), *Can. J. Phys.* **45**, 2581.
Kunsch, P. L., and Boursey, E. (1979), *J. Chem. Phys.* **70**, 731.
Lagerqvist, A, and Miescher, E. (1958), *Helv. Phys. Acta* **31**, 221.
Lee, Y.-P., and Pimentel, G. C. (1978), *J. Chem. Phys.* **69**, 3063.
Lee, Y.-P., and Pimentel, G. C. (1979), *J. Chem. Phys.* **70**, 692.
Levy, D. H. (1972), *J. Chem. Phys.* **56**, 5493.
Li Li, and Field, R. W. (1983), *J. Phys. Chem.* **87**, 3020.
Lofthus, A. (1957), *Can. J. Phys.* **35**, 216.
Maeda, M., and Stoicheff, B. P. (1984), "Proc. Laser Techniques in the Extreme Ultraviolet" (S. E. Harris and T. B. Lucatorto, eds.), American Institute of Physics, New York.
McKellar, A. R. W., Bunker, P. R., Sears, T. J., Evenson, K. M., Saykally, R. J., and Langhoff, S. R. (1983), *J. Chem. Phys.* **79**, 5251.
Miladi, M., le Falher, J.-P., Roncin, J.-Y., and Damany, H. (1975), *J. Mol. Spectrosc.* **55**, 81.
Miladi, M., Roncin, J.-Y., and Damany, H. (1978), *J. Mol. Spectrosc.* **69**, 260.
Miller, T. A. (1973), *J. Chem. Phys.* **58**, 2358.
Miller, T. A., and Freund, R. S. (1974), *J. Chem. Phys.* **61**, 2160.
Miller, T. A., and Freund, R. S. (1975), *J. Chem. Phys.* **63**, 256.
Miller, T. A., Freund, R. S., and Zegarski, B. R. (1974), *J. Chem. Phys.* **60**, 3195.
Moehlmann, J. G., Hartford, Jr., A., and Lombardi, J. R. (1972), *J. Chem. Phys.* **57**, 4764.
Mukamel, S. (1979), *Chem. Phys. Lett.* **60**, 310.
Mulliken, R.S. (1964), *J. Am. Chem. Soc.* **86**, 3784.
Nedelec, O., and Dufayard, J. (1984), *Chem. Phys.* **84**, 167.
Noda, C., and Zare, R. N. (1982), *J. Mol. Spectrosc.* **95**, 254.
Parker, A. E. (1933), *Phys. Rev.* **44**, 84.
Pratt, D. W., and Broida, H. P. (1969), *J. Chem. Phys.* **50**, 2181.
Provorov, A. C., Stoicheff, B. P., and Wallace, S. (1977), *J. Chem. Phys.* **67**, 5393.
Radford, H. E. (1961), *Phys. Rev.* **122**, 114.
Radford, H. E. (1962), *Phys. Rev.* **126**, 1035.
Radford, H. E. (1964), *Phys. Rev. A* **136**, 1571.
Radford, H. E., and Broida, H. P. (1962), *Phys. Rev.* **128**, 231.
Radford, H. E., and Broida, H. P. (1963), *J. Chem. Phys.* **38**, 644.
Raoult, M., LeRouzo, H., Raseev, G., and Lefebvre-Brion, H., (1983), *J. Phys. B.* **16**, 4601.
Renhorn, I. (1980), *Mol. Phys.* **41**, 469.
Rossetti, R., and Brus, L. (1979), *J. Chem. Phys.* **71**, 3963.
Schadee, A. (1978), *J. Quant. Spectrosc. Radiat. Transfer* **19**, 451.
Schamps, J. (1973), Thesis, Université des Sciences et Techniques de Lille, Lille, France.
Schmid, R., and Gerö, L. (1935), *Z. Phys.* **94**, 386.
Schweda, H. S., Renn, A., and Hese, A. (1980), *Symp. Mol. Spectrosc. 35th, Columbus, Ohio.* Talk MF 8.

Schweda, H. S., Renn, A., Büsener, H., and Hese, A. (1985), *Chem. Phys.* **98**, 157.
Silvers, S., Bergeman, T., and Klemperer, W. (1970), *J. Chem. Phys.* **52**, 4385.
Slanger, T. G., and Black, G. J. (1971), *J. Chem. Phys.* **55**, 2164.
Solarz, R., and Levy, D. H. (1973), *J. Chem. Phys.* **58**, 4026.
Stahel, D., Leoni, M., and Dressler, K., (1983), *J. Chem. Phys.* **79**, 2541.
Steimle, T. C., Brazier, C. R., and Brown, J. M. (1982), *J. Mol. Spectrosc.* **91**, 137.
Tanaka, T., Field, R. W., and Harris, D. O. (1974), *J. Chem. Phys.* **61**, 3401.
Tatum, J. B. (1967), *Astrophys, J. Suppl.* **14**, 21.
Tilford, S. G., and Wilkinson, P. G. (1964), *J. Mol. Spectrosc.* **12**, 231.
Tellinghuisen, J. (1984), *J. Mol. Spectrosc.* **103**, 455.
Townes, C. H., and Schawlow, A. L. (1955), "Microwave Spectroscopy", McGraw-Hill, New York.
Vichon, D., Hall, R. I., Gresteau, F., and Mazeau, J. (1978), *J. Mol. Spectrosc.* **69**, 341.
Vigué, J., Broyer, M., and Lehmann, J. C. (1974), *J. Phys. B* **7**, L158.
Vigué, J., Broyer, M., and Lehmann, J. C. (1975), *J. Chem. Phys.* **62**, 4941.
Watson, J. K. G. (1968), *Can. J. Phys.* **46**, 1637.
Watson, W. W. (1932), *Phys. Rev.* **41**, 378; **42**, 509.
Wayne, F. D., and Colbourn, E. A. (1977), *Mol. Phys.* **34**, 1141; erratum in (1984) **51**, 531.
White, M. G., Seaver, M., Chupka, W., and Colson, S. D. (1982), *Phys. Rev. Lett.* **49**, 28.
Whiting, E. E., and Nicholls, R. W. (1974), *Astrophys. J. Suppl.* **27**, 1.
Whiting, E. E., Schadee, A., Tatum, J. B., Hougen, J. T., and Nicholls, R. W. (1980), *J. Mol. Spectrosc.* **80**, 249.
Wilkinson, P. G., and Houk, N. B. (1956), *J. Chem. Phys.* **24**, 528.
Wolniewicz, L., and Dressler, K. (1982), *J. Mol. Spectrosc.* **96**, 195.
Wood, R. W., and Hackett, F. E. (1909), *Astrophys. J.* **30**, 339.
Yoshino, K., Freeman, D. E., and Tanaka, Y. (1979), *J. Mol. Spectrosc.* **76**, 153.

Chapter 6

Predissociation

6.1 Introduction to Predissociation

A predissociation manifests itself by decomposition of the molecule when it is excited into a state that is quasi-bound with respect to the dissociation continuum of the separated atoms:

$$AB + h\nu \rightarrow AB^* \rightarrow A + B$$

This quasi-bound state, AB^*, is called a resonance or a resonant state.

Two cases of predissociation may be distinguished in diatomic molecules.

1. Predissociation by rotation. The non rotating molecule can dissociate only if it is excited above the highest bound vibrational level. This is normal dissociation. If the molecule rotates, a centrifugal potential is added to the electronic potential. This can result in a centrifugal barrier to dissociation whereby the molecule is quasi-bound in rotation–vibrational levels at energies above the rotationless dissociation limit but below the top of the barrier. Such quasi-bound levels can dissociate through the centrifugal rotational barrier by a tunnelling effect. This is called *predissociation by rotation* (see Fig. 6.1) or tunnelling-predissociation, and is described in detail in Herzberg's book (1950, p. 425).† The theory of this phenomenon will not be discussed here (however, see Section 6.6.3).

2. Electronic predissociation. This type of predissociation is a form of perturbation. A molecule in a vibrational level v_1 of a bound state, which lies at lower energy than the dissociation limit of this state, can dissociate into atoms by mixing with the continuum "level" v_E that belongs to another potential curve. These two potential curves may or may not cross. For the crossing curve case, see Fig. 6.2. This two-electronic-state process is the type of predissociation discussed in this chapter. Different curve-crossing cases are distinguished in Section 6.4. The experimental characteristics of predissociation are presented

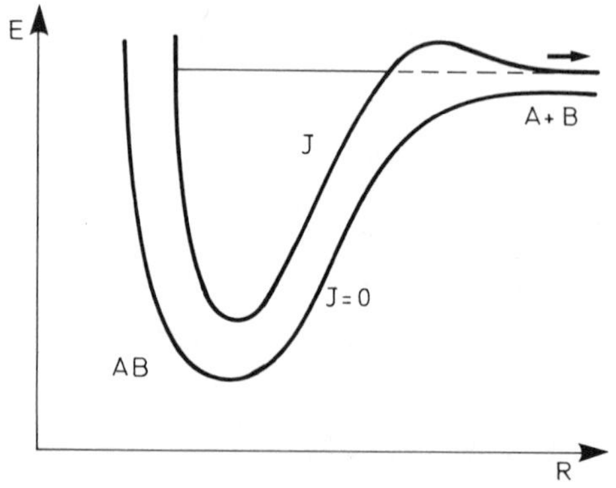

Fig. 6.1 Predissociation by rotation. The $J \neq 0$ potential is obtained by adding the centrifugal energy, $\hbar^2 J(J+1)/2\mu R^2$, to the $J = 0$ potential.

† When there is a barrier in the rotationless potential, there will also be quasi-bound rotation–vibration levels that dissociate by tunnelling through the combined intrinsic and centrifugal barrier (Herzberg, 1950, p. 429).

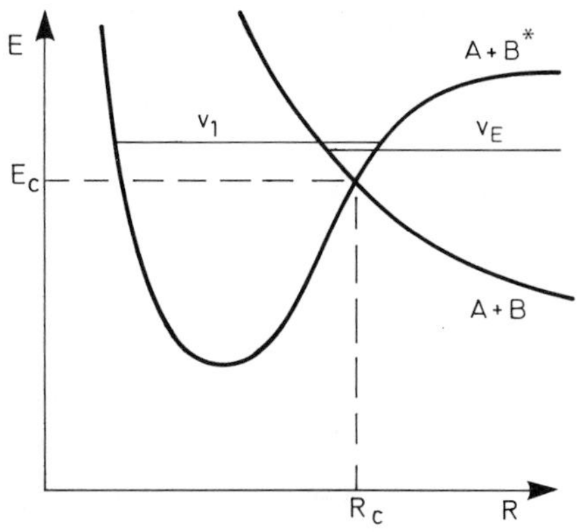

Fig. 6.2 Predissociation. The bound v_1 vibrational level of an electronic state dissociating into excited atoms (A + B*) is predissociated by the continuum (v_E) of another state dissociating to ground state atoms. The two potential energy curves intersect at E_C, R_C.

first. This is followed by a discussion of the sorts of information that can be deduced from experimental data, pertaining to both the nature of the repulsive state and the mechanism of the predissociation.

6.2 Experimental Aspects of Predissociation

There are two complementary ways of looking at the phenomenon of predissociation: *time-dependent* and *time-independent* pictures.

1. Time-dependent picture. Experimental observations can be discussed in terms of a competition between two processes: the radiative process (absorption or emission), characterized by a radiative rate τ_r^{-1}, and a nonradiative process, predissociation, characterized by a nonradiative rate τ_{nr}^{-1}. There are two possible decay pathways for the excited state,

$$AB^* \rightarrow AB + h\nu \qquad \text{radiative}$$

or

$$AB^* \rightarrow A + B \qquad \text{nonradiative.}$$

Predissociation is a nonradiative process because some of the internal energy of the predissociated state is transformed into kinetic energy of the fragment atoms rather than into radiation.

Only the total lifetime τ of a level can be measured. τ is related to the rate of decrease of the number of molecules initially in a given level via both radiative and nonradiative routes. Let k_r be the radiative rate constant (the probability per unit time that a molecule will leave the level as a result of emission of a quantum of light) and k_{nr} the predissociation rate (the dissociation probability per unit time). Recall that the pressure is assumed to be low enough that the rates are unimolecular. The number of molecules leaving the initial state during the time interval dt is given by

$$dN = -k_r N\,dt - k_{nr} N\,dt,$$

with the solution

$$N(t) = N_0 e^{-(k_r + k_{nr})t},$$

N_0 being the initial population, and the emitted radiation has intensity proportional to $N(t)$. The lifetime is defined by

$$N(t) = N_0 e^{-t/\tau}, \tag{6.2.1}$$

and thus

$$1/\tau = k_r + k_{nr} = 1/\tau_r + 1/\tau_{nr}. \tag{6.2.2}$$

Assume that the continuum state involved in the predissociation is not radiatively coupled to the initial state of the quasi-bound ← bound transition (a restriction that is relaxed in Section 6.7). This is a very common situation that often occurs because the continuum-state turning point is located outside of the Franck–Condon range of allowed vertical transitions out of the initial bound state. The continuum ← bound optical transition may also be forbidden by electric dipole selection rules. If the molecule is excited by a pulse of light of shorter duration than τ_{nr}, it can be shown that the system is prepared in the bound *basis* function. Owing to the perturbation interaction between bound and continuum basis states, the prepared nonstationary state decays nonradiatively into the continuum. The rate of this bound → continuum radiationless transition, $1/\tau_{nr}$, may be measured in a time-resolved experiment.

A relative magnitude for the predissociation rate can be obtained by measuring the fluorescence quantum yield,

$$\phi_f = k_r/(k_r + k_{nr}) = \tau/\tau_r. \tag{6.2.3}$$

The dissociative quantum yield is

$$\phi_d = 1 - \phi_f = \tau/\tau_{nr}. \tag{6.2.4}$$

Since experiments generally determine k_{nr} relative to k_r, it is natural to think of the strength of a predissociation relative to τ_r.

2. Time-independent picture. The opposite extreme from short-pulse excitation involves the use of nearly monochromatic radiation. Practically, this means that the interaction between molecule and radiation field is of longer

duration than τ_{nr}. In this limit, the quantity measured is the absorption lineshape. It will be shown below that the linewidth observed in an energy-resolved experiment is related in a very simple way to the predissociation lifetime in the time-resolved experiment.

The effect of predissociation on spectral features depends on whether it is the initial or final state of the transition that is predissociated. Predissociation can be detected either by direct measurements of lifetimes (τ), linewidths (Γ), or level shifts (δE), or indirectly by observation of fragments. Table 6.1 surveys the range of predissociation rates sampled by different methods. Erman (1979) has reviewed the experimental methods for characterizing predissociation phenomena.

6.2.1 MEASUREMENT OF LIFETIMES

Consider first predissociation of the initial state of a transition (i.e., the upper state in emission or the lower state in absorption).

If there is only a barely observable weakening of the lines in emission, the nonradiative lifetime has the same order of magnitude as the radiative lifetime. For an allowed transition, the usual order of magnitude for τ_r is 10^{-8} s; thus, a weakening of the emission can be seen if predissociation corresponds to τ_{nr} around 10^{-8} s, as for example in the predissociation of the N_2^+ $C^2\Sigma_u^+$ state (see Table 6.2). If the transition is weak or nominally forbidden, much weaker predissociations can be detected. In the case of the $I_2 B^3\Pi_{0_u^+}-X^1\Sigma_g^+$ transition ($\Delta S = 1$), with a lifetime of 10^{-6} s, weak predissociations ($\tau_{nr} = 10^{-6}$–10^{-7} s) have been observed. This also depends on the sensitivity of the technique. Brzozowski *et al.* (1976) have detected a very small decrease in the measured lifetimes of CH $A^2\Delta$ for rotational levels above the dissociation limit. This decrease corresponds to a predissociation rate of 8×10^5 s^{-1} (Fig. 6.3). Measurements of the lifetime of a level are useful for detecting weak predissociation, but only if a reliable model exists for the v, J-dependence of the radiative lifetime, τ_r.

The mechanism of a predissociation may be characterized by measurements of the lifetime, τ_v, of each vibrational level or, even better, τ_{vJ}, of every rotational level, carefully extrapolated to zero pressure. When the two unknown rates that appear in Eq. (6.2.2) have similar magnitudes, it is necessary to partition the observed total decay rate into τ_r and τ_{nr}. If the radiative lifetime is known for a non-predissociated level of the same electronic state, τ_r can be calculated for predissociated levels assuming an R-independent value for the electronic transition moment. The nonradiative lifetime is then

Table 6.1
Order of Magnitude of Predissociation Rates

τ_{nr} (s)	τ_r (s)	Γ_{nr} (cm^{-1})	Molecule	State	Ref.[a]	v'	J' or N'	Detection technique
				Measurements of τ				
10^{-5}	5×10^{-8}	5×10^{-7}	K_2	$C^1\Pi_u$	(1)	2 & 10		Selectively detected laser-induced fluorescence (K atom $4^2P_{3/2} \rightarrow 4^2S$)
1.2×10^{-6}	5×10^{-6}	4×10^{-6}	CH	$A^2\Delta$	(2)	1	16	High-frequency deflection
10^{-9}	10^{-8}	5×10^{-3}	N_2^+	$C^2\Sigma_u^+$	(3)	6		Fluorescence quantum yield
				Measurements of Γ				
0.5×10^{-9}		10^{-2}	O_2^+	$b^4\Sigma_g^-$	(4)	5	9	Laser photofragment spectroscopy
2×10^{-10}–4×10^{-11}	10^{-6}	0.025–0.125	ICl	$B^3\Pi_{0^+}$	(5)			Fabry–Perot interferometric spectroscopy
1.8×10^{-11}		0.3	H_2	$B''^1\Sigma_u^+$	(6)	2	2	V.U.V. laser spectroscopy
5×10^{-12}		1.0	O_2	$B^3\Sigma_u^-$	(7)	4		Photoelectrically recorded absorption spectroscopy
2×10^{-14}		300	N_2	$C^3\Pi_u$	(8)	6		Electron energy loss spectroscopy
10^{-14}		400	HBr^+	$^2\Sigma^+$	(9)	4		Photoelectron spectroscopy

[a] References: (1) Meiwes and Engelke (1982); (2) Brzozowski *et al.* (1976); (3) Govers *et al.* (1975); (4) Carré *et al.* (1980) and references therein; (5) Olson and Innes (1976); (6) Rothschild *et al.* (1981); (7) Lewis *et al.* (1980); (8) Mazeau, private communication; (9) Turner (1970).

Table 6.2
Well-Characterized Weak Predissociations

Molecule	Ref.[a]	Predissociated state	Predissociating state	τ (s)	Electronic matrix element (calc.)	Origin of the Interaction
OH	(1)	$A^2\Sigma^+$	$^4\Sigma^-$	2×10^{-7} ($v = 2$)	13 cm^{-1}	Spin–orbit (configuration-interaction effect)
N_2^+	(2)	$C^2\Sigma_u^+$	$B^2\Sigma_u^+$	10^{-9} ($v = 6$)	3.6 $Å^{-1}$	Nuclear kinetic energy operator
I_2	(3)	$B^3\Pi_{0_u^+}$	$^1\Pi_{1_u}$	10^{-6} ($v = 9, J = 0$) 7×10^{-7} ($v = 9, J = 20$)	3×10^{-2} cm^{-1} 3×10^{-4} dimensionless	Hyperfine Gyroscopic (second-order effects)
O_2^+	(4)	$b^4\Sigma_g^-$	$d^4\Sigma_g^+$	10^{-10} ($v = 5, N = 9$)	11 cm^{-1}	Spin–orbit (configuration-interaction)

[a] References: (1) Bergeman *et al.* (1981) (overlap with tail of the discrete vibrational wavefunction); (2) Roche and Tellinghuisen (1979) (factorization of the vibrational part not possible due to strong R-variation of H^e); (3) Broyer *et al.* (1976); (4) Carré *et al.* (1980).

Table 6.3
Well-Characterized Strong Predissociations

Molecule	Ref.[a]	Predissociated state	Predissociating state	Γ_{max} (cm^{-1}) Experimental (except Se_2, BeH)[b]	$\langle \chi_v \vert \chi_E \rangle^2$ Differential Franck–Condon factor (cm)	H^e Electronic matrix element (calc.)	Origin of the interaction
H_2	(1)	$D^1\Pi_u$	$B'^1\Sigma_u^+$	4 ($v = 4$, $J = 1$)		$(2)^{1/2}$, (dimensionless)	Gyroscopic
OH	(2)	$A^2\Sigma^+$	$^4\Pi$	1.47 ($v = 6$)	7.5×10^{-5}	57 cm^{-1}	Spin–orbit
O_2	(3)	$B^3\Sigma_u^-$	$^5\Pi_u$	3 ($v = 4$)	1.1×10^{-4}	65 cm^{-1}	Spin–orbit
Se_2	(4)	$B0_u^+$	0_u^+	180 ($v = 24$)	2×10^{-4}	373 cm^{-1}	Spin–orbit
NO	(5)	$I^2\Sigma^+$	$A'^2\Sigma^+$	9 ($v = 6$)	3.6×10^{-5}	200 cm^{-1}	Electrostatic
NO	(6)	$Q^2\Pi$	$B^2\Pi$	8 ($v = 2$)	3.5×10^{-6}	660 cm^{-1}	Electrostatic
BeH	(7)	$^2\Sigma^+$ (Rydberg)	$^2\Sigma^+$ (valence)	250 ($v = 0$)	1.1×10^{-4}	600 cm^{-1}	Electrostatic

[a] References: (1) Herzberg (1971); (2) Czarny *et al.* (1971); (3) Julienne and Krauss (1975); (4) Atabek and Lefebvre (1972); (5) Gallusser (1976); (6) Miescher (1976); (7) Lefebvre-Brion and Colin (1977).

[b] Se_2 from ref. (4); BeH from ref. (7).

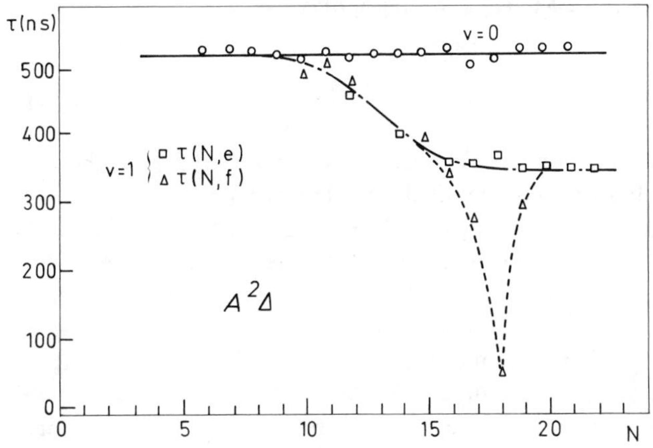

Fig. 6.3 Measured lifetimes of the CH $A^2\Delta$ state. The decay of the $v = 0$ level is purely radiative. The lifetime of the $v = 1$ level indicates a weak predissociation which has been attributed to the continuum of the $X^2\Pi$ state. The shorter lifetimes for the *f*-levels of the F_1 component probably reflect an accidental predissociation (see Section 6.11) through the $B^2\Sigma^-$ state. [From Elander *et al.* (1979) based on data of Brzozowski *et al.* (1976).]

deduced by subtraction of $1/\tau_r$ from the experimental $1/\tau$ value as follows:

$$\frac{1}{\tau(\text{expt})} - \frac{1}{\tau_r(\text{calc})} = \frac{1}{\tau_{nr}}.$$

To obtain the exact variation with v and J of τ_{nr}, the dependence of the electronic transition moment on R must be taken into account in the calculation of τ_r. Note that the lifetime of a state generally varies very little with v. A rapid decrease of the lifetime with v reveals either an unusually large variation of the electronic transition moment with R or, more likely, the presence of predissociation.

A direct measure of the ratio of the radiative $(1/\tau_r)$ and dissociative $(1/\tau_{nr})$ rates is possible when the nonradiative channel exclusively forms an electronically excited atom (see Section 6.2.4).

If the nonradiative lifetime becomes much shorter than the radiative lifetime, lines that are quite sharp and strong in an absorption spectrum are entirely absent from an emission spectrum (Fig. 1.3*a* and *b*). The loss of emission intensity can occur quite abruptly, and the rotational lines or a vibrational progression appear to be cut off. The molecule follows a dissociative route rather than a radiative one. The weakness or the absence of such strongly predissociated emission lines prevents direct measurement of their lifetimes. Only an upper limit for the nonradiative lifetime can then be obtained.

6.2.2 MEASUREMENT OF LINEWIDTHS

Consider now predissociation of the final state of a transition (upper state in absorption or lower state in emission). It is often the upper state of an absorption transition that is predissociated, but the effect of the predissociation on spectral linewidths is identical to the case when the lower state of an emission transition is predissociated (for example, the $B^2\Sigma^+ \rightarrow A^2\Sigma^+$ system of OH where the A state is strongly predissociated). In either case, the spectrum consists of diffuse lines (Fig. 6.4). The predissociated state obtains its diffuse character from the continuum state.

To treat predissociation quantitatively, precise measurements of the width (full width at half maximum, FWHM, or Γ, often misleadingly referred to as the half-width) of each rotational level are necessary. In the absence of Doppler broadening, the lineshape $\sigma_a(E)$ is usually Lorentzian,

$$\sigma_a(E) = \frac{\sigma_{max}(\Gamma/2)^2}{(E - E_r)^2 + (\Gamma/2)^2}, \tag{6.2.5}$$

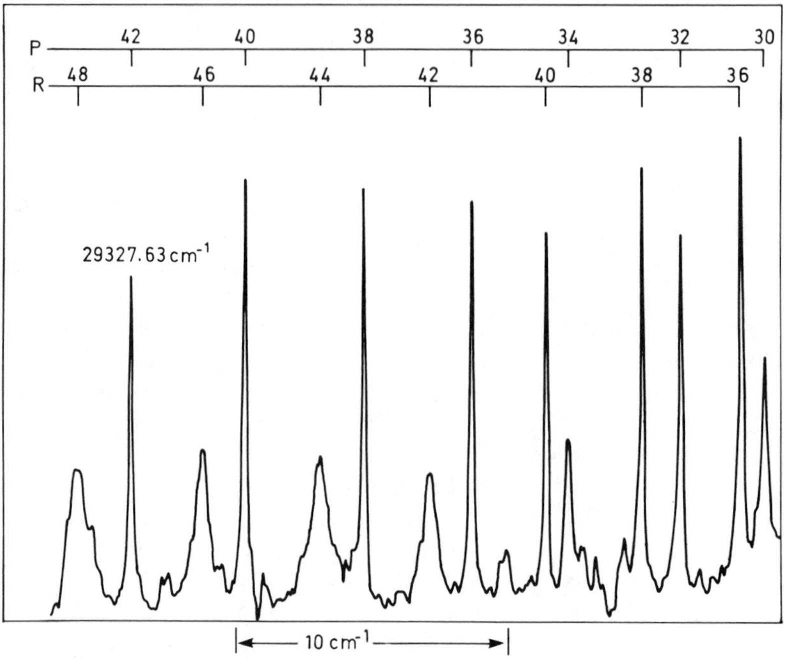

Fig. 6.4 The diffuse 15–0 Band of the B 0_u^+–X 0_g^+ system of $^{78}Se_2$ observed in absorption. [Courtesy R.F. Barrow from Barrow *et al.* (1966).]

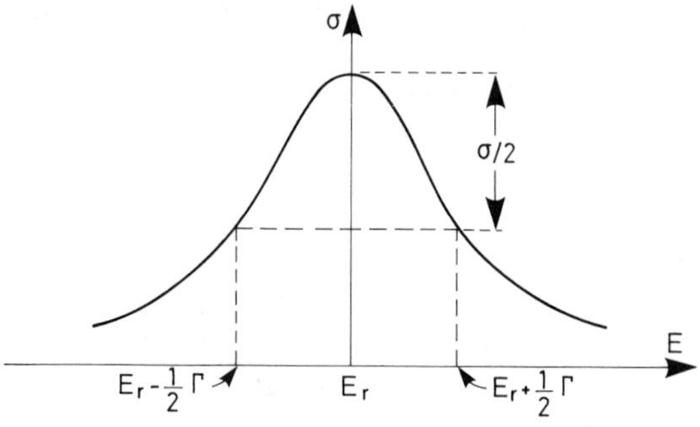

Fig. 6.5 A Lorentzian lineshape.

where Γ is related to the lifetime of the state (see Section 6.3) by

$$\Gamma(\text{cm}^{-1}) = (2\pi c\tau)^{-1} = 5.3 \times 10^{-12}/\tau(\text{s}). \tag{6.2.6}$$

For $E = E_r$ (the energy including the predissociation-induced energy shift, called the resonance energy), the absorption has its maximum value, $\sigma_a(E_r) = \sigma_{\text{max}}$. For $E = E_r \pm \Gamma/2$, the absorption is $\sigma_{\text{max}}/2$ (see Fig. 6.5). Asymmetric line profiles will be discussed in Section 6.7.

In typical experiments, only strong predissociations that result in a nonradiative lifetime much shorter than the radiative lifetime can be detected by line broadening, since the usual radiative lifetime of 10^{-8} s corresponds to a width of only 5×10^{-4} cm^{-1}. By new Doppler-free spectroscopic techniques, it has become possible to measure extremely small predissociation linewidths (Carré *et al.*, 1980; Carrington *et al.*, 1978).

For a nonradiative lifetime of 10^{-13} s, the linewidth will be 50 cm^{-1}, causing the rotational structure to disappear and the band to become diffuse. For a shorter nonradiative lifetime, 10^{-15} s, the vibrational structure disappears and the spectrum becomes similar to a continuous spectrum. As spectral lines become broader and broader, their intensity is spread out and becomes lost in the background. However, low-resolution techniques, such as photoelectron spectroscopy or electron impact spectroscopy, can enable detection of strongly predissociated bands (see Table 6.1).

Doppler broadening has a Gaussian lineshape, and its convolution with the Lorentz natural lineshape yields a Voigt profile. In typical experiments, this effect can be neglected since the Doppler width is usually much smaller than the resolution of the apparatus. Collisional line broadening is also Lorentzian, and the Lorentzian component of measured lines must be carefully extrapolated to zero pressure.

Measurements by photographic photometry require careful calibration due to the nonlinear response of photographic plates: saturation effects can lead to erroneous values. Line profiles can be recorded photoelectrically, if the stability of the source intensity and the wavelength scanning mechanism are adequate. Often individual rotational lines are composed of incompletely resolved spin or hyperfine multiplet components. The contribution to the linewidth from such unresolved components can vary with J (or N). In order to obtain the FWHM of an individual component, it is necessary to construct a model for the observed lineshape that takes into account calculated level splittings and transition intensities. An average of the widths for two lines corresponding to predissociated levels of the same parity and J-value (for example the P and R lines of a $^1\Pi$–$^1\Sigma^+$ transition) can minimize experimental uncertainties. A theoretical Lorentzian shape is assumed here for simplicity, but in some cases, as explained in Section 6.7, interference effects with the continuum can result in asymmetric Fano-type lineshapes.

6.2.3 ENERGY SHIFTS

In addition to line broadening, the predissociation process can cause line shifts. Each discrete or diffuse level can be shifted by its interaction with the entire continuum of the predissociating state, but this effect is considerably smaller than level shifts caused by interactions between discrete levels. The orders of magnitude of predissociation-induced level shifts and linewidths are comparable.

Thus, in classical spectroscopy, only predissociations observable in absorption, having a width of at least 10^{-1} cm^{-1}, can be accompanied by detectable level shifts. Despite the common practice of ignoring predissociative level shifts, the shifts can occasionally be larger than the corresponding level widths. However, determination of level shifts requires a good model for the unshifted line positions. This effect is best detected by studying, not the level shift of the band origin, but the second vibrational difference $\Delta^2 G_v$, where the presence of irregularities may be due to this effect. An example is given in Fig. 6.6 from the very careful study of the predissociation of the O_2 B $^3\Sigma_u^-$ state by Julienne and Krauss (1975).

In principle, rotational constants and fine structure parameters of the predissociated state may also be affected by the predissociating state, as in the case of perturbations. Levels below the disssociation limit, which therefore cannot be predissociated, will also be shifted by their interaction with the continuum. In the case of strong predissociation, only this level shift affecting bound (sharp) levels will be seen, because the levels above the crossing point

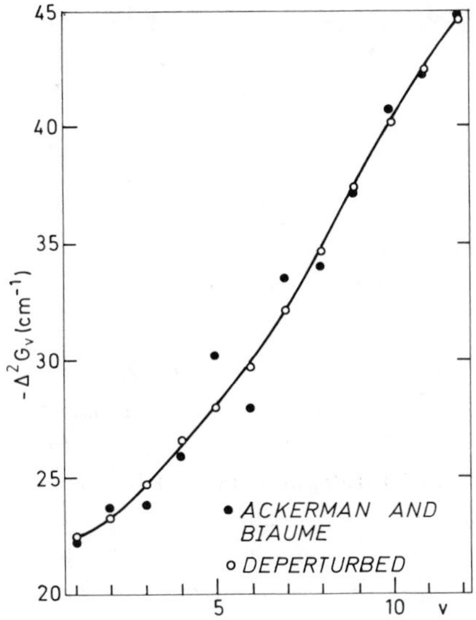

Fig. 6.6 $\Delta^2 G(v)$ differences in the O_2 $B^3\Sigma_u^-$ State. The filled circles show the measured values of Ackerman and Biaume (1970). The open circles show the deperturbed values obtained by subtracting the calculated level shifts from the observed G_v values. [From Julienne and Krauss (1975).]

become too diffuse to be observed. This is the situation for the homogeneously predissociated $Se_2\, B0_u^+$ (Atabek and Lefebvre, 1972) and NO $G^2\Sigma^-$ states (Ben-Aryeh, 1973).

6.2.4 DETECTION OF FRAGMENTS

Laser photofragment spectroscopy is a sensitive indirect method for characterizing certain predissociation processes. A beam of molecular ions is excited by a laser while the intensity variation of a mass-selected photofragment ion is monitored versus laser frequency. By measuring the kinetic energy (i.e., time-of-flight) of each photofragment, the electronic states of the dissociation products may be identified. Extremely small predissociation rates may be detected, since only dissociation can result in a signal at the selected fragment–ion mass. Structure in the photodissociation cross-section versus laser frequency reflects the vibrational (and possibly rotational) structure of the predissociated state. An example is given by the predissociation of the $b^4\Sigma_g^-$ state of O_2^+ (Carrington *et al.*, 1978; Moseley *et al.*, 1979; Carré *et al.*, 1980).

The photofragment angular distribution can indicate the electronic symmetry of the predissociating state (Zare, 1972).

Another possibility involves monitoring the light emitted by a specific electronically excited atomic dissociation product versus laser excitation frequency (Borrell *et al.*, 1977; Meiwes and Engelke, 1982).

6.3 Theoretical Expressions for Widths and Level Shifts

Mixing of the bound $\Psi_{1,v,J}$ and continuum $\Psi_{2,E,J}$ states is governed by an interaction matrix element,

$$H_{v,J;E,J} = \langle \Psi_{1,v,J} |\mathbf{H}| \Psi_{2,E,J} \rangle = \langle \phi_1(r,R)\,\chi_{v,J}(R)\,|\mathbf{H}|\,\phi_2(r,R)\chi_{E,J}(R)\rangle. \tag{6.3.1}$$

The vibrational energy of the continuum state is not quantized; consequently, the vibrational wave functions for energies above the dissociation limit are labeled by the good quantum number, E. These continuum functions are "energy-normalized," rather than "space-normalized" as are the bound vibrational wavefunctions.

One way to understand the necessity for energy normalization and the relationship between energy- and space-normalized functions is to artificially convert the dissociation continuum to a discrete (quasi-continuous) spectrum by adding a vertical and infinite barrier to dissociation at very large internuclear distance. If the location of this outer wall is moved to larger and larger internuclear distances, the amplitude of each discrete, space-normalized wavefunction becomes smaller and smaller as the density of discrete levels, ρ (inverse of the energy interval between two successive levels), becomes larger and larger. The amplitude decrease occurs because the wavefunction occupies an increasingly large region of space and is normalized so that there is unit probability of finding the system in this spatial region. As the barrier is moved to infinity, the amplitude becomes zero and the density of states becomes infinite (the spectrum becomes continuous). Fortunately, the product of the amplitude times the density of states goes to a constant, finite value as the barrier goes to infinite internuclear distance.

The idea of energy normalization can be extended to levels in the discrete region of the spectrum as follows:

$$\chi_E(R) = [\rho(E)]^{1/2}\chi_v(R)$$

or

$$R^{-1}\xi_E(R) = [\rho(E)]^{1/2}R^{-1}\xi_v(R) \tag{6.3.2}$$

where the density, ρ, is the inverse of the separation between two successive vibrational levels or, more rigorously, dv/dE. The space-normalized (with respect to dR) function, $\xi_v(R)$, has the dimensionality of $R^{-1/2}$, and ρ has that of E^{-1}; thus the energy-normalized function, ξ_E of Eq. (6.3.2) must have the same dimensionality as the energy-normalized continuum function, namely,

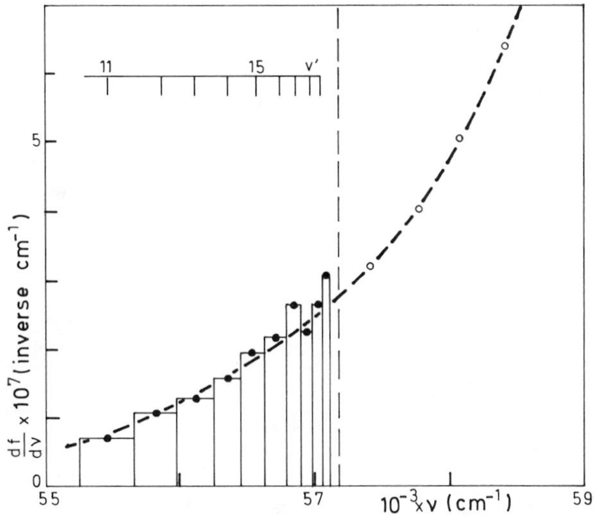

Fig. 6.7 Differential oscillator strengths for the $v''=0$ progression of the $B^3\Sigma_u^- \leftarrow X^3\Sigma_g^-$ system of O_2. Data points corresponding to continuous (Hudson *et al.*, 1966) and discrete (Farmer *et al.*, 1969) transitions are represented by open and filled circles, respectively. [From Child (1980b).]

$E^{-1/2}R^{-1/2}$. Matrix elements of this type of function can only be compared to the corresponding matrix elements of the continuum function. For example, it is well known that the differential oscillator strength, df/dv, varies continuously through a dissociation limit (Allison and Dalgarno, 1971; Smith, 1971). Figure 6.7 shows the continuous variation for the O_2 Schumann–Runge system. Below the dissociation limit, the differential oscillator strength for absorption from initial v'' to final discrete v',

$$\rho(E_{v'})f_{v'v''} = (dv'/dE)\, f_{v'v''}, \tag{6.3.3}$$

is plotted versus E; above the limit, the corresponding continuum quantity,

$$df_{E,v''}/dE,$$

is plotted.

Fano's theory (1961) relates the width of a predissociated level to the interaction matrix element of Eq. (6.3.1). It can be shown that the discrete state amplitude, $a(E)$, in the continuum eigenfunction

$$\Psi_{E,J} = a(E)\Psi_{1,v,J} + \int dE'\, b_{E'}(E)\Psi_{2,E',J} \tag{6.3.4}$$

is given by

$$a(E) = \frac{H_{v,J;E,J}}{E - E_r + i\pi|H_{v,J;E,J}|^2}, \tag{6.3.5}$$

where $H_{v,J;E,J}$ is the matrix element between discrete and *energy*-normalized continuum wavefunctions. When $H_{v,J;E,J}$ varies slowly with energy, $|a(E)|^2$ is a Lorentzian function [cf. Eq. (6.3.5)] with FWHM

$$\Gamma_{v,J} = 2\pi|H_{v,J;E,J}|^2 \tag{6.3.6a}$$

with the prescription that the coupling is calculated with E equal to the energy of the discrete state. Another equivalent expression for Γ explicitly includes $\rho(E)$,

$$\Gamma_{v,J} = 2\pi\rho(E)\,|H_{E,J;v,J}|^2, \tag{6.3.6b}$$

but this expression implies a matrix element of **H** evaluated between space-normalized wavefunctions. This expression for $\Gamma_{v,J}$ is the well-known Fermi–Wentzel Golden Rule.

Similarly, Fano's theory requires that the continuum levels in the interval E' to $E' + dE'$, where $E' \neq E$, contribute an energy shift

$$dF(E') = \frac{|H_{v,J;E,J}|^2}{E^0_{v,J} - E'}\,dE'. \tag{6.3.7}$$

This formula is analogous to second-order perturbation theory for the case of bound states, but the total level shift, $\delta E_{v,J}$, is obtained by replacing the summation over discrete perturbers by an integration.

$$E_{v,J} = E^0_{v,J} + \delta E_{v,J} = E^0_{v,J} + \mathscr{P}\int dF(E'). \tag{6.3.8}$$

Care must be taken to avoid the singularity in the integration at $E = E'$ [for example, by analytic integration in the region of the singularity (Atabek and Lefebvre, 1972)], where $\mathscr{P}$ means "principal part."

If one views predissociation as a time-dependent process, then it is possible to derive the relationship between the FWHM of the predissociated line (assuming no oscillator strength from the continuum basis functions) and the predissociation rate. At time $t = 0$, let the system be prepared with unit amplitude in the discrete state,

$$A(0) = \langle\Psi_{1,vJ}|\Psi(t=0)\rangle = 1.$$

Then, at time t, the amplitude of the discrete state component is

$$A(t) = \langle\Psi_{1,vJ}|\Psi(t)\rangle = \int_{-\infty}^{\infty} dE\,\exp[-(i/h)Et]\,|a(E)|^2, \tag{6.3.9}$$

where $a(E)$ has been given in Eq. (6.3.5). By the definition of $A(t)$, $|a(E)|^2$ and $A(t)$ are related by a Fourier transform and (Fontana, 1982, p. 30)

$$A(t) \propto \exp[-\Gamma t/2\hbar] \exp[-(i/\hbar)E_r t].$$

The time evolution of the discrete state *probability* is given by

$$|A(t)|^2 \propto \exp(-\Gamma t/h),$$

and, by comparison with Eq. (6.2.1),

$$1/\tau = \Gamma/\hbar \qquad \text{or} \qquad \Gamma\tau = \hbar,$$

which is the minimum linewidth compatible with the Heisenberg uncertainty principle. Achievement of this minimum Γ (or maximum τ) is a consequence of assuming either a perfect Lorentzian lineshape or perfect single-exponential decay and level shift of zero. This ideal level shift of zero would occur if H were independent of E and the continuum extended to infinity above and below E.

If the electronic matrix element in Eq. (6.3.1) is assumed to be independent of R, or to vary linearly with R, the matrix element $H_{v,J;\,E,J}$ can be factored into two parts:

$$\begin{aligned} H_{v,J;\,E,J} &= \langle \phi_1(r,R)\,|\mathbf{H}|\,\phi_2(r,R)\rangle\langle \chi_{v,J}(R)|\chi_{E,J}(R)\rangle \\ &= H^e\langle \chi_{v,J}(R)|\chi_{E,J}(R)\rangle, \end{aligned} \qquad (6.3.10)$$

where H^e is either the constant electronic interaction or its value taken at the R-centroid defined by

$$\bar{R}_{v,E} = \frac{\langle \chi_{v,J}(R)\,|R|\,\chi_{E,J}(R)\rangle}{\langle \chi_{v,J}(R)|\chi_{E,J}(R)\rangle}.$$

When two potential curves cross, the value of the R-centroid between isoenergetic levels is, as for bound-state perturbations, the internuclear distance at the curve crossing point, R_C. This approximation predicts a FWHM of

$$\Gamma_{v,J} = 2\pi\,|H^e|^2\,\langle \chi_{v,J}|\chi_{E,J}\rangle^2. \qquad (6.3.11)$$

$\langle \chi_{v,J}|\chi_{E,J}\rangle^2$ is a differential Franck–Condon factor because the continuum function $\chi_{E,J}$ is energy-normalized. It has the dimensionality of E^{-1}, whereas the usual Franck–Condon factor is dimensionless. Contrary to the slow variation of H^e with energy, the differential Franck–Condon factor often varies rapidly and in an oscillatory manner with energy.

In the case of J-dependent $\Delta\Omega \neq 0$ interaction matrix elements, the same approximation can be made, giving the following expression for the width:

$$\Gamma_{v,J} = 2\pi(H^e)^2\langle \chi_{v,J}|\hbar^2/2\mu R^2|\,\chi_{E,J}\rangle^2[J(J+1) - \Omega(\Omega+1)] \qquad (6.3.12)$$

and

$$\langle \chi_{v,J}|\hbar^2/2\mu R^2|\,\chi_{E,J}\rangle \simeq (\hbar^2/2\mu R_C^2)\langle \chi_{v,J}|\chi_{E,J}\rangle = B_C\langle \chi_{v,J}|\chi_{E,J}\rangle. \qquad (6.3.13)$$

Predissociation effects may be treated in either the diabatic (crossing potential curves) or adiabatic (noncrossing curves) representation. Criteria for choosing the more convenient representation were discussed in Section 2.3.4. When the adiabatic representation is appropriate, predissociative interactions between states of the same symmetry result from the nuclear kinetic energy operator. The matrix elements of this operator cannot be factored in the same way as Eq. (6.3.10) because the electronic element

$$W_{12}^e(R) = \langle \Phi_1 | \partial / \partial R | \Phi_2 \rangle$$

can vary rapidly with R. In fact, for a simple two-state avoided crossing, $W_{12}^e(R)$ can be shown to be a Lorentzian function of R [Section 2.3.3, Eq. (2.3.14)]. The $W_{12}^e(R)$ function cannot be factored outside of the vibrational integral

$$H_{v_1,J;E_2,J} = \left\langle \chi_{v_1,J} \middle| W_{12}^e(R) \frac{d}{dR} \middle| \chi_{E_2,J} \right\rangle, \tag{6.3.14}$$

and the magnitude of the adiabatic interaction matrix element is very sensitive to the nodal structure of $\chi_{v_1,J}$ and $\chi_{E_2,J}$ in the R region near the maximum of $W_{12}^e(R)$. An example, the N_2^+ $B^2\Sigma_u^+ \sim C^2\Sigma_u^+$ interaction is discussed using the adiabatic picture in Section 6.9.2.

Calculations of predissociation level shifts in the diabatic representation are simplified by the Eq. (6.3.10) factorization,

$$\delta E_{v,J} = E_{v,J} - E_{v,J}^0 = (H^e)^2 \mathscr{P} \int \frac{|\langle \chi_{vJ} | \chi_{E'J} \rangle|^2}{E_{v,J}^0 - E'} dE'. \tag{6.3.15}$$

One need only consider the energy dependence of the differential Franck–Condon factor. For small H^e, this factorization may be justified in the same way as in the case of second-order energy shifts of bound states. The variation of Γ and δE with v and J provides information about the initially unknown shape of the repulsive potential curve.

In general the bound (predissociated) potential curve is much better characterized experimentally than the repulsive (predissociating) curve. Predissociation linewidths and shifts are usually the best available experimental information about the repulsive state. Indeed, as for bound $\sim$ bound interactions, the vibrational variation of the overlap factor is related to the relative locations of the nodes of the bound and continuum vibrational wavefunctions near R_C (the point of stationary phase of the product $\chi_{v,J}^* \chi_{E',J}$, which is where $\chi_{v,J}^* \chi_{E',J}$ oscillates most slowly). The $\chi_{v,J}$ and $\chi_{E,J}$ functions are solutions of the following nuclear Schrödinger equations (Section 2.6) expressed in atomic units:

$$\left\{-\frac{1}{2\mu}\frac{d^2}{dR^2} + V_1(R) + \frac{1}{2\mu R^2}[J(J+1)] - E\right\}\chi_{v,J}(R) = 0$$
$$\left\{-\frac{1}{2\mu}\frac{d^2}{dR^2} + V_2(R) + \frac{1}{2\mu R^2}[J(J+1)] - E\right\}\chi_{E,J}(R) = 0. \quad (6.3.16)$$

In these equations, $V_1(R)$ is the potential energy curve of the predissociated state and $V_2(R)$ is the potential curve of the predissociating state. Note that these equations determine χ_v and χ_E independently from each other. These equations are uncoupled equations. For a coupled equation approach, see Section 6.10.

6.4 Mulliken's Classification of Predissociations

Mulliken (1960) has identified several classes of predissociation based on the location of the curve crossing (E_C, R_C) relative to the dissociation energy (D^0) and the R_e of the predissociated state. Cases b, a, and c correspond to $E_C < D^0$, $E_C = D^0$, and $E_C > D^0$, respectively. The superscripts $-$, i, $+$, and 0 specify the cases $R_C < R_e$, $R_C = R_e$, $R_C > R_e$, and noncrossing potentials, respectively. For the present discussion, it is useful to retain only Mulliken's $-$, $+$, and 0 classifications. It is immaterial whether the predissociating curve is partly attractive or completely repulsive. These three classes of predissociation manifest profound differences in the variation of linewidth versus the vibrational quantum number of the predissociated level.

1. If the potential curves cross, the crossing can be on the right side of the bound potential (attractive branch). This is an outer crossing or, in Mulliken's notation, a case a^+, c^+, or b^+ predissociation. In this case, the vibrational dependence of the linewidth and level shift exhibits large fluctuations. An example is given in Fig. 6.8, which shows the expected rapidly oscillatory behavior. One even encounters significant variation with J within a given v. Figure 6.9 shows this J-variation for the OD $A^2\Sigma^+$ state. Figure 6.10 is a pictorial representation of similar variations. The energies of rotational levels are plotted versus $J(J+1)$ as solid lines with slope B_v. Dashed and dotted lines, respectively, connect linewidth minima (Δ) and maxima ($\bigcirc$). The slopes of these lines are found to be $\hbar^2/2\mu R_C^2$, where R_C is the location of the curve crossing. Since, at any R-value, the centrifugal energy is identical for all potential curves, R_C does not change with J. Consequently, the J-dependent

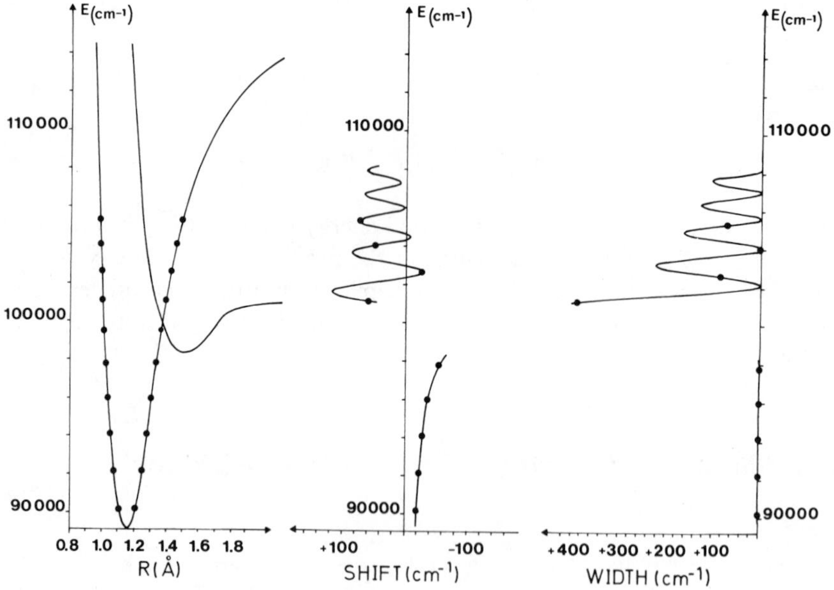

Fig. 6.8 Example of an outer-limb curve crossing. An electrostatic predissociation of the N_2 $C^3\Pi_u$ state by the continuum of the $C'^3\Pi_u$ state. The curves relate the values of level shifts calculated by the coupled equations approach [Eqs. (6.10.1)] for energies below the dissociation limit of the $C'^3\Pi_u$ state to the level shifts and level widths obtained by the semiclassical method [Eqs. (6.5.3) and (6.5.10)] above the dissociation limit. The points shown on the level shift and width curves correspond to the vibrational energies of the $C^3\Pi_u$ state (indicated by the turning points on the potential curve). If all of these values had been observed, they would have been insufficient to suggest the actual shape of the δE and Γ curves. [Courtesy of J. M. Robbe from data of Robbe (1978).]

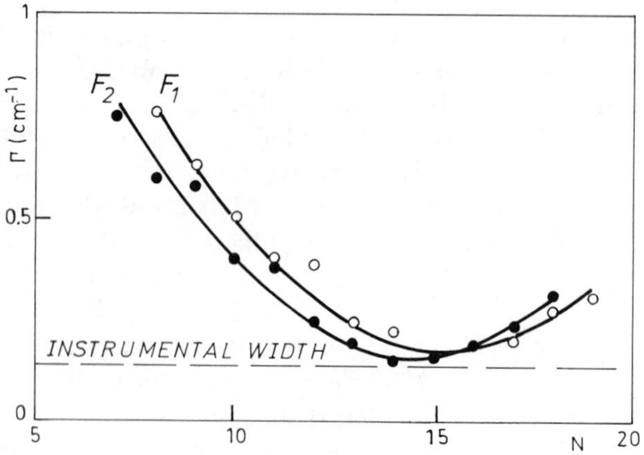

Fig. 6.9 Variation of linewidth versus N in the OD $A^2\Sigma^+$ state. The resolved spin components of $v = 11$ are shown. [Data from Czarny *et al.* (1971).]

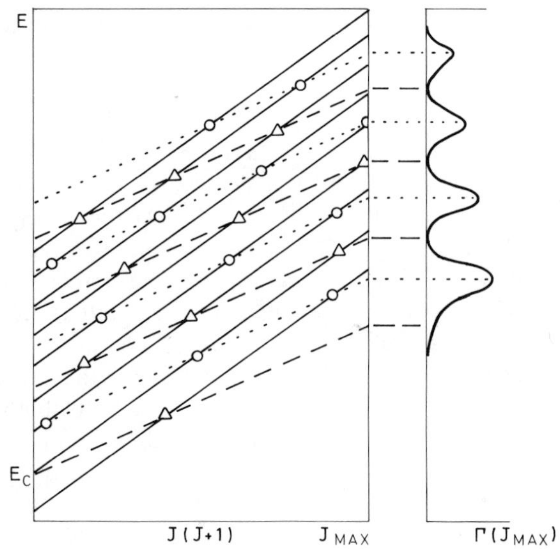

Fig. 6.10 Schematic illustration of the variation of Γ versus v and J. The dashed and dotted lines correspond, respectively, to the minima and maxima of the level widths, except for the first dashed line, which corresponds to the inflection points for the first maximum of the linewidth variation. The Γ curve depicts the variation of the linewidth at J_{MAX} versus energy. The solid lines correspond to rotational levels of the bound state. The points, △ and ○, are the observed locations of linewidth minima and maxima. [Courtesy of M. S. Child.]

energy of the curve crossing is given by

$$\begin{aligned} E_C(J) &= E_C(J=0) + \frac{\hbar^2}{2\mu R_C^2} J(J+1) \\ &= E_C(0) + B_C J(J+1). \end{aligned} \tag{6.4.1}$$

At the right of Fig. 6.10 is a plot of the calculated variation of Γ versus E at a constant J-value, designated as J_{MAX}. The oscillations in Γ versus J within a given v-level are rapid for hydrides because of their large rotational constants. For heavier molecules such as O_2, the linewidths show only a slight J-dependence (Lewis *et al.*, 1980). The linewidth maximum occurs at an energy just above E_C. The curve of the energy of the first maximum in Γ versus $J(J+1)$ (the lowest energy dotted line on Fig. 6.10) is similar to the "limiting curve of dissociation" from which the dissociation energy may be inferred in the case of dissociation by rotation (cf. Herzberg, 1950, p. 428).

To a good approximation, the vibrational overlap integral is determined in the region near R_C, the point of stationary phase. This means that $\langle \chi_{vJ} | \chi_{EJ} \rangle$

will be a function of the energy difference, $E_{vJ} - E_C(J)$, and will be largely insensitive to the values of v and J. However,

$$E_{vJ} - E_C(J) = E_{v0} - E_C(0) + (B_v - B_C)J(J+1),$$

which implies that if there is a relative maximum in $\langle \chi_{vJ} | \chi_{EJ} \rangle$ for the v, J-level, a series of corresponding maxima will occur at the v, J-values shown on Fig. 6.10 connected by each line with slope B_C. At the energy of an outer-wall curve crossing, $E_{vJ} = E_C(J)$, the first maximum in $|\chi_{vJ}(R)|$ occurs at $R_{MAX} < R_C$. As $E_{vJ} - E_C(J)$ increases, R_{MAX} is swept through R_C, causing $\langle \chi_{vJ} | \chi_{EJ} \rangle$ to pass through its absolute maximum value. As $E_{vJ} - E_C(J)$ continues to increase, the vibrational overlap will continue to oscillate. One might naively expect each successive relative maximum to be smaller than the previous one because of the decrease in the amplitudes of both $\chi_{vJ}(R_C)$ and $\chi_{EJ}(R_C)$ as the kinetic energy at R_C increases, but this effect is reduced by the increase in continuum density of states. The net result is the production of secondary maxima.

The rates of oscillation of the linewidth and level shift versus $E_{vJ} - E_C(J)$ depend on the difference in the slopes of the two potentials at R_C [Eq. (6.5.3)]. Murrell and Taylor (1969) describe many examples of the variation of Γ with v. Several exceptions to the expected strong fluctuation in Γ for outer limb crossing will be discussed in Section 6.12.

2. For a crossing on the left side (repulsive branch) of the bound potential, an *inner* crossing or one of the a$^-$, c$^-$, or b$^-$ predissociation cases, a single oscillation of Γ_v will extend over many vibrational levels so that, in practice, only a small variation with v (and J) can be sampled (see Fig. 6.11). The reason

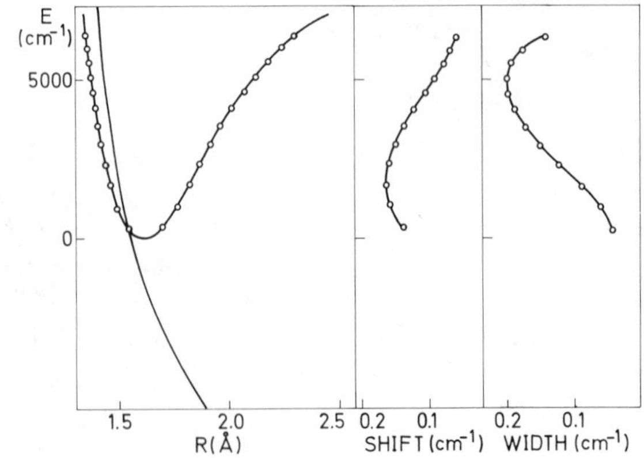

Fig. 6.11 Example of an inner-limb curve crossing: a spin-orbit predissociation of the $O_2 B^3\Sigma_u^-$ state by the continuum of the $^3\Pi_u$ state. [From data of Julienne and Krauss (1975).]

for this is that the slopes of the bound and repulsive curves at R_C are much larger and more nearly equal for an inner rather than an outer wall crossing. Consequently, as $E_{vJ} - E_C$ increases, the innermost maximum of $\chi_{vJ}(R)$ moves to smaller R much more slowly than the outermost maximum sweeps to larger R.

3. For the case of noncrossing or parallel curves (case a^0, b^0, or c^0 of Mulliken), the oscillatory character of Γ is difficult to detect (Durmaz and

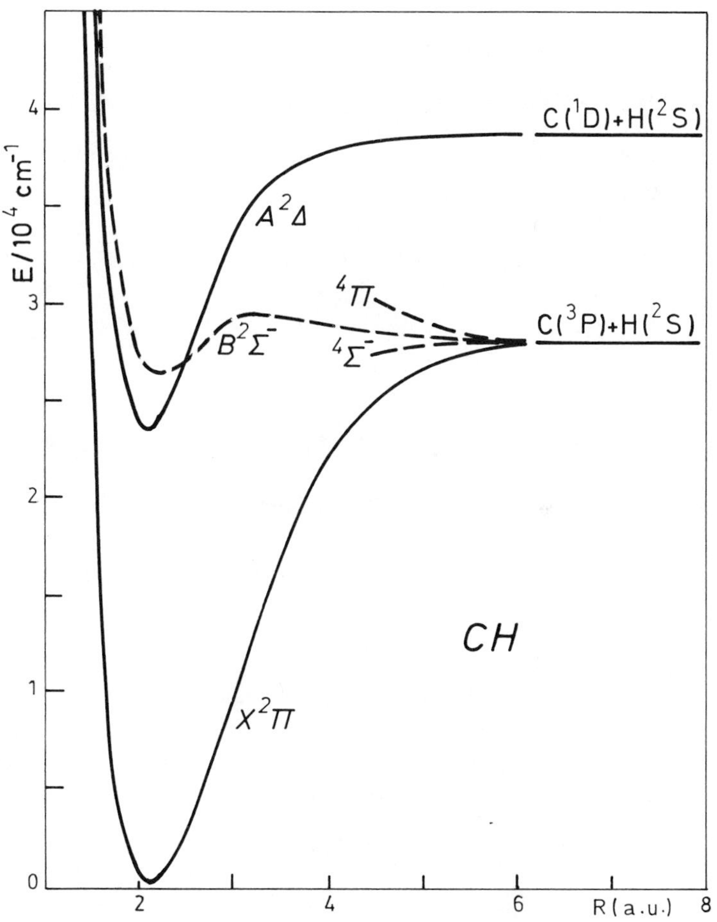

Fig. 6.12 The potential energy curves of CH. Although the $A^2\Delta$ potential is not crossed by that of the $X^2\Pi$ state, the $A^2\Delta$ state is weakly predissociated by the continuum of $X^2\Pi$. [Reprinted courtesy of Brzozowski *et al.* (1976) and the *Astrophysical Journal*, published by the University of Chicago Press; © 1976 The American Astronomical Society.]

Murrell, 1971). This is actually a limiting case of an inner crossing. This situation results typically when the predissociating states are the continua (inner walls) of the lowest-energy bound states, for example, the ground state. The differential Franck–Condon factor between χ_{vJ} and χ_{EJ} vibrational functions is typically about 10^{-2} to 10^{-6} times smaller than in the case of crossing curves. For example, the predissociation of the CH $A^2\Delta$ states probably results from coupling with the dissociative continuum of the $X^2\Pi$ ground state (see Figs. 6.3 and 6.12).

6.5 The Vibrational Factor

Spectroscopically determined potential energy curves for bound electronic states may be extrapolated into the repulsive region by an expression $V_2(R) = aR^{-n} + b$ where $n = 12$ is typically chosen. (A repulsive wall with $n > 12$ would cause the overlap factor to be even smaller than for $n = 12$ because the amplitude of the continuum function near the turning point would decrease.) One of two analytical forms is usually chosen to model repulsive potentials:

$$V_2(R) = D_e + Ae^{-bR} \tag{6.5.1}$$

or

$$V_2(R) = D_e + A'R^{-n'}. \tag{6.5.2}$$

As long as the two parameters A and b or A' and n' are chosen so that the repulsive curve crosses the bound curve at the same point, R_C, and with the same slope, overlap factors obtained from either form of the potential are nearly identical (Julienne and Krauss, 1975). In practice, the two parameters of Eq. (6.5.2) are varied until optimal agreement with the experimental vibration-rotation dependence of Γ (or τ) is obtained.

Analytic formulas can be very useful if linewidths have been measured for many vibrational levels (for example, Child, 1974). It is convenient to represent χ_{vJ} and χ_{EJ} in a uniform semiclassical approximation (Section 4.1.1). As for the bound ~ bound case, the overlap integral between bound and continuum wavefunctions can be expressed as an Airy function (taking into account the proper normalization factor). The linewidth is then

$$\Gamma_{v,J} = \frac{4\pi\hbar\bar{\omega}_{v,J}|H^e|^2}{\hbar \underset{\sim}{v}\,\Delta F}\xi_{v,J}^{1/2}\mathrm{Ai}(-\xi_{v,J})^2 = \pi\Gamma_{v,J}^0\xi_{v,J}^{1/2}\mathrm{Ai}(-\xi_{v,J})^2, \tag{6.5.3†}$$

† Equation (6.5.3) may be rewritten in the usual spectroscopic units as

$$\Gamma\ (\mathrm{cm}^{-1}) = \frac{1.5307|H^e\ (\mathrm{cm}^{-1})|^2[\mu\ (\text{Aston units})]^{1/2}\bar{\omega}_{v,J}\ (\mathrm{cm}^{-1})\xi_{v,J}^{1/2}\mathrm{Ai}(-\xi_{v,J})^2}{[E\ (\mathrm{cm}^{-1}) - V_J(R_C)]^{1/2}\Delta F}$$

where, in the calculation of ΔF, V and R are expressed in reciprocal centimeters and angstroms, respectively.

where ΔF is the difference between the slopes of the two potentials at R_C,

$$\Delta F = \left(\frac{dV_1}{dR}\right)_{R=R_C} - \left(\frac{dV_2}{dR}\right)_{R=R_C}; \tag{6.5.4}$$

$\bar{\omega}_{v,J}$ is the local vibrational spacing,

$$\hbar\bar{\omega}_{v,J} \simeq \frac{dE_{v,J}}{dv} = \frac{E_{(v+1)J} - E_{(v-1)J}}{2}; \tag{6.5.5}$$

$\underset{\sim}{v}$ is the classical velocity at R_C,

$$\underset{\sim}{v} = \frac{\hbar k(R_C)}{\mu}, \tag{6.5.6}$$

where $k(R_C)$ is the common value of the wavenumber for the two wavefunctions at the point of stationary phase, R_C,

$$k_1(E_{v,J},R_C) = k_2(E_{E,J},R_C) = \frac{(2\mu)^{1/2}}{\hbar}[E_{v,J} - V_J(R_C)]^{1/2}, \tag{6.5.7}$$

with

$$V_{J1}(R_C) = V_{J2}(R_C) = V_J(R_C);$$

and

$$\xi_{v,J} = [\tfrac{3}{2}\phi(E_{v,J})]^{2/3}.$$

$\phi(E_{v,J})$ is a phase difference, which, for the case of an outer wall crossing (Fig. 6.13a), is

$$\phi(E_{v,J}) = \int_{a_2}^{R_C} dR\, k_2(E_{v,J}, R) + \int_{R_C}^{b_1} dR\, k_1(E_{v,J}, R) \tag{6.5.8}$$

where a and b are inner and outer turning points and the subscripts 1 and 2 refer to the bound and repulsive potentials, respectively. ϕ is one-half the difference in phase between two classical paths for the nuclear motion at energy E_v. Path I is from a_1 to R_C on V_1 and from R_C to ∞ on V_2; path II is from a_1 to b_1 on V_1, b_1 to R_C on V_1, R_C to a_2 on V_2, and a_2 to ∞ on V_2 (Schaefer and Miller, 1971). When $\xi_v > 1.5$, the Airy function of Eq. (6.5.3) becomes a sine function and

$$\Gamma_{v,J} = \Gamma^0_{v,J}\sin^2[\phi(E_{v,J}) + \pi/4], \tag{6.5.9}$$

which is the result obtained using the JWKB approximation (Child, 1980b, Appendix).

Figure 6.13a is a pictorial description of $\phi(E_v)$ for an outer wall curve crossing. In phase space, bound motion in the vth vibrational level appears as an ellipse in the harmonic approximation. Motion on a linear unbound potential is represented as a parabola. The shaded area is $2\phi(E_v)$. As v increases, the

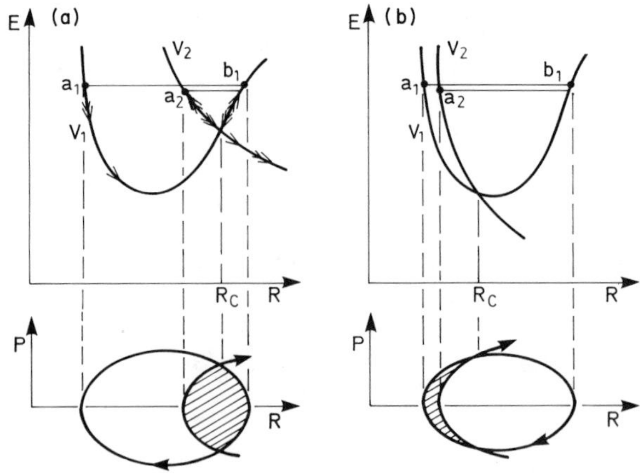

Fig. 6.13 Pictorial descriptions of the phase difference between bound and continuum vibrational wavefunctions. The top part of the figure shows the crossing bound and repulsive potential curves and the two paths between which the phase shift is to be determined. The lower part of the figure represents the classical phase-space trajectories for motion on V_1 (ellipse) and V_2 (parabola). The shaded area is the phase difference between the two paths. (*a*) Outer wall crossing. Path I (single arrows) is the most direct dissociation path: a_1 to R_C on V_1, R_C to ∞ on V_2. Path II (double arrows) is the shortest indirect path: a_1 to b_1 to R_C on V_1, R_C to a_2 to ∞ on V_2. (*b*) Inner wall crossing. The phase difference is between the shortest (b_1 to R_C on V_1, R_C to a_2 to ∞ on V_2) and next longer (b_1 to a_1 to R_C on V_1, R_C to ∞ on V_2) path.

area of the ellipse increases [the unbound motion parabola shifts to the left so that the minimum value of R at $P = 0$ occurs at $V_2(R_{\min}) = E_v$]; consequently, ϕ increases with v for $E > E_C$. Whenever the value of $\phi(E_v) = (2v + 1)\pi/4$, the predissociation linewidth has its maximum value [except for the first maximum, $v = 0$, at which Eq. (6.5.9) is invalid] (Child, 1980b).

If V_1 and the form of $\phi(E_v)$ are known, the repulsive curve V_2 can be determined by an RKR-like method that computes individual turning points (Child, 1973, 1974). This method is useful for obtaining an initial approximation for the repulsive potential. However, if only a few experimental Γ_v-values are known, it is difficult to unambiguously identify the oscillatory frequency of Γ versus v. For example, in Fig. 6.8 the number of vibrational levels sampled is insufficient to determine the actual shape of Γ_v.

The predissociation level shift, δE_{vJ}, for $E_{vJ} > E_C$ has also been treated semiclassically (Bandrauk and Child, 1970):

$$\delta E_{vJ} = 1/2\Gamma^0_{v,J} \sin\left[\phi(E_{vJ}) + \frac{\pi}{4}\right] \cos\left[\phi(E_{vJ}) + \frac{\pi}{4}\right] + \Delta_{vJ} \quad (6.5.10)$$

where Δ_{vJ} is a correction term arising from the nonlinearity of the potentials and is nearly independent of v and J. Note that a relative maximum of $|\delta E_{vJ}|$ will occur between each maximum and minimum value of Γ_{vJ}. δE_{vJ} oscillates at the same frequency as Γ_{vJ} but leading in phase by $\pi/2$ (Fig. 6.8). For $E < E_C$, δE is always negative for an outer crossing and positive for an inner crossing because the continuum χ_{EJ} functions having maximum overlap with χ_{vJ} lie above and below E_{vJ}, respectively, for outer and inner crossings. Level shifts are even observable below the dissociation limit (i.e., for $E < D^0 < E_C$), where it is impossible for line broadening to occur (because $E < D^0$).

Alternatively, the level shift for $E < E_C$ can be understood in terms of a noncrossing curve (adiabatic) representation. The outer- (inner-) crossing case corresponds to an adiabatic curve that is wider (narrower) below E_C than the diabatic curve. Broadening (narrowing) the potential curve has the effect of shifting each v, J-level to lower (higher) energy.

The difference between the behavior of Γ at inner versus outer wall crossings can be understood by examining Eq. (6.5.9). For an outer crossing, the phase difference, $\phi(E)$, varies rapidly with v and the $\sin^2 \phi(E)$ function will oscillate rapidly. For an inner crossing, the phase difference does not change very much because the potential curves are nearly parallel. This is evident from the definition of $\phi(E)$,

$$\phi(E) = \int_{a_1}^{R_C} k_1\, dR - \int_{a_2}^{R_C} k_2\, dR, \tag{6.5.11}$$

and Fig. 6.13b. Thus $\phi(E)$ and $\sin^2 \phi(E)$ are nearly independent of energy. The level shift varies slowly; consequently, it will be difficult to detect inner-crossing level shifts experimentally from the $\Delta^2 G(v)$ variation. For the third case, noncrossing or parallel potential curves, the variation of the phase difference is even slower and the function $\sin^2 \phi(E)$ becomes very flat. (See also Section 6.12.)

6.6 The Electronic Interaction Strength

It is insufficient, having obtained an approximate shape for the dissociative curve from the study of the *relative* variation of Γ (or τ), for the interpretation of the predissociation to be complete. An equally important but often quite difficult step consists of accounting for the *absolute* value of Γ. The mere presence of a crossing by a repulsive curve is not sufficient to cause predissociation of bound levels. A value for the electronic interaction H^e may be deduced from an experimental value for $\Gamma_{v,J}$ if the calculated value of

$\langle \chi_{v,J} | \chi_{E,J} \rangle^2$ is introduced into Eq. (6.3.11). This semiexperimental value for H^e may be compared to a calculated or estimated value. The few cases that have been completely interpreted are listed in Tables 6.2 and 6.3.

The coupling operator, **H**, for predissociation in Eq. (6.3.1) has exactly the same origin as for perturbations (see Table 2.2). It can be an electronic, spin–orbit, or rotational operator. It is easiest to calculate the electronic part of the predissociation matrix element in the case (a) basis, for which $\mathbf{\Omega}$ is a good quantum number. The case (b) matrix elements can be obtained from the case (b) → case (a) transformations [for example, Eqs. (2.2.19) and (2.2.20)]. Examples of case (b) predissociation matrix elements are given by Julienne and Krauss (1975) for triplet states and by Carrington *et al.* (1978) for quartet states.

The selection rules and v, J-dependence of predissociation effects depend on the identity of the operator responsible for the predissociation. From knowledge of the selection rules, qualitative information can immediately be obtained from the variation of the total interaction with v or J. For example, if lines from low-J levels are missing in emission, the predissociation is certainly not due to a gyroscopic ($\Delta\Omega \neq 0$) interaction, which would be zero for $J = 0$, but must arise from a homogeneous interaction.

When a predissociation is weak, its interpretation is often difficult: small first-order effects can be masked by second-order effects. If only a few lines are missing or weakened, it is necessary to consider the possibility of an accidental predissociation, or, in other words, a three-state interaction involving a local perturbation by a weakly predissociated level (See Section 6.10). Predissociation of normally long-lived (metastable) states detected in emission may originate from very small interactions such as spin–spin or hyperfine interaction, as is the case for the I_2 $B^3\Pi_{0^+u}$ state (Broyer *et al.*, 1976).

Some general conclusions may be drawn from Tables 6.2 and 6.3.

6.6.1 ELECTROSTATIC PREDISSOCIATION

There are two ways of interpreting the interactions responsible for homogeneous predissociations between states of the same symmetry and multiplicity. The N_2^+ $B^2\Sigma_u^+ \sim C^2\Sigma_u^+$ interaction must be interpreted using an adiabatic model (noncrossing curves); thus the coupling is attributed to the d/dR operator (see Section 6.9.2) and the predissociation will be called nonadiabatic. Rydberg ~ non-Rydberg predissociations in NO are represented by a diabatic model (crossing curves), and consequently the coupling is due to the $1/r_{12}$ electrostatic interaction operator. The NO $^2\Pi$ Rydberg states are strongly predissociated (Miescher, 1976). These predissociations are J-independent and clearly due to the $B^2\Pi$ valence state, which is also responsible for perturbations of the lowest Rydberg states below the dissociation limit (Giusti-Suzor and Jungen, 1984) (Fig. 6.14).

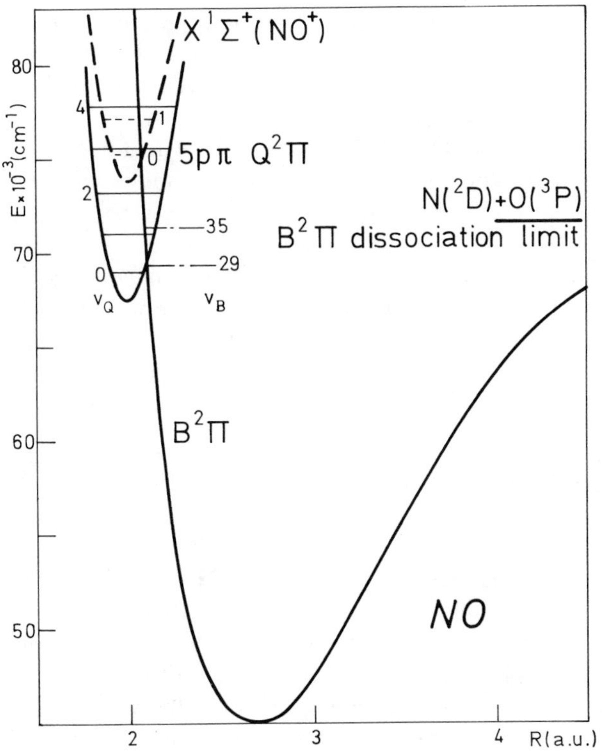

Fig. 6.14 Example of $^2\Pi \sim {}^2\Pi$ Rydberg$\sim$valence electrostatic interaction in NO. The $v = 0$ and 1 levels of the $5p\pi$ $Q^2\Pi$ Rydberg state are perturbed by bound levels ($v = 29$ and 35) of the $B^2\Pi$ valence state, but the $v \geq 2$ levels of $Q^2\Pi$ are predissociated by the continuum of the same $B^2\Pi$ state. (Adapted from Gallusser and Dressler (1982).]

For light molecules, the most important predissociation mechanism is via electrostatic interaction. The affected levels often completely disappear in the absorption spectrum. For example, the N_2 $C^3\Pi_u(v = 6)$ level (cf. Fig. 6.8) is observed only by electron impact spectroscopy (Mazeau, private communication). Its width is found to be $\Gamma = 300$ cm^{-1}.

6.6.2 SPIN–ORBIT PREDISSOCIATION

The spin–orbit interaction, especially between states of different spin, is the *most frequent* cause of predissociation. In Table 6.2, five out of twelve cases result from this interaction. This means that the $\Delta S = 0$ selection rule so often invoked for predissociation is *totally invalid.* The direct spin–orbit interaction causes predissociations, the strengths of which increase with the molecular weight. Compare predissociations of corresponding states for the isovalent molecules OH and HBr^+ ($\Gamma_{max} = 1.5\ cm^{-1}$, Table 6.3, and $400\ cm^{-1}$, Table 6.1) or those of O_2 and Se_2 ($\Gamma_{max} = 3\ cm^{-1}$ and $180\ cm^{-1}$). The difference in the widths is mainly a manifestation of the difference between the spin–orbit constant of a light atom ($\zeta_O = 150\ cm^{-1}$) and that of a heavy atom ($\zeta_{Br^+} = 2500\ cm^{-1}$ or $\zeta_{Se} = 1600\ cm^{-1}$). A spin–orbit interaction due to a slight departure from the single-configuration limit is exemplified by the predissociation of the $O_2^+\ b^4\Sigma_g^-$ state by a ${}^4\Sigma_g^+$ state (Carré *et al.*, 1980) (see Section 6.7.1) and that of the first vibrational levels of the OH $A^2\Sigma^+$ state by the ${}^4\Sigma^-$ state (Bergeman *et al.*, 1981).

6.6.3 ROTATIONAL OR GYROSCOPIC PREDISSOCIATION

Predissociations induced by terms from $\mathbf{H}^{ROT}$ $[\mathbf{B}(\mathbf{J}^+\mathbf{L}^- + \mathbf{J}^-\mathbf{L}^+)]$ are designated as gyroscopic predissociations, to avoid confusion with predissociation by rotation (Fig. 6.1). Experimentally, it is difficult to distinguish between a gyroscopic interaction and predissociation by rotation, because both exhibit linewidths which increase monotonically with J. For example, tunnelling through a barrier in the *rotationless* potential [an intrinsic rather than centrifugal barrier as in the $Na_2\ B^1\Pi_u$ state; see Vedder *et al.*, (1984)] can be confused with a predissociation that involves gyroscopic effects; in both cases, Γ varies with J (see Fig. 5 of Helm *et al.*, 1980). Only careful study of predissociation effects in several vibrational levels can eliminate this ambiguity.

Gyroscopic coupling only connects states of the same multiplicity and can be important, especially at high J in light molecules such as H_2, but its importance for heavy molecules decreases in proportion to the rotational constant. Note, however, that thermal access to extremely high J-levels for heavy molecules can overcome the effect of small $\langle v, J|\mathbf{B}(R)|E, J\rangle$ matrix elements.

6.6.4 HYPERFINE PREDISSOCIATION

The first example of hyperfine predissociation appears simultaneously with gyroscopic predissociation in the $I_2\ B^3\Pi_{0u}^+$ state (Broyer *et al.*, 1976). The predissociation due to gyroscopic coupling is very small in this particular case.

Taking into account its effect, a residual effective radiative lifetime (nonzero for $J = 0$) has been found that shows a strong variation with v. This is actually a predominantly nonradiative lifetime, weakly dependent on J, due to hyperfine interaction between the $^3\Pi O_u^+$ and $^1\Pi_u$ states. This hyperfine predissociation rate depends on the quantum number $\mathbf{F} = \mathbf{I} + \mathbf{J}$ ($\mathbf{I}$ is the nuclear spin angular momentum), but its interpretation goes beyond the scope of this book. The determination of the predissociation rate due to the hyperfine interaction has been made possible by separate observations of the lifetime of individual hyperfine components (Vigué *et al.*, 1977, 1981; Pique *et al.*, 1983). Another example of hyperfine predissociation has been observed in Br_2 (Koffend *et al.*, 1983).

6.7 Interference Effects

As in the case of perturbations between two states of the same multiplicity, when the interacting basis states, 1 and 2, both have nonzero transition moments to a common state, 0, then an interference effect can occur in transitions from the $1 \sim 2$ mixed levels into level 0 (see Section 5.3). Bound ~ bound interference effects usually appear as a transfer of intensity from one line to another, the direction of the transfer depending on the *relative* signs of transition moments and perturbation matrix elements. The predissociation counterpart manifests itself as an asymmetric line broadening, characterized by Fano's index q_v. The absorption cross-section has the form

$$\sigma_a(\varepsilon) = \sigma_i \frac{(q + \varepsilon)^2}{1 + \varepsilon^2}, \tag{6.7.1}$$

where

$$\varepsilon(E) = (E - E_r)/(\Gamma/2) \tag{6.7.2}$$

is the dimensionless energy-offset from line-center and σ_i corresponds to the dissociative continuum cross-section. A negative q-value results in a "window" on the high-frequency side of an absorption line, $q > 0$ is associated with a window on the low-frequency side of the line, and $q = 0$ corresponds to an absorption feature that looks as if it were an emission line but is actually an absorption window superimposed on a continuous absorption background. Figure 6.15 displays the different types of Fano line profiles. $q \to \infty$ is the most frequently encountered situation, and occurs when the continuum state has zero oscillator strength from the initial state (level 0). Since $q^2\sigma_i \propto \sigma_{10}$ [q is defined in Eq. (6.7.5) and σ_{10} is the absorption cross-section for the bound

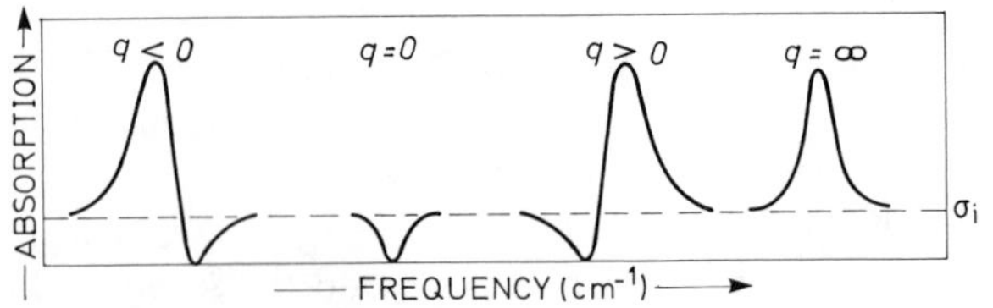

Fig. 6.15 Fano profiles for $q < 0$, $q = 0$, $q > 0$, and $q = \infty$. The dotted line corresponds to the σ_i continuous background absorption in the absence of a predissociated level.

$\leftarrow$ bound $1 \leftarrow 0$ transition], Eq. (6.7.1) takes the form of Eq. (6.2.5) (with $\sigma_{\max} = \sigma_{10}$) as $q \to \infty$ and one finds a Lorentzian line of FWHM $= \Gamma$.

Let the bound and continuum Born–Oppenheimer upper-state basis functions be $\phi_1\chi_{vJ}$ and $\phi_2\chi_{EJ}$ and the bound lower state $\phi_0\chi_0$. The mixed upper-state wavefunction is then

$$\Psi_{E,J} = a(E)\phi_1\chi_{vJ} + \int dE'\, b_{E'}(E)\phi_2\chi_{E'J}, \tag{6.7.3}$$

and the $\Psi_E \leftarrow \phi_0\chi_0$ transition moment is

$$\langle\Psi_{E,J}|\boldsymbol{\mu}|\phi_0\chi_0\rangle$$
$$= a^*(E)\langle\phi_1|\boldsymbol{\mu}|\phi_0\rangle\langle\chi_{vJ}|\chi_0\rangle + \langle\phi_2|\boldsymbol{\mu}|\phi_0\rangle\int dE' b^*_{E'}(E)\langle\chi_{E'J}|\chi_0\rangle. \tag{6.7.4}$$

Fano (1961) showed that

$$|\langle\Psi_{E,J}|\boldsymbol{\mu}|\phi_0\chi_0\rangle|^2$$

has the form of Eq. (6.7.1) if q is defined as

$$q_{vJ} = \frac{1}{\pi}\frac{\langle\phi_1\chi_{vJ}|\boldsymbol{\mu}|\phi_0\chi_0\rangle}{\langle\phi_2\chi_{EJ}|\boldsymbol{\mu}|\phi_0\chi_0\rangle}\frac{1}{H_{vJ;EJ}}. \tag{6.7.5}$$

If the $\boldsymbol{\mu}$ and $H_{vJ;EJ}$ and the electronic matrix element of $\boldsymbol{\mu}$ are R-independent, then Eq. (6.7.5) can be factored,

$$q_{vJ} = \frac{1}{\pi}\frac{\langle\phi_1|\boldsymbol{\mu}|\phi_0\rangle\langle\chi_{vJ}|\chi_0\rangle}{\langle\phi_2|\boldsymbol{\mu}|\phi_0\rangle\langle\chi_{EJ}|\chi_0\rangle}\frac{1}{\langle\phi_1|\mathbf{H}|\phi_2\rangle\langle\chi_{vJ}|\chi_{EJ}\rangle}. \tag{6.7.6}$$

Note that q is dimensionless because $\langle\chi_{EJ}|\chi_0\rangle$ and $\langle\chi_{vJ}|\chi_{EJ}\rangle$ have units of $(\text{energy})^{-1/2}$ and $\langle\phi_1|\mathbf{H}|\phi_2\rangle$ has units of $(\text{energy})^{+1}$. Again, as for bound $\sim$ bound perturbations, the sign of the interference index, q, does not depend

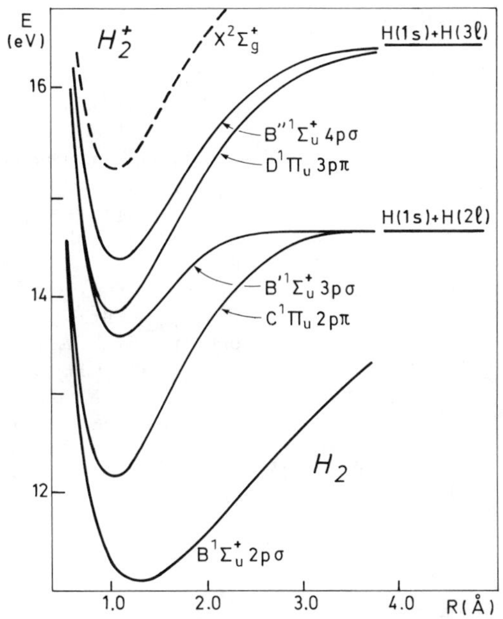

Fig. 6.16 The potential energy curves for some excited states of H_2. The zero of energy is the $v = 0$, $J = 0$ level of the $X^1\Sigma_g^+$ ground state. [From Sharp (1970).]

on arbitrary phase choices for ϕ and χ functions. The definition of q contains each of the three relevant electronic and vibrational wavefunctions exactly twice.

The only observations of Fano lineshapes in predissociation have been in the spectrum of the H_2 molecule (Fig. 6.16). The $v = 3$, $J = 1$ level of H_2 $D^1\Pi_u$ is predissociated by the $B'^1\Sigma_u^+$ state (Herzberg, 1971). Also, Fano profiles are found in $v = 1$ of $B''^1\Sigma_u^+$, which also mixes with the continuum of $B'^1\Sigma_u^+$. Since Franck–Condon factors for absorption from $X^1\Sigma_g^+$ ($v = 0$) into the continuum of $B'^1\Sigma_u^+$ are favorable, the overlap term ($\langle \chi_{EJ} | \chi_0 \rangle$) in the denominator of Eq. (6.7.6) is nonzero and q has a finite value. q is positive for $D^1\Pi_u$ and negative for $B''^1\Sigma_u^+$.

Owing to photographic plate saturation effects, precise q-values cannot be measured from Herzberg's spectra. Figure 6.17 shows a saturation-free Fano profile for H_2 $D^1\Pi_u$ ($v = 5$) obtained using synchrotron radiation by simultaneously recording the absorption spectrum and the H-atom Lyman-α ($2P \rightarrow 1S$) fluorescence excitation spectrum resulting from the $H_2^* \rightarrow H(1S) + H(2P)$ dissociation process (Glass-Maujean *et al.*, 1979).

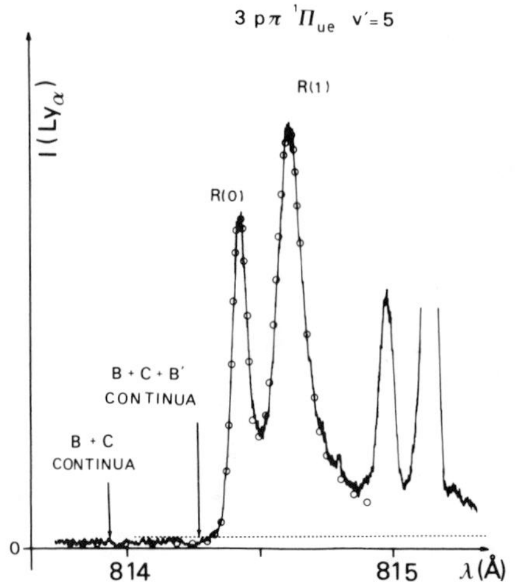

Fig. 6.17 Fano lineshape in H_2. The predissociation of the $N = 2$ [$R(1)$ line] and $N = 1$ [$R(0)$ line] levels of the $D^1\Pi_{ue}$ ($v = 5$) state by the continuum of $B'^1\Sigma_u^+$ is detected by monitoring the Lyman-α emission from one of the fragment atoms. The dots represent the lineshape calculated from the Fano formula [Eq. (6.7.1)] with parameter values $\Gamma(N = 2) = 14.5 \text{ cm}^{-1}$, $q(N = 2) = -9$; $\Gamma(N = 1) = 4.8 \text{ cm}^{-1}$, $q(N = 1) = -18$. These lineshapes should be compared to the symmetric profile of Fig. 6.5 ($q = \infty$). The horizontal dotted line separates the interacting continuum σ_i from the noninteracting continua [σ_d of Eq. (7.8.1)]. From Glass-Maujean *et al.* (1986).]

6.8 Isotope Effects

Even more so than for perturbations, the isotopic dependence of predissociations is useful for identifying the electronic symmetry of the unbound state. The naive assertion that the lightest isotopic molecule is predissociated most rapidly must be examined with caution. It is valid only in the cases of gyroscopic predissociation and predissociation by rotation through a centrifugal barrier. In Eq. (6.3.12), the matrix element

$$\langle \chi_{vJ} | \hbar^2/2\mu R^2 | \chi_{EJ} \rangle$$

is approximately equal to $B_C\langle \chi_{vJ}|\chi_{EJ}\rangle$, where B_C is the value of the rotational constant at the crossing point R_C. The ratio of the linewidths for two isotopic species is thus proportional to the square of the ratio of their rotational constants or inversely proportional to their μ^2 ratio. In the case of hydrides, for example,

$$\frac{\Gamma(\mathrm{XD})}{\Gamma(\mathrm{XH})} = \frac{\tau(\mathrm{XH})}{\tau(\mathrm{XD})} \simeq \left[\frac{\mu(\mathrm{XH})}{\mu(\mathrm{XD})}\right]^2 \left(\frac{\langle \chi_v|\chi_E\rangle_{\mathrm{XD}}}{\langle \chi_v|\chi_E\rangle_{\mathrm{XH}}}\right)^2 \approx \frac{1}{4}. \tag{6.8.1}$$

If isotopic effects on the differential Franck–Condon factors are neglected, it is indeed found that the hydride is four times more strongly predissociated than the deuteride, but only when the predissociation is due to a gyroscopic interaction.

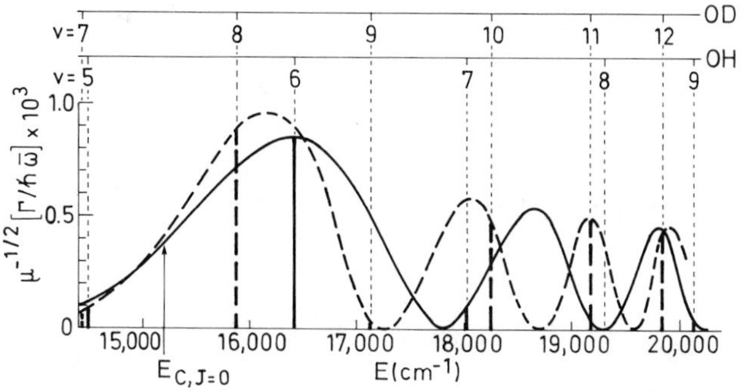

Fig. 6.18 Reduced mass dependence of the reduced (dimensionless) linewidth. The solid and dashed vertical lines depict the reduced linewidths for the $A^2\Sigma^+$ state of OH and OD, respectively, as calculated by Czarny *et al.* (1971). The curves represent the values calculated using Eqs. (6.5.8) and (6.5.9). [From Child (1974).]

The isotope effect on differential Franck–Condon factors has been investigated by Child (1974) (see Fig. 6.18), and it can be predicted that the absolute maximum value of Γ varies roughly as $\mu^{1/6}$. This $\mu^{1/6}$ dependence for the magnitude of Γ is weaker and in the opposite sense to the μ^{-2} dependence for gyroscopic predissociations. The oscillation frequency of $\Gamma(E)$ versus E is also sensitive to the reduced mass. Since the phase difference, $\phi(E_{v,J})$ [Eq. (6.5.8)] increases approximately in proportion to the area,

$$\Delta_v = \int \hbar k(R)\, dR = (v + \tfrac{1}{2})h/2,$$

of the bound-state circle in Fig. 6.13b,

$$\frac{d\phi(E_{v,J})}{dE} \simeq \frac{d\Delta_v}{dv}\frac{dv}{dE} = \left(\frac{1}{\bar{\omega}_{v,J}}\right) \propto \mu^{1/2}, \tag{6.8.2}$$

one expects that successive maxima in $\Gamma(E)$ will be closer together in energy for the heavier isotopic molecule. The $\Gamma(E)$ functions for light and heavy isotopic molecules start out in phase at E_C. Γ_{light} will lag farther and farther behind Γ_{heavy} as $E - E_C$ increases. Eventually, far above E_C, one isotopic species will exhibit a minimum in Γ at the same energy that the other has a linewidth maximum (Fig. 6.18).

Anomalous isotope effects occur at accidental or indirect predissociations, which are discussed in Section 6.11. The accidentally predissociated v, J-level is perturbed by a v, J-level that is directly predissociated by a third (unbound) state. The accidentally predissociated level, having acquired an admixture of the perturber's wavefunction, borrows part of the characteristics of the perturber, in particular the linewidth. Line broadening at an accidental predissociation depends, in a complex way, on both the magnitudes of coupling matrix elements and accidental near-degeneracies. Consequently, the observed linewidths will be strongly influenced by isotopic substitution (Lorquet and Lorquet, 1974; Lefebvre-Brion and Colin, 1977).

Another type of anomalous isotope effect, which is observed in the N_2^+ $C^2\Sigma_u^+$ state, is discussed in the following section.

6.9 Examples of Predissociation

6.9.1 SPIN–ORBIT PREDISSOCIATION

If a predissociation is strong, the observed widths can often be explained in terms of a single-configuration picture for the two states. Consider, for example, the OD $A^2\Sigma^+$ state (see Fig. 6.19). The $v > 7$ vibrational levels of the $A^2\Sigma^+$ state of OD are strongly predissociated. The rotational lines of the $B^2\Sigma^+ - A^2\Sigma^+$ emission system are broadened by predissociation of $A^2\Sigma^+$, the lower state of the transition, and the linewidth is found to vary rapidly with v and N. Figure 6.8 shows this variation for the $v = 11$ level of OD $A^2\Sigma^+$. This rapid variation of Γ means that the outer wall of the $A^2\Sigma^+$ potential is crossed by a repulsive state. By adjusting the location of the crossing point R_C and the slope of the repulsive curve at the crossing, so that the computed and observed linewidth variations match, a repulsive curve has been shown to cross the $A^2\Sigma^+$ potential between $v = 7$ and 8. The experimental value for Γ_{max} is 1.65 cm^{-1} ($v = 8$, $N = 5$). The calculated value of $\langle \chi_{v,N} | \chi_{E,N} \rangle^2$ is 1.20×10^{-4} cm. From

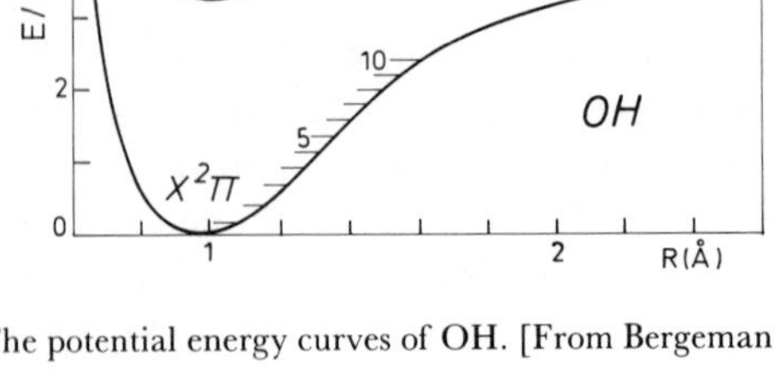

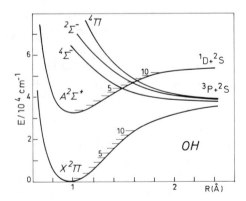

Fig. 6.19 The potential energy curves of OH. [From Bergeman *et al.* (1981).]

Eq. (6.3.11), an electronic matrix element of 47 cm^{-1} is obtained (Czarny *et al.*, 1971). The Wigner–Witmer rules require that four electronic states originate from the $O(^3P) + H(^2S)$ dissociation limit: $X^2\Pi$, $^4\Sigma^-$, $^2\Sigma^-$, and $^4\Pi$. The $X^2\Pi$ potential does not cross $A^2\Sigma^+$ and can be disregarded. The $^4\Sigma^-$ and $^2\Sigma^-$ states arise mainly from the $\sigma^2\pi^2\sigma^*$ configuration, which differs by two spin–orbitals from the predominant $\sigma\pi^4$ configuration of the $A^2\Sigma^+$ state. However, the $^4\Pi$ state arises from the $\sigma\pi^3\sigma^*$ configuration, which differs by only one spin–orbital from the configuration of the $A^2\Sigma^+$ state. Consequently, with the wave functions

$$\Psi(^4\Pi_{1/2}) = 3^{-1/2}[|\sigma\beta\,\sigma^*\beta\,\pi^-\alpha\,\pi^+\alpha\,\pi^+\beta| + |\sigma\beta\,\sigma^*\alpha\,\pi^-\beta\,\pi^+\alpha\,\pi^+\beta|$$
$$+|\sigma\alpha\sigma^*\beta\pi^-\beta\pi^+\alpha\pi^+\beta|] \tag{6.9.1}$$

$$\Psi(^2\Sigma^+_{1/2}) = |\sigma\alpha\,\pi^-\alpha\,\pi^-\beta\,\pi^+\alpha\,\pi^+\beta|, \tag{6.9.2}$$

the spin–orbit interaction is (Section 2.4.2)

$$\langle A^2\Sigma^+_{1/2}|\mathbf{H}^{SO}|^4\Pi_{1/2}\rangle = 3^{-1/2}[0 + 0 + \tfrac{1}{2}a_+]$$

where

$$a_+ = \langle\pi|\hat{a}\mathbf{l}^+|\sigma^*\rangle \simeq \langle 2p\pi_O|\hat{a}\mathbf{l}^+|2p\sigma_O\rangle = -2^{1/2}A(X^2\Pi)$$
$$= 2^{1/2}(140\ cm^{-1}) = 197\ cm^{-1}. \tag{6.9.3}$$

The hypothesis that the σ^* and π orbitals are in a pure precession (Section 4.5) relationship to each other is supported by *ab initio* calculations (Czarny *et al.*, 1971), which find dominant $2p_O$ character for these orbitals in the $A^2\Sigma^+$ and $^4\Pi$ states at the internuclear distance, R_C, of the curve crossing. The pure precession value of the $^4\Pi_{1/2} \sim A^2\Sigma^+$ interaction is $3^{-1/2}a_+/2 = 57\ cm^{-1}$, in satisfactory agreement with the value of 47 cm^{-1} obtained from the observed linewidths.

Sometimes a predissociation cannot be understood in such a simple single-configuration picture. An example is the predissociation of the O_2^+ $b^4\Sigma_g^-$ state (see Fig. 6.20). This predissociation is very weak but has been detected as a line broadening by a Doppler-free laser technique. The width of the upper state of the $b^4\Sigma_g^-$–$a^4\Pi_u$ transition in absorption is 0.05 cm^{-1} for $v = 5$, $N = 9$, a factor of 25 smaller than the width in OD. As the oxygen atom spin–orbit constant is responsible for both of these predissociations, a direct mechanism for O_2^+ seems improbable. The $b^4\Sigma_g^-$ state is most likely predissociated by a $^4\Sigma_g^+$ state (Carré *et al.*, 1980). The dominant configuration of this $^4\Sigma_g^+$ state, $3\sigma_g^2\ 1\pi_u^3\ 1\pi_g\ 3\sigma_u$, differs by two spin–orbitals from the main configuration of the $b^4\Sigma_g^-$, which is $3\sigma_g\ 1\pi_u^4\ 1\pi_g^2$. The interaction between these two states is explained by the fact that the electronic wave function of $b^4\Sigma_g^-$, in the region of the curve crossing, contains a small functional admixture (6×10^{-3}) of the configuration $3\sigma_g^2\ 1\pi_u^3$

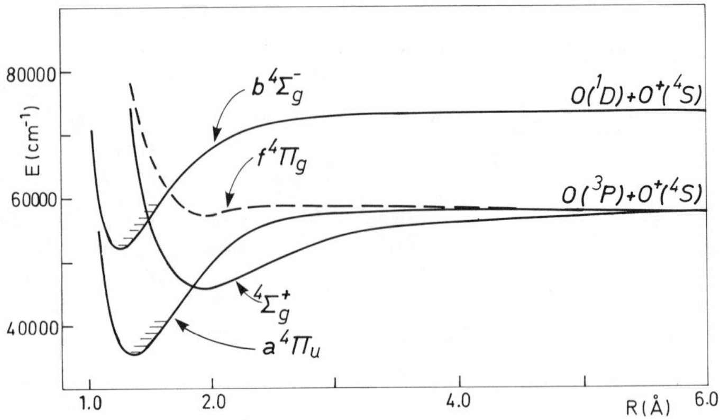

Fig. 6.20 The potential energy curves of O_2^+. [Calculations by A. L. Roche from Carré *et al.* (1980).]

$1\pi_g\ 3\sigma_u$, the dominant configuration of the $^4\Sigma_g^+$ state. The electronic matrix element is

$$\langle {}^4\Sigma_g^+ |\mathbf{H}^{\mathrm{SO}}| \mathrm{b}^4\Sigma_g^- \rangle = 0.99\langle \sigma_g^2\pi_u^3\pi_g\sigma_u |\mathbf{H}^{\mathrm{SO}}| \sigma_g\pi_u^4\pi_g^2 \rangle$$
$$+ 0.08\langle \sigma_g^2\pi_u^3\pi_g\sigma_u |\hat{a}\mathbf{l}_z\mathbf{s}_z| \sigma_g^2\pi_u^3\pi_g\sigma_u \rangle. \quad (6.9.4)$$

The first matrix element is zero and the second is calculated to be 140 cm^{-1}. Thus the overall electronic matrix element has a value of about 11 cm^{-1}, which is sufficient to partly explain the observed width.

6.9.2 NONADIABATIC PREDISSOCIATION

The N_2^+ $C^2\Sigma_u^+$ ($v \geq 3$) levels are subject to very weak predissociation. The predissociation rate is comparable to the purely radiative decay rate. Figure 6.21 illustrates the strong isotopic dependence of the observed decay rates. Roche and Tellinghuisen (1979) have shown that this effect results mainly from an interaction with the continuum of the $B^2\Sigma_u^+$ state. The *ab initio* potential energy curves for the $B^2\Sigma_u^+$ and $C^2\Sigma_u^+$ states (Fig. 6.22) reflect the very strong, two-state avoided curve crossing predicted many years ago by Douglas (1952). Consequently, the adiabatic picture is the ideal starting point for analyzing the $B \sim C$ predissociation.

In the adiabatic picture, the electronic coupling matrix element involves the nuclear kinetic energy operator. The wavefunctions for the two adiabatic $^2\Sigma_u^+$

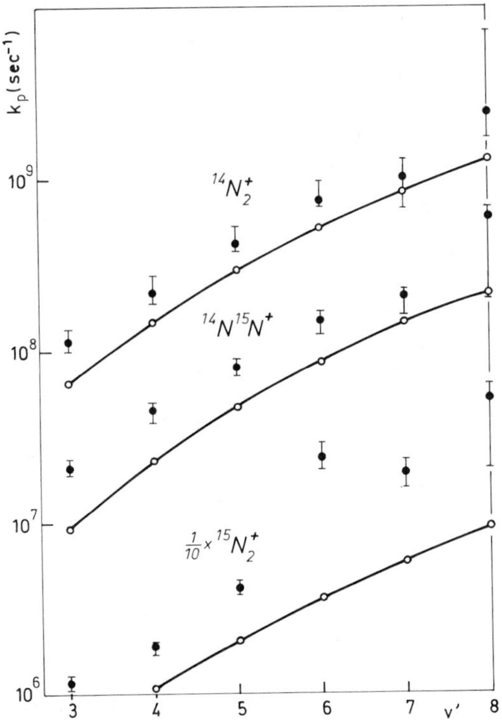

Fig. 6.21 Vibration dependence of the observed (solid circles with error bars) and calculated (open circles) predissociation rates for the $C^2\Sigma_u^+$ state of $^{14}N_2^+$, $(^{14}N^{15}N)^+$, and $^{15}N_2^+$. [From the data given in Table 2 of Roche and Tellinghuisen (1979).]

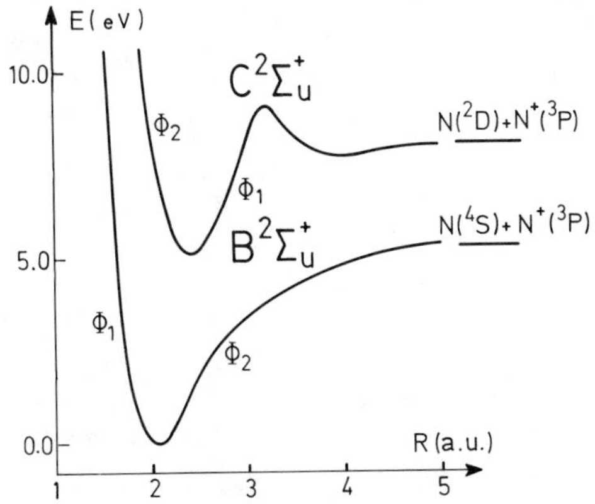

Fig. 6.22 Calculated adiabatic potential energy curves for the $B^2\Sigma_u^+$ and $C^2\Sigma_u^+$ States of N_2^+. The curves are labeled by the dominant character, ϕ_1 versus ϕ_2, of the wavefunctions. [From Roche and Tellinghuisen (1979).]

states are mixtures of two dominant configurations,

$$\phi_1 = 2\sigma_u 3\sigma_g^2 1\pi_u^4$$

$$\phi_2 = 2\sigma_u^2 3\sigma_g 1\pi_u^3 1\pi_g,$$

which differ by two orbitals. (ϕ_2 actually represents two isoconfigurational $^2\Sigma_u^+$ states derived from the $\pi_u^3\pi_g\, {}^1\Sigma_u^+$ and $^3\Sigma_u^+$ parents.) The B and C state wavefunctions are

$$\Psi_B = C_{1B}(R)\phi_1 + C_{2B}(R)\phi_2$$

$$\Psi_C = C_{1C}(R)\phi_1 + C_{2C}(R)\phi_2,$$

where the C-coefficients vary slowly with R. At small R, Ψ_B is dominated by ϕ_1, whereas at large R it is dominated by ϕ_2 (vice versa for Ψ_C).† The electronic matrix element (Section 2.3.3) is

$$\left\langle \Psi_B \left| \frac{d}{dR} \right| \Psi_C \right\rangle_r = W^e_{BC}(R)$$

$$W^e_{BC}(R) = C_{1B}\frac{d}{dR}C_{1C} + C_{2B}\frac{d}{dR}C_{2C}.$$

There are no $C_1(d/dR)C_2$ cross terms because ϕ_1 and ϕ_2 are mutually orthogonal. The matrix elements of the form

$$\langle \phi_1 | d/dR | \phi_2 \rangle$$

must vanish since ϕ_1 and ϕ_2 differ by two orbitals (Section 2.3.2).

The $W^e(R)$ function, resulting from a simple two-state interaction, has a Lorentzian form [Eq. (2.3.14)]. For the N_2^+ B$\sim$C interaction, $W^e(R)$ has its maximum in an R-region of numerous constructive and destructive interferences between the vibrational wavefunctions. Thus, slight changes in the vibrational functions resulting from isotopic substitution can drastically alter the vibronic matrix element,

$$H_{BE;Cv} = \langle \chi_E^B(R) \,|\, W^e_{BC}(R)\, d/dR |\, \chi_v^C(R) \rangle. \tag{6.9.5}$$

Figure 6.23a shows the electronic factor $W^e_{BC}(R)$ (dotted line) superimposed on the $^{14}N_2^+$ vibrational wavefunctions χ_E^B and $\chi_{v=4}^C$ (solid lines).

Figure 6.23b shows how the value of the vibronic integral [Eq. (6.9.5)] accumulates as the integration proceeds from $R = 0$ to larger R. Results for the $^{14}N_2^+$ and $^{15}N_2^+$ isotopic molecules are shown. After rapid, large-amplitude

† The R-value where ϕ_1 and ϕ_2 are equally mixed is where the diabatic potential curves cross. Note that in the diabatic picture the dominant configurational character of a given state is identical on either side of R_C, whereas in the adiabatic picture the configurational parentage reverses.

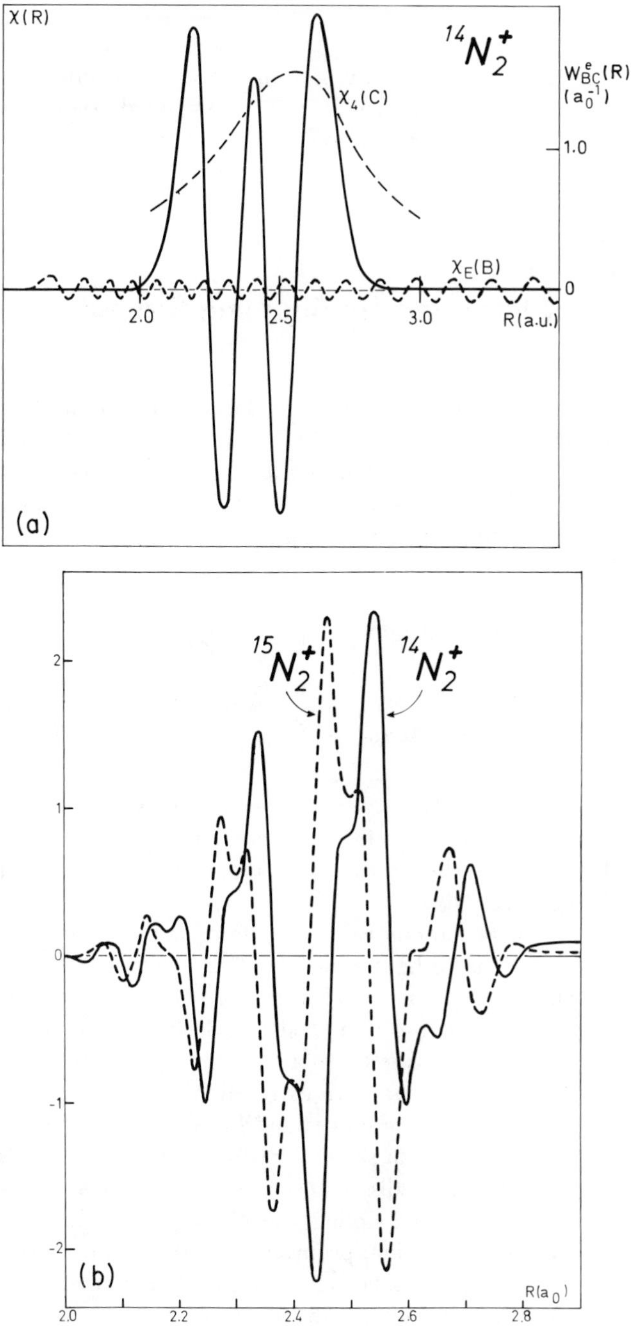

Fig. 6.23 (*a*) Vibrational wavefunctions χ_4^{C} and χ_E^{B} for the N_2^+ $\mathrm{C}^2\Sigma_u^+$ $v = 4$ level and the isoenergetic continuum of $\mathrm{B}^2\Sigma_u^+$. The electronic matrix element, $W^e_{\mathrm{BC}}(R)$, is drawn as a dotted line. (*b*) Isotopic dependence in the accumulation of $\langle \chi_E^{\mathrm{B}} | W^e_{\mathrm{BC}}(\mathrm{R}) | d/dR | \chi_4^{\mathrm{C}} \rangle$ vibronic integral [Eq. (6.9.5)] [from Roche and Tellinghuisen (1979).]

oscillation, the integral eventually stabilizes to a very small value at an internuclear distance where $\chi_{v=4}^{C}$ is zero. The calculated predissociation rates are displayed on Fig. 6.21 and agree well with the experimental isotopic dependence.

6.10 Case of Intermediate Coupling Strength

The Golden Rule formula (6.3.11) and Eq. (6.3.8) for the level shift are expressed in terms of the unperturbed vibrational wavefunctions. For strong predissociations, this approximation becomes untenable. Exact methods exist that can determine both the linewidth and level shift. One method consists of numerically solving the following coupled equations (Lefebvre-Brion and Colin, 1977; Child and Lefebvre, 1978):

$$\left[-(1/2\mu)\frac{d^2}{dR^2}+V_1(R)+(1/2\mu R^2)J(J+1)-E\right]\chi'_{1,J}(R)=H^e(R)\chi'_{2,J}(R) \tag{6.10.1}$$

$$\left[-(1/2\mu)\frac{d^2}{dR^2}+V_2(R)+(1/2\mu R^2)J(J+1)-E\right]\chi'_{2,J}(R)=H^e(R)\chi'_{1,J}(R)$$

where the χ' vibrational functions are unknown exact solutions. Note that the Golden Rule formula uses vibrational wavefunctions that are solutions of the uncoupled Eqs. (6.3.16) where the terms on the right-hand side of Eqs. (6.10.1) have been disregarded.

Very strong predissociations can result from homogeneous interactions: electrostatic interactions for light molecules ($I^2\Sigma^+$ state of NO) or spin–orbit interactions for heavy molecules ($B^3\Pi_{0^+}$ state of IBr). Above the energy of the curve crossing, the expectation that levels should belong to one or the other potential curve seems completely unsound. One finds only numerous very broad levels, arranged without obvious rotation–vibration structure, interspersed with a smaller number of sharp lines. These sharp lines are the key to an understanding of such a case, as shown by Child (1976), who has developed a method for treating strong predissociations by a semiclassical approach.

Even for this case of strong coupling, the adiabatic picture of two potential curves that avoid crossing is inappropriate. Child has introduced an intermediate coupling picture that takes advantage of both diabatic and adiabatic characteristics. The diabatic curve of the predissociated state is displayed (solid lines) in Fig. 6.24a. The corresponding diabatic vibrational levels, E_d, are plotted versus $J(J+1)$ (solid lines) in Fig. 6.24b.

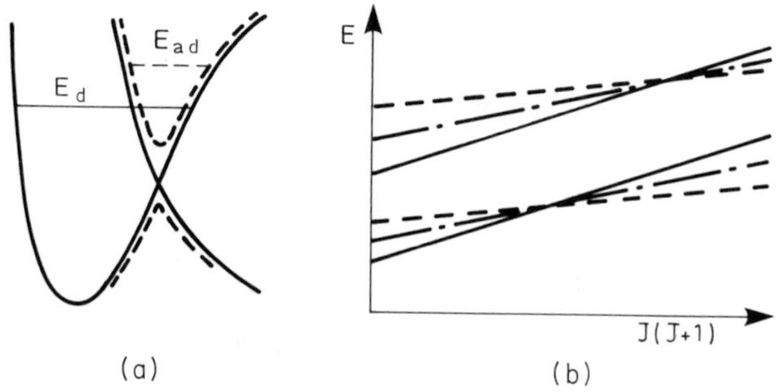

Fig. 6.24 Predissociation for a pair of states intermediate between adiabatic and diabatic coupling limits. [From Child (1980b).] (*a*) Diabatic (solid line) and adiabatic (dashed line) potential curves and the corresponding vibrational levels. (*b*) Term values plotted versus $J(J+1)$: diabatic (solid line), adiabatic (dashed line), and actual term values (dotted-and-dashed line).

The adiabatic picture provides an alternative point of view. In Fig. 6.24a, the adiabatic curves are plotted as dashed lines. The upper adiabatic potential supports a set of adiabatic levels, $E_{\mathbf{ad}}$. As the rotational constants for the levels of the adiabatic curve are smaller than those for the corresponding levels of the diabatic curve, the adiabatic levels (dashed lines on Fig. 6.24b) will cross the diabatic levels. (Note that there is no relationship between the numbering of $v_{\mathbf{d}}$ and $v_{\mathbf{ad}}$.)

By Child's semiclassical approach, it has been shown (Child, 1976) that the actual levels E (dotted–dashed lines in Fig. 6.24b) in the region of the curve crossing lie intermediate in energy between the two sets of approximate levels[†]

$$E = (E_{\mathbf{d}} + xE_{\mathbf{ad}})/(1 + x), \tag{6.10.2}$$

where x is a dimensionless coupling-strength parameter,

$$x = (\bar{\omega}_{\mathbf{d}}/\bar{\omega}_{\mathbf{ad}})(\lambda^{-2} - 1)$$

($\bar{\omega}_{\mathbf{d}}$ and $\bar{\omega}_{\mathbf{ad}}$ are the local vibrational spacing for the diabatic and adiabatic potentials, respectively, and $(\lambda^{-2} - 1)$ is the Landau–Zener coupling strength parameter (see for example, Preston *et al.*, 1974) with

$$\lambda^2 = \exp[-2\pi(H^e)^2/\hbar \underset{\sim}{v} \Delta F]$$

and v and ΔF are defined in Eqs. (6.5.6) and (6.5.4), respectively. Similarly, the values of the observed rotational constants are

$$B = (B_{\mathbf{d}} + xB_{\mathbf{ad}})/(1 + x). \tag{6.10.3}$$

[†] A small correction must be added to the diabatic and adiabatic levels to satisfy this equation. This correction for the diabatic levels is the term $\Delta_{v,J}$ of Eq. (6.5.10). For adiabatic levels this shift is, at most, one quarter of the adiabatic vibrational interval, $\Delta G_{\mathbf{ad}}(v)/4$.

Figure 6.24b shows that there exists certain J-values for which the energies of the three levels—diabatic, adiabatic, and exact—are identical. For this value, the linewidth is predicted and found to be zero. For the neighboring levels, the linewidths are shown to vary as

$$\Gamma = 2\pi x(a + x)(E_{\mathrm{d}} - E_{\mathrm{ad}})^2/\hbar\bar{\omega}_{\mathrm{ad}}(1 + x)^3, \tag{6.10.4}$$

where $a = \bar{\omega}_{\mathrm{d}}/\bar{\omega}_{\mathrm{ad}}$, with $\bar{\omega}_{\mathrm{d}}$ and $\bar{\omega}_{\mathrm{ad}}$ the local vibrational spacing for diabatic and adiabatic potentials, respectively. This semiclassical approach has been found to be in excellent agreement with exact numerical solution of the coupled Eqs. (6.10.1) (Child and Lefebvre, 1978).

In order to make practical use of Child's semiclassical method, it is necessary to have absolute rotational assignments for the sharp lines. This is a nontrivial problem since the usual rotation–vibration branch and band structure is shattered beyond recognition. However, the sharpest lines will have appreciable fluorescence quantum yields, thus enabling rotational assignments to be made on the basis of resolved $R(J-1) - P(J+1) = B''(4J' + 2)$ splittings in laser-excited fluorescence spectra. It is fortunate that the theoretical analysis is based on the sort of assignment information that is most accessible in the simplest tunable-laser experiment.

Once rotational constants are obtained from the sharpest lines, trial diabatic and adiabatic potentials are refined until the rotational constants, B_{d} and B_{ad}, calculated from them satisfy Eq. (6.10.3). Figure 6.25 shows that the exact solutions of the coupled equations give a width that is not proportional to $(H^e)^2$ when the coupling becomes very strong. One can understand this by varying the electronic matrix element. As H^e increases, the adiabatic level shifts. When it has shifted into coincidence with the diabatic level, the width is zero according to Eq. (6.10.4).

For very large widths, slight deviations from a Lorentzian lineshape are predicted (Child and Lefebvre, 1978).

It becomes necessary to solve coupled equations in more complicated cases. Here are two examples.

1. Several states cross the dissociated state (multichannel problem). In general, the total width is assumed to be the sum of the partial widths arising separately from interactions with individual states, but this is valid only if the continua of the dissociative states do not interact with each other. If the continua are coupled to each other by an interaction, I (I is a dimensionless parameter), the width is

$$\Gamma = \frac{\Gamma_1 + \Gamma_2}{1 + \pi^2 I^2}, \tag{6.10.5}$$

where Γ_1 and Γ_2 are the partial widths from each separate dissociative state

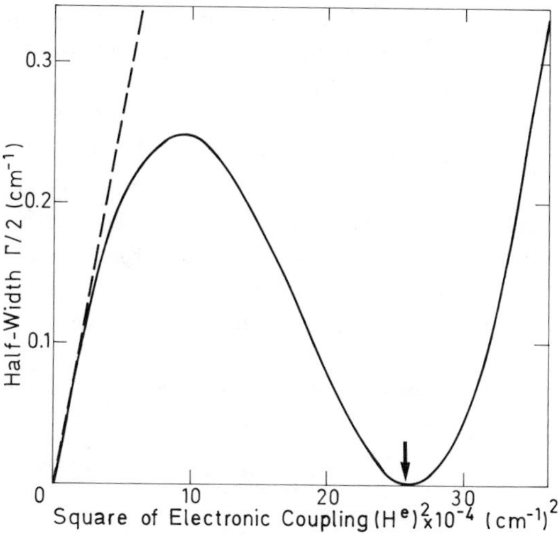

Fig. 6.25 Variation of Γ versus $(H^e)^2$. The full line corresponds to the $v=21$ diabatic level. The arrow indicates the value of H^e for which the adiabatic $v=0$ level coincides with the diabatic $v=21$ level. The dotted line gives the Γ-values expected from the Golden Rule. [Adapted from data of Child and Lefebvre (1978) and unpublished results.]

(Beswick and Lefebvre, 1973). I^2 will be, in general, small. The interaction strength between a discrete state and a continuum can be extrapolated to the continuum ~ continuum interaction by dividing it by ΔG, the separation between two successive vibrational levels of the discrete state [see Eq. (6.3.2)]. Then

$$\pi^2 I^2 = \frac{\pi\Gamma}{2}\frac{1}{\Delta G};$$

assuming $\Gamma = 10\ \text{cm}^{-1}$ and $\Delta G = 1500\ \text{cm}^{-1}$, $\pi^2 I^2 \simeq 0.01$.

2. Indirect or accidental predissociation, which is treated in the following section.

6.11 Indirect (Accidental) Predissociation

In Section 6.4 the possibility of predissociation of isolated lines was mentioned. This is usually called accidental predissociation and can be interpreted as perturbation of a rotational level by a diffuse level that lies nearby in energy for this value of J. This type of predissociation should more generally be called indirect predissociation, since the predissociation takes place through an intermediate state.

Let H_{12} be the coupling between level 1 and level 2, where level 2 is predissociated by the continuum of a third state and has a linewidth Γ_2. Direct predissociation of level 1 by state 3 is assumed to be negligible.

Kovács and Budó (1947) have argued that one should first compute the mixing of the two discrete states. Then, the width of level v_1 of state 1, Γ_{1,v_1}, is borrowed from state 2 in proportion to the mixing coefficient $a_{1v_1,2v_2}$, which expresses the level 2 character mixed into the nominal level 1 eigenstate, $'\Psi'_{1,v1}$:

$$'\Psi'_{1,v_1} = a_{1,v_1;2,v_2}\Psi_{1,v_1} + (1 - a^2)^{1/2}\Psi_{2,v_2}$$

The width is then

$$\Gamma_{1,v_1} = a^2_{1,v_1;2,v_2}\Gamma_{2,v_2}. \tag{6.11.1}$$

If the energy separation between the two discrete states, ΔE_{12}, is large compared to the interaction matrix element, H_{12}, this mixing coefficient is given by the perturbation formula,

$$a_{1,v_1;2,v_2} = \frac{H_{12}}{\Delta E_{12}}.$$

However, if state 1 borrows its width from several vibrational levels of state 2, Eq. (6.11.1) must be replaced by

$$\Gamma_{1,v_1} = 2\pi\left[\sum_{v_j} a_{1,v_1;2,v_j}H_{2v_j;3}\right]^2, \tag{6.11.2}$$

where $H_{2,v_j;3}$ is related to the width of level v_j by

$$\Gamma_{2,v_j} = 2\pi H^2_{2v_j;3}.$$

This formula shows how interference between terms in the sum over predissociated perturbers can increase or decrease the accidental width (Lefebvre-Brion and Colin, 1977).

If levels 1 and 2 are degenerate before interaction, the linewidth Γ might be expected to be divided equally between them, and

$$\Gamma_{1,v_1} = \Gamma_{2,v_2} = \Gamma.$$

Atabek *et al.* (1980) have pointed out that this conclusion is incorrect when $H_{12} \ll \Gamma_{2,v_2}$. Then it would not be proper to prediagonalize the interaction between the two discrete states. Indeed, level 2, after its interaction with the continuum of state 3, must be considered as a pseudo-continuum with respect to state 1. In the weak coupling ($H_{12} \ll \Gamma_{2,v_2}$) limit, the width of the accidental predissociation is then given by

$$\Gamma_{1,v_1} = \frac{H_{12}^2 \Gamma_{2,v_2}}{\Delta E_{12}^2 + (\Gamma_{2,v_2}/2)^2}. \tag{6.11.3}$$

If Γ_{2,v_2} is small compared to ΔE_{12}, this formula is equivalent to Eq. (6.11.1) when levels 1 and 2 are far from coincidence. However, if $\Delta E = 0$, then

$$\Gamma_{1,v_1} = \frac{4H_{12}^2}{\Gamma_{2,v_2}} \quad \text{for} \quad \Gamma_{2,v_2} \gg H_{12}. \tag{6.11.4}$$

Paradoxically, the accidental width is found to decrease as Γ_{2,v_2} increases. This resonance narrowing phenomenon corresponds to an accidental decoupling of level 1 from the continuum of state 3, when $\Delta E_{12} \ll \Gamma_{2,v_2}/2$.

An accidental predissociation occurs in the N_2 $b^1\Pi_u$ state (Robbe, 1978).

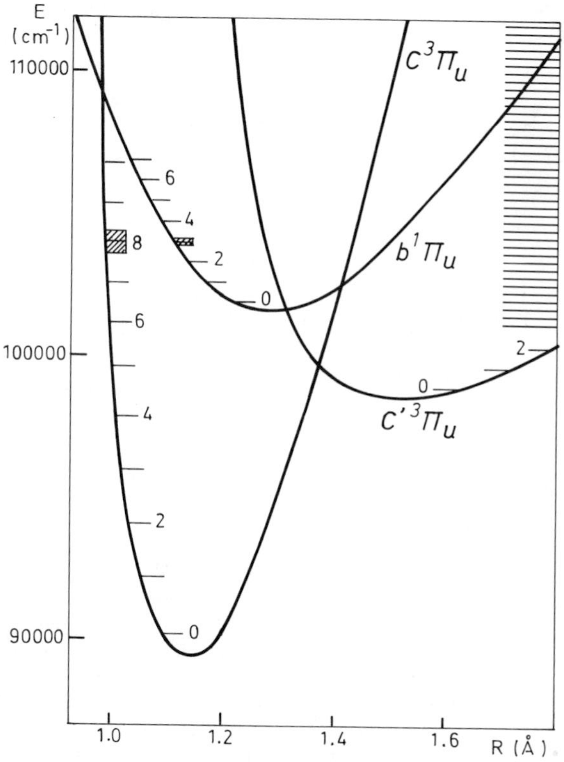

Fig. 6.26 The potential energy curves of some states of N_2. The indirect predissociation of $b^1\Pi_u$ ($v = 3$) by the $C'^3\Pi_u$ continuum via the $C^3\Pi_u$ ($v = 8$) level is shown. The widths of the directly and indirectly predissociated levels are indicated.

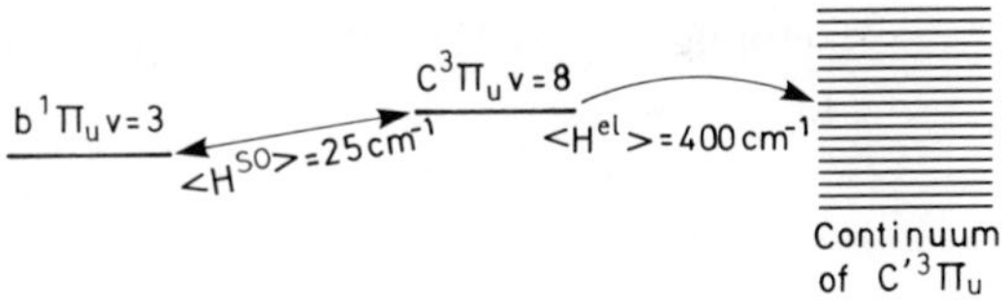

Fig. 6.27 Schematic mechanism of the indirect or accidental predissociation described in Fig. 6.26.

Consider the $b^1\Pi_u(v = 3)$ level shown in Fig. 6.26. Its coupling with the continuum of the $C'^3\Pi_u$ state is negligible, but it is degenerate with $C^3\Pi_u(v=8)$, which is strongly predissociated by the continuum of the $C'^3\Pi_u$ state. The situation is represented schematically in Fig. 6.27. Figure 6.28 shows the difference between the value of $\Gamma_{1,v}$ calculated using the Kovács–Budó model and the exact value, obtained by solution of coupled equations. If the coupling between the C and C′ states is increased, the $\Gamma_C \gg H_{bC}$ limiting case is reached and the Γ_{b,v_b} value is given by Eq. (6.11.4).

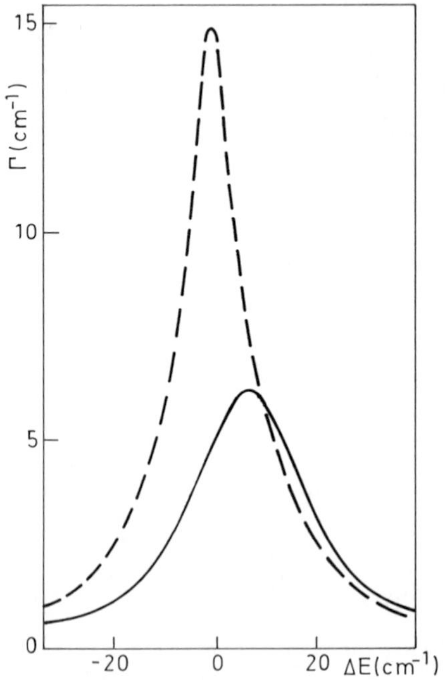

Fig. 6.28 Comparison between the Budó–Kovács (dashed line) and coupled equations (solid line) models for indirect predissociation. The width of the indirectly predissociated N_2 $b^1\Pi_u$ $(v = 3)$ level is plotted versus the energy separation from the $C^3\Pi_u$ $(v=8)$ intermediate predissociated perturbing level. The electronic coupling between the C and C′ states is taken as 400 cm^{-1}. [Courtesy J. M. Robbe from data of Robbe (1978).]

6.12 Some Recipes for Interpretation

This chapter concludes with a brief outline of the information crucial to selection and testing of a mechanistic model for an observed predissociation.

It is frequently possible to eliminate all but a few plausible candidates for the electronic identity of the predissociating state simply by considering the symmetries (Wigner–Witmer rules) of the electronic states associated with the lowest energy dissociation limit. For predissociation of very high-lying electronic states, it is often necessary to consider all of the atomic dissociation asymptotes that lie at lower energy than the predissociated level. Knowledge of the configurational character of the predissociated and plausible predissociating states can be valuable because, in addition to the usual electronic selection rules for perturbations, orbital selection rules and estimates of relevant orbital matrix elements can frequently eliminate all but one candidate.

Once a hypothesis about the symmetry and configurational parentage of a predissociating state is made, a variety of qualitative and quantitative tests must be made. The approximate size of the electronic interaction factor may be derived from an experimental Γ (or τ) value and an estimate of the vibrational factor. Some differential Franck–Condon factors for crossing potential curves are listed in Table 6.3, column 6. Recall that the isotopic dependence of the differential Franck–Condon factor is $\propto \mu^{1/6}$, and it is unusual to find a value larger than $\mu^{1/6} \times 10^{-4}$ cm. If the electronic interaction is of spin–orbit origin, it is a simple matter to estimate its magnitude from the spin–orbit parameters of the constituent atoms (Section 4.3). If the predissociation is gyroscopic, then the electronic factor will be approximately $B_C[l(l+1)]^{1/2}[J(J+1)]^{1/2}$. (Strictly speaking, the electronic factor is only $[l(l+1)]^{1/2}$, as listed in Table 6.2.)

If experimental data concerning the variation of Γ with v, J, and μ are available, then a detailed interpretation of the predissociation may be attempted making use of computer programs for calculating vibrational factors. These calculations may involve either semiclassical formulas [Eq. (6.5.3)] or numerical integration of vibrational wavefunctions.

Greater caution is required when dealing with weak predissociations. In general, the weaker the effect, the larger the number of plausible mechanisms. As for perturbations, second-order effects (especially spin–orbit) can give rise to weak interactions between states differing by $\Delta\Lambda = \pm 2$ (and other spin–spin selection rules). Second-order predissociation will have very small electronic matrix elements. On the other hand, very small vibrational factors occur for interactions between noncrossing potential curves. Table 6.2 summarizes several examples of weak predissociations.

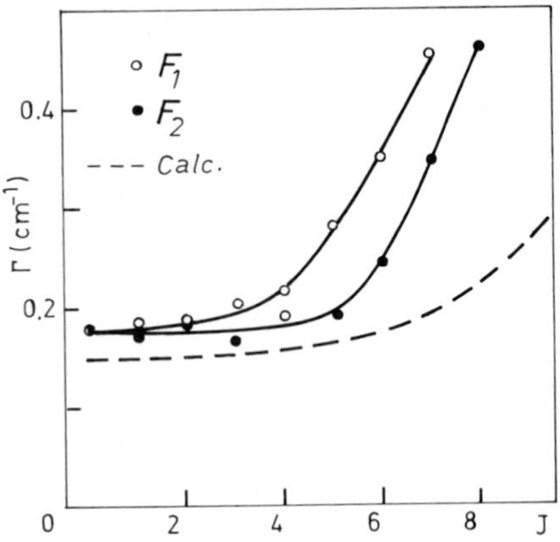

Fig. 6.29 Linewidth variation in the OH $A^2\Sigma^+$ ($v = 8$) level for the resolved F_1 and F_2 spin components. The solid lines represent the experimental data of Czarny and Felenbok (1968). The dotted line shows the calculated linewidths from Czarny *et al.* (1971).

There are several pitfalls to be avoided in formulating a predissociation mechanism. A linear variation of the linewidth with $J(J + 1)$ can be caused by either gyroscopic interaction or penetration through a centrifugal barrier. Another ambiguous indication is given by the J-dependence of the vibrational factor for the case of homogeneous, outer-crossing predissociations. This is illustrated (Fig. 6.29) by the OH $A^2\Sigma^+$ ($v = 8$) level, which, when predissociated by the $^4\Pi$ state (Section 6.9.1), shows a J-variation of the linewidth similar to that expected for a gyroscopic interaction.

When the linewidth exhibits no oscillations, this suggests the occurrence of an inner crossing, but two cases exist where an outer crossing is shown to display no linewidth oscillation. The first example concerns the OD molecule. Below the energy of the curve crossing, the bound $\sim$ free vibrational overlap comes only from the tail of the discrete wave function (tunnelling). The nonradiative decay rate is very slow, but it increases smoothly with J [predissociation of the OD $A^2\Sigma^+$ ($v = 0$–2) levels by the $^4\Sigma^-$ state (Bergeman *et al.*, 1981)].

A second example is given by the heterogeneous predissociation of the Br_2 $B^3\Pi_{0_u^+}$ state by the $^1\Pi_{1_u}$ state (Child 1980a). Oscillations in the linewidth disappear for high vibrational levels because, in this energy region, the repulsive curve is nearly parallel to the bound potential curve and the outer crossing case behaves as a noncrossing case.

These ambiguities can be eliminated if levels belonging to successive values of v or J are examined. Isotope effects are also useful for confirming the nature of a predissociation. In the absence of detailed information, one must be cautious about inferring the origin of observed predissociation effects.

References

Ackerman, M., and Biaume, F. (1970), *J. Mol. Spectrosc.* **35**, 73.
Allison, A. C., and Dalgarno, A. (1971), *J. Chem. Phys.* **55**, 4342.
Atabek, O., and Lefebvre, R. (1972), *Chem. Phys. Lett.* **17**, 167.
Atabek, O., Lefebvre, R., and Requena, A. (1980), *J. Mol. Spectrosc.* **82**, 364.
Bandrauk, A. D., and Child, M. S. (1970), *Mol. Phys.* **19**, 95.
Barrow, R. F., Chandler, G. G., and Meyer, C. B. (1966), *Philos. Trans. R. Soc. London, Ser. A* **260**, 395.
Ben-Aryeh, Y. (1973), *J. Quant. Spectrosc. Radiat. Transfer* **13**, 1441.
Bergeman, T., Erman, P., Haratym, Z., and Larsson, M. (1981), *Phys. Scr.* **23**, 45.
Beswick, J. A., and Lefebvre, R. (1973), *Mol. Phys.* **29**, 1611.
Borrell, P., Guyon, P. M., and Glass-Maujean, M. (1977), *J. Chem. Phys.* **66**, 818.
Broyer, M., Vigué, J., and Lehmann, J. C. (1976), *J. Chem. Phys.* **64**, 4793.
Brzozowski, J., Bunker, P., Elander, N., and Erman, P. (1976), *Astrophys. J.* **207**, 414.
Carré, M., Druetta, M., Gaillard, M. L., Bukow, H. H., Horani, M., Roche, A. L., and Velghe, M. (1980), *Mol. Phys.* **40**, 1453.
Carrington, A., Milverton, D. R. J., and Sarre, P. (1978), *Mol. Phys.* **35**, 1505.
Child, M. S. (1973), *J. Mol. Spectrosc.* **45**, 293.
Child, M. S. (1974), *in* "Molecular Spectroscopy", Vol. 2 (R. F. Barrow, D. A. Long, and D. J. Millen, eds.), p. 466. Chemical Soc., London, Specialist Periodical Report.
Child, M. S. (1976), *Mol. Phys.* **32**, 1495.
Child, M. S. (1980a), *J. Phys. B.* **13**, 2557.
Child, M. S. (1980b), *in* "Semi-Classical Methods in Molecular Scattering and Spectroscopy" (M. S. Child, ed.) p. 127 Reidel Publ. Dordrecht, Holland.
Child, M. S., and Lefebvre, R. (1978), *Chem. Phys. Lett.* **55**, 213.
Czarny, J., and Felenbok, P. (1968), *Ann. Astrophys.* **31**, 141.
Czarny, J., Felenbok, P., and Lefebvre-Brion, H. (1971), *J. Phys. B* **4**, 124.
Douglas, A. E. (1952), *Can. J. Phys.* **30**, 302.
Durmaz, S., and Murrell, J. N. (1971), *Trans. Faraday Soc.* **67**, 3395.
Elander, N., Hehenberger, M., and Bunker, P. R. (1979), *Phys. Scr.* **20**, 631.
Erman, P. (1979), *in* "Molecular Spectroscopy", Vol. 6 (R. F. Barrow, D. A. Long, and J. Sheridan, eds.), p. 174. Chemical Soc., London, Specialist Periodical Report.
Fano, U. (1961), *Phys. Rev.* **124**, 1866.
Farmer, A. J. D., Fabian, W., Lewis, B. R., Lokan, K. H., and Haddad, G. N. (1969), *J. Quant. Spectrosc. Radiat. Transfer* **8**, 1739.
Fontana P. R. (1982), "Atomic Radiative Processes", Academic Press, New York.
Gallusser, R. (1976), thesis, Physical Chemistry Laboratory, ETH Zurich, Switzerland.
Gallusser, R., and Dressler, K. (1982), *J. Chem. Phys.* **76**, 4311.
Giusti-Suzor, A., and Jungen, C. (1984), *J. Chem. Phys.* **80**, 986.
Glass-Maujean, M., Breton J., and Guyon, P. M. (1979), *Chem. Phys. Lett.* **63**, 591.
Glass-Maujean, M., Guyon, P. M., and Breton, J. (1986), *Phys. Rev.* To be published.
Govers, T. R., van de Runstraat, C. A., and de Heer, F. J. (1975), *Chem. Phys.* **9**, 285.
Herzberg, G. (1950), "Spectra of Diatomic Molecules," Van Nostrand, Princeton.
Herzberg, G. (1971), *in* "Topics in Modern Physics," Colorado Assoc. Univ. Press p. 191, Boulder, Colo.
Helm, H., Cosby, P. C., and Huestis, D. L. (1980), *J. Chem. Phys.* **73**, 2629.
Hudson, R. D., Carter, V. L., and Stein, J. A. (1966), *J. Geophys. Res.* **7**, 2295.
Julienne, P. S., and Krauss, M. (1975), *J. Mol. Spectrosc.* **56**, 270.

Koffend, J. B., Bacis, R., Churassy, S., Gaillard, M. L., Pique, J. P., and Hartmann, F. (1983), *Laser Chem.* **1**, 185.
Kovács, I., and Budó, A. (1947), *J. Chem. Phys.* **15**, 166.
Lefebvre-Brion, H., and Colin, R. (1977), *J. Mol. Spectrosc.* **65**, 33.
Lewis, B. R., Carver, J. H., Hobbs, T. I., McCoy, D. G., and Gies, H. P. F. (1980), *J. Quant. Spectrosc. Radiat. Transfer* **24**, 365.
Lorquet, A. J., and Lorquet, J. C. (1974), *Chem. Phys. Lett.* **26**, 138.
Meiwes, K. H., and Engelke, F. (1982), *Chem. Phys. Lett.* **85**, 409.
Miescher, E. (1976), *Can. J. Phys.* **54**, 2074.
Mulliken, R. S. (1960), *J. Chem. Phys.* **33**, 247.
Moseley, J. T., Cosby, P. C., Ozenne, J. B., and Durup, J. (1979), *J. Chem. Phys.* **70**, 1974.
Murrell, J. N., and Taylor, J. M. (1969), *J. Mol. Spectrosc.* **16**, 609.
Olson, C. D., and Innes, K. K. (1976), *J. Chem. Phys.* **64**, 2405.
Pique, J. P., Bacis, R., Hartmann, F., Sadeghi, N., and Churassy, S. (1983), *J. Phys. Colloq (Orsay, Fr.)* **44**, 347.
Preston, R. K., Sloane, C., and Miller, W. H. (1974), *J. Chem. Phys.* **60**, 4961.
Robbe, J. M. (1978), Ph.D. thesis, Université des Sciences et Techniques de Lille, Lille, France, unpublished.
Roche, A. L., and Tellinghuisen, J. (1979), *Mol. Phys.* **38**, 129.
Rothschild, M., Egger, H., Hawkins, R. T., Bokor, J., Pummer, H., and Rhodes, C. K. (1981), *Phys. Rev. Sect. A* **23**, 206.
Schaefer, H. F., III, and Miller, W. H. (1971), *J. Chem. Phys.* **55**, 4107.
Sharp, T. E. (1970), *At. Data* **2**, 119.
Smith, A. L. (1971), *J. Chem. Phys.* **55**, 4344.
Turner, D. W. (1970), "Molecular Photoelectron Spectroscopy", Wiley (Interscience), New York.
Vedder, H. J., Chawla, G. K., and Field, R. W. (1984), *Chem. Phys. Lett.* **111**, 303.
Vigué, J., Broyer, M., and Lehmann, J. C. (1977), *J. Phys. B.* **10**, L379.
Vigué, J., Broyer, M., and Lehmann, J. C. (1981), *J. Phys. Colloq. (Orsay, Fr.)* **42**, 937.
Zare, R. N. (1972), *Mol. Photochem.* **4**, 1.

Chapter 7

Autoionization

7.1 Experimental Aspects

Above ionization limits, lines in the absorption spectrum are often broad or diffuse. This diffuse character results from an interaction of a very highly excited neutral molecule state, AB**, with the continuum of an ionized molecule, AB^+, plus an electron. This continuum reflects the fact that the electron can be ejected from the molecule over a continuous range of kinetic energy:

$$AB + h\nu \longrightarrow AB^{**} \longrightarrow AB^+ + e \quad (KE) \tag{7.1.1}$$

Level AB**, often called the "superexcited" state, is an autoionized or resonance state. Autoionization is called preionization by Herzberg. This can be justified by the analogy between preionization and predissociation. In predissociation, the interaction of a discrete state with the vibrational continuum of the nuclei allows this discrete state a finite probability of dissociation. In preionization, it is the mixing of a discrete state with the electronic continuum that provides a finite ionization probability.

Equation (7.1.1) is not the only mechanism governing the decay of AB**. Indeed, three main mechanisms compete:

$$AB^{**} \begin{cases} \nearrow AB^+ + e & \text{autoionization} \\ \rightarrow A + B^* & \text{dissociation} \\ \searrow AB + h\nu & \text{radiation} \end{cases}$$

We consider in the following the case of a single isolated "resonance," far removed from other resonances. If an absorption line has a Lorentzian profile, the linewidth,† Γ, is related to the total lifetime, τ, of the state by the usual relation: $\Gamma \propto \tau^{-1}$ [cf. Eq. (6.2.6)]. This width includes contributions from each of these three different types of decay. If interactions between continua of the electron, nuclei, and radiation field are neglected, one can write

$$\Gamma \simeq \Gamma_a + \Gamma_d + \Gamma_r \tag{7.1.2}$$

or

$$\tau^{-1} \simeq \tau_{nr}^{-1} + \tau_r^{-1}, \tag{7.1.3}$$

where

$$\tau_{nr}^{-1} = \tau_a^{-1} + \tau_d^{-1}. \tag{7.1.4}$$

These partial widths, Γ_a, Γ_d, and Γ_r, associated with autoionizing, dissociative, and radiative decay, respectively, and the corresponding partial lifetimes τ_a, τ_d, and τ_r, have no physical meaning. Only the total width Γ can be observed experimentally. This width is identical whether the line is observed by absorption, photoionization, or fluorescence spectroscopy, but the associated cross-sections can be different. Here we address only the problem of autoionization in absorption spectroscopy.

The orders of magnitude of these partial lifetimes can be very different. The radiative lifetime is greater than 10^{-8} S for most diatomic molecule excited states. Generally, the rates of nonradiative processes above an ionization limit are much faster than radiative decay, and consequently no emission lines are easily detected above such limits. An exception is found in the spectrum of H_2, where Q-branch lines of the first members of the $^1\Pi_u$ Rydberg series are seen in emission (Larzillière *et al.*, 1980) above the ionization limit. The emitting levels of these Q-lines, which correspond to f-parity levels of $^1\Pi_u$ states, are very weakly predissociated. In contrast, the e-levels (upper levels of the P and R branches) are strongly predissociated by $^1\Sigma_u^+$ Rydberg states ($^1\Sigma^+$ states here are exclusively e-parity with perturbation selection rule $f \not\leftrightarrow e$). These f-levels, in addition to being weakly predissociated, are also very weakly autoionized. In many other cases, decay by predissociation can be very fast, particularly for electrostatic predissociations (see Tables 6.2 and 6.3). For example, in the

† Width here means the homogeneous part of the width, in the absence of any pressure effect.

spectrum of the NO molecule, most $^2\Pi$ Rydberg states are predissociated by the vibrational continuum of the $B^2\Pi$ valence state so rapidly that autoionization cannot compete (Giusti-Suzor and Jungen, 1984).

In general, complex features may arise in the absorption spectrum due to competition between the two processes, autoionization and predissociation. For example, as will be shown in Section 7.7, the partial width due to electronic autoionization is independent of v for a given $nl\lambda$ Rydberg state, but the partial width due to predissociation can vary considerably from one vibrational level to another because of oscillatory variations of the vibrational part of the interaction with the predissociating state.

Another problem that affects any interpretation of a diffuse peak is instrumental resolution. Recall that the resolution is given by

$$\Delta\lambda \quad (\text{Å}) = 10^{-8}\lambda^2\,\Delta\nu \quad (\text{cm}^{-1}).$$

For example, at 800 Å, a very good resolution of $\Delta\lambda = 0.02$ Å corresponds to $\Delta\nu = 3\ \text{cm}^{-1}$. Such resolution is sufficient to resolve individual rotational lines of only the H_2 molecule.

When autoionized rotational lines are broad, it is difficult to extract from the observed band contour the fundamental quantity of interest, the width of a single rotational line. The width of the rotational contour is, for example, about $40\ \text{cm}^{-1}$, for the $n = 13$ member of the Worley–Jenkins Rydberg series of N_2 (Carroll, 1973) at liquid nitrogen temperature (77 K). Figure 7.1 shows the band contour obtained from individual lines of a given width. When the width of a single line ($25\ \text{cm}^{-1}$) is of the order of magnitude of the rotational *envelope*, the observed band contour ($65\ \text{cm}^{-1}$) masks the true linewidth. Experiments using supersonic molecular beams, where the rotational temperature is lowered to a few kelvins, will reduce this effect of rotational congestion.

The band contour depends not only on the temperature but also on Hund's coupling cases. Jungen (1980) has pointed out that the rotational structure becomes narrower as n increases due to a change from Hund's case (b) to case (d). This envelope-narrowing effect must not be construed as a variation of autoionization widths with effective quantum number n^*. Fortunately, double-resonance techniques in the vacuum ultraviolet region are beginning to isolate individual rotational line profiles. (See also Section 5.5.7.)

Another pernicious effect of limited resolution is a reduction of the peak height. Theoretically (i.e., at infinite instrumental resolution), the maximum peak height is obtained at $E = E_r$ and its value is proportional to μ^2/Γ [see Eq. (6.2.5)], where μ is the transition moment. Since both μ^2 and Γ vary as $(n^*)^{-3}$, the maximum peak height is independent of n^* along an autoionized Rydberg series. However, because the linewidth, Γ, decreases as $(n^*)^{-3}$, the apparent peak height begins to be affected by limited resolution as n^* increases. The narrower the peak, the more its appearent peak height is reduced.

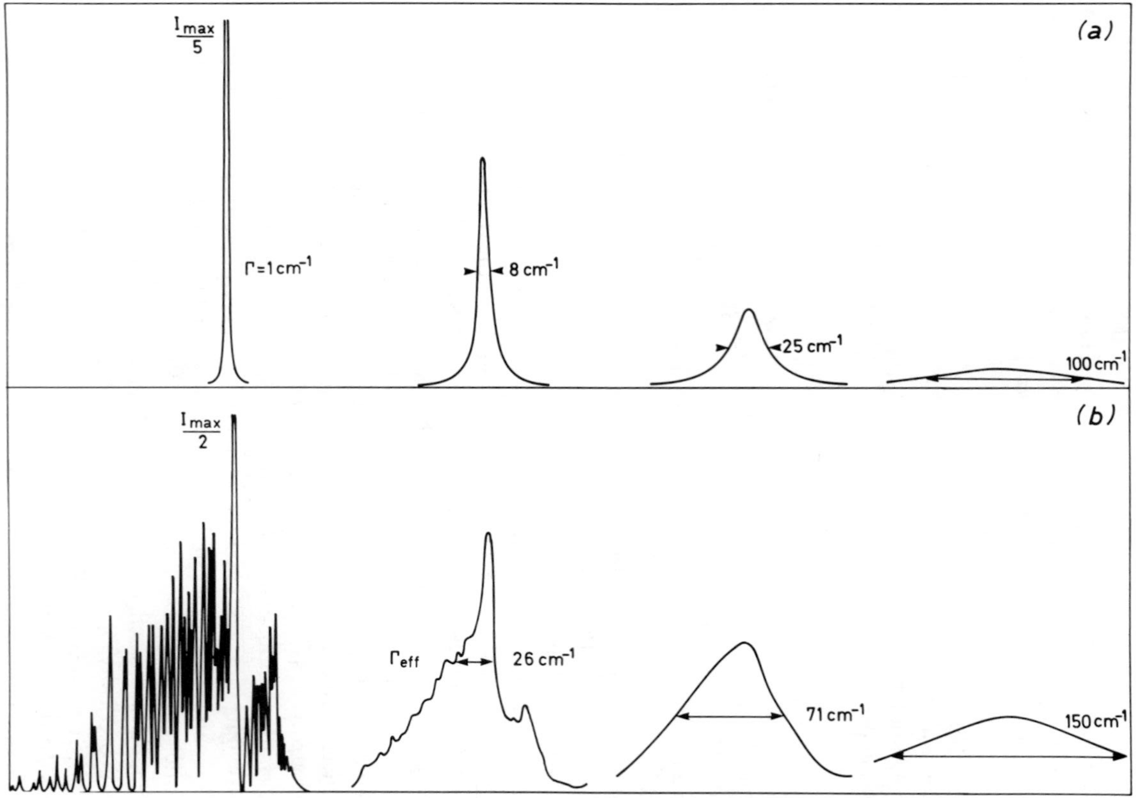

Fig. 7.1 Relationship between full width at half maximum (FWHM) of individual lines and band contours. (*a*) Single lines with FWHM consistent with band contour shown below. (*b*) Band contours are constructed by convolution of individual lines (each with the lineshape shown above) with an instrumental resolution of 0.01 cm^{-1}. The spectrum is calculated for a $^2\Pi$ [case (b)] $\leftarrow X^2\Pi_{1/2}$ [case (a)] transition of the NO molecule observed in absorption at 78 K. [From Giusti-Suzor and Jungen (1984).]

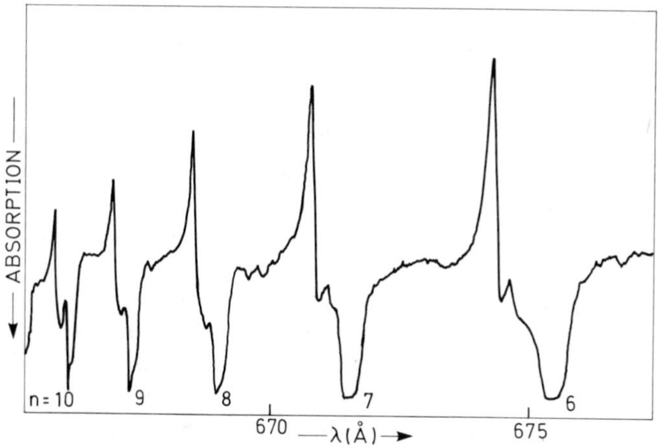

Fig. 7.2 Absorption spectrum showing the autoionization of the N_2 Hopfield Rydberg series. (Courtesy K. Yoshino.)

Consequently, the experimental peak height appears to decrease as $(n^*)^{-3}$.† This resolution effect is clearly evident in Fig. 7.2.

Spectra with higher resolution will soon become available with the development of new experimental techniques such as tunable, vacuum ultraviolet lasers, but in many cases the linewidth resulting from autoionization or predissociation exceeds the interval between two successive rotational lines and prevents resolution of the rotational structure regardless of instrumental resolution. Only double-resonance techniques such as two-color resonantly enhanced multiphoton ionization will help to obtain useful information on true linewidths.

7.2 The Nature of Autoionized States

Autoionized states are Rydberg states converging to an ionization limit that lies above the lowest ionization threshold. The general formula for the energy of Rydberg states is

$$E_n = E_\infty - \frac{\mathcal{R}}{(n^*)^2} = E_\infty - \frac{\mathcal{R}}{(n-a)^2} \tag{7.2.1}$$

† Note that the integrated intensity of a line $[\sim(\mu^2/\Gamma)\Gamma]$ decreases as $(n^*)^{-3}$, regardless of instrumental resolution.

where n and n^* are the principal and effective quantum numbers, E_∞ is the ionization limit, $\mathscr{R}$ the Rydberg constant (109,737.318 cm^{-1}), and a the quantum defect, which is characteristic of a particular Rydberg series. This quantum defect is about 1.0 for $ns\sigma$ orbitals of molecules composed of atoms between Li and Ne, since these orbitals "penetrate" into the molecular core (cf. Mulliken, 1964) and consequently are strongly modified by the presence of inner electrons of the same $s\sigma$ symmetry. The quantum defect is of the order 0.7 for $np\sigma$ and $np\pi$ orbitals, which are less modified by the core, and nearly zero for $nd\sigma$, $nd\pi$, and $nd\delta$ orbitals, which have no "precursors" in the core.

Rydberg series can converge to excited vibrational levels of the ground state of the ion or to excited electronic states of the ion. Each Rydberg state can be described as one external electron in the field of the ion core state that forms the limit of the corresponding $nl\lambda$ Rydberg series. Autoionization of a Rydberg state involves formation of a state of the ion that lies lower than the Rydberg series limit of the autoionized state, plus an ejected electron, which carries away the excess energy.

Since autoionization involves a nonradiative change in energy of the core ion, the ejected electron obtains its kinetic energy from the superexcited (autoionized) state by conversion of

1. rotational energy,
2. vibrational energy,
3. spin–orbit energy,
4. electronic energy,

depending on which processes are energetically possible. The first three types of core-to-ejected-electron energy transfer are observed for autoionization of discrete states by the continuum that belongs to the same electronic state but to a different vibrational, rotational, or spin–orbit state of the core ion. Approximate selection rules exist for vibrational transfer. The last type of core-to-electron energy transfer occurs between a discrete state built on an electronically excited core and a continuum associated with a lower-energy electronic state of the ion.

The first two types of interaction are peculiar to molecules. The last two also appear in atoms, but in the case of electronic autoionization, nuclear motion introduces a level of complexity beyond that of the atomic case.

Some general rules for the autoionization process are derived next. These rules are based on the following approximations. The feature in which autoionization effects are manifested must be a single, well-resolved line associated with a well-characterized rovibronic level, for which the autoionization process can be described to a good approximation by a single predominant mechanism. The consequences of the breakdown of these approximations are discussed in Section 7.8.

7.3 Autoionization Widths

The observed autoionization width is the result of several partial decays into different continua. Therefore, it is useful for calculations to define a quantity, the partial autoionization width, in which the initial and final states of the autoionization process are defined perfectly. One can then describe the total width, the only experimentally observable autoionization width, by summing over all possible final states. The initial state is specified by 1, n, and v, the final state by 2, v_+ of the resultant ion, and ε of the ejected electron. The partial autoionization width can be expressed by the Golden Rule formula:

$$\Gamma_{1,n,v;\,2,\varepsilon,v_+} = 2\pi|H_{1,n,v;\,2,\varepsilon,v_+}|^2. \tag{7.3.1}$$

The interaction H_{12} can often be factored into electronic and vibrational parts, as in the case of predissociation, but here it is the electronic part that has the continuous character. In some cases, the vibrational part is simply an overlap factor and

$$H_{1,n,v;\,2,\varepsilon,v_+} = \langle\psi_{1n,v}|\mathbf{H}|\psi_{2\varepsilon,v_+}\rangle = \langle\phi_{1n}|\mathbf{H}|\phi_{2\varepsilon}\rangle\langle v|v_+\rangle, \tag{7.3.2}$$

where the function $\phi_{2\varepsilon}$ for the continuum electron is energy-normalized.

The interactions responsible for autoionization have the same origins as those for perturbations between discrete states. A useful relationship between perturbation and autoionization matrix elements can be obtained as follows: the autoionized state is the nth member of a given Rydberg series converging to a limit higher in energy than the lowest ionization limit. If the first members of this Rydberg series lie below the lowest ionization limit, they can perturb a higher n' member of the Rydberg series that converges to the lowest ionization limit.

See Fig. 7.6, where the $n = 3$ member of a Rydberg series that converges to the vibrationally excited H_2^{*+} state of the ion perturbs the $n' = 10$ member of a Rydberg series that converges to the ground state of the ion.

The interaction between two Rydberg states belonging to different series comes from the part of the Rydberg electron wave functions near the ion core and, for light molecules, is restricted to a distance within about 2 Å from the center of mass. In this region, the first lobes (i.e., the region within the first nodal surface) of successive Rydberg orbitals of a given series are very similar in form (because they must be orthogonal to the lowest-energy, nodeless orbital), but differ only by a scale factor (see Fig. 7.3 and Slater, 1960, p. 223, Fig. 9.2). Indeed, as n increases, the amplitude of this inner part of the Rydberg electron wavefunction decreases proportionally with the normalization factor, $(n^*)^{-3/2}$, of the wavefunction.

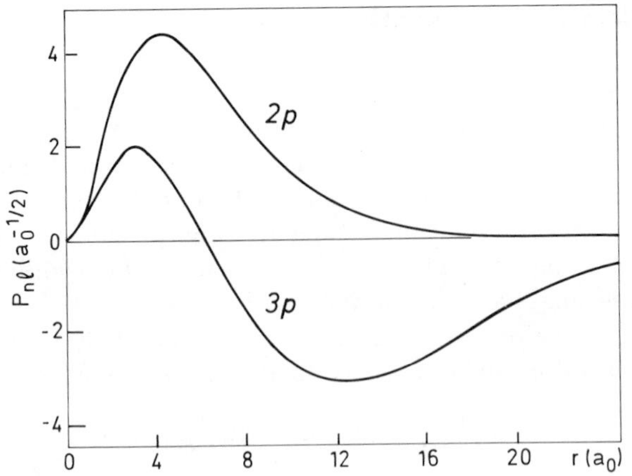

Fig. 7.3 Hydrogenic radial wavefunctions for successive $2p$ and $3p$ Rydberg orbitals; r is the distance of the Rydberg electron from the molecular center of mass.

The interaction between two Rydberg states of the same n, which belong to two different series, ϕ_{1n} and ϕ_{2n}, will be derived using a convenient Hund's case basis set and by expressing the interaction as part of a residual *electronic* interaction of the phenomenological form

$$H^e_{1n,2n} = \langle \phi_{1n}|\mathbf{H}|\phi_{2n}\rangle = \frac{2\mathscr{R}}{(n^*)^3} I, \tag{7.3.3}$$

where $(n^*)^3$ plays the role of a scale factor and I is a dimensionless parameter that, to a good approximation, is independent of n and is characteristic of the type of interaction. This interaction will be generalized to states of different n by the formula

$$H^e_{1n,2n'} = \langle \phi_{1n}|\mathbf{H}|\phi_{2n'}\rangle = \frac{2\mathscr{R}}{(n^*)^{3/2}(n'^*)^{3/2}} I, \tag{7.3.4}$$

where I is the same dimensionless parameter and is, to a good approximation, independent of n and n'. This formula is valid even if state 2 is a valence state. In Section 4.2, Table 4.5 shows how the "experimental" value of the $(n^*)^{3/2} H^e_{1n,2}$ electronic parameter is found to be nearly constant for different Rydberg states, $1n$, of the same series interacting with successive vibrational levels of the same valence electronic state, 2.

The electronic part $\langle \phi_{1n}|\mathbf{H}|\phi_{2n'}\rangle$ of the perturbation interaction between two discrete states cannot be compared directly with the electronic part

$\langle \phi_{1n}|\mathbf{H}|\phi_{2\varepsilon}\rangle$ of the autoionization matrix element. Only the energy-normalized perturbation interaction (see Section 6.3), $\rho_{n'}^{1/2}\langle \phi_{1n}|\mathbf{H}|\phi_{2n'}\rangle$, which has the same dimensionality as $\langle \phi_{1n}|\mathbf{H}|\phi_{2\varepsilon}\rangle$, can be compared to this last quantity and should have the same magnitude.

For Rydberg states, the density of states ρ_n is given by the inverse separation between two successive states:

$$\frac{1}{\rho_{n'}} = \Delta E_{n'} = \frac{\mathcal{R}}{(n'^*)^2} - \frac{\mathcal{R}}{(n'^* + 1)^2} \simeq \frac{2\mathcal{R}}{(n'^*)^3}. \tag{7.3.5}$$

For a Rydberg state of a given n, the quantity I from Eq. (7.3.4), assumed previously to be constant for all n' members of a second Rydberg series, will now be assumed constant in the limit $n' \to \infty$, which corresponds to a continuum electron with zero kinetic energy. Thus, by combining Eqs. (7.3.4) and (7.3.5),

$$\langle \phi_{1n}|\mathbf{H}|\phi_{2\varepsilon=0}\rangle = \rho_{n'}^{1/2}\langle \phi_{1n}|\mathbf{H}|\phi_{2n'}\rangle = \frac{(2\mathcal{R})^{1/2}}{(n^*)^{3/2}} I, \tag{7.3.6}$$

the partial width becomes

$$\Gamma_{n,v;\varepsilon=0,v_+} = 2\pi \frac{2\mathcal{R}}{(n^*)^3} I^2 \langle v|v_+\rangle^2. \tag{7.3.7}$$

With the assumption that I is independent of n and therefore also independent of the energy of the ejected electron, Eq. (7.3.7) is valid for any ε energy. If I varies smoothly with the energy of the ejected electron, it must be replaced by $I(\varepsilon)$ in Eq. (7.3.7).

Two conclusions may be drawn from this general formula.

1. For successive autoionized states of the same series, the widths decrease as $(n^*)^{-3}$. This law is general regardless of the type of electronic interaction and assumes that I is independent of (or weakly dependent on) the energy of the continuum electron. Expressions for I, valid for different types of autoionization, are discussed in the following sections.

2. The magnitude of the width depends on the interval between ionization thresholds. Indeed, autoionization into a given continuum is possible only if the superexcited state lies above the ionization threshold of that continuum,

$$E_n = E(\mathrm{AB}^{*+}) - \frac{\mathcal{R}}{(n^*)^2} \geq E(\mathrm{AB}^+). \tag{7.3.8}$$

Equation (7.3.8) sets a lower bound on the value of n^* required for autoionization:

$$n^* \geq \left(\frac{\mathcal{R}}{\Delta E}\right)^{1/2}, \tag{7.3.9}$$

Table 7.1
Minimum *n*-Values Required for Different Types of Autoionized States[a]

Interaction	Molecule	Ionization thresholds	Interval ΔE between the thresholds (cm^{-1})	n_{min}
Electronic	N_2^+	$X^2\Sigma_g^+$ $A^2\Pi_u$	8950	4
	NO^+	$X^1\Sigma^+$ $b^3\Pi$	58,793	2
Spin–orbit	N_2^+	$A^2\Pi_{3/2}$ $A^2\Pi_{1/2}$	80	37
	HI^+	$^2\Pi_{3/2}$ $^2\Pi_{1/2}$	5352	5
Vibrational	H_2^+	$X^2\Sigma_g^+$ ($v = 0$) $X^2\Sigma_g^+$ ($v = 1$)	2300	7
Rotational	*ortho*-H_2^+	$X^2\Sigma_g^+$ ($v = 0$, $N = 0$) $X^2\Sigma_g^+$ ($v = 0$, $N = 2$)	180	25
	ortho-N_2^+	$X^2\Sigma_g^+$ ($v = 0$, $N = 0$) $X^2\Sigma_g^+$ ($v = 0$, $N = 2$)	12	96

[a] From an original idea of Jungen (private communication).

where $\Delta E = E(AB^{*+}) - E(AB^+)$. This in turn, when inserted into Eq. (7.3.7), sets an upper limit for the autoionization width. Table 7.1 summarizes several such bounds on n associated with different types of autoionization.

7.4 Rotational Autoionization

In rotational autoionization, the coupling matrix element is J-dependent and has the same origin as the coupling that occurs for heterogeneous rotational interactions where the selection rule is $\Delta\Omega = \pm 1$ (cf. Section 2.5.3). Since rotational transitions into very high Rydberg states cannot generally be resolved, especially when they are broadened by autoionization, rotational autoionization has been studied only in the H_2 molecule (Herzberg and Jungen, 1972; Jungen and Dill, 1980).

Let us first consider the rotational interaction between discrete Rydberg states. For example, the $^1\Pi_u$ and $^1\Sigma_u^+$ levels of an np-complex of H_2 are mixed by the $\mathbf{l}$-uncoupling term in $\mathbf{H}^{ROT}$. Since $\Delta\Sigma = 0$, the selection rule $\Delta\Omega = \pm 1$ implies that $\Delta\lambda = \pm 1$. The 2×2 $^1\Pi \sim {}^1\Sigma^+$ perturbation secular determinant for the Jth rotational levels, expressed in the case (a) basis, is, for $l = 1$,

$$\begin{vmatrix} E_{n\Sigma} + B[J(J+1)+2] - E & -2B[J(J+1)]^{1/2} \\ -2B[J(J+1)]^{1/2} & E_{n\Pi} + B[J(J+1)] - E \end{vmatrix} = 0. \quad (7.4.1)$$

The $\Delta J = 0$ energy separation between the Π and Σ states of the $l = 1$ complex is proportional to $(n^*)^{-3}$; thus, at high values of n, the Π and Σ levels of the same rotational quantum number become nearly completely mixed. At the limit ($n = \infty$), the energy separation between the Π and Σ levels is $2B$, because

$$E_{(n=\infty)\Sigma} = E_{(n=\infty)\Pi} = E_{\infty}.$$

In this case, the eigenvalues of Eq. (7.4.1) are

$$\begin{aligned} E_1 &= [B + B(2J+1)] + E_{\infty} + BJ(J+1) \\ E_2 &= [B - B(2J+1)] + E_{\infty} + BJ(J+1), \end{aligned} \quad (7.4.2)$$

which correspond to the two ionization limits associated with the $N = 2$ and 0 ion core rotational levels. The $\Delta J = 0$ energy separation between these two series of $N = J + 1$ (E_1) and $N = J - 1$ (E_2) rotational levels is $2B(2J + 1)$. The corresponding eigenfunctions in this complete $\Pi \sim \Sigma$ mixing limit,

$$\begin{aligned} \Psi_1 &= \left(\frac{J+1}{2J+1}\right)^{1/2} \psi(^1\Sigma) - \left(\frac{J}{2J+1}\right)^{1/2} \psi(^1\Pi) \\ \Psi_2 &= \left(\frac{J}{2J+1)}\right)^{1/2} \psi(^1\Sigma) + \left(\frac{J+1}{2J+1}\right)^{1/2} \psi(^1\Pi), \end{aligned} \quad (7.4.3)$$

are the case (d) basis functions. The mixing coefficients in Eq. (7.4.3) form the transformation matrix, which is convenient for describing high values of n [which are intermediate between cases (d) and (a)]. Using this coefficient transformation matrix, the residual $\Delta J = 0$ interaction between the different levels of identical n becomes, for large n, a consequence of the small electronic energy separation between Σ and Π basis components of the $l = 1$ complex, which vanishes at the case (d) limit. The off-diagonal matrix element, evaluated in the case (d) limit, is

$$\begin{aligned} \langle \Psi_{1n} | \mathbf{H} | \Psi_{2n} \rangle &= \frac{[J(J+1)]^{1/2}}{2J+1} (\langle n^1\Sigma | \mathbf{H} | n^1\Sigma \rangle - \langle n^1\Pi | \mathbf{H} | n^1\Pi \rangle) \\ &\quad + \frac{(J+1) - J}{J+1} \langle n^1\Sigma | \mathbf{H} | n^1\Pi \rangle \end{aligned} \quad (7.4.4)$$

or, since $\langle n^1\Sigma|\mathbf{H}|n^1\Pi\rangle = -2B[J(J+1)]^{1/2}$,

$$\langle\Psi_{1n}|\mathbf{H}|\Psi_{2n}\rangle = \frac{[J(J+1)]^{1/2}}{2J+1}(E_{n\Sigma} + 2B - E_{n\Pi} - 2B)$$

$$= \frac{[J(J+1)]^{1/2}}{2J+1}(E_{n\Sigma} - E_{n\Pi}). \tag{7.4.5a}$$

Similarly,

$$\langle\Psi_{1n}|\mathbf{H}|\Psi_{1n}\rangle = \frac{E_{n\Sigma} + E_{n\Pi}}{2} - \frac{E_{n\Pi} - E_{n\Sigma}}{2(2J+1)} + \frac{4BJ(J+1)}{2J+1} \tag{7.4.5b}$$

$$\langle\Psi_{2n}|\mathbf{H}|\Psi_{2n}\rangle = \frac{E_{n\Sigma} + E_{n\Pi}}{2} + \frac{E_{n\Pi} - E_{n\Sigma}}{2(2J+1)} - \frac{4BJ(J+1)}{2J+1}. \tag{7.4.5c}$$

Now, in order to reexpress Eq. (7.4.5a) in terms of quantum defects, let

$$E_{n\Pi} = E_\infty - \frac{\mathcal{R}}{(n-a_\pi)^2} \quad \text{and} \quad E_{n\Sigma} = E_\infty - \frac{\mathcal{R}}{(n-a_\sigma)^2}, \tag{7.4.6}$$

where a_π and a_σ are the quantum defects of the unperturbed Π and Σ series, respectively. Expressed in terms of the very small difference of quantum defects, $a' = \frac{1}{2}(a_\pi - a_\sigma)$, Eq. (7.4.6) becomes

$$E_{n\Pi} = E_\infty - \frac{\mathcal{R}}{(n^* - a')^2} \quad \text{and} \quad E_{n\Sigma} = E_\infty - \frac{\mathcal{R}}{(n^* + a')^2} \tag{7.4.7}$$

where n^* is defined as a mean effective principal quantum number for the two series:

$$n^* \equiv n - \tfrac{1}{2}(a_\pi + a_\sigma).$$

Using a truncated Taylor expansion,

$$\frac{\mathcal{R}}{(n^* \pm a')^2} \simeq \frac{\mathcal{R}}{(n^*)^2} \mp \frac{2\mathcal{R}}{(n^*)^3}a', \tag{7.4.8}$$

Eq. (7.4.5a) becomes

$$\langle\Psi_{1n}|\mathbf{H}|\Psi_{2n}\rangle = \frac{[J(J+1)]^{1/2}}{2J+1}\frac{2\mathcal{R}}{(n^*)^3}2a' = \frac{[J(J+1)]^{1/2}}{2J+1}\frac{2\mathcal{R}}{(n^*)^3}(a_\pi - a_\sigma). \tag{7.4.9}$$

Further, by analogy with Eq. (7.3.4), the interaction ($\Delta J = 0$, $\Delta N = \pm 2$) between two states, one belonging to an np complex, the other to an $n'p$

complex, is

$$\langle \Psi_{1np} | \mathbf{H} | \Psi_{2n'p} \rangle = \frac{2\mathscr{R}}{(n^*)^{3/2}(n^{*\prime})^{3/2}} I \tag{7.4.10}$$

with

$$I = \frac{[J(J+1)]^{1/2}}{2J+1}(a_\pi - a_\sigma). \tag{7.4.11}$$

As this interaction occurs between states that converge to the same electronic state of the ion (but to different N levels), their vibrational functions are nearly identical. Thus, the vibrational factor is almost unity for $\Delta v = 0$ and zero for $\Delta v \neq 0$.

For autoionization of the Jth level, which belongs to the series converging to the second ($N^+ = 2$) ionization limit and autoionized by the continuum of $H_2^+(N^+ = 0)$, Eq. (7.3.7) gives (see Herzberg and Jungen, 1972)

$$\Gamma_{n,J} = 2\pi \frac{2\mathscr{R}}{(n^*)^3} \frac{J(J+1)}{(2J+1)^2}(a_\pi - a_\sigma)^2, \tag{7.4.12}$$

which, for $J = 1$, becomes

$$\Gamma_{n,J} = 2\pi \frac{2\mathscr{R}}{(n^*)^3}\left(\frac{2}{9}\right)(a_\pi - a_\sigma)^2. \tag{7.4.13}$$

Thus, it is possible to view rotational autoionization as arising from the departure of the molecular-ion core from spherical symmetry ($a_\pi \neq a_\sigma$), which causes the *np* complex to split into Π and Σ levels. Figure 7.4 illustrates rotational autoionization in H_2. The right side of Fig. 7.5 displays lines of the $v = 0$, $J = 0$ series converging to the $N^+ = 2$ limit of the ion (called *np*2) that are autoionized by the continuum of $H_2^+\ X^2\Sigma_g^{-1}(v_+ = 0,\ N^+ = 0)$ resulting in the appearance of "emission windows" (see Section 7.8).

The measured difference between the quantum defects for Σ and Π Rydberg series is, for $H_2(v_+ = 1)$: $a_\sigma - a_\pi = 0.230 - (-0.085) = 0.315$. From this and Eq. (7.3.4), $I = 0.148$ for $J = 1$ (see Table 7.2). For *para*-H_2, the interval between the $N^+ = 2$ and $N^+ = 0$ ionization limits is equal to $6B = 108\ \text{cm}^{-1}$ for $J = 1$, where B is the rotational constant of H_2^+. Thus, the smallest possible value of n (for $J = 1, \Delta N = 2$ rotational autoionization) is 25 (see Table 7.1). For the rotational line 26*p*2, the width is 2.3 cm^{-1} (Table 7.2). The f-parity levels cannot be autoionized by this mechanism because there is only one such level in an *np* complex for each value of J.

The interval $2B(2J + 1)$ between the two ionization limits becomes small for heavy molecules and low values of J. Therefore, limited resolution prevents observation of rotational autoionization peaks for molecules heavier than H_2.

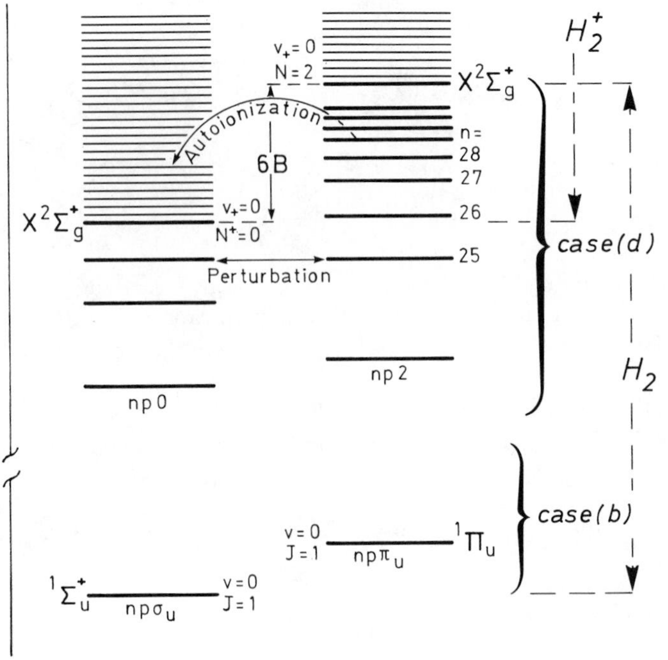

Fig. 7.4 Schematic illustration of rotational autoionization in *para*-H_2 showing the transition from case (b) to case (d). The electronic continua associated with H_2^+ $X^2\Sigma_g^+$ ($v_+ = 0$, $N^+ = 0$ and 2) are depicted by light horizontal lines. The stable and autoionizing levels of the H_2 *np* Rydberg series with $v = 0$ and $J = 1$ are shown as heavy horizontal lines.

For example, for N_2, using data from Table 7.1 and the value of I from H_2, the maximum width for a rotationally autoionized line is predicted to be $0.03\ \text{cm}^{-1}$.

7.5 Vibrational Autoionization

Autoionizing interaction between high Rydberg levels converging to one vibrational level of the ion and the continuum associated with a lower vibrational level of the same ion-core electronic state are more easily observable than rotational autoionization. This process is illustrated schematically in Fig. 7.6 for $v_+ = 0$ and $v_+ = 1$. At sufficiently high n, the $v = 1$ Rydberg levels converging to the ion-core $v_+ = 1$ level of the electronic ground state are embedded in the electronic continuum of the $v_+ = 0$ level of this ground state.

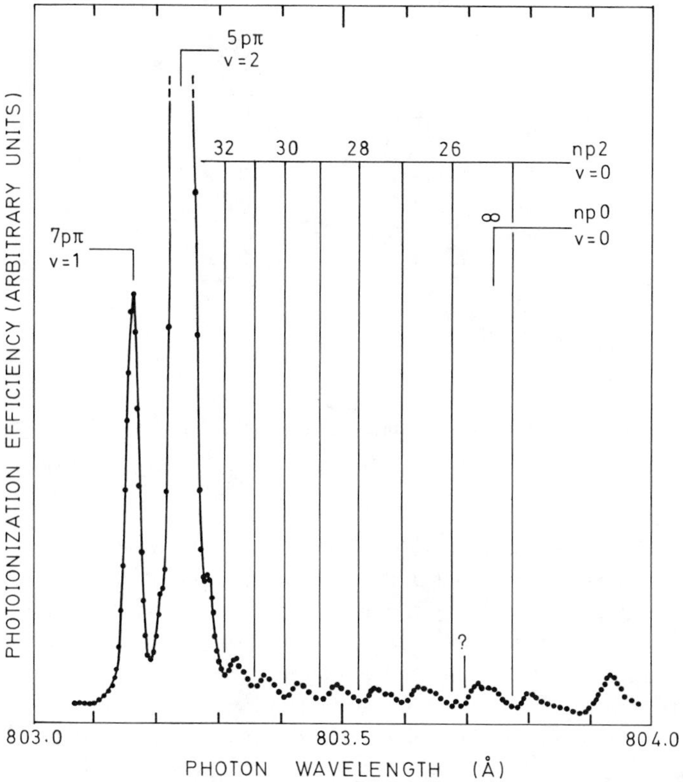

Fig. 7.5 Relative photoionization cross section of *para*-H_2, at 78 K, in the region of the ionization threshold, recorded at a wavelength resolution of 0.016 Å. On the right-hand side of the figure, the rotationally autoionized lines of the $np\hat{2}$ series appear as emission windows. The large peaks on the left are the result of vibrational autoionization (see Section 7.5). [From Dehmer and Chupka (1976).]

Table 7.2
Typical Examples of Autoionization Linewidths

Type of autoionization	Molecule	State	Γ (cm^{-1}) (obs)	I (calc)	Vibrational selection rules
Rotational	H_2	$26p2$, $v = 1$, $J = 1$	2.3	0.148	$\Delta v = 0$
Vibrational	H_2	$8p\sigma$, $v = 2$, $J = 1$	9.7	0.05	$\Delta v \simeq 1$
Spin–orbit	HI	$n = 6$	370	(0.24)[a]	$\Delta v = 0$
Electronic	N_2	$^1\Sigma_u^+$ $(n = 5)$	(2)[b]	0.05	any Δv (Franck–Condon
Electronic	NO	$(b^3\Pi)\,3p\pi$	500	(0.05)[a]	factors)

[a] Deduced from experimental value of Γ and Eqs. (7.6.7), (7.6.8), (7.7.1).
[b] Calculated by Duzy and Berry (1976).

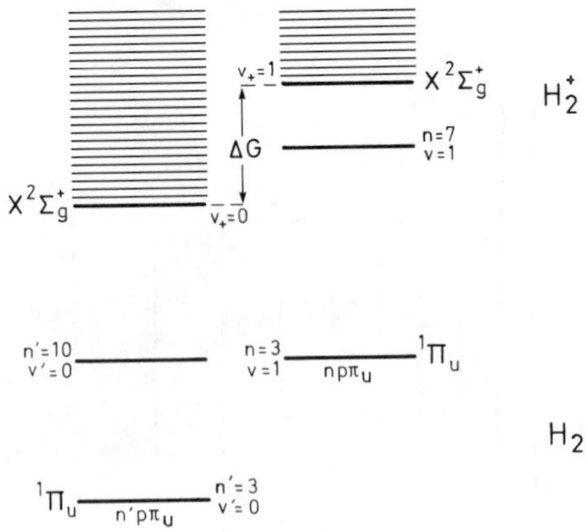

Fig. 7.6 Schematic illustration of vibrational autoionization of the H_2 $np\pi_u\,{}^1\Pi_u$ $(v = 1)$ Rydberg series by the continuum of H_2^+ $X^2\Sigma_g^+$ $(v_+ = 0)$.

Just as for rotational autoionization, this vibrational interaction can be related to homogeneous perturbations between discrete Rydberg levels.

Because these states have the same symmetry and identical angular momentum (Λ), spin, and core electronic wavefunctions, but different vibrational wavefunctions, the nuclear kinetic energy operator, which depends on R, is the only operator that can couple these two vibronic states (see Section 2.3). This interaction can be understood using a very simple model (Herzberg and Jungen, 1972). The potential energy curve $U_{n\Lambda}(R)$ of a Rydberg state is not strictly parallel to that of the ion $U^+(R)$, and this deviation, which reflects the minor bonding/antibonding effect of the Rydberg electron, can be expressed by an R-dependent quantum defect as follows (cf. Mulliken, 1969, Fig. 1):

$$U_{n\Lambda}(R) = \langle \phi_{n\Lambda}|\mathbf{H}|\,\phi_{n\Lambda}\rangle = U^+(R) - \mathscr{R}/[n - a_\lambda(R)]^2 \qquad (7.5.1)$$

where $a_\lambda(R)$ varies slowly with R and can be represented by a truncated Taylor expansion about the equilibrium internuclear distance, R_e^+, of the ion:

$$a_\lambda(R) = a_{\lambda,(R=R_e^+)} + \left(\frac{da_\lambda}{dR}\right)_{R=R_e^+}(R - R_e^+) + \frac{1}{2}\left(\frac{d^2a_\lambda}{dR^2}\right)_{R=R_e^+}(R - R_e^+)^2 + \cdots \qquad (7.5.2)$$

In this expression, da_λ/dR reflects the difference in the R_e values of the ion and Rydberg state potential curves and d^2a_λ/dR^2 reflects the difference in the harmonic constants (ω_e) or in the dissociation energies of these two curves. If one defines $n^* = n - a_{\lambda(R=R_e^+)}$, using a truncated Taylor expansion as in

Eq. (7.4.8) and keeping only the linear term of Eq. (7.5.2), one obtains

$$U_{n\Lambda}(R) = U^{+}(R) - \frac{\mathscr{R}}{(n^*)^2} - \frac{2\mathscr{R}}{(n^*)^3}\frac{da_\lambda}{dR}(R - R_e^+). \tag{7.5.3}$$

Converting to the usual off-diagonal form [Eq. (7.3.4)] of Rydberg ~ Rydberg matrix elements between members of the same $l\Lambda$ series but with different vibrational factors, the matrix element can be written as

$$\langle \Psi_{n\Lambda,v}|\mathbf{H}|\Psi_{n'\Lambda,v'}\rangle = \frac{2\mathscr{R}}{(n^*)^{3/2}(n^{*\prime})^{3/2}}\frac{da_\lambda}{dR}\langle v|R - R_e^+|v'\rangle. \tag{7.5.4}$$

In the harmonic approximation, the only nonzero matrix elements of $(R - R_e^+)$ are $\Delta v = v' - v = -1$ (see, for example, Wilson *et al.*, 1980, Appendix III),

$$\langle \Psi_{n\Lambda,v}|\mathbf{H}|\Psi_{n'\Lambda,v-1}\rangle = \frac{2\mathscr{R}}{(n^*)^{3/2}(n^{*\prime})^{3/2}}\frac{da_\lambda}{dR}\left[\frac{h}{8\pi^2\mu\omega c}\right]^{1/2}(v)^{1/2} \tag{7.5.5}$$

where μ is the reduced mass and ω the harmonic vibrational frequency. The linewidth for the n^*, v-level autoionized by the continuum of the $v - 1$ level is obtained from Eq. (7.5.5) and the Golden Rule (Eq. 7.3.1), eliminating the $n^{*\prime}$ dependence by inserting the density of states factor,

$$\left(\frac{dE}{dn}\right)^{-1} = \left[\frac{2\mathscr{R}}{(n^{*\prime})^3}\right]^{-1}, \tag{7.5.6}$$

to obtain (Herzberg and Jungen, 1972)

$$\Gamma_{n_v,v-1}\ (\text{cm}^{-1}) = 2\pi\frac{2\mathscr{R}}{(n^*)^3}\left[\frac{da_\lambda}{dR}\ (\text{Å}^{-1})\right]^2 \frac{16.8576}{\mu\ (\text{amu})\ \ \omega\ \ (\text{cm}^{-1})}v, \tag{7.5.7}$$

where v and $v - 1$ are, respectively, the vibrational quantum numbers of the nth autoionized Rydberg state and the product ion.

Vibrational autoionization corresponds to an exchange between the kinetic energy of the ejected electron and the vibrational energy of the core ion. This process occurs because the potential curves of the ion and Rydberg states are slightly different. Equation (7.5.7) can be derived from a more sophisticated model using multichannel quantum defect theory (Raoult and Jungen, 1981).

Equation (7.5.7) shows that the widths (1) decrease as $(n^*)^{-3}$ and (2) are proportional to v if only the first derivative of the quantum defect is retained. The approximate selection rule, $\Delta v = -1$, known as the propensity rule of Berry (1966), is well verified experimentally, at least for H_2. The width, for $\Delta v = -2$ in H_2, is typically two orders of magnitude smaller than for $\Delta v =$

-1. The effect of the second derivative is neglected for H_2, but second-order effects proportional to $\langle v-1|v\rangle \langle v|v+1\rangle$ (and higher-order effects) can produce small widths for any Δv value (Dehmer and Chupka, 1976).

For H_2, the width is larger for Σ states than for Π states because the quantum defect a_σ varies more rapidly with internuclear distance than a_π (see Fig. 2 of Jungen and Atabek, 1977). This, in turn, can be understood by the fact that, at $R = \infty$ the $3p\sigma$ state of H_2 dissociates into H(2p) + H(1s) atoms. The value of n is therefore "promoted" by one unit on going from the separated atom to the united atom limit. Consequently, a_σ must vary rapidly with R. This does not occur for Π Rydberg states. A typical autoionization width is given in Table 7.2 for the level $8p\sigma$ $v = 2$ autoionized by the $v_+ = 1$ continuum of H_2^+ $X^2\Sigma_g^+$. In Fig. 7.5, one can see a peak that corresponds to $R(0)$ $7p\pi$ ($v = 1$), for which the calculated width (without convolution with the experimental resolution) is only 0.15 cm^{-1}. The peak corresponding to the level $5p\pi$ ($v = 2$), autoionized by the $v_+ = 0$ continuum of H_2^+, illustrates a deviation from Berry's propensity rule. This deviation is due to interference between rotational and vibrational autoionization (Jungen and Dill, 1980). The proportionality of the widths to v is often well verified in pure vibrational autoionization cases (see, for example, Dehmer and Chupka, 1976).

In HF (Guyon *et al.*, 1976) vibrational autoionization peaks are clearly seen in the threshold photoelectron spectrum. For heavy molecules, the da_λ/dR factors appear to be small, as suggested by the calculations for N_2 by Duzy and Berry (1976) and verified in the photoionization spectrum of N_2 (Fig. 7.7), where the members of the series converging to the $v_+ = 1$ level of the $X^2\Sigma_g^+$ state are very weakly autoionized by the continuum of the X state ($v_+ = 0$) level (see, for example, $n = 11$ or 12). For the NO molecule (Ng *et al.*, 1976), fine

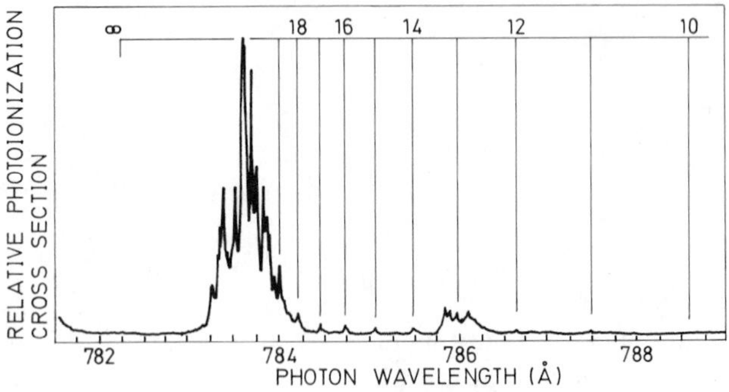

Fig. 7.7 Photoionization cross-section of N_2 taken at 77 K with a resolution of 0.016 Å (from Dehmer *et al.*, 1984). The first members of the $v = 1$ series are vibrationally autoionized by the continuum $v_+ = 0$ of N_2^+ $X^2\Sigma_g^+$. For $n \geq 14$, anomalies result from indirect electrostatic autoionization. (From Giusti-Suzor and Lefebvre-Brion, 1984.)

structure is also visible in the photoionization curve. An apparent deviation from the $\Delta v = -1$ propensity rule is observed that cannot be explained simply by a nonnegligible second derivative term. This anomalous line broadening has been shown to be induced by predissociation (Giusti-Suzor and Jungen, 1984).

7.6 Spin–Orbit Autoionization

This type of autoionization can affect Rydberg levels that lie between the ionization limits associated with different multiplet components of a given ion-core electronic state. A well-known example concerns the Rydberg states of the argon atom. This type of autoionization has not yet been demonstrated in molecules. An expression for the autoionization width can be deduced in a similar fashion to that for rotational autoionization (Jungen, private communication). Consider, for example, the Rydberg states converging to the N_2^+ $A^2\Pi_{1/2}$ substate, which can be autoionized by the continuum of the $A^2\Pi_{3/2}$ substate. The $A^2\Pi$ state is inverted, which means that $E(^2\Pi_{3/2}) < E(^2\Pi_{1/2})$. The spin–orbit interaction will appear between discrete Rydberg states of different series having the same Ω value (see Section 2.4.2).

Consider the $(p\pi)^3(ns\sigma)$ $^3\Pi_1$ and $^1\Pi_1$ Rydberg series of N_2 converging to the $A^2\Pi$ state of N_2^+, as described by the Rydberg formulas

$$\begin{aligned} E[(ns\sigma)\,^1\Pi] &= E(A^2\Pi) - \frac{\mathcal{R}}{(n-a_1)^2} = E(A^2\Pi) - \frac{\mathcal{R}}{(n^* + a')^2} \\ E[(ns\sigma)\,^3\Pi] &= E(A^2\Pi) - \frac{\mathcal{R}}{(n-a_3)^2} = E(A^2\Pi) - \frac{\mathcal{R}}{(n^* - a')^2}. \end{aligned} \tag{7.6.1}$$

$E(A^2\Pi)$ is the electronic energy in the absence of spin–orbit splitting:

$$E(A^2\Pi) = \tfrac{1}{2}[E(A^2\Pi_{1/2}) + E(A^2\Pi_{3/2})]$$

$$a' = \tfrac{1}{2}(a_3 - a_1) \qquad \text{and} \qquad n^* = n - \tfrac{1}{2}(a_3 + a_1).$$

The spin–orbit interaction associated with the unpaired $p\pi$ core electron mixes the $^3\Pi_1$ and $^1\Pi_1$ components of the same n value ($\Delta\Omega = 0$). In the case (a) basis, the secular equation is [see Eq. (2.4.14)]

$$\begin{vmatrix} E[(ns\sigma)\,^3\Pi_1] - E & A(^3\Pi) \\ A(^3\Pi) & E[(ns\sigma)\,^1\Pi_1] - E \end{vmatrix} = 0. \tag{7.6.2}$$

The difference between the triplet and singlet energies is approximately $2K_n$, where K_n is an exchange integral proportional to $(n^*)^{-3}$. As discussed in Section 2.4.1, the mixing between $^3\Pi_1$ and $^1\Pi_1$ increases with n, and when n becomes infinite the states are completely mixed and are described by a case (c) basis that corresponds to the limiting solutions,

$$\begin{aligned} E_1 &= E(A^2\Pi) + A(^3\Pi) \qquad \Psi_1 = 2^{-1/2}(^1\Pi_1 + {}^3\Pi_1) \\ E_2 &= E(A^2\Pi) - A(^3\Pi) \qquad \Psi_2 = 2^{-1/2}(^1\Pi_1 - {}^3\Pi_1). \end{aligned} \tag{7.6.3}$$

These case (a) → (c) transformation coefficients will now be used to describe Rydberg states for high values of n. Two new series are obtained. One of the two $\Omega = 1$ series converges to the $A^2\Pi_{1/2}$ component of the ion and the other to the $A^2\Pi_{3/2}$ component, with energies

$$\begin{aligned} E_{1n} &= E(A^2\Pi_{1/2}) - \frac{\mathcal{R}}{(n - a_{1/2})^2} \\ E_{2n} &= E(A^2\Pi_{3/2}) - \frac{\mathcal{R}}{(n - a_{3/2})^2}. \end{aligned} \tag{7.6.4}$$

From Eqs. (7.6.3) and (7.4.8) and noting that $E(A^2\Pi_{3/2}) - E(A^2\Pi_{1/2}) = 2A(^3\Pi)$ it is easy to show that

$$a_{1/2} = a_{3/2} = \tfrac{1}{2}(a_1 + a_3).$$

The residual interaction between these same-n Rydberg levels, expressed in the case (c) basis, but intermediate between cases (c) and (a), can now be viewed as a consequence of the electronic part of the Hamiltonian, since the case (c) basis functions are exact eigenfunctions only when the energy difference between isoconfigurational triplet and singlet states is zero. The off-diagonal matrix element in the case (c) basis is

$$\begin{aligned} &\langle 2^{-1/2}[(ns\sigma)^1\Pi_1 + (ns\sigma)^3\Pi_1]|\mathbf{H}|2^{-1/2}[(ns\sigma)^1\Pi_1 - (ns\sigma)^3\Pi_1]\rangle \\ &= \tfrac{1}{2}\{E[(ns\sigma)^1\Pi_1] - E[(ns\sigma)^3\Pi_1]\} \\ &= K_n \simeq \frac{\mathcal{R}}{(n^*)^3}2a' = \frac{\mathcal{R}}{(n^*)^3}(a_3 - a_1). \end{aligned} \tag{7.6.5}$$

The approximate equality is obtained from a Taylor expansion of Eq. (7.6.4) analogous to that of Eq. (7.4.8). The interaction between the nth term of the series converging to the $A^2\Pi_{1/2}$ limit and the (n')th term of the $A^2\Pi_{3/2}$ series can be written, by analogy with Eq. (7.3.4), as

$$\langle \Psi_{1n}|\mathbf{H}|\Psi_{2n'}\rangle = \frac{\mathcal{R}}{(n^*)^{3/2}(n'^*)^{3/2}}(a_3 - a_1). \tag{7.6.6}$$

From the general result (Eq. 7.3.4), the I parameter may be identified here as

$$I = \frac{a_3 - a_1}{2}. \tag{7.6.7}$$

To a good approximation, the vibrational wavefunctions of the two series, which converge to the same vibrational level of the $^2\Pi_{1/2}$ and $^2\Pi_{3/2}$ limits, are identical, leading to a $\Delta v = 0$ propensity rule and a vibrational factor equal to one. Finally, the autoionization width given by Eq. (7.3.7) is

$$\Gamma_n = 2\pi \frac{2\mathscr{R}}{(n^*)^3}\left(\frac{a_3 - a_1}{2}\right)^2. \qquad (7.6.8)$$

For N_2, Worley's third Rydberg series is clearly $^1\Pi$ while the series observed by Ogawa and Tanaka (1962) belongs to the corresponding $^3\Pi$ states. The difference between the two quantum defects for $v = 0$ ($a_1 \simeq 1.04$ and $a_3 \simeq 1.18$) gives $I \simeq 0.07$. The interval between the N_2^+ $A^2\Pi_{3/2}$ and $A^2\Pi_{1/2}$ limits is small (80 cm^{-1}). Consequently, the minimum value of n for autoionization obtained from Eq. (7.3.9) is very large, $n_{\min} \geq 37$; thus the maximum value of Γ will be about 0.12 cm^{-1}. Figure 7.8 schematically illustrates $N_2 \rightarrow N_2^+$ $(A^2\Pi)$ spin–orbit autoionization. Actually, electronic (or, more precisely, electrostatic) autoionization is considerably stronger and conceals this spin–orbit autoionization mechanism (see Fig. 7.9).

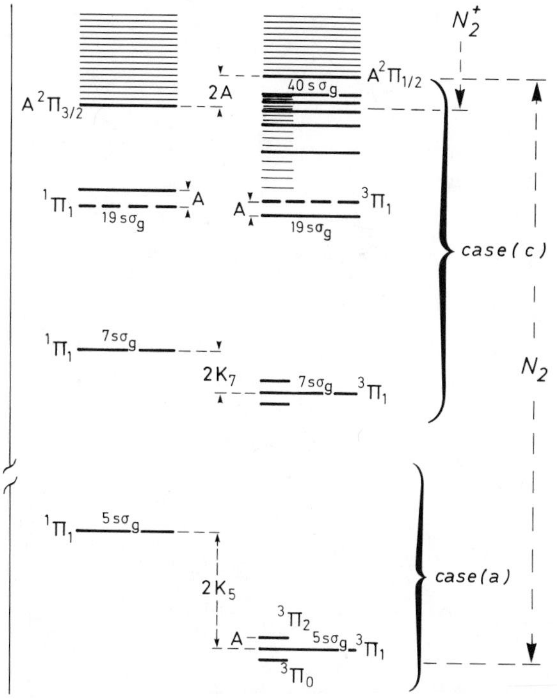

Fig. 7.8 Schematic illustration of spin–orbit autoionization for the Π_u Rydberg states converging to the $A^2\Pi_u$ state of N_2^+.

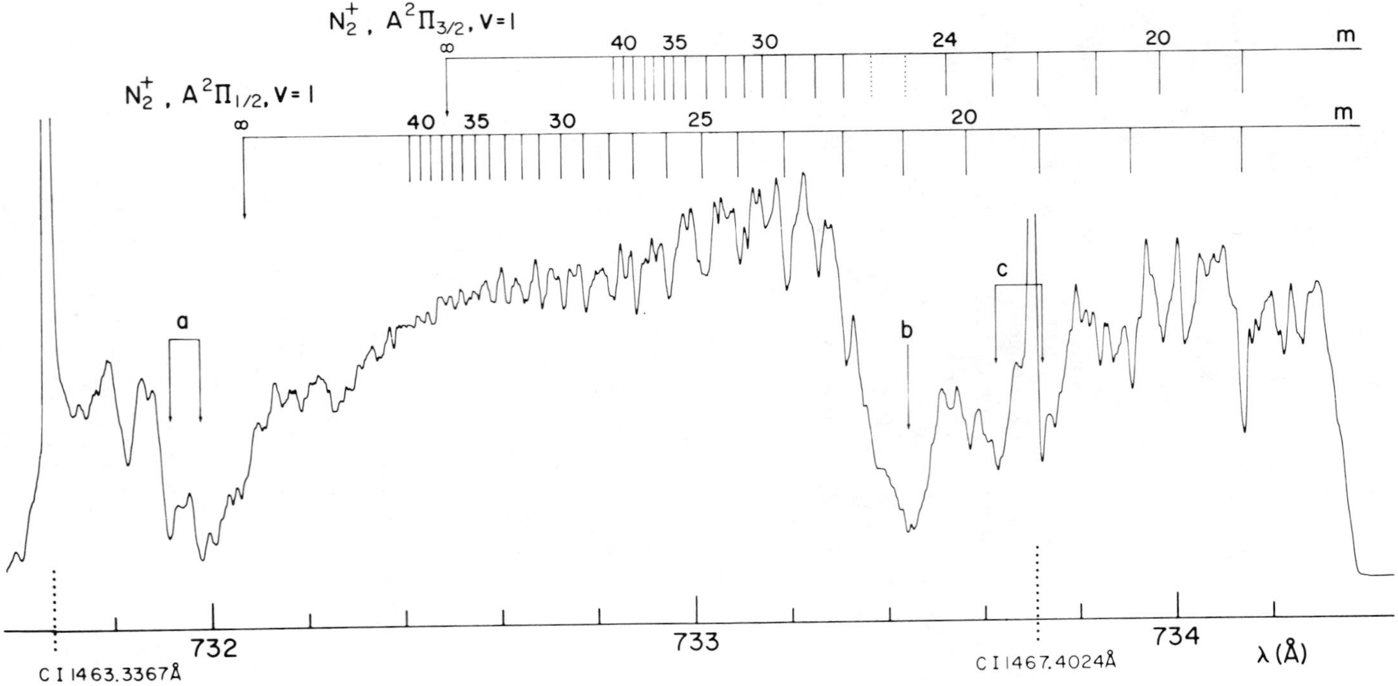

Fig. 7.9 Densitometer trace (absorption downward) at liquid N_2 temperature of the higher members of the A–X Rydberg series of N_2, which converge to the N_2^+ $A^2\Pi_{1/2}$ and $A^2\Pi_{3/2}$ $v_+ = 1$ limits. The strong bands indicated by arrows belong to other series: a, $m = 8$ of $A^2\Pi_{1/2}$, $v_+ = 2$; b, progression (3), $v_+ = 2$; c, $m = 4$ of $A^2\Pi_{3/2}$, $v_+ = 5$. (From Yoshino *et al.*, 1976).

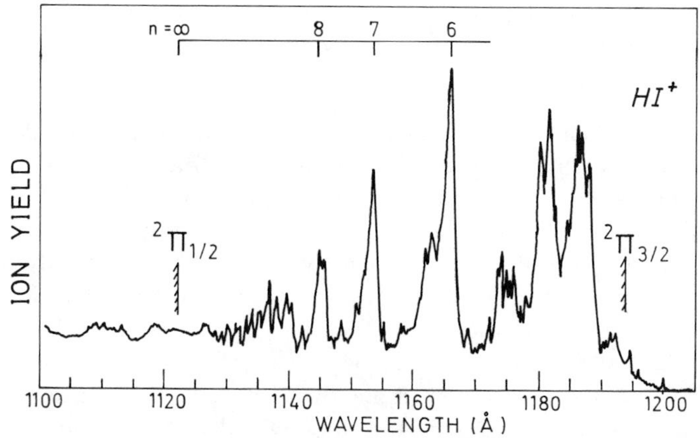

Fig. 7.10 Photoionization of HI at 0.14 Å resolution showing the spin–orbit autoionization between the two $\Omega = \frac{3}{2}$ and $\Omega = \frac{1}{2}$ thresholds of the ground state of the ion. [From Eland (1980).]

For molecules containing an atom heavier than N, the spin–orbit splitting of the ion will be larger and n_{min} can become small. Consequently, the spin–orbit interaction could then be the most important cause of autoionization, as it is already well known to be for predissociation. The HI photoionization spectrum (Eland and Berkowitz, 1977) is a good example. Figure 7.10 shows the Rydberg levels as they converge to the $^2\Pi_{1/2}$ limit. (For HI the $^2\Pi$ state of the ion is inverted.) The width of the $n = 6$ member of the series is estimated from the spectrum to be 370 cm^{-1}. As this series is certainly very perturbed even at low n by spin–orbit interaction, it is difficult to extract deperturbed values for a_1 and a_3 and, from them, I. However, using the experimental value for Γ, Eq. (7.6.8) gives $I = 0.24$. This value is similar to the value that can be deduced for the Xe atom from the widths of autoionized levels between the $^2P_{1/2}$ and $^2P_{3/2}$ thresholds.

The magnitude of the spin–orbit splitting of the ion-core state affects the autoionization width most importantly through the minimum value of n, not through the size of I. This result is not surprising, since the present treatment was based on case (c) basis functions, which are eigenfunctions of the spin–orbit Hamiltonian (i.e., which diagonalize the spin–orbit part of the total Hamiltonian). (See Lefebvre-Brion *et al.* (1985) for a theoretical treatment of the HI spin–orbit autoionization.)

7.7 Electronic (or Electrostatic) Autoionization

Electronic autoionization occurs between the nth Rydberg state converging to an excited electronic state of the ion in its v vibrational level and the continuum of a lower electronic state of the ion in its v_+ vibrational level. In

Eq. (7.3.7), the general expression for the width,

$$\Gamma_{nv;\varepsilon v_+} = 2\pi \frac{2\mathscr{R}}{(n^*)^3} I^2 \langle v | v_+ \rangle^2 \tag{7.7.1}$$

the vibrational functions, χ_v and χ_{v_+}, correspond to different potential curves of the ion. There is thus no selection rule for the vibrational factor, and the overlap-squared factor is analogous to a Franck–Condon factor. In the electronic part, the parameter I corresponds to an electrostatic interaction between states of identical symmetry (same values of Λ, S, and Σ). In a simple picture, these two states must be derived from electronic configurations which differ by at most two orbitals.

The unknown parameter, I, can be related to perturbation interactions between two Rydberg states converging to different states of the ion, provided that homogeneous perturbations are experimentally observed in the discrete spectrum below the ionization limit. Figure 7.11 illustrates such a situation for N_2. The first two members of two different $^1\Pi_u$ Rydberg series, the $[(X^2\Sigma_g^+)3p\pi_u]c^1\Pi_u$ and the $[(A^2\Pi_u)3s\sigma_g]o^1\Pi_u$ states, which converge, respectively, to the N_2^+ $X^2\Sigma_g^+$ ($v_+ = 0$) and $A^2\Pi_u$ ($v_+ = 0$), states are found at similar energies and perturb each other. These homogeneous perturbations have been studied by Stahel *et al.* (1983). They have deduced a value of about 120 cm^{-1} for the effective electronic parameter, from which Eq. (7.3.4) gives

$$I = (n^*)^{3/2}(n'^*)^{3/2} \times 120/2\mathscr{R} \simeq 0.01.$$

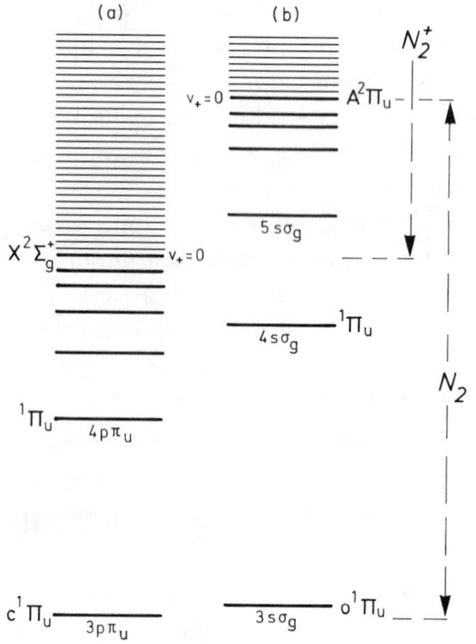

Fig. 7.11 Schematic illustration of the electrostatic autoionization of the $^1\Pi_u$ Rydberg states converging to the $A^2\Pi_u$ state of N_2^+ by the continuum of the $X^2\Sigma_g^+$ state. (a) Worley–Jenkins Rydberg series, (b) Worley's (third) Rydberg series.

Table 7.1 shows that the separation between the two ionization limits, $E(A^2\Pi_u) - E(X^2\Sigma_g^+) = 9000\ cm^{-1}$, implies $n_{min} = 4$ [Eq. (7.3.9)]. The $n = 5$ member of the $(A^2\Pi_u)\ ns\sigma_g$ series lies above the first ionization limit $(X^2\Sigma_g^+, v = 0)$ and appears strongly in the ionization spectrum (Fig. 7.7). Unfortunately, the peak is much broader than expected from $I \approx 0.01$ ($\Gamma_{obs} \approx 80\ cm^{-1}$, $\Gamma_{calc} \approx 0.6\ cm^{-1}$). This results from an enhancement by indirect electronic autoionization (analogous to accidental predissociation) of the higher members of the Rydberg series converging to the $v = 1$ level of the $X^2\Sigma_g^+$ state (Giusti-Suzor and Lefebvre-Brion, 1984).

For NO, the separation between the first two excited states of NO^+ is considerably larger than for N_2^+, giving $n_{min} \geq 2$. A linewidth of 500 cm^{-1} has been observed in the ultraviolet spectrum of NO (Takezawa, 1977) for a $3p\pi$ Rydberg state converging to the $NO^+\ b^3\Pi$ excited state. If that width is caused exclusively by electronic autoionization and not by predissociation, it corresponds to a value of about 0.05 for the I parameter.

In conclusion, Table 7.2 shows that the strongest autoionization process is electronic autoionization for light molecules and spin–orbit autoionization for heavy molecules.

Nothing has been said yet about level shifts. As the coupling responsible for autoionization is even less energy-dependent than that for predissociation, the level shift should be less important. Autoionization level shifts have not yet been experimentally detected for molecules, but the possible occurrence of such shifts should not be forgotten.

7.8 Validity of the Approximations

In the preceding sections, we have assumed that an absorption line has a Lorentzian shape. If this is not true, then the linewidth cannot be defined as the full width at half maximum intensity. Transitions from the ground state of a neutral molecule to an ionization continuum often have appreciable oscillator strength, in marked contrast to the situation for ground state to dissociative continuum transitions. The absorption cross-section near the peak of an autoionized line can be significantly affected by interference between two processes: (1) direct ionization or dissociation, and (2) indirect ionization (autoionization) or indirect dissociation (predissociation). The line profile must be described by the Beutler–Fano formula (Fano, 1961):

$$\sigma_a = \sigma_d + \sigma_i \frac{(q + \varepsilon)^2}{1 + \varepsilon^2}, \tag{7.8.1}$$

where σ_d is the cross-section for the portion of the ionization or dissociative continuum that, by symmetry or selection rules, does not interact with the resonance state, and σ_i corresponds to the part of the ionization or dissociative continuum that interacts with this state. The line shape is characterized by two parameters, ε and q:

1. $\varepsilon = (E - E_r)/(\Gamma/2)$, where E_r and Γ are the energy and full width of the resonance state (and ε must not be confused with the kinetic energy of the electron).
2. q is defined similarly to Eq. (6.7.6):

$$q = \frac{1}{\pi} \frac{\langle \phi_r | \boldsymbol{\mu} | \phi_0 \rangle \langle v | v_0 \rangle}{\langle \phi_i | \boldsymbol{\mu} | \phi_0 \rangle \langle v_+ | v_0 \rangle \langle \phi_r | \mathbf{H} | \phi_i \rangle \langle v | v_+ \rangle}, \tag{7.8.2}$$

where $\langle \phi_r | \boldsymbol{\mu} | \phi_0 \rangle$ and $\langle \phi_i | \boldsymbol{\mu} | \phi_0 \rangle$ are the electronic transition moments from the ground state ϕ_0 in its vibrational level v_0 into the resonance and continuum states, respectively.†

Figure 7.2 reproduces Hopfield's Rydberg series in the N_2 absorption spectrum, consisting (in absorption) of a series with a very large q and an apparent "emission" series with $q \simeq 0$. This means that the bound Rydberg states of the latter series have near-zero transition moments from the ground state of the N_2 molecule. This series appears because its intensity is borrowed from the continuum [see Raoult *et al.* (1983) for a theoretical interpretation]. Such transitions into states with zero nominal transition moments will appear less strongly below the ionization limit, since they can only borrow intensity from nearby discrete states (see Chapter 5).

In the preceding sections, the case of one resonance isolated from the others and interacting with a single continuum of a specific nature has been discussed. A brief treatment of more general situations follows.

If the resonance state interacts with several continua, Eq. (7.8.1) remains valid provided that one defines q by summing over each of these continua in Eq. (7.8.2). In the case where multiple continua are involved—for example, those corresponding to all vibrational levels of a specific electronic state of the ion—it can be shown (Smith, 1970) that the q and Γ parameters become independent of the vibrational quantum number of the Rydberg state. Note that the lineshape asymmetry parameter, q, does not depend on the quantum number n; in contrast, the width Γ depends on $(n^*)^{-3}$.

This generalized form of Eq. (7.8.1) can be applied only when *a single discrete state* interacts with one or several continua *that do not interact with each other*. This situation is typical for a predissociated state. Unfortunately, it does not often occur for an autoionized Rydberg state. When there are several resonances that

† Since ϕ_i is an energy-normalized continuum function, q is dimensionless, as is ε.

have large widths relative to their separations, Fano's formula must be replaced by the formula given by Mies (1968):

$$\sigma = \sigma_{\mathrm{d}} + \sigma_{\mathrm{i}} \frac{\left[1 + \sum_n (q_n/\varepsilon_n)\right]^2}{1 + \left(\sum_n 1/\varepsilon_n\right)^2}, \tag{7.8.3}$$

where q_n and ε_n are defined as previously for a single resonance, n. This formula takes into account interference between several near-degenerate resonances.

Typically, each resonance is coupled to several continua, and these continua interact with each other. In the case where two continua are coupled by an interaction I [I is a dimensionless parameter analogous to that in Eq. (7.3.4)], the width of the resultant Lorentzian peak is given by the same formula as in Eq. (6.10.5) (Beswick and Lefebvre, 1973):

$$\Gamma = \frac{\Gamma_1 + \Gamma_2}{1 + \pi^2 I^2}, \tag{7.8.4}$$

where Γ_1 and Γ_2 represent the widths arising from each independent continuum. This formula is valid regardless of whether the interacting continua belong to different electronic states of the ion or correspond to ionization and dissociative continua.

In reality, one encounters overlapping multiple resonances interacting with multiple continua. The analytical expressions given here represent an idealized situation. Several examples involving interferences between different autoionization processes [for example, in H_2, interference between rotational and vibrational autoionization (Fig. 7.2); in N_2, interference between vibrational and electronic autoionization (Fig. 7.7)] have been discussed. Multichannel quantum defect theory (MQDT) is capable of treating such problems without introducing the concept of a width (Dill and Jungen, 1980).

The basic concepts of MQDT are unfamiliar to most molecular spectroscopists. The crucial ideas of frame transformation and phase shift are briefly explained in the next two paragraphs.

In this chapter on autoionization processes several examples have been given where the members of a Rydberg series follow one Hund's case when the principal quantum number, n, is small and another Hund's case when n is large. For spin–orbit autoionization, the small- and large-n Rydberg states are well represented by Hund's cases (a) and (c), respectively. For rotational autoionization (of H_2, for example), the small- and large-n Rydberg states are well represented by Hund's cases (b) and (d), respectively. One of the basic concepts of MQDT is that the space in which the outer (Rydberg) electron moves may be divided into two parts: the *inner region* near the nuclei, where the maximum in the radial probability distribution for a low-n Rydberg orbital occurs, and the *outer region*, far from the core, where large-n Rydberg orbitals have their maximum radial probability. In the inner region the Rydberg electron is strongly affected by the field of the molecule; exchange interactions between the Rydberg and core electrons are particularly important. In the outer region, exchange effects become negligible and the Rydberg electron becomes weakly coupled to the core electrons; the molecular ion core appears to the outer electron as a definite $|\Lambda^+, S^+, J^+, \Omega^+\rangle$ [case (a) inner region → case (c) outer region] or $|\Lambda^+, S^+, N^+, J^+\rangle$ [case (b) inner region → case (d) outer region] state. Thus, the appropriate Hund's coupling case depends on the instantaneous coordinate of the electron in the Rydberg orbital. When the electron moves from the inner to the outer region, a *frame transformation* is performed to adapt the total electronic wavefunction (core plus Rydberg electron) to the appropriate Hund's case.

A second crucial idea of MQDT is the relationship between the *quantum defect*, a, of a Rydberg series and the *phase shift* of the radial part of the outer electron wavefunction. At large r, the radial wavefunction of the external electron would be a pure Coulomb function if the potential were of the hydrogenic atom form

$$V(r) = -eZ/r.$$

The asymptotic form of a Coulomb function is a sine function. For a nonhydrogenic atom or any molecule, the external electron wavefunction is more complicated than a Coulomb function. However, at large r this wavefunction can be well approximated by a Coulomb radial function with an asymptotic form that is shifted in phase by ϕ with respect to the hydrogenic atom wavefunction. Figure 7.12 illustrates the effect of this phase shift on the

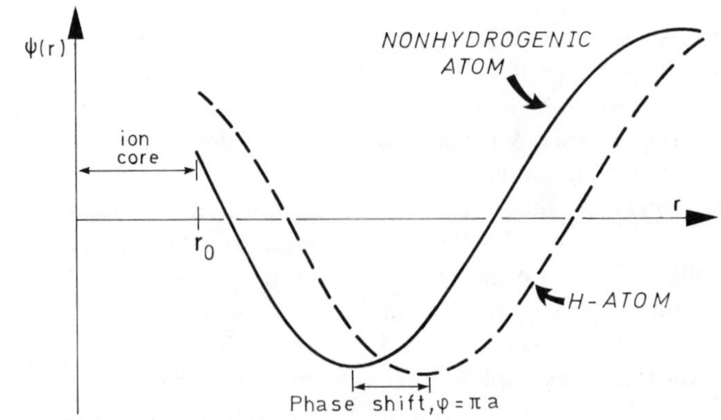

Fig. 7.12 Diagram representing the phase shift induced by the electric field of the core ion in a nonhydrogenic atom; $r > r_0$ denotes the asymptotic region. [From Fano (1975).]

Coulomb wavefunction of a nonhydrogenic atom. The phase shift summarizes the net effect of all interactions experienced by the external electron when it enters the inner region. These interactions are strong relative to the binding energy of the external electron. Therefore the phase shift, ϕ, varies slowly as the energy of the outer electron is increased, even when an ionization threshold is crossed and the electron energy changes from negative (bound electron) to positive (continuum electron). For a bound electron, the quantum defect, a, can be related to the phase shift, ϕ, of the radial wavefunction by

$$\phi = \pi a,$$

and ϕ is, in general, nearly independent of n. For a molecule, ϕ can depend on internuclear distance, and it is the $a(R)$ variation that is the critical factor controlling vibrational autoionization (see Section 7.5).

MQDT uniformly treats all members of a Rydberg series with given l,† λ, and v and the adjoining continuum to which this series converges as a single entity, a "channel." Based on relationships such as Eqs. (7.3.4) and (7.3.6), MQDT replaces separate treatments of the many interactions between individual states by a global treatment of interchannel interactions. The underlying unity of the phenomena of perturbations, predissociation, and autoionization, which has been the central theme of this book, appears to be optimally expressed by MQDT (Fano, 1970; Jungen and Atabek, 1977; Jungen, 1982; Giusti-Suzor and Jungen, 1984; Green and Jungen, 1985).

References

Berry, R. S. (1966), *J. Chem. Phys.* **45**, 1228.

Beswick, J. A., and Lefebvre, R. (1973), *Mol. Phys.* **29**, 1611.

Carroll, P. M. (1973), *J. Chem. Phys.* **58**, 3597.

Dehmer, P. M., and Chupka, W. A. (1976), *J. Chem. Phys.* **65**, 2243.

Dehmer, P. M., Miller, P. J., and Chupka, W. A. (1984), *J. Chem. Phys.* **80**, 1030.

Dill, D., and Jungen, C. (1980), *J. Phys. Chem.* **84**, 2116.

Duzy, C., and Berry, R. S. (1976), *J. Chem. Phys.* **64**, 2431.

Eland, J. H. D. (1980), *J. Chim. Phys. Phys. Chim. Biol.* **77**, 613.

Eland, J. H. D., and Berkowitz, J. (1977), *J. Chem. Phys.* **67**, 5034.

Fano, U. (1961), *Phys. Rev.* **124**, 1866.

Fano, U. (1970), *Phys. Rev. A* **2**, 353.

Fano, U. (1975), *J. Opt. Soc. Am.* **65**, 979.

Giusti-Suzor, A., and Jungen, C. (1984), *J. Chem. Phys.* **80**, 986.

Giusti-Suzor, A., and Lefebvre-Brion, H. (1984), *Phys. Rev. A* **30**, 3057.

† l can never be a rigorously good quantum number in a molecule.

Greene, C. H., and Jungen, C. (1985), *Adv. At. Mol. Phys.* **21**, 51.
Guyon, P. M., Spohr, R., Chupka, W. A., and Berkowitz, J. (1976), *J. Chem. Phys.* **65**, 1650.
Herzberg, G. (1950), "Spectra of Diatomic Molecules," Van Nostrand-Reinhold, Princeton, New Jersey.
Herzberg, G., and Jungen, C. (1972), *J. Mol. Spectrosc.* **41**, 425.
Jungen, C. (1980), *J. Chim. Phys. Phys. Chim. Biol.* **77**, 27.
Jungen, C. (1982), *in* "Physics of Electronic and Atomic Collisions" (S. Datz, ed.), p. 455. North-Holland-Publ., Amsterdam.
Jungen, C., and Atabek, O. (1977), *J. Chem. Phys.* **66**, 5584.
Jungen, C., and Dill, D. (1980), *J. Chem. Phys.* **73**, 3338.
Larzillière, M., Launay, F., and Roncin, J. Y. (1980), *J. Phys. (Orsay, Fr.)* **41**, 1431.
Lefebvre-Brion, H., Giusti-Suzor, A., and Raçeev G. (1985), *J. Chem. Phys.* **83**, 1557.
Mies, F. H. (1968), *Phys. Rev.* **175**, 164.
Mulliken, R. S. (1964), *J. Am. Chem. Soc.* **86**, 3183.
Mulliken, R. S. (1969), *J. Am. Chem. Soc.* **91**, 4615.
Ng, C. Y., Mahan, B. H., and Lee, Y. T. (1976), *J. Chem. Phys.* **65**, 1956.
Ogawa, M., and Tanaka, Y. (1962), *Can. J. Phys.* **40**, 1593.
Raoult, M., and Jungen, C. (1981), *J. Chem. Phys.* **74**, 3388.
Raoult, M., LeRouzo, H., Raçeev, G., and Lefebvre-Brion, H. (1983), *J. Phys. B* **16**, 4601.
Slater, J. C. (1960), "Quantum Theory of Atomic Structure," McGraw-Hill, New York.
Smith, A. L. (1970), *Philos. Trans. R. Soc. London. A* **268**, 169.
Stahel, D., Leoni, M., and Dressler, K. (1983), *J. Chem. Phys.* **79**, 2541.
Takezawa, S. (1977), *J. Mol. Spectrosc.* **66**, 121.
Wilson, E. B., Decius, J. C., and Cross, P. C. (1980), "Molecular Vibrations," Dover, New York.
Yoshino, K., Ogawa, M., and Tanaka, Y. (1976), *J. Mol. Spectrosc.* **61**, 403.

Index

D

E

F

G

H

PF, forbidden transition $^3\Sigma^- - {}^1\Sigma^+$, 288